T0310246

SYNTHESIS AND APPLICATIONS OF COPOLYMERS

SYNTHESIS AND APPLICATIONS OF COPOLYMERS

Edited by

ANBANANDAM PARTHIBAN

Institute of Chemical and Engineering Sciences,
Agency for Science, Technology and Research,
Singapore

Published by John Wiley & Sons, Inc., Hoboken, New Jersey
Published simultaneously in Canada

Library of Congress Cataloging-in-Publication Data:

Synthesis and applications of copolymers / edited By Andanandam Parthiban, Institute of Chemical and Engineering Sciences, Agency for Science, Jurong Island, Singapore.
pages cm
Includes bibliographical references and index.
ISBN 978-1-118-05746-9 (cloth)
1. Polymerization. 2. Copolymers. 3. Copolymers–Industrial applications. I. Parthiban, Andanandam, editor of compilation.
TP156.P6S95 2014
668.9′2–dc23

2014002692

Printed in the United States of America

ISBN: 9781118057469

10 9 8 7 6 5 4 3 2 1

CONTENTS

PREFACE

Natural polymers such as cellulose and rubber are intertwined with technological advances at various ages of human activity. Also, many technological developments in the space age are centered around the development of polymers. Both light weight and processability have been the key factors that advanced and broadened the use of polymers. The frontiers of technology are constantly pushed forward because of factors such as environmental regulations and depletion of resources, which make development of high performance materials a constant affair.

In the modern era, polymers are widely used in everyday life in a variety of forms such as films, fibers, foams, molded articles, and sheets. Polymers are also used as adhesives, binders, fillers and in various other forms. In some of these applications, polymers are used in the dispersed form either in a solid matrix or in a liquid form. It is often the case that polymers that are made up of single monomers do not possess the desired properties in terms of processability, thermomechanical properties and surface compatibility. Polymerization of one monomer with other monomers, commonly referred to as copolymerization, is one of the strategies often used to improve various properties of polymers such as adhesion, film-forming tendency, high temperature performance, processability, solvent resistance and wettability. Reactivity of monomers becomes crucial in determining the outcome of such polymerization processes. Since the physicochemical characteristics of each monomer are unique, their polymerization nature also varies with the monomer. Hence, it is important to understand the reactivity of monomers and reaction conditions in order to make copolymers, which are completely or predominantly free of homopolymers so that the newly formed polymeric materials exhibit properties expected of them.

Reports on the synthesis and properties of copolymers remain scattered in journal publications and conference proceedings. The focus of some review articles and dedicated conference proceedings also tend to be relatively narrow in scope, revolving around one key area of application. Unlike all previously published articles and materials available in public domain, this book aims to be broad in scope so that it may appeal to a wider section of professionals, from students to research scientists, in academia and industry. By keeping this in mind, a broad range of subjects have been covered.

This book is divided into two sections with the first section covering synthesis of copolymers and the second section discussing various applications of copolymers. One unique characteristic of polymers is their ability to form films; because of which they find application in areas as far apart as conducting materials to structural resins. The ability to form thin films in polymers has been widely exploited in areas such as antifouling surfaces, capacitors, conducting coatings, dielectric materials, lubricants, photosensitive materials, semiconductors and solar cells. The final application of a polymer is determined by the backbone composition which in turn is decided by the nature of monomers used during polymerization. These monomers differ in functionality and hence the mode of polymerization differs as well. These complexities bring in challenges not only for design and synthesis of monomers but also for polymerization of these monomers. As discussed in Chapter 1, synthetic techniques for making polymers are constantly evolving and the trend is more and more toward controlling the chain length, chain-end functionality, composition, arrangement of monomer (sequence) in the chain, and so on. Polyolefins are well known for over half a century at present. There have been many interesting developments of materials and mechanisms over the years such as ultrahigh molecular weight (UHMW) polymers and chain shuttling, respectively. However, as described in Chapter 2, some developments like copolymerization with functional monomers are still elusive. Chapter 2 also gives a detailed mechanistic account of olefin polymerization. Polymerization of vinyl monomers constitutes one of the most important, high volume industrial activities. Property tuning by copolymerization is a vital process and the reactivity of monomers is an important parameter that determines the copolymer formation as well as composition of copolymers. These are the subjects covered as part of Chapter 3. Quite apart from introducing comonomers during polymerization in order to tune the properties of polymers, it is also quite possible to influence polymer properties by performing post-polymerization reactions. Polymers possessing reactive functional groups that are otherwise inert during polymerization are useful for this purpose. The cyclic carbonate group, with its ability to undergo nucleophilic addition with amines and alcohols, is one such suitable functionality. Synthesis and its specific properties of such polymers are discussed in detail in Chapter 4. Depletion of fossil fuels as well as efforts to curtail the emission of green house warming gases has increased the focus in the direction of renewable processes and materials. Chapter 5 summarizes the state of the art as well as the possibility of making various monomers and polymers by renewable processes. One of the recent developments in polymer synthesis has been forming polymers by reacting monomers bearing multiple functional groups. Depending on the symmetry and reactivity of monomers involved, ladder type, networked, and hyper-branched polymers are formed, as described in

Chapters 6 and 7. Chapter 6 exhaustively analyzes the synthesis, nature, and type of pores formed and applications of microporous organic polymers. Chapter 7 provides a broad account of various synthetic strategies and type of monomers used for making dendritic copolymers.

Fixing carbon dioxide (CO_2), a global warming gas whose emission increased proportionately with various industrial activities of humans, is one of the aims of the global research community. Chapter 8 describes a potential application of copolymers obtained by fixing CO_2 in the area of polymer electrolytes. The key feature of copolymers in particular block copolymers is its ability to self-assemble. The nanopatterns formed as a result of such self-assembly have been proposed in various applications in electronic industry as discussed in Chapter 9. Polymers that exhibit solubility characteristics dependent on external stimuli, such as temperature owing to structural changes that accompany temperature, are useful for many applications such as controlled release of active ingredients and injectable drug delivery. Chapter 10 gives a detailed account of stimuli-responsive polymers, that is, various stimulants and applications of the stimuli-responsive polymers. Historically, many biocompatible and biodegradable polymers have been used as coating materials for drugs. Chapter 11 provides a detailed summary of such pharmaceutical polymers. It is a recent trend in the biomedical field to covalently link drugs and polymers, commonly called polymer conjugates, in order to either increase the solubility of drugs or slow down the rate of absorption of drugs by the human body. Such delayed release not only maintains a desired concentration of drugs in the body but also helps to overcome the toxicity possessed by some drugs. Chapter 12 gives an overview of polymer–drug conjugates and other related up-to-date developments.

CONTRIBUTORS

Ajazuddin Department of Pharmaceutics, Rungta College of Pharmaceutical Sciences & Research, Chattisgarh, India

João A. S. Bomfim Public Research Center "Henri Tudor", Department of Advanced Materials and Structures, Hautcharage, Luxembourg; Mondo Luxembourg SA, R&D Department, Rue de l'Industie, Foetz, Luxembourg

Hideki Ichikawa Faculty of Pharmaceutical Sciences, Kobe Gakuin University, Chuo-ku, Kobe, Japan

Satyasankar Jana Institute of Chemical and Engineering Sciences, Agency for Science, Technology and Research (A*STAR), Singapore

Parijat Kanaujia Institute of Chemical and Engineering Sciences, Agency for Science, Technology and Research (A*STAR), Singapore

Sivashankar Krishnamoorthy Institute of Materials Research, Research and Engineering, Agency for Science Technology and Research (A*STAR), Singapore; Science et Analysis des Materiaux (SAM), Centre de Recherche Public-Gabriel Lippmann, Belvaux, Luxembourg

Fabio di Lena Public Research Center "Henri Tudor", Department of Advanced Materials and Structures, Hautcharage, Luxembourg; Empa, Swiss Federal Laboratories for Materials Science and Technology, Laboratory for Biomaterials, Gallen, Switzerland

Anbanandam Parthiban Institute of Chemical and Engineering Sciences, Agency for Science, Technology and Research (A*STAR), Singapore

Bien Tan School of Chemistry and Chemical Engineering, Huazhong University of Science and Technology, Wuhan, China

He Tao Institute of Chemical and Engineering Sciences, Agency for Science, Technology and Research (A*STAR), Singapore

Yoichi Tominaga Department of Organic and Polymer Materials Chemistry, Tokyo University of Agriculture and Technology, Tokyo, Japan

Alex van Herk Institute of Chemical and Engineering Sciences, Agency for Science, Technology and Research, Singapore

Natarajan Venkatesan Chicago College of Pharmacy, Midwestern University, IL, USA

Srinivasa Rao Vinukonda Central Institute of Plastics Engineering & Technology, Phase II, Cherlapally Industrial Area, Hyderabad, India

Shujun Xu School of Chemistry and Chemical Engineering, Huazhong University of Science and Technology, Wuhan, China

SECTION I

SYNTHESIS OF COPOLYMERS

1

TRENDS IN SYNTHETIC STRATEGIES FOR MAKING (CO)POLYMERS

ANBANANDAM PARTHIBAN

1.1 BACKGROUND AND INTRODUCTION

Polymers have been an inherent part of human life for well over half a century at present. In spite of the comparably poor mechanical properties of polymers with that of metals, polymers encompass the applications of materials ranging from metals to glass and have replaced them in many applications. Light weight in combination with ease of processing as compared to that of metals and glasses are two of the most favorable characteristics of polymers. These characteristics are of great significance in the present circumstances as efforts are being made to lower energy consumption in various processes, and thus such lesser energy consumption would also accompany lesser emissions of CO_2. Polymeric materials with enhanced properties are required in order to meet the ever improving technological needs in various fields. In addition to the demand in technological improvements, health concerns, predominantly about monomers, also bring in legislative changes leading to the disappearance of polymers from the markets, if not completely, in selected sectors of application where, in fact, these polymers were in use for many decades. One recent example is the polycarbonate derived from Bisphenol A. Owing to the suspected nature of Bisphenol A as an endocrine disruptor, its use in drinking water bottles has been banned recently in some of the developed countries. There has been as immense pressure to replace Bisphenol A in other applications as well. Indeed, until now, Bisphenol A is one of the widely employed monomers for making linear polymers like polycarbonates as well as thermosetting resins and adhesives based on bisepoxy compound (Figure 1.1).

Synthesis and Applications of Copolymers, First Edition. Edited by Anbanandam Parthiban.

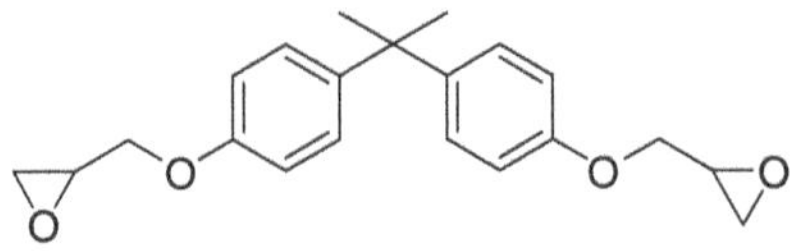

Figure 1.1 Chemical structure of bisepoxy compound.

Although the use of Bisphenol-A-based polymers is likely to face continued intense scrutiny, there are interesting developments with some other polymers that date farther back such as polyethylene. Polyethylene is an interesting case. For a long time, it is the largest volume of synthetic polymer falling under the category of commodity plastic. However, there are recent trends that expand the application of polyethylene into selected specialty areas. The efforts for making ultrahigh molecular weight polyethylene (UHMWPE) and make use of these materials in offshore applications are of particular significance in this regard. With the development of processes for converting bioethanol to ethylene, a green label is being attached to polyethylene in addition to other claims like lower carbon footprint in comparison to other polymers. It is interesting to note that the nondegradable nature of polyolefins and in particular polyethylene is of immense concern for a long time because it is the largest volume of synthetic polymer and thus constitutes a major component in landfill.

Synthesis and the development of synthetic methodologies are the life blood of new materials. The challenges posed by changing and ever demanding technologies and also health and environmental concerns can be met by synthetic methods that evolve with time. It is an objective of this chapter to give an overview of interesting synthetic methods developed in the recent past. In published literature, newer methods and polymers made therefrom abound. However, any new development has to meet many if not all of the following requirements in order for the process and/or product to reach industrial scale manufacturing processes and subsequently the market:

1. Reagents and catalysts employed for making new monomers and polymers should be, preferably, available on industrial scale.
2. Requirement of low consumption of energy, which means that the process be of low or moderately high temperature and pressure in nature.
3. Processes that demand unusual machinery, very high rate of mixing, exotic reagents, solvents, and reaction conditions should be avoided.
4. Methods of purification should be simple and straightforward, preferably free from techniques like column chromatography.
5. Use of nontoxic and recyclable solvents is an important criterion to be considered for industrial scale processes.
6. It is also preferable to have processes in which lesser number of steps is involved to make the desired product.

7. Atom economical and high yielding processes are generally preferred.
8. It is also desirable to have processes that do not generate lot of waste water and do not make use of highly corrosive reagents or catalysts.

The developments discussed in following sections may be looked at by keeping the above requirements for a successful process.

Interestingly, some of the stated objectives of the synthetic methodologies developed in the recent past are as follows:

1. To make polymers of well-defined or predetermined molecular weights and low polydipersities.
2. Control over (co)polymer architectures.
3. Sequence-regulated polymerization.
4. Formation of reversible covalent bonds.
5. Self-healing or self-repairing polymers.
6. Recyclable thermoset polymers.
7. Chain-shuttling polymerization.

Although many of the abovementioned developments have generated immense interest only among academic communities and thus resulted in enormous volumes of publications, these are nevertheless worth noting on account of very interesting material characteristics achieved through these developments.

1.2 SIGNIFICANCE OF CONTROL OVER ARRANGEMENT OF MONOMERS IN COPOLYMERS

Developments in controlled radical polymerization had led to the formation of polymers of varying structures such as block copolymers, cylindrical brushes, gradient copolymers, graft copolymers, hyperbranched polymers, macrocycles, and miktoarm stars. Each of these polymers possessed unique characteristics that were absent in the corresponding linear polymers, although in terms of chemical composition they were alike. An interesting case is the gradient copolymers whose physical properties differed considerably from the corresponding block and random copolymers of similar chemical composition as given in Table 1.1 [1].

1.3 CHAIN-GROWTH CONDENSATION POLYMERIZATION

Among the various polymerization techniques, step-growth or condensation polymerization has its own place in making polymeric materials with unique properties. A large majority of condensation polymers are engineering thermoplastics well known for their high temperature properties, crystallinity, excellent mechanical properties, and so on. Polyesters as represented by poly(ethylene and butylene terephthalate)s, aromatic and aliphatic polyamides, polyimides, a wide

TABLE 1.1 Comparison of Block and Gradient Copolymers of Poly(styrene-*co*-methylacrylate)

Type of Copolymer	T_g, °C	Dynamic Mechanical Analysis Results
PS-*b*-PMA	~100, ~10	Two distinct segmental relaxation processes
PS-*ran*-PMA	75	
PS-*grad*-PMA	>10 (rapidly cooled)	Single, extremely broad segmental relaxation
PS-*grad*-PMA	>10, ~70 (annealed above 50 °C)	

variety of poly(arylene ether)s such as polyether ether ketone (PEEK) and other poly(ether ketone)s, poly(ether sulfone), and poly(benzimiazole)s are some of the well-known examples of polymers formed by condensation polymerization. Condensation polymerization that typically involves AA- and BB-type monomers or AB-type monomers, where A and B represent different reacting functionalities during polymerization, generally yields polymers with polydispersity of 2 or more. However, recently, Yokozawa et al. [2] have introduced a new concept termed as *chain-growth* condensation polymerization whereby the molecular weights of condensation polymers such as polyamides, polyesters, and polyethers have been controlled and polydispersity of these polymers is well below the theoretically predicted 2. Some special *para*-substituted AB-type aromatic monomers were employed for this purpose (Scheme 1.1). By introducing an activated functional group in the AB-type monomer, a preferred reaction site was created that resulted in sequential addition of monomers.

1.3.1 Sequential Self-Repetitive Reaction (SSRR)

Dai et al. [3] reported a sequential self-repetitive reaction by which the condensation of diisocyanate with diacid in the presence of a carbodiimide catalyst like 1,3-dimethyl-3-phospholene oxide (DMPO) led to the formation of polyamide (Scheme 1.2). The reaction is so called because of the occurrence of repetitive reactions sequentially by the following three steps:

1. Condensation of two isocyanates to yield carbodiimide.
2. Addition of carboxylic acid to carbodiimide leading to the formation of *N*-acyl urea.
3. Thermal fragmentation of *N*-acyl urea that results in amide and isocyanate fragments in half or the original molar quantities.

Carbodiimide formation and the addition of carboxylic acid to carbodiimide took place in a facile manner at an ambient temperature. On the contrary, fragmentation of *N*-acyl urea occurred above 140 °C. In order to form the

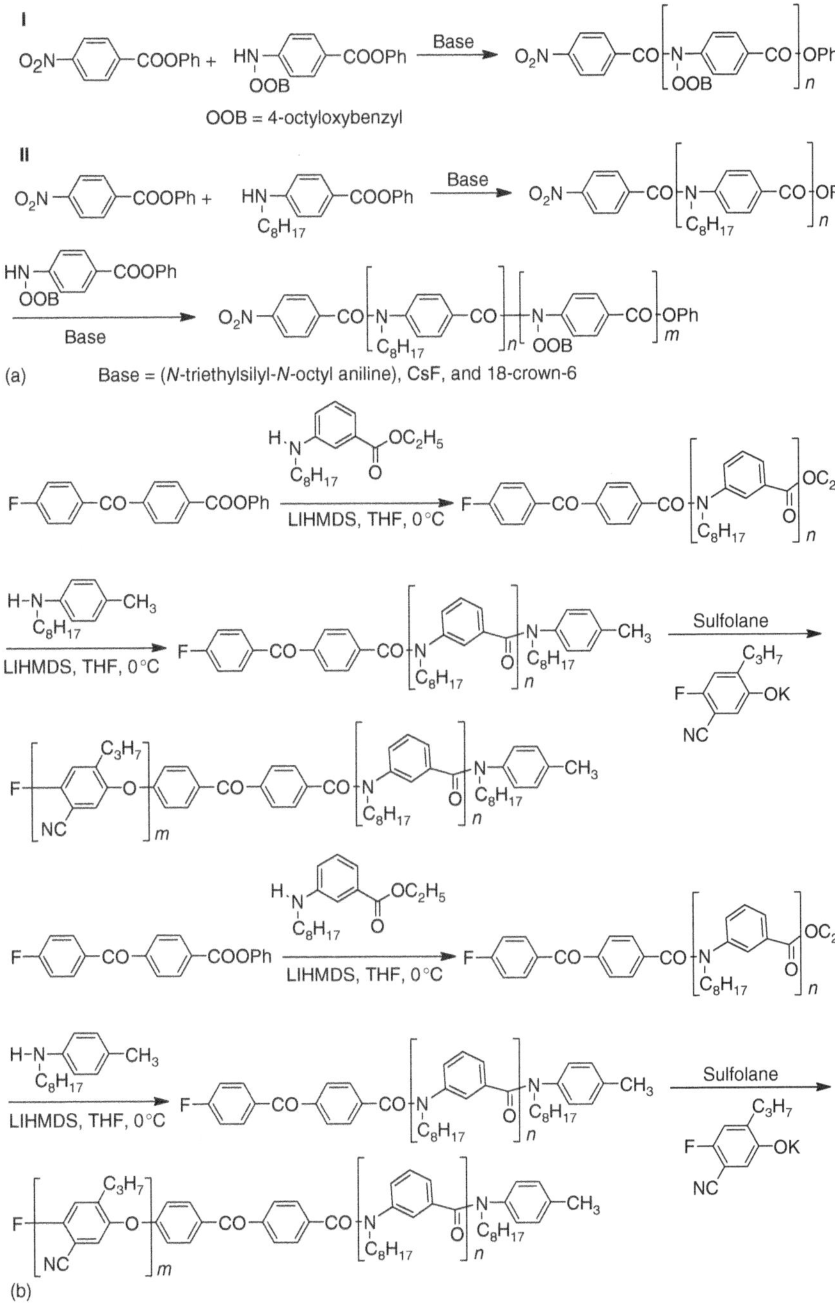

I

O_2N–C6H4–COOPh + HN(OOB)–C6H4–COOPh —Base→ O_2N–C6H4–CO[N(OOB)–C6H4–CO]$_n$OPh

OOB = 4-octyloxybenzyl

II

O_2N–C6H4–COOPh + HN(C_8H_{17})–C6H4–COOPh —Base→ O_2N–C6H4–CO[N(C_8H_{17})–C6H4–CO]$_n$OPh

HN(OOB)–C6H4–COOPh, Base → O_2N–C6H4–CO[N(C_8H_{17})–C6H4–CO]$_n$[N(OOB)–C6H4–CO]$_m$OPh

(a) Base = (*N*-triethylsilyl-*N*-octyl aniline), CsF, and 18-crown-6

H-N(C_8H_{17})–C6H4–CO–OC_2H_5; LIHMDS, THF, 0 °C; H-N(C_8H_{17})–C6H4–CH_3; Sulfolane; C_3H_7, F, OK, NC; CO; OC_2H_5; F; CO; COOPh; *n*; *m*

(b)

Scheme 1.1 (a) Polyamides (I) and block copolyamides (II) prepared by step-growth polymerization. (Reprinted with permission from [2l]. Copyright © 2002 American Chemical Society.) (b) Preparation of diblock copolymers, poly(amide-*block*-ether) by chain-growth condensation polymerization. (Reprinted with permission from [2m]. Copyright © 2009 Wiley Periodicals Inc.)

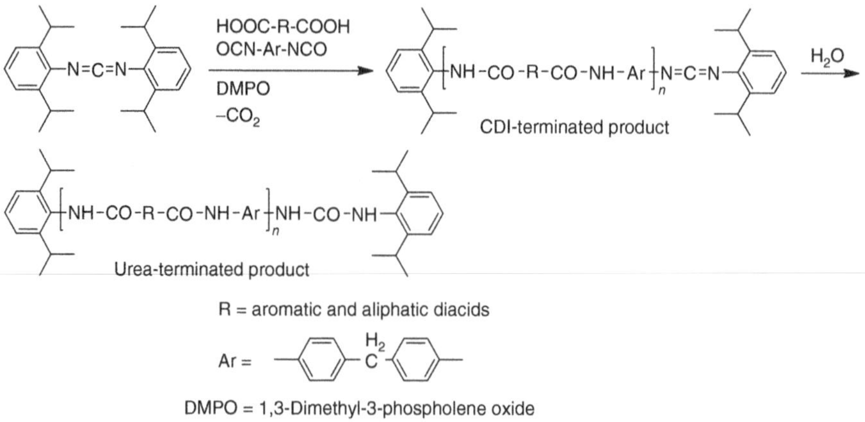

HOOC-R-COOH
OCN-Ar-NCO
DMPO
$-CO_2$
CDI-terminated product
H_2O
Urea-terminated product

R = aromatic and aliphatic diacids

DMPO = 1,3-Dimethyl-3-phospholene oxide

Scheme 1.2 Synthesis of polyamides by sequential self-repetitive reaction. (Reprinted with permission from [3a]. Copyright © 2002 American Chemical Society.)

polyamide, the aforementioned three steps were carried out by heating the reaction mixture with sufficient acid.

1.3.2 Poly(phenylene Oxide)s by Chain-Growth Condensation Polymerization

Kim et al. [4] reported the synthesis of well-defined poly(phenylene oxide)s bearing trifluoromethyl ($-CF_3$) groups through nucleophilic aromatic substitution (S_NAr) reaction (Scheme 1.3). For the chain-growth condensation polymerization to proceed in a controlled manner, the correct choice of initiator was a prerequisite. If the initiator chosen was far more reactive, chain transfer reactions occurred during the course of polymerization. As a result of this, multimodal gel permeation chromatograms were observed indicating the inhomogeneous nature of polymers that were produced by this process. For example, when 4-nitro-3-(trifluoromethyl)benzonitrile was employed as an initiator in the chain-growth condensation polymerization of 4-fluoro-3-(trifluoromethyl)potassium phenolate, transetherification prevailed in the reaction because of the more activated nature of initiator. However, the polymerization proceeded in a controlled manner when 2-nitrobenzotrifluoride was used as an initiator. Gel permeation chromatographic analysis of polymers of the reaction between 2-nitrobenzotrifluoride and 4-fluoro-3-(trifluoromethyl)potassium phenolate revealed that the molecular weight increased linearly with conversion and the observed molecular weight was close to the theoretically estimated molecular weight. The polydispersity lowered with conversion and the polydispersity of final polymers were relatively narrow.

1.3.3 Hydroxybenzoic Acids as AA′ Type Monomer in Nucleophilic Aliphatic Substitution Polymerization

Hydroxybenzoic acids such as 4-hydroxybenzoic acid are well-known AB-type monomers capable of undergoing self-polymerization yielding, in some cases,

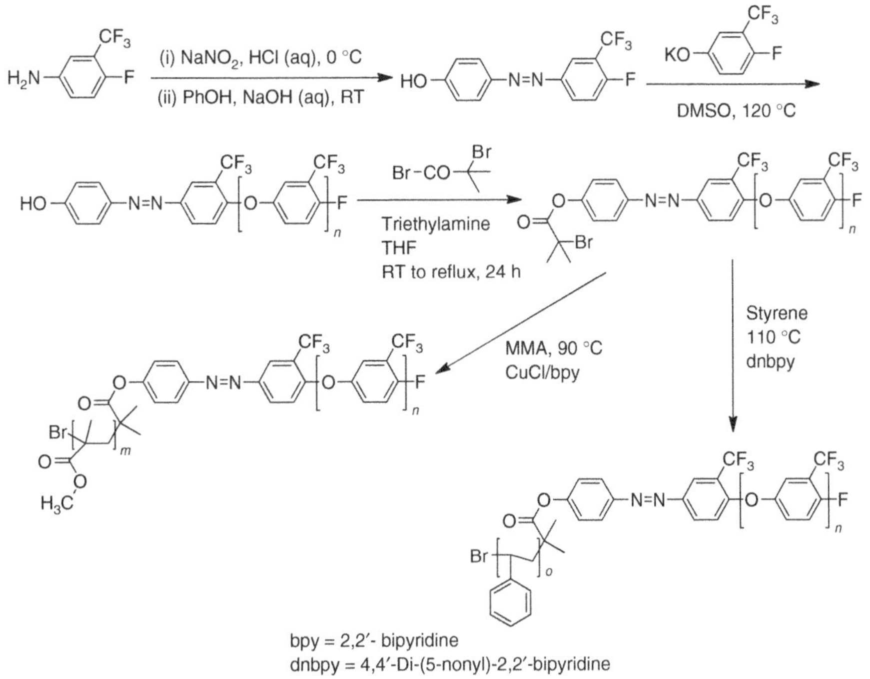

bpy = 2,2′- bipyridine
dnbpy = 4,4′-Di-(5-nonyl)-2,2′-bipyridine

Scheme 1.3 Synthesis of rod–coil block copolymers by chain-growth condensation polymerization. (Reprinted with permission from [4a]. Copyright © 2010 Wiley Periodicals Inc.)

polymers with interesting and useful characteristics such as liquid crystallinity. However, this AB-type monomer has been reported to function as AA′-type monomer in the reaction with *trans*-1,4-dibromo-2-butene, yielding a new class of materials [5] (Scheme 1.4) with high molecular weights (Table 1.2). Both phenoxide and carboxylate anions of hydroxybenzoic acid took part in the nucleophilic substitution nearly simultaneously. As expected, the reaction between AB_n-type hydroxybenzoic acid and *trans*-1,4-dibromo-2-butene yielded network polymers. Reduction of unsaturated bond in such poly(ether ester)s yielded an interesting class of polymers whose chemical structure was comparable to that of poly(butylene terephthalate), except that one of the ester linkages of latter polymer was replaced by an ether (–O–) linkage in the former. Such substitution led to substantial reduction in melting transitions, about 100 °C, but also induced crystallization during heating as well as upon cooling.

1.4 SEQUENCE-CONTROLLED POLYMERIZATION

Nature's way of making polymers, in particular polypeptides, is often envied because of the control with which such polymers are made precisely in terms of

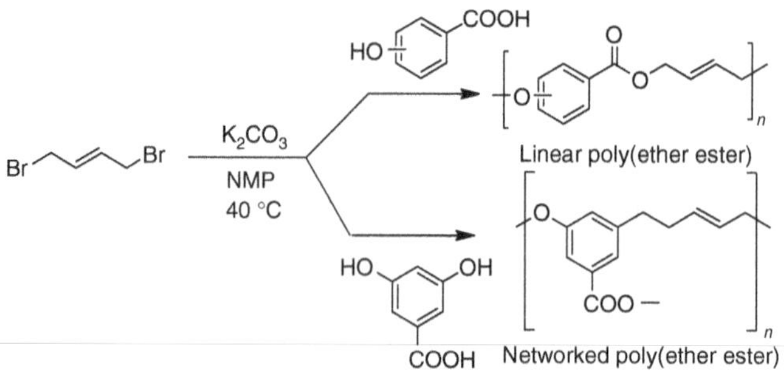

Scheme 1.4 Linear and networked polymers formed by using hydroxybenzoic acid as AA′-type comonomer. (Reprinted with permission from [5a]. Copyright © 2011 Wiley Periodicals Inc.)

TABLE 1.2 Hydroxy Benzoic acid as AA′-Type Comonomer in Aliphatic Nucleophilic Substitution Reaction[a]

HBA monomer (solvent)	Reaction time (h) and temperature (°C)	⟨M_n⟩	⟨M_w⟩	PD
4-HBA (DMA)	48, 40	22,200	35,850	1.60
4-HBA (NMP)	64, RT	10,800	17,300	1.60
4-HBA (NMP)	24, RT	8,300	12,120	1.50
4-HBA (NMP)	48, 40	22,530	33,750	1.50
4-HBA (NMP)	64, 40	26,460	39,220	1.50
4-HBA (NMP)	48, 50	9,970	15,590	1.60
3-HBA (NMP)	48, 40	25,550	38,080	1.50
HNA (NMP)	48, 40	Insoluble		

4-HBA, 4-hydroxy benzoic acid; 3-HBA, 3-hydroxy benzoic acid; HNA, 6-hydroxy-2-naphthoic acid; NMP, *N*-methyl-2-pyrrolidinone; DMA, *N,N*-dimethyl acetamide. [a][5].

sequence and tacticity [6]. In such natural process, it is very common that polypeptides are made up of 20 or more amino acids in repeat unit sequence. Developments reported thus far and the following description need to be considered cautiously, because in comparison to the natural processes, the polymeric systems reported are too simple and also are restricted to just two different monomer sequences.

1.4.1 Sequence-Controlled Copolymers of N-Substituted Maleimides

The reactivity difference between styrene and various N-substituted maleimides has been utilized for making copolymers (Scheme 1.5) [7].

1.4.2 Alternating Copolymers by Ring-Opening Polymerization

By making use of bis(phenolate) group 3 metal complexes as initiators, highly alternating copolymers were prepared by ring-opening polymerization of a mixture of

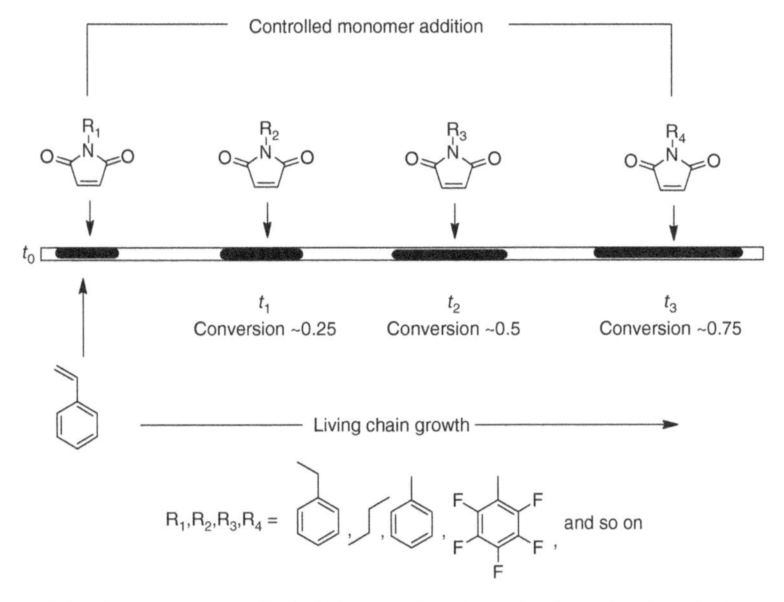

Scheme 1.5 Sequence-controlled chain-growth polymerization. (Reprinted with permission from [6]. Copyright © 2010 Royal Society of Chemistry.)

enantiomerically pure but different monomers (Scheme 1.6) [8]. The co-catalyzed carbonylation of epoxides yielded optically active β-lactone. The β-lactone in presence of yttrium initiators formed alternating copolymers.

1.4.3 Selective Radical Addition Assisted by a Template

A template-dependent recognition was noticed in the copolymerization of sodium methacrylate (NaMA) and methacryloyloxyethyl trimethylammonium chloride (ACMA) (Scheme 1.7) [9].

1.4.4 Alternating AB-Type Sequence-Controlled Polymers

A naphthalene template was used for making alternating copolymer of methyl methacrylate and acrylate [10]. The monomeric units methacrylate and acrylate were anchored spatially close to each other in the *peri*-position (1,8-position) of naphthalene to form the A–B templated divinyl monomer that was polymerized under dilute conditions by metal-catalyzed living radical polymerization. After polymerization, the naphthalene template was removed by hydrolysis, and the polymers were methylated again to yield alternating AB-type copolymer of poly(methyl methacrylate-*alt*-acrylate).

1.4.5 Metal-Templated ABA Sequence Polymerization

A palladium template structure was designed to achieve selective, intramolecular, and directional double cyclopolymerization at the template. ABA-type

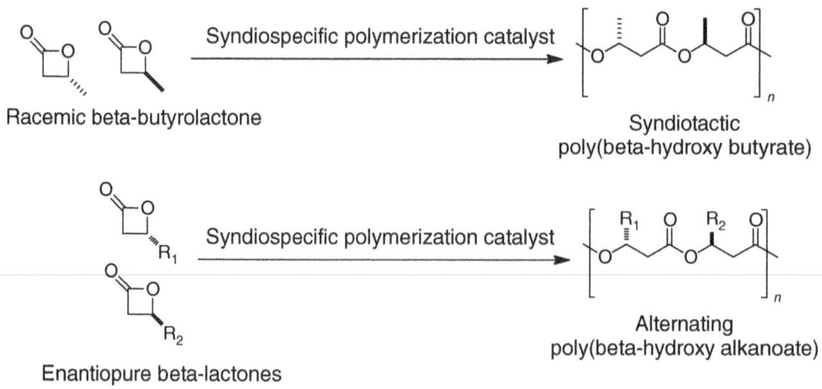

Racemic beta-butyrolactone
Syndiospecific polymerization catalyst
Syndiotactic poly(beta-hydroxy butyrate)
Enantiopure beta-lactones
Syndiospecific polymerization catalyst
Alternating poly(beta-hydroxy alkanoate)

Scheme 1.6 Sequence-controlled ring-opening polymerization. (Reprinted with permission from [8]. Copyright © 2009 American Chemical Society.)

Ru catalyst
Radical addition
Radical initiating site
Size-selective monomer recognition
Highly selective addition

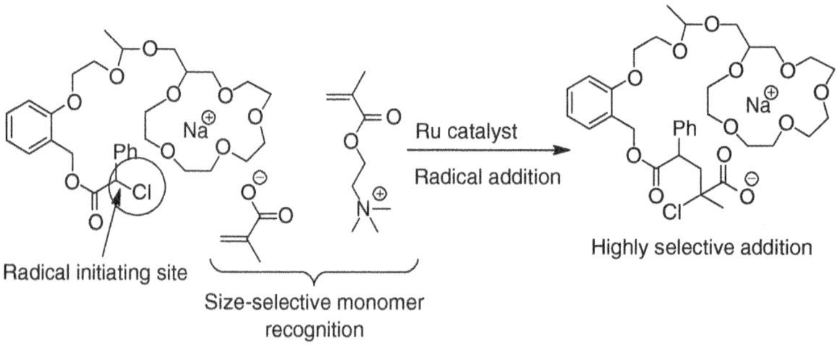

Scheme 1.7 Template-assisted size-selective monomer recognition. (Reprinted with permission from [9]. Copyright © 2010 American Chemical Society.)

alternating copolymers of styrene-4-vinylpyridine-styrene were prepared by this approach (Scheme 1.8) [11]. It is important to note that the polymerization was not straightforward and insoluble polymers were obtained generally due to cross-linking reactions. Polymerization in a bulky fluoroalcohol like 1,1,1,3,3,3-hexafluorophenyl-2-propanol (HFFP) at temperatures −5 to −60 °C proceeded smoothly. However, the reaction time required was very long ranging from 48 to 120 h.

1.4.6 Sequence-Controlled Vinyl Copolymers

In an atom transfer radical polymerization reaction, the chain end was made inactive by atom transfer radical addition of allyl alcohol to yield a primary alcohol. The chain end was again activated by the oxidation of primary alcohol to carboxylic acid followed by esterification to induce polymerization again. These sequence of reactions were repeated with the introduction of different side groups

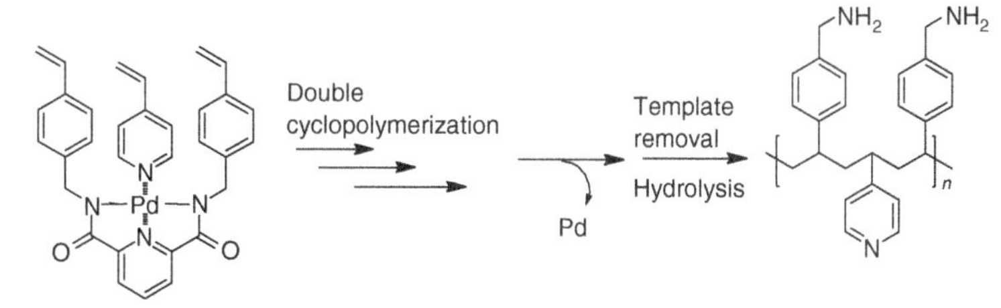

Scheme 1.8 Sequence-regulated metal-templated polymerization of ABA monomer. (Reprinted with permission from [11]. Copyright © 2011 Wiley Periodicals Inc.)

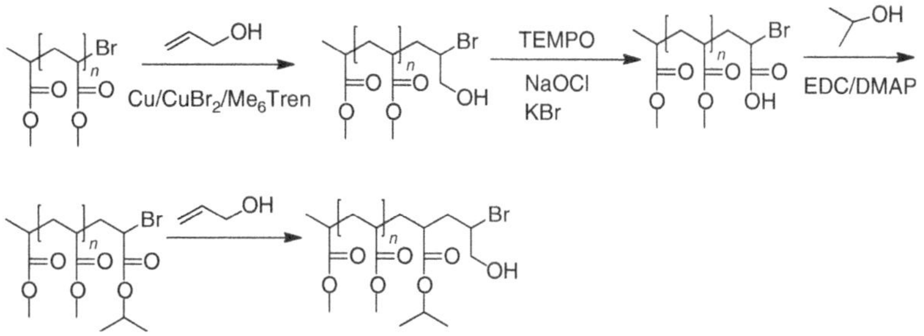

Scheme 1.9 Sequence-controlled vinyl copolymers. (Reprinted with permission from [12]. Copyright © 2011 Royal Society of Chemistry.)

at each esterification step (Scheme 1.9) [12]. Atom transfer radical addition also led to loss of bromo-end groups although oxidation and esterification were nearly quantitative.

1.4.7 Sequence-Regulated Polymerization Induced by Dual-Functional Template

By a suitable design of template having cationic and radical initiating sites, it was possible to polymerize, preferably, only the vinyl monomer that interacted with the template over the noninteracting vinyl monomer (Scheme 1.10) [13].

1.5 PROCESSING OF THERMOSET POLYMERS: DYNAMIC BOND FORMING PROCESSES AND SELF-HEALING MATERIALS

Temperature-induced flow is one of the distinguishing features of thermoplastics from thermosets. This behavior of thermoplastics makes them easily processable, thereby allowing thermoplastics to be extruded, injection molded, thermally formed into fibers, films, filaments, pellets, and so on. Thermosets have many interesting, useful, and unique characteristics that are as follows: superior dimensional

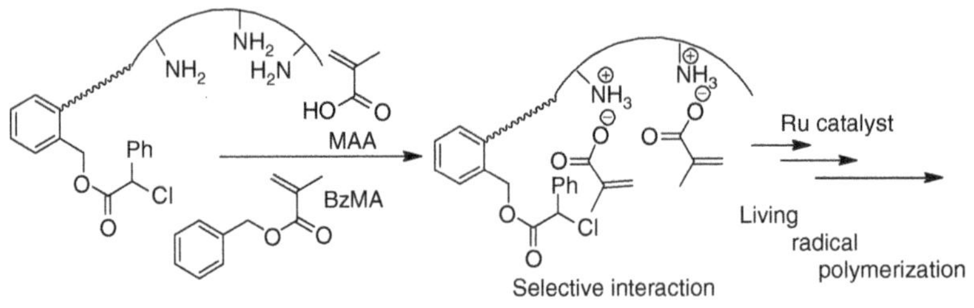

Scheme 1.10 Template-assisted sequence-regulated polymerization. (Reprinted with permission from [13]. Copyright © 2011 Wiley Periodicals Inc.)

stability, ability to perform at high temperatures, solvent resistance, creep resistance, high fracture strength, and high modulus. These properties make thermosets attractive for applications in a variety of fields such as adhesives, coatings, electrical insulation, printed circuit boards, and rubbers. Unlike thermoplastics that are reprocessable, thermoset formation is an irreversible process. However, some of the recently evolved concepts challenge this age-old phenomenon. There are instances, where, by making the covalent linkages reversible, concepts such as repairability and processability could be introduced into thermosets. For this purpose, it is necessary to form networks with chemical bonds that are labile, that is, bond breaking and bond making could be induced, for example, heat, catalysts, light, and reagents.

1.5.1 Plasticity of Networked Polymers Induced by Light

A photomediated reversible backbone cleavage in a networked structure without any degradation of mechanical properties was achieved by addition–fragmentation chain transfer process involving allyl sulfides. Initially, reaction of a radical with an in-chain functionality leads to the formation of an intermediate. This intermediate in turn fragments thereby reforming the initial functionality and the radical. As a result of the addition–fragmentation process, the topology of the network was changed; however, the nature of network remained unchanged. The network strands were unaffected, provided there were no side reactions or radical termination processes. Under such conditions, the number of allyl sulfide groups also remained unchanged. The fragmentation and reformation process facilitated the stress relaxation in each bond. It may be noted that cleavage and reformation reactions occurred in a facile manner because of the rubbery nature of network with very low T_g, −25 °C [14a]. Since these polymers were unsuitable as structural materials, thiol–yne networks were proposed as suitable alternatives [14b].

1.5.2 Radically Exchangeable Covalent Bonds

Alkoxyamine units were utilized as thermodynamic covalent cross-linking system. Heating of a network polymer induced a state of equilibrium between dissociation

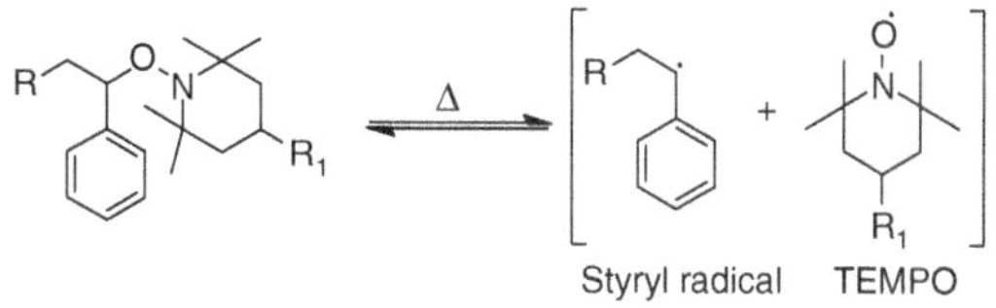

Styryl radical TEMPO

Scheme 1.11 Radical exchange reaction of an alkoxyamine derivative. (Reprinted with permission from [15]. Copyright © 2006 American Chemical Society.)

and association at the point of cross-linking. Poly(methacrylic ester)s possessing alkoxyamine as pendant groups underwent radical exchange reaction upon heating. The cross-linked structure was quantitatively decross-linked under stoichiometric control (Scheme 1.11) [15].

1.5.3 Self-Repairing Polyurethane Networks

Polyurethane networks exhibiting self-repairing characteristics when exposed to UV light were prepared by introducing chitosan substituted with oxetane groups in two component polyurethane. Any mechanical damage induced the ring-opening reaction of four-membered oxetanes, which resulted in two reactive ends. These reactive ends underwent cross-linking reaction with fragments formed from chitosan upon exposure to UV light thus repairing the network. These materials have been reported to self-repair in less than an hour and are proposed for applications as wide as transportation, packaging, fashion, and biomedical industries [16].

1.5.4 Temperature-Induced Self-Healing in Polymers

Polyketones with carbonyl groups in 1,4-arrangement were obtained by alternating co- or terpolymerization of carbon monoxide, ethylene, and propylene using homogeneous Pd-based catalysts. By Paul–Knorr reaction with furfurylamine, these 1,4-arranged polyketones were converted to furan derivatives that were subsequently subjected to Diels–Alder reaction with bismaleimide leading to the formation of cross-linked polymers. At elevated temperatures, these cross-linked polymers underwent retro Diels–Alder reaction. Thus, by making use of heat as external stimulus, these polymers could be subjected to many cycles of cross-linked–decross-linked structures. Dynamic mechanical analysis and three-point bonding tests demonstrated that this cycle is repeatable 100% for multiple times (Scheme 1.12) [17].

1.5.5 Diels–Alder Chemistry at Room Temperature

One of the interesting aspects of Diels–Alder reaction is that it is an addition process, and thus it is an atom economical process. Materials formed by Diels–Alder reactions have been termed as *dynamers*, which in turn is defined as a class

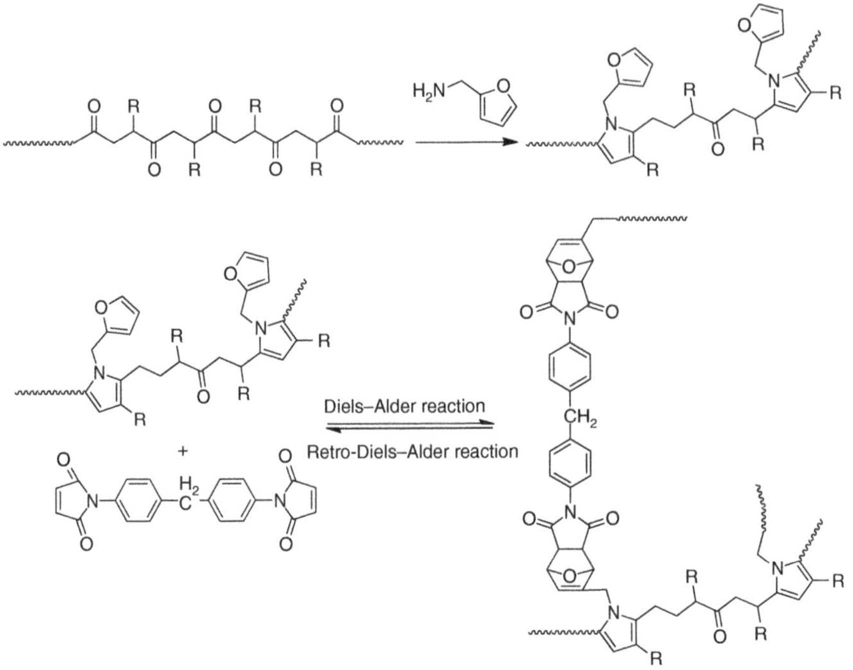

Scheme 1.12 Temperature-induced self-healing polymers. (Reprinted with permission from [17]. Copyright © 2009 American Chemical Society.)

of polymers that are formed by linking monomers in a reversible process. The reversible nature of the process allows continuous scrambling of polymer chain sequences. By employing bisfulvene dienes and bis(tricyanoethylene carboxylate) or bis(dicyanofumarate) as dienophile, a room temperature Diels–Alder process was reported (Scheme 1.13) [18].

1.5.6 Trithiocarbonate-Centered Responsive Gels

Trithiocarbonate units flanked between dimethacrylate terminal functionalities through linkers have been shown to exhibit dynamic covalent chemical characteristics. The network formed by this monomer undergoes reorganization either in the presence of CuBr/ligand or in the presence of radical initiators like AIBN (Scheme 1.14) [19].

1.5.7 Shuffling of Trithiocarbonate Units Induced by Light

The photoresponsive nature of trithiocarbonate units makes it undergo shuffling reaction when exposed to UV irradiation. Poly(n-butyl acrylate) cross-linked using trithiocarbonate was synthesized by radical addition–fragmentation chain transfer

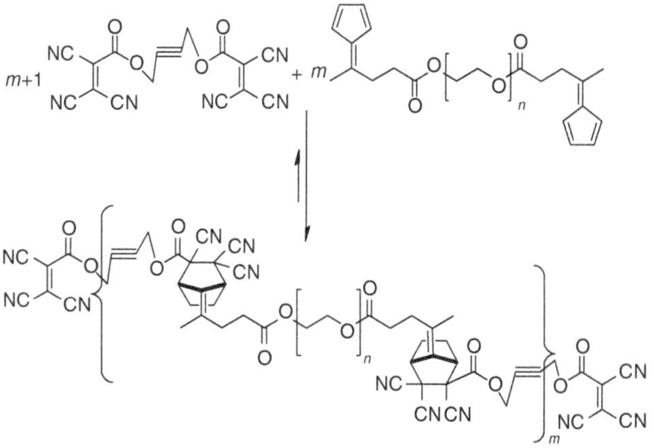

Scheme 1.13 Diels–Alder addition at room temperature. (Reprinted with permission from [18]. Copyright © 2009 Wiley Periodicals Inc.)

(RAFT) polymerization. Poly(*n*-butyl acrylate) was chosen as matrix due to its low T_g (−50 °C) because of which the chain mobility is high at room temperature. The tensile modulus of fresh sample was 69 ± 6 kPa and the same for self-healed polymers in the presence of solvent was 65 ± 11 kPa. The self-healing in bulk was incomplete due to the restricted chain mobility (Scheme 1.15) [20].

1.5.8 Processable Organic Networks

Networks formed by classical epoxy chemistry such as the reaction between diglycidyl ether of Bisphenol A and glutaric anhydride having epoxy/acyl ratio of 1 : 1 in the presence of 5 or 10 mol% zinc acetyl acetonate behave like processable glasses. Broken or ground samples of such networks, in spite of being cross-linked well above gel point, have been reported to be reprocessed by injection molding. These cross-linked networks behave like an elastomer at room temperature and are confirmed as networks by dissolution experiments since they display swelling tendency but do not dissolve even in good solvents upon prolonged immersion at high temperatures. The fact that these networked systems were able to relax stresses completely at high temperature and tend to flow was confirmed by rheology and birefringence studies [21].

1.6 MISCELLANEOUS DEVELOPMENTS

1.6.1 Atom Transfer Radical Polymerization (ATRP) Promoted by Unimolecular Ligand-Initiator Dual-Functional Systems (ULIS)

Within the last few decades, three major developments took place in the free-radical polymerization of vinyl monomers, namely, nitroxide-mediated free-radical

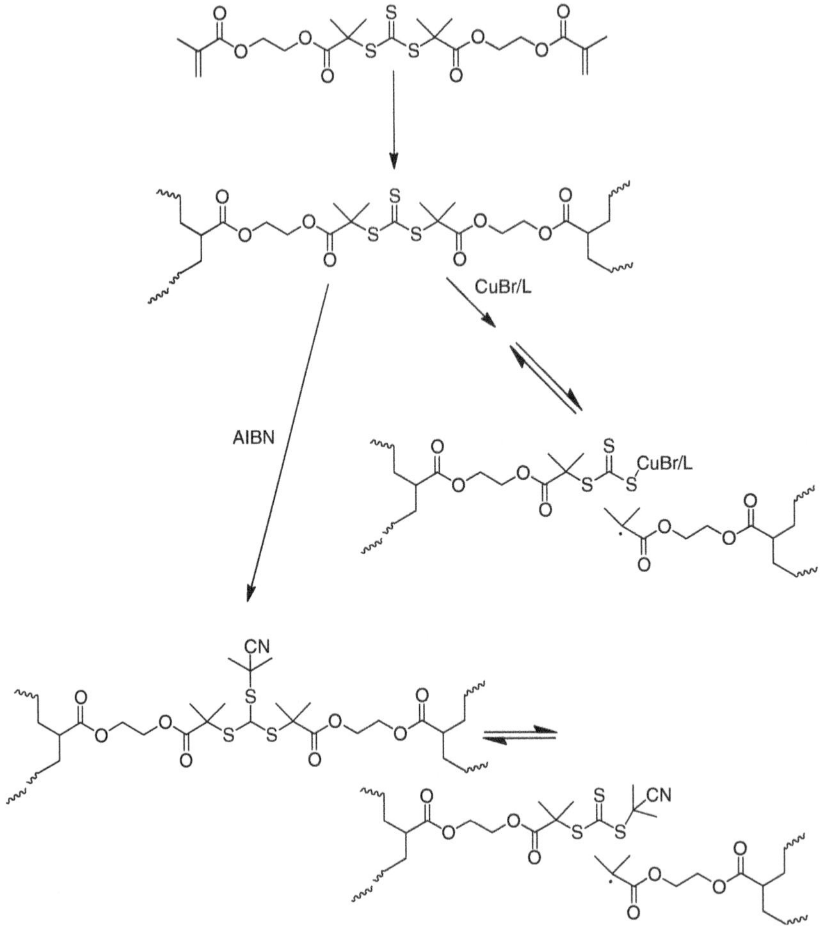

CuBr/L

AIBN

CuBr/L

CN

CN

Scheme 1.14 Trithiocarbonate-centered responsive gels. (Reprinted with permission from [19]. Copyright © 2010 American Chemical Society.)

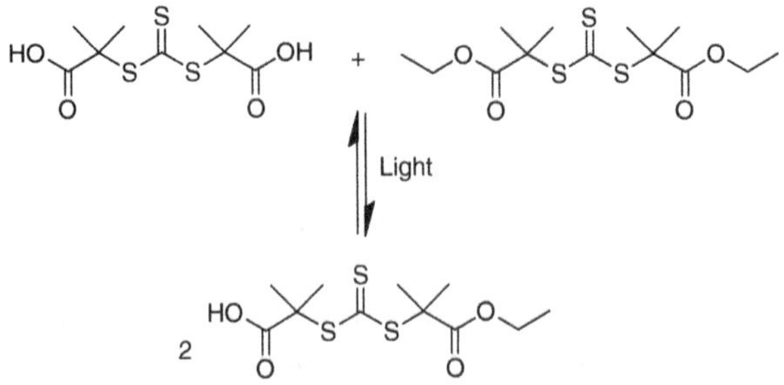

Light

Scheme 1.15 Light-induced shuffling of trithiocarbonate units. (Reprinted with permission from [20]. Copyright © 2011 Wiley Periodicals Inc.)

polymerization (NMP) [22], RAFT polymerization [23], and ATRP [24]. Each of these techniques, in spite of many advantages, has inherent deficiencies that prevent it from becoming a major commercial process. Among these three techniques, ATRP is somewhat more convenient to practice. Two of the major problems associated with ATRP and worthy of mentioning are follows: presence of high residual metal impurities in the form of copper and its salts; and the inability to homo or copolymerize acidic monomers like acrylic and methacrylic acids in the free acid form. The residual metal impurity poses many challenges such as resulting in colored polymers such as black, blue, dark blue, brown and green. In this context, it may be noted that vinyl polymers such as polystyrene and poly(methyl methacrylate) are bright white solids, yielding highly transparent materials upon processing in the absence of any impurities. The presence of metal residue could affect the thermooxidative stability of polymers as well as pose health concerns because of the toxic nature of residual metals. Copolymers bearing acrylic and methacrylic acids are important for imparting variety of characteristics such as improved adhesion and stimuli responsive behavior, and for dispersing in aqueous media.

In ATRP, copper in lower oxidation state such as Cu(I)Br undergoes a redox transition during polymerization in cycles described as active and dormant, which represent chain growth and dead stages, respectively. Typically, tertiary amines are used for complexing the copper salt. Also, initiation process involves the abstraction of halogen atom such as bromine or chlorine from alkyl bromides or chlorides, which result in the formation of alkyl radical. The alkyl radical thus formed initiates polymerization by transferring the radical to vinyl monomer.

A modified ATRP process that overcomes the aforementioned deficiencies has been reported recently [25,26]. The ligand that coordinates with the metal salt and the alkyl halide were covalently linked to form ligand initiators (Figure 1.2) [25,27]. In this modified process, reducing the concentration of copper salt from 1000s of ppm to 10s of ppm did not have much influence on the rate of polymerization [26]. The polymerization was highly influenced by reaction conditions such as nature of solvent employed and polymerization temperature. Through this ULIS-promoted ATRP, very high molecular weight polymers as well as block copolymers were obtained [28]. The residual metal content of polymers obtained by ATRP promoted by ULIS was theoretically estimated to be two orders of magnitude lower. Analysis of polymers by techniques like inductively coupled plasma (ICP) also confirmed that the metal residue present in polymers precipitated in methanol without passing through alumina column to be 5 ppm or lower, well below than that present in purified polymers obtained by conventional ATRP [29]. It was also possible to homo- and copolymerize acrylic and methacrylic acids directly through the modified process [26,30,31]. Terpolymers derived by using acrylic acid as one of the comonomers have also been reported by ULIS-promoted ATRP [30]. The end group fidelity of polymers obtained by the modified process has been verified by chain extension reactions as well as by elemental analysis of macroinitiators [26,28,29,31]. This modified process potentially takes ATRP one step closer to downstream.

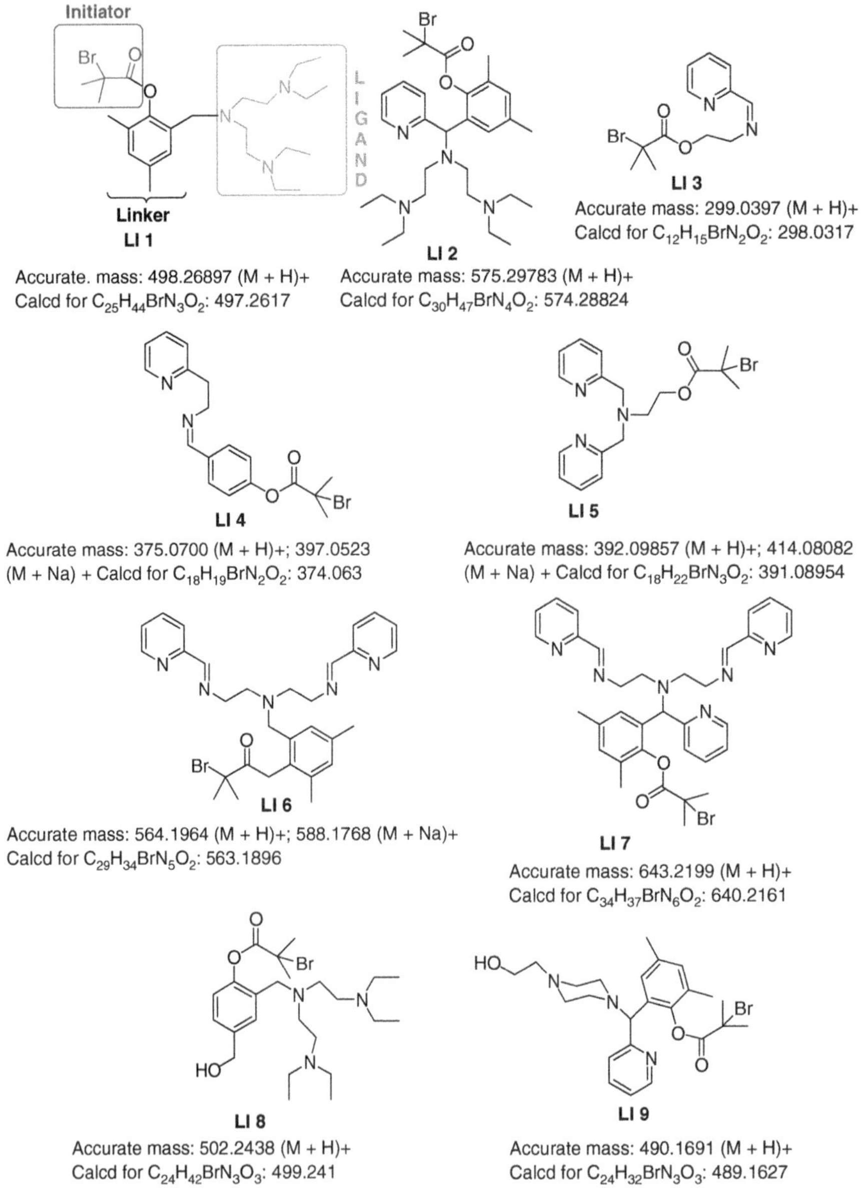

Initiator
Linker
LIGAND

LI 1
Accurate. mass: 498.26897 (M + H)+
Calcd for $C_{25}H_{44}BrN_3O_2$: 497.2617

LI 2
Accurate mass: 575.29783 (M + H)+
Calcd for $C_{30}H_{47}BrN_4O_2$: 574.28824

LI 3
Accurate mass: 299.0397 (M + H)+
Calcd for $C_{12}H_{15}BrN_2O_2$: 298.0317

LI 4
Accurate mass: 375.0700 (M + H)+; 397.0523 (M + Na) + Calcd for $C_{18}H_{19}BrN_2O_2$: 374.063

LI 5
Accurate mass: 392.09857 (M + H)+; 414.08082 (M + Na) + Calcd for $C_{18}H_{22}BrN_3O_2$: 391.08954

LI 6
Accurate mass: 564.1964 (M + H)+; 588.1768 (M + Na)+
Calcd for $C_{29}H_{34}BrN_5O_2$: 563.1896

LI 7
Accurate mass: 643.2199 (M + H)+
Calcd for $C_{34}H_{37}BrN_6O_2$: 640.2161

LI 8
Accurate mass: 502.2438 (M + H)+
Calcd for $C_{24}H_{42}BrN_3O_3$: 499.241

LI 9
Accurate mass: 490.1691 (M + H)+
Calcd for $C_{24}H_{32}BrN_3O_3$: 489.1627

Figure 1.2 Chemical structure of unimolecular ligand initiator systems [25,27].

1.6.2 Unsymmetrical Ion-Pair Comonomers and Polymers

Ion-pair comonomers are those, which as the name imply, composed of anionic and cationic vinyl monomers existing in pairs through the attraction of opposite charges. Symmetrical ion pairs are those where the vinyl functionality is chemically similar, for example, ion-pair monomers derived from methacrylamides [32] and methcrylates [33]. Because of the chemical similarity of vinyl functionality, the reactivity toward polymerization is also similar. Polymers of this type where opposite charges prevail upon two adjacent monomeric units in the polymer chain are called as *polyampholytes* [34] contrary to zwitterions where the opposite charges are present within the same monomer. Unsymmetrical ion pairs (Figure 1.3) [35] are those where chemical nature of vinyl functionalities is different. Owing to this difference, the rate of polymerization of anionic and cationic components of ion-pair monomers could also vary. However, the mobility of monomers during polymerization would be governed by charge neutralization. Thus, even though the individual components of monomer pair may polymerize at different rates, the oppositely charged entities can be expected to be together whether polymerized or in the monomeric form. Indeed, this was found to be the case, when unsymmetrical ion pairs composed of *N*-alkyl-1-vinyl imidazole and styrene-4-sulfonate were polymerized under RAFT. Under conventional free-radical polymerization, this unsymmetrical ion-pair monomer yielded completely insoluble polymers. However, under RAFT-mediated polymerization process, soluble polymers were obtained. The nuclear magnetic resonance (NMR) analysis of this polymer indicated the presence of unreacted *N*-alkyl-1-vinyl imidazole monomer accompanying poly(styrene-4-sulfonate) even after dialysis in order to compensate the excess negative charge of polymer chain [35]. Such monomer pair has been found to be useful for making ionically cross-linked poly(methyl methacrylate) even at a concentration of about 5 mol% (Scheme 1.16) [35].

1.6.3 Imidazole-Derived Zwitterionic Polymers

Zwitterionic polymers are potentially useful in many applications such as antifouling membranes, enhanced oil recovery, and low temperature precipitation of proteins. Zwitterionic polymers are typically derived from metharylamides and methacrylates. Owing to the hydrolytic instability of amide and ester linkages, zwitterionic methacrylamides and methacrylates undergo hydrolysis to varying degrees even during polymerization. To avoid this hydrolysis, zwitterions free of hydrolytically unstable linkages such as vinylimidazole and benzimidazole based

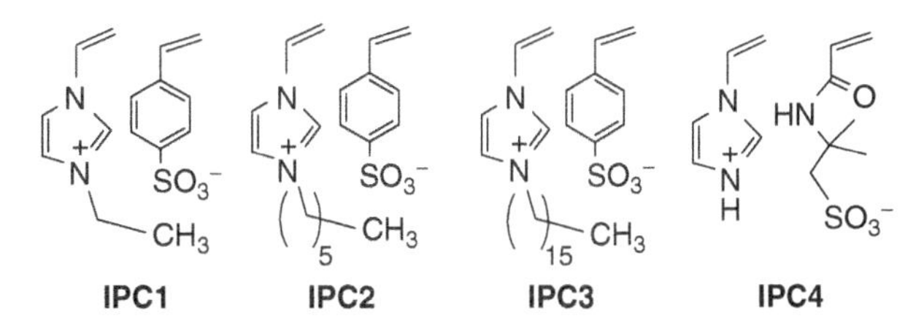

Figure 1.3 Chemical structure of unsymmetrical ion-pair comonomers. (Reprinted with permission from [35]. Copyright © 2013 Wiley Periodicals Inc.)

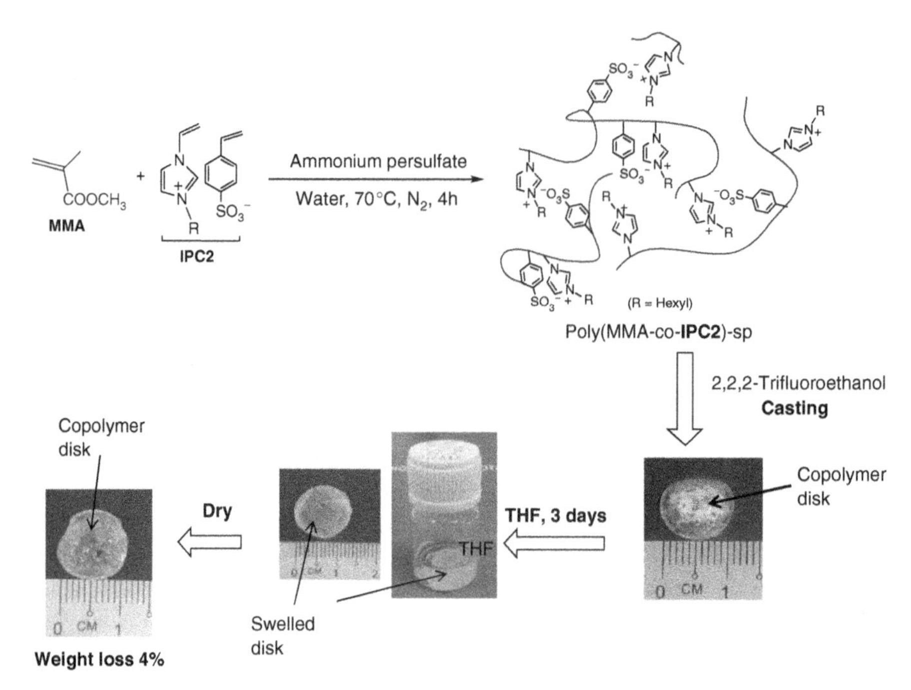

Scheme 1.16 Preparation of ionically cross-linked poly(methyl methacrylate) (PMMA) using unsymmetrical ion-pair comonomer. (Reprinted with permission from [35]. Copyright © 2013 Wiley Periodicals Inc.)

have been proposed recently (Scheme 1.17) [36]. These zwitterionic polymers showed very interesting solubility characteristics such as insoluble but swelling tendency in deionized water and solubility in concentrated brine solution. It also showed upper critical solution temperature (UCST) behavior as well as gel–sol behavior in brine solution. These zwitterionic polymers also showed non-Newtonian flow characteristics. Increased hydrophobicity with increased $\pi-\pi$ interaction and intermolecular association between charged species are responsible for the aforementioned characteristics of these novel zwitterionic polymers. Owing to the unique solubility characteristics, these polymers are potentially useful in applications such as enhanced oil recovery and low temperature precipitation of proteins.

1.6.4 Post-Modification of Polymers Bearing Reactive Pendant Groups

Polymers bearing reactive pendant groups are useful for post-polymerization modification reactions. Such modifications can be used for improving many of the properties of polymers such as compatibility, solubility, thermal stability, and processability. A copolymer of maleic anhydride was converted to water and organosoluble through one such modification reaction (Scheme 1.18) [37]. The copolymer that was water insoluble became amphiphilic upon modification

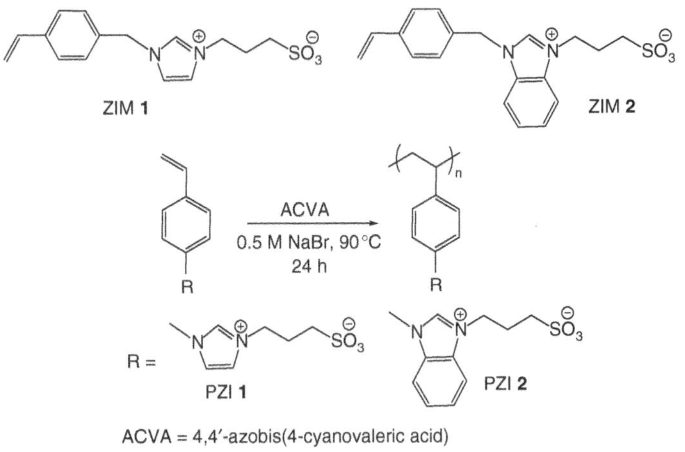

Scheme 1.17 Zwitterionic polymers free of hydrolyzable linkages. (Reprinted with permission from [36]. Copyright © 2013 Royal Society of Chemistry.)

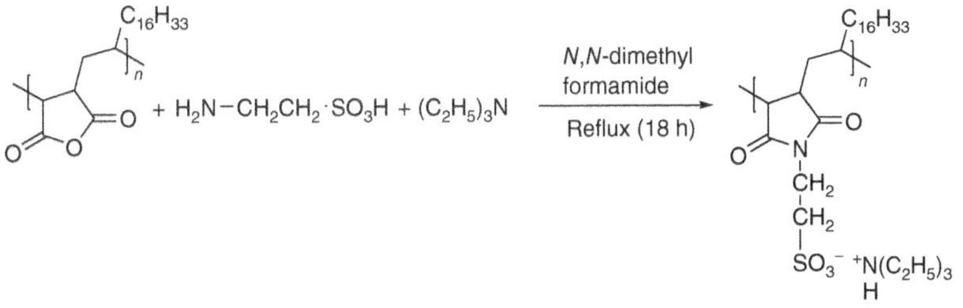

Scheme 1.18 Preparation of amphiphilic polymer by post-polymerization modification reaction involving reactive pendant group [37].

whereby it turned soluble in deionized water as well as in organic solvents. The modification reaction also enhanced the thermal stability of starting polymer (Figure 1.4) [37]. The aqueous solution of modified, amphiphilic polymer was clear and low in viscosity as compared to aqueous solutions of polymers like polyvinyl alcohol. These polymers are potentially useful as elastic coatings for protecting glass, metal, and plastic during transport, in cosmetics, and as flocculants in water purification.

1.7 CONCLUSION

It may be noted that the field of polymer science is applied in nature. As discussed earlier, even though, new developments in published literature abound, not many of these developments have been translated into commercial processes. As there has been a tremendous increase in the number of people involved in research particularly so in applied research fields like polymer science, any new development

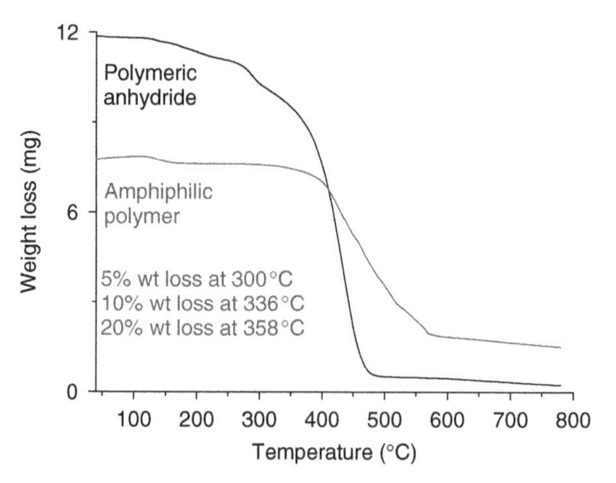

Figure 1.4 Comparison of thermal stability of unmodified and modified polymers by thermogravimetric analysis [37].

attracts the attention of vast array of researchers resulting in the generation of enormous body of published literature within a very short span of time. One good example, in the recent past, for this trend, is the development in controlled radical polymerization processes. Each one of these processes has inherent weaknesses that prevent these processes from being practiced as large-scale industrial process. In the history of chemical research, such occurrences are not uncommon. It is common for challenges to arise from many fronts in any new development. Successful outcome relies on overcoming these challenges. Opportunity for newer developments constantly arises due to many factors like legislation-induced banning of products on account of environment and health concerns, development of new processes particularly from bio-based renewable processes and products, demanding applications, and increasing environmental awareness and concerns of general public on chemicals. Nowadays, the economic cycles also are a major determining factor.

REFERENCES

1. Matyjaszewski, K., Ziegler, M. J., Arehart, S. V., Greszta, D., Pakula, T. (2000). Gradient copolymers by atom transfer radical copolymerization. *J. Phys. Org. Chem.*, *13*, 775–786.
2. (a) Yoshino, K., Yokoyama, A., Yokozawa, T. (2009). Well-defined star-shaped poly(p-benzamide) via chain-growth condensation polymerization: use of tetra-functional porphyrin initiator to optimize star polymer formation. *J. Polym. Sci. Part A: Polym. Chem.*, *47*, 6328–6332; (b) Yokozawa, T., Ajioka, N., Yokoyama, A. (2008). Reaction control in condensation polymerization. *Adv. Polym. Sci.*, *217*, 1–77; (c) Miyakoshi, R., Yokoyama, A., Yokozawa, T. (2008). Development of catalyst-transfer condensation polymerization. Synthesis of π-conjugated polymers with controlled molecular weight and low polydispersity. *J. Polym. Sci. Part A: Polym. Chem.*, *46*, 753–765;

(d) Yokozawa, T., Yokoyama, A. (2007). Chain-growth polycondensation: the living polymerization process in polycondensation. *Prog. Polym. Sci.*, *32*, 147–172; (e) Sugi, R., Yokoyama, A., Yokozawa, T. (2003). Synthesis of well-defined telechelic aromatic polyamides by chain-growth polycondensation: application to the synthesis of block copolymers of polyamide and poly(tetrahydrofuran). *Macromol. Rapid Commun.*, *24*, 1085–1090; (f) Ajioka, N., Suzuki, Y., Yokoyama, A., Yokozawa, T. (2007). Synthesis of well-defined polystyrene-b-aromatic polyether using an orthogonal initiator for atom transfer radical polymerization and chain-growth condensation polymerization. *Macromolecules*, *40*, 5294–5300; (g) Kim, S., Kakuda, Y., Yokoyama, A., Yokozawa, T. (2007). Synthesis of controlled rod-coil block copolymers by a macroinitiator method: chain-growth polycondensation for an aromatic polyamide from a polystyrene macroinitiator. *J. Polym. Sci. Part A: Polym. Chem.*, *45*, 3129–3133; (h) Yokozawa, T., Yokoyama, A. (2009). Chain-growth condensation polymerization for the synthesis of well-defined condensation polymers and π-conjugated polymers. *Chem. Rev.*, *109*, 5595–5619; (i) Yokoyama, A., Yokozawa, T. (2007). Converting step-growth to chain-growth condensation polymerization. *Macromolecules*, *40*, 4093–4101; (j) Yokozawa, T., Suzuki, H. (1999). Condensative chain polymerization in solid-liquid phase. Synthesis of polyesters with a defined molecular weight and a narrow molecular weight distribution by polycondensation. *J. Am. Chem. Soc.*, *121*, 11573–11574; (k) Yokozawa, T., Maeda, D., Hiyama, N., Hiraoka, S. (2001). Chain-growth polycondensation in solid-liquid phase with ammonium salts for well-defined polyesters. *Macromol. Chem. Phys.*, *202*, 2181–2186; (l) Yokozawa, T., Ogawa, M., Sekino, A., Sugi, R., Yokoyama, A. (2002). Chain-growth polycondensation for well-defined aramide. *Synthesis of unprecedented block copolymer containing aramide with low polydispersity. J. Am. Chem. Soc.*, *124*, 15158–15159; (m) Yokoyama, A., Masukawa, T., Yamazaki, Y., Yokozawa, T. (2009). Successive chain-growth condensation polymerization for the synthesis of well-defined diblock copolymers of aromatic polyamide and aromatic polyether. *Macromol. Rapid Commun.*, *30*, 24–28; (n) Ohishi, T., Masukawa, T., Fujii, S., Yokoyama, A., Yokozawa, T. (2010). Synthesis of core cross-linked star polymers consisting of well-defined aromatic polyamide arms. *Macromoleules*, *43*, 3206–3214; (o) Yoshino, K., Hachiman, K., Yokoyama, A., Yokozawa, T. (2010). Chain-growth condensation polymerization of 4-aminobenzoic acid esters bearing tri(ethylene glycol) side chain with lithium amide base. *J. Polym. Sci. Part A: Polym. Chem.*, *48*, 1357–1363; (p) Yokozawa, T., Asai, T., Sugi, R., Ishigooka, S. Hiraoka, S. (2000). Chain-growth polycondensation for nonbiological polyamides of defined architecture. *J. Am. Chem. Soc.*, *122*, 8313–8314; (q) Yoshino, K., Yokoyama, A., Yokozawa, T. (2011). Synthesis of a variety of star-shaped polybenzamides via chain-growth condensation polymerization with tetrafunctional porphyrin initiator. *J. Polym. Sci. Part A: Polym. Chem.*, *49*, 986–994.

3. (a) Chen, A.-L., Wei, K.-L., Jeng, R.-J., Lin, J.-J., Dai, A.A. (2011). Well-defined polyamide synthesis from diisocyanates and diacids involving hindered carbodiimide intermediates. *Macromolecules*, *44*, 46–59; (b) Wei, K.-L., Wu, C. H., Huang, W. H., Lin, J.-J., Dai, S. A. (2006). *N*-Aryl acylureas as intermediates in sequential self-repetitive reactions to form poly(amide–imide)s. *Macromolecules*, *39*, 12–14; (c) Chen, C. W., Cheng, C. C., Dai, S. A. (2007). Reactive macrocyclic ether–urethane carbodiimide (MC–CDI): synthesis, reaction, and ring-opening polymerization (ROP). *Macromolecules*, *40*, 8139–8141.
4. (a) Kim, Y. J., Seo, M., Kim, S. Y. (2010). Synthesis of well-defined rod-coil block copolymers containing trifluoromethylated poly(phenylene oxide)s by chain-growth condensation polymerization and atom transfer radial polymerization. *J. Polym. Sci.*

Part A: Polym. Chem., *48*, 1049–1057; (b) Kim, Y. J., Kakimoto, M., Kim, S. Y. (2006). Synthesis of hyperbranched poly(arylene ether) from monomer containing nitro group: kinetically controlled growth of polymer chain through dynamic exchange of end functional groups. *Macromolecules*, *39*, 7190–7192; (c) Lee, M. S., Kim, S. Y. (2005). Synthesis of poly(arylene ether)s containing triphenylamine units via nitro displacement reaction. *Macromolecules*, *38*, 5844–5845; (d) Kim, Y. J., Chung, I. S., Kim, S. Y. (2003). Synthesis of poly(phenylene oxide) containing trifluoromethyl groups via selective and sequential nucleophilic aromatic substitution reaction. *Macromolecules*, *36*, 3809–3811; (e) Chung, I. S., Kim, S. Y. (2001). Meta-activated nucleophilic aromatic substitution reaction: poly(biphenylene oxide)s with trifluoromethyl pendent groups via nitro displacement. *J. Am. Chem. Soc.*, *123*, 11071–11072; (f) Chung, I. S., Kim, S. Y. (2000). Poly(arylene ether)s via nitro displacement reaction: synthesis of poly(biphenylene oxide)s containing trifluoromethyl groups from AB type monomers. *Macromolecules*, *33*, 9474–9476.

5. (a) Parthiban, A., Ming Choo, F., Chai, C. L. L. (2011). Linear and networked polymers formed by the near simultaneous occurrence of etherification and esterification under mild reaction conditions. *Polym. Int.*, *60*, 1624–1628; (b) Parthiban, A., Venugopal, M. (2006). Modifiable polyunsaturated polymers and their preparation. *US patent*, 20060189774; (c) Parthiban, A., Babu Rao, T. R. (2007). Unsaturated polyesters prepared under mild reaction conditions by aliphatic nucleophilic substitution reactions. *Polym. Prepr.*, *48(1)*,541–542; (d) Parthiban, A., Hung, N. Y., Babu Rao, T. R., Yang, L. L. (2006). Novel polyunsaturated poly(ether ester) made by consecutive formation of ether and ester bonds under one pot conditions. *Polym. Prepr.*, *47(2)*, 580–581; (e) Parthiban, A., Yu, H., Chai, C. L. L. (2008). Unsaturated polyimide prepared under mild reaction conditions by nucleophilic substitution reaction through C-N bond formation. *Polym. Prepr.*, *49(2)*, 844–845; (f) Parthiban, A., Ming Choo, F. (2012) A new addition to the family of polyesters—synthesis and characterization of novel (alternating) poly(ether ester)s, poly(butylene 3-/4-hydoxy benzoate)s. *J. Polym. Res.*, *19*, 1–7; (g) Parthiban, A., Yu, H., Chai, C. L. L. (2009). Etherification vs esterification – superior reactivity of carboxylate anion over phenoxide in aliphatic nucleophilic substitution reaction with 1,4-dibromo-2-butene. *Polym. Prepr.*, *50(2)*, 479–480; (h) Parthiban, A., Yu, H., Chai, C. L. L. (2009). Unsaturated copolyesters of 4,4′-oxybis(benzoic acid) and terephthalic acid prepared under mild reaction conditions by aliphatic nucleophilic substitution reaction with 1,4-dibromo-2-butene. *Polym. Prepr.*, *50(2)*, 729–730.
6. Lutz, J-F. (2010). Sequence-controlled polymerizations: the next holy grail in polymer science? *Polym. Chem.*, *1*, 55–62.
7. Pfeifer, S., Lutz, J-F. (2008). Development of a library of N-substituted maleimide for the local functionalization of linear polymer chains. *Chem. Eur. J.*, *14*, 10949–10957.
8. Cramer, J. W., Treitler, D. S., Dunn, E. W., Castro, P. M., Roisnel, T., Thomas, C. M., Coates, G. W. (2009). Polymerization of enantiopure monomers using syndiospecific catalysts: a new approach to sequence control in polymer synthesis. *J. Am. Chem. Soc.*, *131*, 16042–16044.
9. Ida, S., Ouchi, M., Sawamoto, M. (2010). Template-assisted selective radical addition toward sequence-regulated polymerization: lariat capture of target monomer by template initiator. *J. Am. Chem. Soc.*, *132*, 14748–14750.
10. Hibi, Y., Tokuoka, S., Terashima, T., Ouchi, M., Sawamoto, M. (2011). Design of AB divinyl "template monomers" toward alternating sequence control in metal-catalyzed living radical polymerization. *Polym. Chem.*, *2*, 341–347.

11. Hibi, Y., Ouchi, M., Sawamoto, M. (2011). Sequence-regulated radical polymerization with a metal-templated monomer: repetitive ABA sequence by double copolymerization. *Angew. Chem. Int. Ed.*, *50*, 7434–7437.
12. Tong, X., Guo, B.-h., Huang, Y. (2011). Toward the synthesis of sequence-controlled vinyl copolymers. *Chem. Commun.*, *47*, 1455–1457.
13. Ida, S., Ouchi, M., Sawamoto, M. (2011). Designer template initiator for sequence regulated polymerization: systems design for substrate-selective metal-catalyzed radical addition and living radical polymerization. *Macromol. Rapid Commun.*, *32*, 209–214.
14. (a) Scott, T. F., Schneider, A. D., Cook, W. D., Bowman, C. N. (2005). Photoinduced plasticity in cross-linked polymers. *Science*, *308*, 1615–1617; (b) Park, H. Y., Kloxin, C. J., Scott, T. F., Bowman, C. N. (2010). Stress relaxation by addition-fragmentation chain transfer in highly cross-linked thiol-yne networks. *Macromolecules*, *43*, 10188–10190.
15. Higaki, Y., Otsuka, H., Takahara, A. (2006). A thermodynamic polymer cross-linking system based on radically exchangeable covalent bonds. *Macromolecules*, *39*, 2121–2125.
16. Ghosh, B., Urban, M. W. (2009). Self-repairing oxetane-substituted chitosan polyurethane networks. *Science*, *323*, 1458–1460.
17. Zhang, Y., Broekhuis, A. A., Picchioni, F. (2009). Thermally self-healing polymeric materials: the next step to recycling thermoset polymers? *Macromolecules*, *42*, 1906–1912.
18. Reutenauer, P., Buhler, E., Boul, P. J., Candau, S. J., Lehn, J.-M. (2009). Room temperature dynamic polymers based on Diels-Alder chemistry. *Chem. Eur. J.*, *15*, 1893–1900.
19. Nicolay, R., Kamada, J., Wassen, A. V., Matyjaszewski, K. (2010). Responsive gels based on dynamic covalent trithiocarbonate cross-linker. *Macromolecules*, *43*, 4355–4361.
20. Amamoto, Y., Kamada, J., Otsuka, H., Takahara, A., Matyjaszewski, K. (2011). Repeatable photoinduced self-healing of covalently cross-linked polymers through reshuffling of trithiocarbonate units. *Angew. Chem. Int. Ed.*, *50*, 1660–1663.
21. Montarnal, D., Capelot, M., Tournilhac, F., Leibler, L. (2011). Silica-like malleable materials from permanent organic networks. *Science*, *334*, 965–968.
22. Nicolas, J., Guillaneuf, Y., Lefay, C., Bertin, D., Gigmes, D., Charleux, B. (2013). Nitroxide mediated polymerization. *Prog. Polym. Sci.*, *38*, 63–235.
23. (a) Moad, G., Rizzardo, E., Thang, Sh. (2008). *Polymer*, *49*, 1079–1131; (b) Perrier, S., Takolpuckdee, P. (2005). *J. Polym. Sci. Part A: Polym. Chem.*, *43*, 5347–5393.
24. Wang, S., Matyjaszewski, K. (1995). *J. Am. Chem. Soc.*, *117*, 5614–5615.
25. Parthiban, A. (2011). Unimolecular ligand-initiator dual functional systems (umlidfs) and use thereof. *PCT Int. Appl. WO 2011040881* (assigned to Agency for Science, Technology and Research, Singapore).
26. Jana, S., Parthiban, A., Ming Choo, F. (2012). Unimolecular ligand-initiator dual functional systems (ULIS) for low copper ATRP of vinyl monomers including acrylic/methacrylic acids. *Chem. Commun.*, *48*, 4256–4258.
27. Parthiban, A. (2012). A modified ATRP process by unimolecular ligand initiator (ULIS) dual functional systems. *Polym. Prepr.*, *53(2)*, 177–178.
28. Parthiban, A. (2012). Preparation of block copolymers and high molecular weight polymers by ATRP using unimolecular ligand initiator (ULIS) dual functional systems. *Polym. Prepr.*, *53(2)*, 245–246.

29. Parthiban, A. (2012). Comparing ATRP with that promoted by unimolecular ligand initiator (ULIS) dual functional systems. *Polym. Prepr., 53(2)*, 241–242.
30. Parthiban, A. (2012). Homo and copolymerization of acrylic acid in the free acid form by ATRP using unimolecular ligand initiator (ULIS) dual functional systems. *Polym. Prepr., 53(2)*, 243–244.
31. Parthiban, A., Jana, S., Ming Choo, F. (2012). Direct polymerization of acrylic and methacrylic acids by ATRP using unimolecular ligand initiator (ULIS) dual functional systems. *Polym. Prepr., 53(2)*, 179–180.
32. Salamone, J. C., Mahmud, N. A., Mahmud, M. U., Nagabushanam, T., Watterson, A. C. (1982). Acrylic ampholytic polymers. *Polymer*, *23*, 843–848.
33. Yang, J. H., John, M. S. (1995). The conformation and dynamics study of amphoteric copolymers, p(sodium-2-methacryloyloxyethanesulfonate-co-2-methacryloyloxy ethyl trimethylammonium iodide), using viscometry, ^{14}N- and ^{23}Na-NMR. *J. Polym. Sci. Part A: Polym. Chem.*, *33*, 2613–2621.
34. Salamone, J. C., Watterson, A. C., Hsu, T. D., Tsai, C. C., Mahmud, M. U. (1977). *J. Polym. Sci. Polym. Lett. Ed.*, *15*, 487–491.
35. Jana, S., Vasantha, V. A., Stubbs, L. P., Parthiban, A., Vancso, J. G. (2013). Vinylimiazole-based asymmetric ion pair comonomers: synthesis, polymerization studies and formation of ionically crosslinked PMMA. *J. Polym. Sci. Part A: Polym. Chem.*, *51*, 3260–3273.
36. Vasantha, V. A., Jana, S., Parthiban, A., Vancso, J. G. (2013). Water swelling, brine soluble imidazole based zwitterionic polymers – synthesis and study of reversible UCST behavior and gel-sol transitions. *Chem. Commun.* DOI: 10.1039/C3CC44407D.
37. (a) Parthiban, A., Yu, H., Chai, C. L. L. (2009). Amphiphilic water and organosoluble grafted copolymers. *Polym. Prepr. 50(2)*, 428–429; (b) Parthiban, A. (2009). Water swellable and water soluble polymers and use thereof. *PCT Int. Appl. WO 2009154568* (assigned to Agency for Science, Technology and Research, Singapore); (c) Parthiban, A. (2011). Water swellable and water soluble polymers and use thereof. *Singapore patent SG 167476* (assigned to Agency for Science, Technology and Research, Singapore).

2

FUNCTIONAL POLYOLEFINS FROM THE COORDINATION COPOLYMERIZATION OF VINYL MONOMERS

Fabio Di Lena
João A. S. Bomfim

2.1 MOLECULAR ASPECTS OF OLEFIN COORDINATION TO METALS

Coordination compounds of alkenes and metals have been known since the early nineteenth century, although their structure remained mysterious until the late twentieth century. The first documented example of a coordination compound of alkenes is the air-stable Zeise's salt, $K[PtCl_3(ethene)]\cdot H_2O$ [1]. Alkenes undergo significant structural changes upon coordination with a metal center: first, the C–C bond distance increases, which indicates a decrease in the bond order. Second, the C–H bonds are bent out of the C=C plane, which can be attributed to a deviation from the sp^2 character of the carbon atoms. These structural changes are strictly related to the very nature of the bond between alkene and metal, which is based on the following aspects: (i) a partial donation of electrons from the π orbital of the olefin to an empty d orbital of the metal (σ-dative component); and (ii) a partial donation of electrons from a d orbital of the metal to the π^* orbital of the olefin (π-retrodative component). Both interactions contribute to weaken the olefinic carbon–carbon double bond, change the hybridization of the carbons, and, consequently, modify the olefin's reactivity. For instance, while free alkenes are known to react with electrophiles such as Br_2, coordinated alkenes become capable of undergoing both

Synthesis and Applications of Copolymers, First Edition. Edited by Anbanandam Parthiban.

intramolecular and intermolecular nucleophilic attacks. Particularly interesting is the case of intramolecular nucleophilic attack by a *cis*-coordinated anionic-like ligand such as a hydride moiety or an alkyl group. Assumed to proceed via a four-centered transition state (Scheme 2.1), the attack results in the insertion of the olefin into the metal–hydrogen or metal–carbon bond with the simultaneous migration of the newly generated alkyl group to the coordination site previously occupied by the olefin (site *epimerization*). This reaction is at the basis of well-known catalytic processes such as, alkene hydrogenation with the Wilkinson's catalyst [2,3] and alkene polymerization with Ziegler catalysts* [4], and it is commonly referred to as *migratory insertion*. Another unsaturated, neutral molecule that can undergo migratory insertion is carbon monoxide, which is "activated" toward nucleophilic attack by metal coordination via a σ-dative/π-retrodative bond scheme similar to that described earlier for alkenes.

Organometallic compounds based on electron-poor, early transition metals such as Ti(III), Ti(IV), Zr(IV), and Hf(IV) as well as those based on electron-rich, late transition metals such as Fe(III), Ni(II), and Pd(II) have been shown to promote alkene polymerization (Figure 2.1) [5]. According to Pearson's [6] *hard* and *soft* (Lewis) acids and bases (HSAB) theory, the first class of compounds are classified as hard Lewis acids, whereas the second class of compounds are classified as soft Lewis acids. Consequently, in the presence of polar olefins such as (meth)acrylates, which contain functional groups (e.g., the carbonyl) that are harder Lewis bases than the C=C bond, early transition metals would rather coordinate the heteroatoms (κ coordination) than the olefinic moiety (π coordination), thus preventing the migratory insertion from taking place. It is worth noting that the κ coordination of polar olefins to an early transition metal center makes them susceptible of both intramolecular and intermolecular nucleophilic attacks as in the coordinative group transfer polymerization (CGTP, *vide infra*). On the other hand, the less basic, late transition metal centers may bind polar olefins also via π coordination [7], thus enabling migratory insertion. This has been observed by single-crystal X-ray diffraction in a number of complexes, especially of platinum and palladium [8,9].

2.2 FUNDAMENTALS OF HOMOPOLYMERIZATION OF ALKENES

Consecutive migratory insertion events of ethene lead to linear polyethylene (PE) if chain-breaking reactions such as β–H transfer to monomer and β–H elimination to metal [10,11], which are supposed to begin with agostic interactions [12],

*There is a debate whether coordination complexes used in coordination polymerizations should be called catalysts or initiators. If the initiation and propagation reactions are fast and predominate over chain transfer processes resulting in one polymer chain per active site (as in living or quasi-living reactions), initiator is usually preferred. In a similar manner, some authors tend to distinguish between catalyst and precatalyst, emphasizing the structural and chemical changes that the starting coordination compounds undergo during the activation phase, for example, with co-catalysts such as organoaluminum compounds. For sake of simplicity, the term catalyst will be used throughout this chapter when a clear distinction between precatalysts and actual catalytic species is not needed.

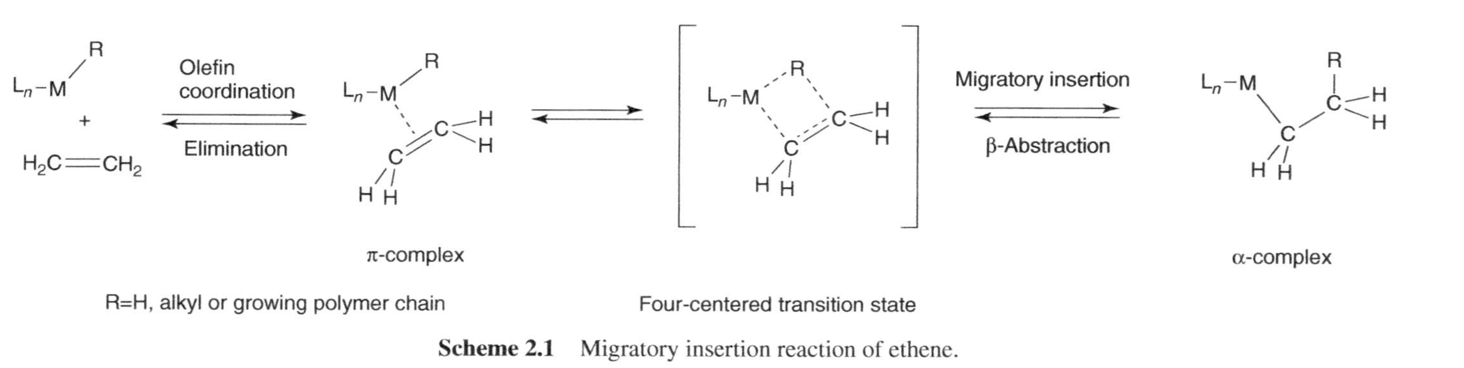

Scheme 2.1 Migratory insertion reaction of ethene.

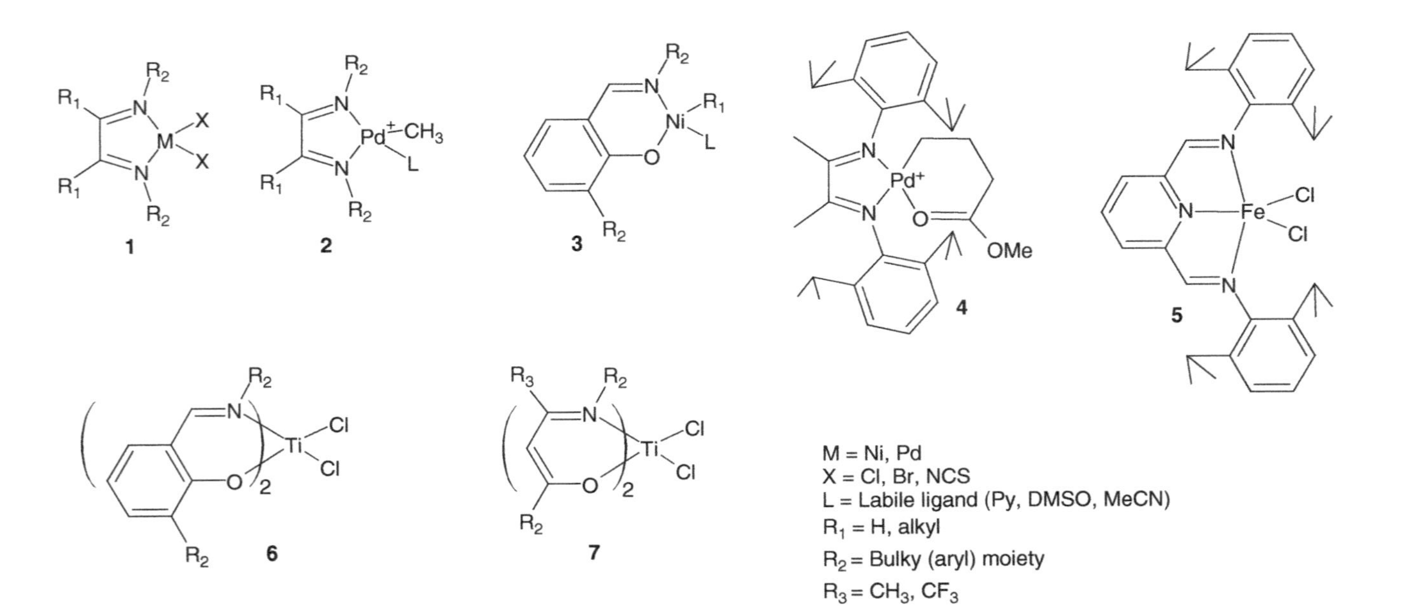

Figure 2.1 Some transition metal catalysts for the migratory insertion polymerization of olefins: group 10 diimine complexes before activation with a co-catalyst (**1**); cationic Pd(II)/diimine complexes (**2**); neutral phenoxyimine nickel(II) catalysts (**3**); cationic Pd(II)/diimine catalyst with an inserted acrylate unit (**4**); iron(II) bis(imino)pyridine catalyst (**5**); titanium(IV) phenoxyimine complex (**6**); and Ti(IV) β-enaminoketonate complex (**7**).

are kinetically noncompetitive with monomer propagation. Otherwise, branched or hyperbranched polymer architectures are obtained [13,14]. This phenomenon, usually referred to as *chain walking* (as if the catalyst *walked* along the chain), has been observed mainly with late transition metal catalysts because of their higher propensity to β–H abstraction [15]. As one may expect, the physical and mechanical properties of linear, branched, and hyperbranched PE are dramatically different. For late transition metal catalysts, the rate ratio isomerization/insertion has been found to increase with (i) increasing steric bulk around the metal center; (ii) increasing temperature; and (iii) decreasing monomer concentration. By changing one of these three parameters during the polymerization, so-called *single monomer copolymers* can be obtained. For example, by alternating high monomer pressure (which disfavors the chain walking and thus affords linear segments) and low monomer pressure (which favors isomerization and thus leads to branched structures), thermoplastic elastomers were prepared from a simple ethene homopolymerization [16,17].

In the case of propene and other α-olefins, the structural regularity, and thus the final properties, of the resulting polymer are connected not only to the linearity of the polymer chain but also to the regio- and stereoselectivity of the insertion reaction [18]. In fact, while ethene is a symmetrical molecule, α-olefins have a "head" and a "tail" and are also prochiral. The former aspect implies that the growing chain may contain head-to-tail, head-to-head, and tail-to-tail sequences depending on which of the two carbon atoms of the olefinic bond is nucleophilically attacked in two consecutive insertion events. On the other hand, the metal coordination of the α-olefin with one or the other enantioface determines the relative stereochemistry of the adjacent chiral centers within the macromolecule. For example, if in two consecutive insertions the α-olefin coordinates to the metal with the same enantioface, a "meso" (*m*) diad is obtained, whereas the coordination of different enantiofaces leads to a "racemic" (*r*) diad. A regular sequence of *m* or *r* diads affords, respectively, an *isotactic* or a *syndiotactic* polymer, whereas a random sequence of *m* and *r* diads within the same chain yields an *atactic* polyolefin. By judicious ligand design and metal selection, it has been possible to finely control the regio- and stereoselectivity of the insertion reaction and thus the final properties of the polyolefin [4,19]. Interestingly, the temperature and the monomer concentration can influence the selectivity of certain catalysts so that "single monomer copolymers" can be prepared also with 1-alkenes. For instance, Coates [20,21] produced various thermoplastic elastomers such as an "isotactic polypropylene-*block*-regioirregular polypropylene-*block*-isotactic polypropylene-*block*-regioirregular polypropylene-*block*-isotactic polypropylene" pentablock copolymer simply by changing the reaction temperature during the homopolymerization of propene in the presence of specifically designed Ni(II) complexes. These copolymers differ from those obtained with Waymouth's [22,23] oscillating catalysts, with which only the stereoselectivity, and not the insertion mode, changes.

The polymerization of cyclic olefins has been demonstrated using Ni and Pd cationic catalysts. Cyclopropene [24] and cyclopentene [25] have been polymerized without ring-opening producing rigid-chain linear polymers.

In contrast, 4-methylcyclopentene undergoes chain walking affording a linear chain of alternating methylene and cyclopentane units [26].

2.3 COPOLYMERIZATION OF ETHENE AND OTHER ALKENES

Although the insertion homopolymerization of ethene with early transition metal catalysts leads to highly linear polymers with high crystallinity and high density, the incorporation of α-olefin comonomers (e.g., 1-octene) introduces short branches in the polymer chain reducing both crystallinity and density. These materials are called *linear low density polyethylenes (LLDPEs)* to distinguish them from low density polyethylene (LDPE) obtained by radical polymerization.

Although homopolymers of cyclic olefins such as norbornene present little interest because of their rigid structure and processability problems, their copolymers with ethene and α-olefins are of great interest as they frequently present good optical properties and transparency as well as high temperature resistance. These materials are traded as cyclic olefin copolymers (COCs). The random copolymerization of ethene and norbornene can be achieved by means of several catalytic systems, including vanadium complexes [27], group 4 metallocenes [28,29], octahedral titanium complexes [30,31], and group 10 diimine complexes [32]. By using sequential monomer addition in the presence of "living" catalysts, for example, some octahedral bis(pyrrolide imine) [33] and bis(β-enaminoketonato) titanium [34] complexes, block copolymers can also be prepared. An interesting example that points out the versatility of octahedral titanium catalyst is the synthesis of diblock copolymers in which each block consists of poly(ethylene-*co*-norbornene) segments with different amounts of norbornene, which can go up to 40 mol% [24]. This molecular architecture leads to blocks with different properties which do not phase segregate.

An interesting strategy for preparing olefin block copolymers is the coordinative chain transfer polymerization (CCTP), also called *chain-shuttling* polymerization. In this process, a chain-shuttling agent (CSA)—typically a main group organometallic compound—is added to a system of (at least) two transition metal catalysts having different selectivities. The role of CSA is to shuttle the growing chains from one catalytic center to another and to act as a reservoir of "living" chains. Owing to the lack of energetically accessible empty orbitals, in fact, organometallic compounds based on main group metals do not undergo β–H abstraction and thus "dead" chains are not produced. If the exchange rate is significantly higher than the rates of monomer propagation, then narrowly dispersed, well-defined block copolymers are obtained even with nonliving catalysts. Recently, researchers at Dow were able to produce olefin block copolymers having both "hard" (PE) and "soft" (poly(1-octene)) segments with two nonmetallocene catalysts and $ZnEt_2$ as CSA [35–37]. Hundreds of poly(ethylene-*b*-1-octene) chains with $M_w/M_n < 2$ per total catalyst were obtained. The CCTP process can be carried out also with a single catalyst through successive monomer addition steps to create diblock copolymers [31]. The main advantage of using successive additions in CCTP is the higher yield, as more chains per catalytic center are produced [32].

Another strategy for preparing functionalized polyolefins is the "reactive olefin" pathway proposed by Chung and Rhubright [38] in the 1990s. The idea behind this approach is to polymerize olefins containing reactive groups that do not poison the catalyst, which can be later converted to the desired functionalities. The main examples are the use of Ziegler–Natta-like catalysts to copolymerize olefins and boranes, which can successively be converted into alcohols and halides [26], or the use of half-metallocenes for the copolymerization of olefins with methylstyrenes, in which the benzyl methyl group can be oxidized to create phenols or carboxylic acids [39].

2.4 COPOLYMERIZATION OF ALKENES AND CARBON MONOXIDE

High molecular weight polyketones have been prepared via the metal-catalyzed alternating copolymerization of a number of alkenes (ethene, propene, styrene, and its derivatives) with carbon monoxide (Figure 2.2) [40–44]. Stereoselective processes as well as alternating terpolymerizations of CO with two different alkenes have also been described. Because of the strong, intermolecular dipolar interactions among the C=O groups, the polymers exhibit high chemical and mechanical resistance over a wide range of temperatures. Polyketones of this type have found industrial applications under commercial names such as Carilon® and Ketonex®.

A number of cationic complexes, mainly based on Pd(II) and Ni(II) centers, have been investigated as catalysts for this reaction. Owing to space limitations and the presence of excellent reviews in the literature, this chapter will not describe this topic in detail; only the most significant mechanistic aspects and structure–reactivity relationships are summarized here. In particular, independent studies have shown the following:

1. The origin of the perfect alternation of the two monomers is both thermodynamic and kinetic. Indeed, while the consecutive insertion of two molecules of CO is thermodynamically disfavored, the insertion of CO into a metal–carbon bond is faster than the insertion of an alkene (i.e., it is kinetically favored). This is possibly because the metallacyclic intermediates that form during the polymerization reaction are opened more easily by the CO than by the olefin.

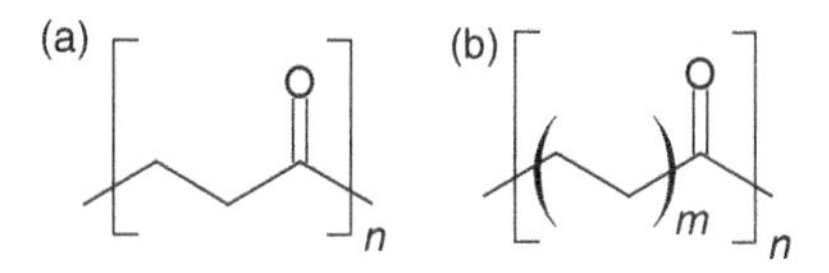

Figure 2.2 Polyketones resulting from the alternating (a) and nonalternating (b) copolymerization of ethene and carbon monoxide.

2. The presence of the metallacycle slows down β–H elimination to metal and β–H transfer to monomer, giving the polymerization the characteristics of a "living" process.
3. Pd(II)-based catalysts have been found to be more active than Ni(II)-based catalysts, irrespective of the ligands employed for complexation. This is probably because, with Pd complexes, the carbonylation step occurs via the migratory insertion of four-coordinated species, whereas, with the Ni complexes, it takes place preferentially via five-coordinated species. The latter produces more stable intermediates in the catalytic cycle, which slows down the reaction.
4. Because of electronic effects, copolymerization of aliphatic olefins (e.g., ethene and propene) usually requires diphosphine-based ligands (P–P), whereas copolymerization of aromatic olefins (e.g., styrene) typically requires either bidentate nitrogen (N–N) or mixed (P–N) ligands.
5. Depending on their nature, P–P ligands promote the opening of the C=C double bonds in a highly 1,2 or 2,1 manner with regio irregularly. In contrast, with N–N ligands, the opening of the olefin double bonds is usually 2,1.
6. C_{2v}-symmetric catalysts usually produce predominantly syndiotactic polymers through chain-end control, whereas C_2-symmetric catalysts usually produce predominantly isotactic polymers via enantiomorphic-site control. The stereospecificity of C_s- and C_1-symmetric catalysts is, on the other hand, hard to predict. Notably, no hemiisotacticity has ever been reported so far.

In 2002, Pugh [45] reported the first example of coordinative, imperfectly alternating copolymerization of ethene and carbon monoxide (Figure 2.2) by using an *in situ* generated, neutral Pd(II) catalysts bearing phosphine–sulfonate ligands (**L1–L3**, Figure 2.3). In methanol at 100–120 °C, high molecular weight (M_n > 30,000) linear polyketones with CO content of 42–49 mol% were prepared in reasonable yields (up to 190 g mmol^{-1} h^{-1}) in circa 2.0 h. Better results in terms of both catalytic activity (up to 600 g mmol^{-1} h^{-1}) and polymer molecular weights (up to ca. 150,000) were obtained with similar metal compounds, which also afforded CO incorporations as little as 1 mol% [46]. Unsurprisingly, the polyketones resulting from the nonalternating copolymerization of ethene and carbon monoxide exhibit

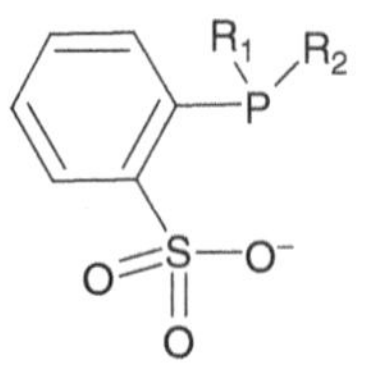

R_1=R_2=*o*-(MeO)C_6H_4 **L1**

R_1=R_2=*o*-(EtO)C_6H_4 **L2**

R_1=R_2=*o*-(^{i}PrO)C_6H_4 **L3**

R_1=R_2=Ph **L4**

Figure 2.3 Typical phosphine–sulfonate ligands used for the *in situ* generation of Pd(II) catalysts for the coordinative copolymerization of polar olefins with ethene or carbon monoxide.

a lower T_m, higher solubility in organic solvents and, thus, a better processability. Positive correlations were found between the degree of nonalternation and the reaction temperature, the P_{ethene}/P_{CO} ratio, and the bulkiness of the ancillary ligand.

2.5 COPOLYMERIZATION OF ALKENES AND POLAR VINYL MONOMERS

2.5.1 Migratory Insertion Polymerization

2.5.1.1 General Aspects A number of metal catalysts have been successfully employed to (co)polymerize, radically, a great variety of polar vinyl monomers [47]. In contrast, the migratory insertion polymerization of (meth)acrylates, vinyl acetate (VA), vinyl ethers (VEs), acrylonitrile (AN), and vinyl halides (VHs), as well as their copolymerization with apolar olefins such as ethene, has been much less fruitful despite the intense research [48–56]. The incorporation of the functionalities borne by these monomers into polyolefins is highly desirable in order to expand the scope of materials. Mechanistic studies have shown that the difficulty of incorporating polar vinyl monomers into polyolefins is due to a number of factors that may also occur concurrently. First, owing to steric reasons, the π coordination to transition metals of the C=C bond of polar olefins may be weaker than that of ethane, making the polymerization of the latter more favorable [44,57]. Second, the κ coordination of the polar olefins to the metal center via their heteroatoms may interfere with the π coordination of the olefinic moiety (*vide supra*) up to the complete catalyst deactivation. Reactivity studies carried out on the same cationic nickel(II) complex have shown that, while approximately equal portions of κ–O and π C=C binding are observed with VA, the κ–N coordination of AN largely predominates over the π coordination of the C=C bond [46,48,58,59]. Third, polar olefins insert into the metal–carbon bond with a rate that is sensibly slower than that of ethene [44,50,60]. Yet, the resulting metal alkyl species form chelates that are difficult to open even at elevated temperature and monomer concentration [44,50]. Moreover, β-atom (e.g., halogen in VHs) or β-group (e.g., OAc in VA) elimination to metal may lead to irreversible catalyst deactivation because of the formation of stable metal–heteroatom bonds [39,41,61–65]. Last but not the least, combinations of electron-poor catalysts and electron-rich olefins may give rise to polymers formed through a cationic mechanism [66,67].

2.5.1.2 Early Examples Recent mechanistic studies have shown that the polymerizations reported in the 1970s [68], in which polar monomers such as AN were polymerized in the presence of late transition metal complexes such as Cy_3PCuMe or $(bpy)_2FeEt_2$, take place anionically rather than via a migratory insertion mechanism as originally stated [69]. Brookhart [26,70] studied the copolymerization of ethene with acrylates and other polar monomers using cationic Pd(II)/diimine complexes. The soft acid character (low oxophilicity) of palladium makes it tolerant toward (meth)acrylic functionalities allowing for π coordination. The metallocycle formed by the growing chain after an acrylate insertion is a catalyst resting state, remaining reactive and able to coordinate incoming monomers. This catalytic system has been studied by Michalak and Ziegler [71] using DFT; an associative displacement mechanism is proposed for the incoming monomer coordination (rate-limiting step). One of the major characteristics of ethane–acrylate copolymers prepared with Pd(II)/diimine complexes is that, while the overall macromolecule has a hyperbranched structure, the acrylate monomers are localized at the branch ends [72].

Compared to that of their acrylic analogs, the insertion of methacrylic monomers is more difficult, even with late transition metal catalysts. After migratory insertion, in fact, MMA monomer creates a much higher steric bulk around the metal center [73]. For instance, cationic MePd(II)(diimine) catalysts can insert an MMA unit affording fairly stable complexes, which do not undergo β-elimination [74]. In a similar manner, attempts to copolymerize MMA and ethene using neutral Ni complexes with P,O ligands, which are fairly active for ethene homopolymerization, lead only to oligomers [75]. New-generation Pd/phosphine–sulfonate complexes, which can copolymerize methyl acrylate and ethene with circa 10% acrylate incorporation (*vide infra*), afford only pristine PE in the presence of ethene and MMA [76].

The first example of incorporation of a VE into a polyolefin through a coordination–insertion mechanism was reported in 2006 by Jordan [77], who copolymerized some silyl VEs with 1-hexene using $[(\alpha\text{-diimine})Pd(II)Me]^+$ $[\alpha\text{-diimine}=(2,6\text{-}^iPr_2\text{-}C_6H_3)N{=}CMeCMe{=}N(2,6\text{-}^iPr_2\text{-}C_6H_3)]$ as the catalyst and dichloromethane as the solvent at 20 °C. The level of $CH_2{=}CHOSiPh_3$ and $CH_2{=}CHOSiMe_3$ incorporation into the poly(1-hexene) backbone was found to be 20 and 11 mol%, respectively. Notably, in the case of $CH_2{=}CHOSiMe_3$, a substantial amount of silyl VE homopolymerization occurring via a cationic mechanism was observed. With both monomers, the obtained copolymers showed a highly branched structure (90–100 branches per 1000 carbons) with comonomer units located mostly at the end of branches. A polymerization mechanism based on subsequent migratory insertion and chain walking events was proposed.

2.5.1.3 Latest Examples In the following years, a number of publications appeared in the literature showing that limited amounts (typically $\leq$10 mol%) of several polar monomers could be randomly incorporated into a PE chain (Figure 2.4) by using palladium(II) catalysts bearing phosphine–sulfonate ligands [78–80]. In all cases, a low to moderate catalyst activity ($<10^3$ TOF) was reported. For instance, Pugh [81], by using an *in situ* generated catalytic system similar to that he employed for the nonalternating copolymerization of ethene and carbon monoxide (**L1**, Figure 2.3), reported the incorporation of circa 10% alkyl acrylate units into a *linear* PE backbone with modest catalytic activity. Between 60 °C and 80 °C as well as 5 h and 15 h of reaction time, the number average molecular weights up to 21,000, $1.6 < \text{PDI} < 2.9$, and circa 1 methyl branch per 10^3 carbon atoms were obtained. Other phosphine–sulfonate ligands were tested with Pd(II), some of which afforded M_n up to 41,200 (**8**, Figure 2.5) [82] and an acrylate incorporation as high as 52 mol% (**16**) [83]. As with Pd(II) α-diimine systems, the

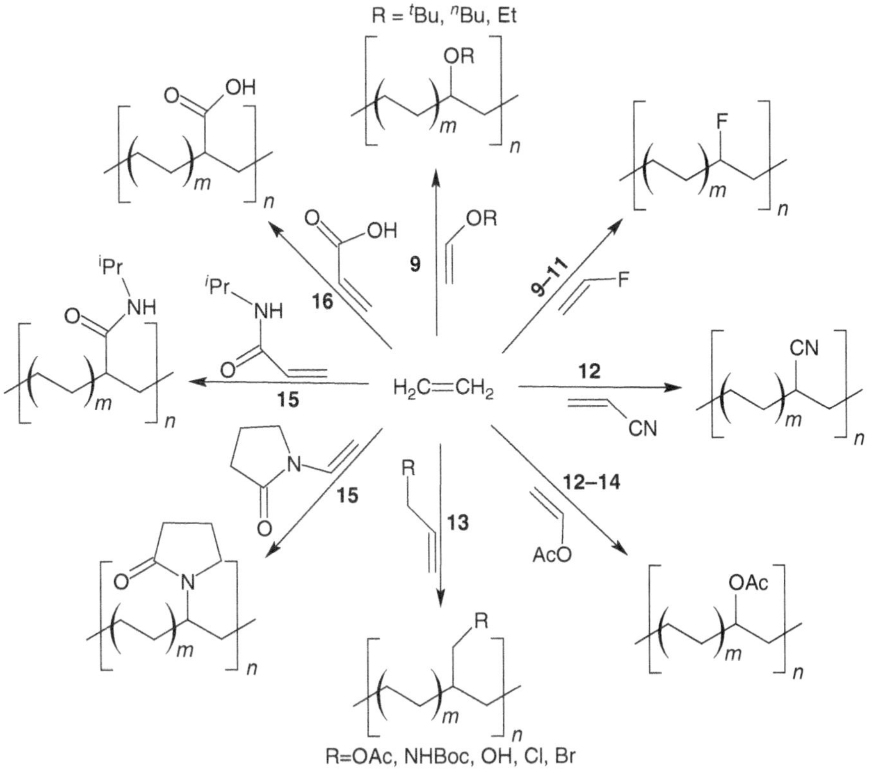

Figure 2.4 Random copolymers of ethene and various polar olefins prepared by means of Pd(II) catalysts bearing phosphine–sulfonate ligands (see Figure 2.5).

incorporation rate and the catalytic activity were found to be, respectively, directly proportional and inversely proportional to the concentration of acrylate.

Jordan [84] described the copolymerization of ethene with various alkyl VEs using the catalyst (**9**) with reaction temperatures between 60 °C and 100 °C and up to 20 bar of ethene. Following circa 20 h of reaction time, random copolymers with $900 < M_n < 4800$, $1.8 < \text{PDI} < 2.0$, less than 10 methyl branches per 10^3 carbon atoms, and an incorporation of approximately 7 mol% VEs were obtained. In another study, the same research group reported the copolymerization of ethene with vinyl fluoride using analogous catalysts (**9–11**) under comparable experimental conditions [85]. Polymers with M_n as high as 14,500, $2.5 < \text{PDI} < 3.0$, less than five methyl branches per 10^3 carbon atoms, and an incorporation of vinyl fluoride in the PE backbone ≤ 0.5 mol% were prepared. Nozaki [86] used the catalyst (**12**) to prepare linear copolymers of ethene and AN at 80–100 °C with a pressure of ethene up to 40 bar. Although the incorporation of AN ranged from 2 to 9 mol%, an M_n up to 12,000 and a PDI as low as 1.5 were measured following a reaction time of 120–270 h. Interestingly, no chain branching could be detected by means

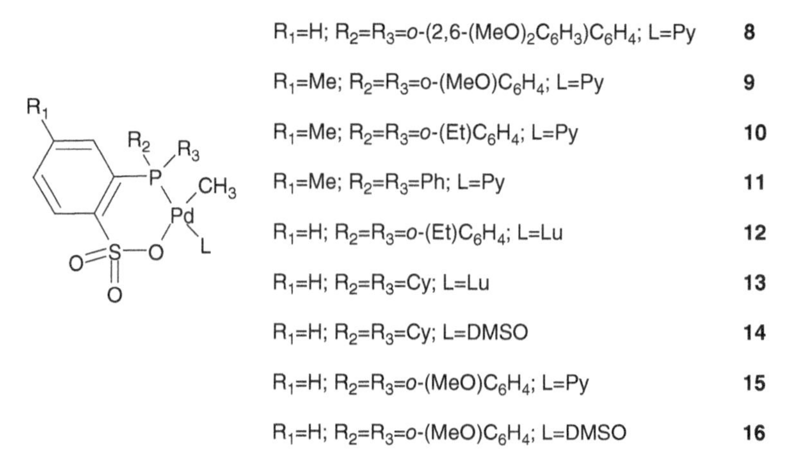

Figure 2.5 Pd(II) catalysts bearing phosphine–sulfonate ligands for the copolymerizations of ethene and polar olefins.

of ^{13}C NMR. In a subsequent publication, Nozaki [87] described the copolymerization of ethene and VA catalyzed by the phosphine–sulfonate complexes **12–14**. At 80 °C and 30 bar of ethene, the resulting polyolefins showed an M_n of circa 5000, a PDI > 2, and contained less than 2 mol% of VAc and 1 methyl branch per 10^3 carbon atoms. Monomers such as allyl acetate and allyl alcohol, as well as various allyl amines and allyl halides, could also be copolymerized with ethene using the catalyst **13** [88]. In all cases, the average number molecular weight (M_n) and the incorporation of allyl monomer into the PE chain did not exceed 7000 and 8 mol%, respectively. The highest catalyst activities were obtained in the presence of allyl acetate, whereas no polymerization occurred with unprotected allyl amine. Similar findings were reported by Claverie [89], who used complex **15** to copolymerize ethene with *N*-vinyl-2-pyrrolidinone or *N*-isopropyl acrylamide at 95 °C. The reaction afforded linear polymers with $M_n \leq 5000$ and a monomer incorporation <5 mol% following a reaction time of up to 72 h. Last but not the least, catalyst **16** was employed by Mecking [90] for the direct synthesis of linear (ca. 3 methyl branches/1000 carbon atoms) ethene-acrylic acid copolymers at 95 °C. Up to 9.6 mol% of acrylic acid was incorporated, the large majority of which was found not in end groups, as in the case of VAc and AN, but in the polymer backbone. Analogous results were obtained by the same group for the copolymerization of vinyl sulfones with catalyst **16** under comparable reaction conditions [91].

2.5.2 Polymerization via a Dual Radical/Migratory Insertion Pathway

Very recently, Monteil [92] reinvestigated the reactivity of two nickel(II) complexes originally designed for the coordination/insertion oligomerization/polymerization of ethene. The authors discovered that, while practically unreactive toward (meth)acrylates when used in the absence of additives, the complexes became able to polymerize *n*-butyl acrylate (nBuA) and MMA in the presence of triphenylphosphine (PPh_3). A mechanism based on a radical polymerization initiated by the thermal homolytic cleavage of the metal–carbon bonds in the complexes, and facilitated by the stabilization of the resulting Ni(I) species upon coordination of PPh_3, was proposed. The picture became very intriguing when the two monomers were copolymerized with nonpolar olefins such as styrene (Sty) or ethene. In fact, not only well-defined copolymers with high incorporation of Sty and ethene were promptly obtained, but the results of all the mechanistic experiments were in favor of a process in which nBuA or MMA were enchained according to a radical-type mechanism whereas Sty or ethene were incorporated via migratory insertion.

2.5.3 Coordinative Group Transfer Polymerization

Since the detailed description of the CGTP is out of the scope of this chapter, in this section only an overview of such a process is given. An extensive treatment of the topic can be found elsewhere [93]. At odds with migratory insertion polymerization, in which the (polar) olefin is π-coordinated to a transition metal center and is nucleophilically and intramolecularly attacked by a cis-coordinated alkyl moiety (Scheme 2.1), in the CGTP the polar olefin is κ-coordinated to an electrophilic metal center (e.g., early transition metals, main group metals, and also lanthanides) and can be nucleophilically attacked, both inter- (Scheme 2.2a) and intramolecularly (Scheme 2.2b), by an anionic moiety such as an enolate [94]. A number of monomers including alkyl (meth)acrylates, (meth)acrylamides, AN, and vinyl ketones have been polymerized through CGTP in good yields as well as stereoselective and "living" fashion affording a variety of polar–polar block copolymers and polar–apolar block copolymers. The preparation of the latter class of macromolecules has been possible by exploiting the ability of several catalysts to mediate both migratory insertion and CGTPs depending on the experimental conditions used. For instance, by polymerizing ethene or propene with conventional metallocene catalysts at temperatures $\leq 0\,°C$, which ensures a "living"-like process, well-defined metal–polymer intermediate species are obtained that can in turn initiate CGTP upon addition of, for example, methyl methacrylate.

2.6 COPOLYMERIZATION OF POLAR VINYL MONOMERS AND CARBON MONOXIDE

The coordination copolymerization of polar monomers such as (meth)acrylates, VA, and AN with carbon monoxide has long been unsuccessful for reasons similar to those reported in the earlier section. Also in this case, the proof-of-principle arrived when neutral Pd(II) catalysts containing phosphine–sulfonate ligands were employed. In 2007, Nozaki [95] reported the alternating copolymerization of VA and CO in bulk at 70 °C with catalysts generated *in situ* by mixing $Pd(dba)_2$ with ligand **L1** or **L4** (Figure 2.3). With a catalytic productivity of up to 3 g $mmol^{-1}$ h^{-1}

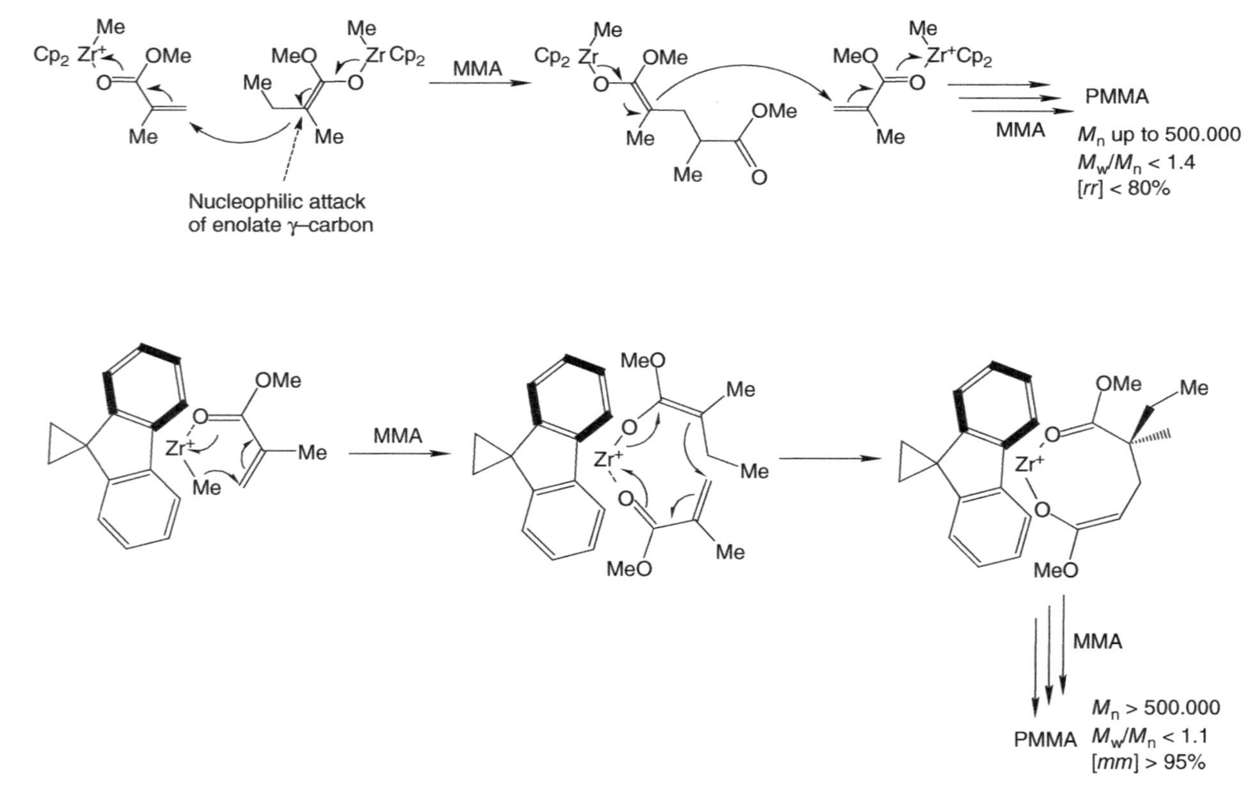

Scheme 2.2 Intermolecular (a) and intramolecular (b) coordinative group transfer polymerization of methyl methacrylate catalyzed by achiral (a) and chiral (b) zirconocenes.

and polymer molecular weights of up to 38,000, linear regio irregular polyketones were prepared. A year later, the same group reported the alternating copolymerization of methyl acrylate with CO using the same catalytic systems in bulk at 70 °C for 20 h [96]. With an equally low catalytic activity (up to 4.4 g mmol^{-1} h^{-1}) and M_n up to 30,000, linear regio regular (highly 2,1) polyketones were obtained. In either polymerization, the intervention of radical-based monomer enchainment was ruled out by means of a series of control experiments.

2.7 WHY ARE PHOSPHINE–SULFONATE LIGANDS SO SPECIAL?

Apart from the steric factors, which act in a way similar to that of other classes of bidentate ligands, the (partial) success of phosphine–sulfonate ligands in the coordinative insertion of polar monomers is due to a combination of different electronic effects. First, being anionic, the phosphine–sulfonate ligand makes the Pd(II) center less electrophilic and thus less reactive toward functional groups such as nitriles and carbonyls. On the one hand, this reduces (i) the poisoning effect of AN via κ–N coordination [97]; (ii) the chances of β-heteroatom elimination to metal with the consequent catalyst deactivation; and (iii) the attraction between the palladium atom and the carbonyl group in the metallacycles generated in the copolymerization with CO, which makes ethene kinetically competitive with carbon monoxide up to the point of allowing its consecutive insertions into the palladium alkyl bond. This explains the formation of polyketones with a CO content lower than 50 mol% [98]. On the other hand, the lower electrophilicity of the neutral metal center, together with the presence of an electron-withdrawing group in the polar olefin, enhances the π-back donation from the palladium to the coordinated olefin. This partial electron transfer strengthens the metal–olefin interaction (enhanced coordination) and, at the same time, weakens (i.e., activates) the C=C bond within the olefinic moiety. Moreover, the weak π-acceptor character of the sulfonate moiety gives an extra contribution to the π-back bonding at its trans position. Last but not the least, since phosphine is a strong σ-donor, it increases the migratory ability of the alkyl moiety at its *trans* position. Consequently, when the polar olefin is *trans* with respect to the sulfonate group and the growing chain is *trans* with respect to the phosphine group, the possibility of having migratory insertion is the highest [97].

As mentioned earlier, a peculiarity of the (co)polymers prepared with palladium catalysts bearing phosphine–sulfonate ligands is their linearity. Experimental and theoretical studies have rationalized this finding as follows [99]. While β–H transfer to monomer seems negligible with this class of catalysts, β–H elimination to metal takes place. However, since the β-hydride elimination is zero order in monomer, whereas the propagation reaction is first order, the former is kinetically noncompetitive at high monomer pressure, which suppresses the chain walking and thus the chain branching.

2.8 TELECHELIC AND END-CAPPED MACROMOLECULES

As discussed earlier, during the copolymerization of olefins with polar monomers, π-type complexes and κ-type complexes may coexist. When there is a chemical equilibrium between these two species, it is possible to copolymerize olefins and polar monomers; however, if the κ complexes are far more stable, the insertion of other monomers is hampered. As a result, polar monomers can be used to quench olefin polymerization while introducing a last polar monomer to the chain in the ω-position. Moreover, the catalyst can be activated with, or already contain, a functional moiety that, after polymerization, can lead to α-functionalized macromolecule. Brookhart's [14,59] Pd metallacycle catalyst is an example of α-functionalization ("head-capping") catalyst, which introduces an acrylate ester as the first monomer in the polymer chain.

Concerning ω-functionalization, a quenching effect is frequently observed when adding polar monomers to polymerization systems with limited tolerance to functional groups. As an example, one may cite the MMA addition to an olefin polymerization with Ni(P,O) catalysts [65]. The quenching strategy has also been employed to prepare β-styrene end-capped polymers with Pd(II)/diimine catalysts [100]. These late transition metal compounds can polymerize ethene in a quasi-living manner (limited chain termination and transfer) and also insert a styrene unit into the growing polymer chain, which in turn makes the resulting benzyl complexes too stable and catalytically inactive.

End-capped macromolecules can also be further functionalized. For instance, Brookhart and Matyjaszewski [101] grafted methacrylate-capped hyperbranched PE onto poly(*n*-butyl acrylate) (PnBuA) by combining Pd-catalyzed, migratory insertion polymerization, and atom transfer radical polymerization (ATRP). The final material possessed a graft-block structure, as determined by SEC and NMR, and self-assembled with PnBuA as continuous phase and PE as cylindrical domains, as shown by atomic force microscopy (AFM).

2.9 ON THE USE OF CHEMOINFORMATICS FOR A MORE RAPID DEVELOPMENT OF THE FIELD

Up to date, the CAS Registry^{SM} contains more than 8×10^7 organic and inorganic chemical substances, which can be taken as a meaningful representation of the *actual* chemical space [102]. However, the stoichiometric combination of all the atoms in the periodic table in all the possible topology isomers—the so-called *virtual* chemical space—affords a much greater number of possible chemical substances, most likely in excess of 10^{60} [103]. Reasoning on a purely statistical basis, it is highly probable that, in such an enormous space, there might be a chemical compound for nearly every purpose. This includes highly active, inexpensive, "green" catalysts for the synthesis of polyolefins having all the properties necessary for any desired application. However, it is unlikely that the conventional trial-and-error and one-variable-at-time approaches may provide all the sought molecules in a reasonable time. Yet, notwithstanding the capability of modern high throughput synthesis and screening techniques to carry out thousands of experiments per day, it has become apparent that the preparation of each one of 10^{60} or more molecules composing the virtual chemical space is materially impossible.

With a clever combination of graph theory, machine learning, data mining, and advanced statistics, as well as chemical topology, molecular mechanics, and quantum mechanics, chemoinformatics offers unique opportunities for a rapid "sifting" of the virtual chemical space [104]. One important goal of chemoinformatics is to provide quantitative and predictive equations of the type $P = f(d_1, d_2, \ldots, d_n)$, where P is a property of interest (e.g., the activity or the selectivity of a catalyst, the half-maximal inhibitory concentration (IC_{50}) of a drug, the Young modulus of a polymer) and $d_1, \ldots, d_n$ are the molecular descriptors. According to Todeschini and Consonni [105], a *molecular descriptor is the final result of a logic and mathematical procedure which transforms chemical information encoded within a symbolic representation of a molecule into a useful number*. With such equations at hand, the performance of a given molecule can be assessed simply by calculating the descriptors $d_1, \ldots, d_n$ prior to any synthesis or experimental assay. The process is dubbed virtual or *in silico* screening. In this way, only the most promising candidates are prepared in the laboratory, saving time, and money.

Although successfully employed in pharmaceutical and agrochemical research for years, chemoinformatics has only started to be applied to the development of catalysts and materials recently [104b]. The first, and up to now, only set of quantitative equations for the *in silico* evaluation of polymerization catalysts' performance has been reported for copper-catalyzed ATRP [106,107]. The mathematical models were in agreement with the generally accepted ATRP mechanistic picture and quantified the roles played in ATRP by energetic and steric factors of both catalysts and initiators as well as by the reaction medium. Yet, the models suggested the existence of long-range interactions that might significantly influence the mutual orientation of the catalyst and the initiator prior to the Brownian collision and the subsequent binding. Importantly, the equations managed to predict the catalytic performance of diverse sets of both Cu complexes and alkyl halide initiators. In the wake of these success stories, chemoinformatics might certainly prove to be a powerful tool for the discovery of new generations of catalysts for the coordination (co)polymerization of vinyl monomers.

2.10 CONCLUSION AND OUTLOOK

Copolymerization reactions combine the possibility of having the chain topologies (e.g., linear, branched, cyclic) that homopolymerization can afford with the opportunity of introducing a multitude of functionalities distributed in various ways (e.g., randomly, alternatingly, blockily) along the polymer backbone. This makes copolymers among the most diverse and tailorable materials available today. Particularly interesting is the case of the copolymers resulting from the insertion polymerization of nonpolar olefins (e.g., ethene, various 1-alkenes, dienes) since they are affordable and of high industrial value. In contrast, and despite the energies squandered, polymer chemists have had much less success in designing metal catalysts for the coordination (co)polymerization of polar olefins. The introduction of Pd(II) catalysts bearing phosphine–sulfonate ligands is only the first step toward the achievement of this objective since many aspects of the process still need to be improved. This includes, for example, the level of incorporation of polar monomer, the catalytic productivity, the minimization of chain-breaking reactions, the stereoselectivity, and the (catalyst-encoded) sequence selectivity when two or more monomers are present. The last three issues need to be addressed also in the case of copolymerization of nonpolar olefins. Hence, despite the age of the field, catalytic olefin copolymerization still offers plenty of room for exciting research.

The development of more advanced catalytic systems that may function simultaneously through different mechanisms (e.g., radical and migratory insertion, also in the presence of CSAs, or migratory insertion and group transfer) together with an increasing use of chemoinformatics may certainly help facing these challenges.

REFERENCES

1. Seyferth, D. (2001). $(C_2H_4)PtCl_3]^{(-)}$, the anion of Zeise's salt, $K[(C_2H_4)PtCl_3]\cdot H_2O$. *Organometallics*, *20*, 2–6.
2. Osborn, J. A., Jardine, F. H., Young, J. F., Wilkinson, G. (1966). The preparation and properties of tris(triphenylphosphine) halogenorhodium(I) and some reactions thereof including catalytic homogeneous hydrogenation of olefins and acetylenes and their derivatives. *J. Chem. Soc. A*, 1711–1732.
3. Halpern, J. (1981). Mechanistic aspects of homogeneous catalytic hydrogenation and related processes. *Inorg. Chim. Acta*, *50*, 11–19.
4. Corradini, P., Guerra, G., Cavallo, L. (2004). Do new century catalysts unravel the mechanism of stereocontrol of old Ziegler-Natta catalysts? *Acc. Chem. Res.*, *37*, 231–241.
5. (a) Britovsek, G. J. P., Gibson, V. C., Wass, D. F. (1999). The search for new-generation olefin polymerization catalysts: life beyond metallocenes. *Angew. Chem. Int. Ed. Eng.*, *38*, 428–447; (b) Gibson, V. C., Spitzmesser, S. K. (2003). Advances in non-metallocene olefin polymerization catalysis. *Chem. Rev.*, *103*, 283–315; (c) Resconi, L., Cavallo, L., Fait, A., Piemontesi, F. (2000). Selectivity in propene polymerization with metallocene catalysts. *Chem. Rev.*, *100*, 1253–1345.
6. (a) Pearson, R. G. (1963). Hard and soft acids and bases. *J. Am. Chem. Soc.*, *85*, 3533–3539; (b) Huheey, J. E., Keiter, E. A., Keiter, R. L. (1993). *Inorganic Chemistry, Principles of Structure and Reactivity*. Harper Collin 4th ed. pp. 344–355.
7. Strömberg, S., Zetterberg, K., Siegbahn, P. E. M. (1997). Trends within a triad: comparison between σ-alkyl complexes of nickel, palladium and platinum with respect to association of ethylene, migratory insertion and β-hydride elimination. A theoretical study. *J. Chem. Soc. Dalton Trans.*, 4147–4152.
8. Fusto, M., Giordano, F., Orabona, I., Ruffo, F. (1997). $[PtR(\eta^2\text{-olefin})(N\text{-}N)]^+$ complexes containing the olefin and the alkyl ligand in a cis arrangement. Preparation, structural characterization, and olefin stereochemistry. *Organometallics*, *16*, 5981–5987.

9. Malinoski, J. M., White, P. S., Brookhart, M. (2003). Structural characterization of $[\kappa^2\text{-(t-Bu)}_2PCH_2C(O)C_6H_5]PdMe(\eta^2\text{-}C_2H_4)^+ BAr'_4{}^-$: a model for the catalyst resting state for ethylene polymerization. *Organometallics*, *22*, 621–623.

10. Britovsek, G. J. P., Bruce, M., Gibson, V. C., Kimberley, B. S., Maddox, P. J., Mastroianni, S., McTavish, S. J., Redshaw, C., Solan, G. A., Stromberg, S., White, A. J. P., Williams, D. J. (1999). Iron and cobalt ethylene polymerization catalysts bearing 2,6- bis(imino)pyridyl ligands: synthesis, structures, and polymerization studies. *J. Am. Chem. Soc.*, *121*, 8728–8740.

11. Gibson, V. C., Redshaw, C., Solan, G. A. (2007). Bis(imino)pyridines: surprisingly reactive ligands and a gateway to new families of catalysts. *Chem. Rev.*, *107*, 1745–1776.

12. Brookhart, M., Green, M. L. H., Parkin, G. (2007). Agostic interactions in transition metal compounds. *Proc. Natl. Acad. Sci. U. S. A.*, *104*, 6908–6914.

13. Crossetti, G. L., Dias, M. L., Queiroz, B. T., Silva, L. P., Ziglio, C. M., Bomfim, J. A. S., Filgueiras, C. A. L. (2004). Ethylene polymerization with imine and phosphine nickel complexes containing isothiocyanate. *Appl. Organomet. Chem.*, *18*, 331–336.

14. Bomfim, J. A. S., Dias, M. L., Filgueiras, C. A. L., Peruch, F., Deffieux, A. (2008). The effect of polymerization temperature on the structure and properties of poly(1-hexene) and poly(1-decene) prepared with a Ni(II)-diimine catalyst. *Cat. Today*, *133–135*, 879–885.

15. Mecking, S., Johnson, L. K., Wang, L., Brookhart, M. (1998). Mechanistic studies of the palladium-catalyzed copolymerization of ethylene and α-olefins with methyl acrylate. *J. Am. Chem. Soc.*, *120*, 888–899.

16. Brookhart, M. S., Johnson, L. K., Killian, C. M. *US patent no. US2004/0102591* (paragraph 715 and examples 063 through 083).

17. Bomfim, J. A. S., Dias, M. L., Filgueiras, C. A. L., Deffieux, A., Peruch, F. (2007). Studies on microstructure and crystallinity control in polyolefins prepared by nickel catalysts. Proceedings of the 9th Brazilian Polymer Congress, n249 (8 pages). Campina Grande, October 10th 2007.

18. Busico, V., Cipullo, R. (2001). Microstructure of polypropylene. *Prog. Polym. Sci.*, *26*, 443–433.

19. Brintzinger, H. H., Fischer, D., Mülhaupt, R., Rieger, B., Waymouth, R. M. (1995). Stereospecific olefin polymerization with chiral metallocene catalysts. *Angew. Chem. Int. Ed. Eng.*, *34*, 1143–1170.

20. Hotta, A., Cochran, E., Ruokolainen, J., Khanna, V., Fredrickson, G. H., Kramer, E. J., Shin, Y.-W., Shimizu, F., Cherian, A. E., Hustad, P. D., Rose, J. M., Coates, G. W. (2006). Semicrystalline thermoplastic elastomeric polyolefins: advances through catalyst development and macromolecular design. *Proc. Natl. Acad. Sci. U. S. A.*, *3*, 15327–15332.

21. Rose, J. M., Deplace, F., Lynd, N. A., Wang, Z., Hotta, A., Lobkovsky, E. B., Kramer, E. J., Coates, G. W. (2008). C_2-symmetric Ni(II) α-diimines featuring cumyl-derived ligands: synthesis of improved elastomeric regioblock polypropylenes. *Macromolecules*, *41*, 9548–9555.

22. Coates, G. W., Waymouth, R. M. (1995). Oscillating stereocontrol: a strategy for the synthesis of thermoplastic elastomeric polypropylene. *Science*, *13*, 217–219.

23. Lin, S., Waymouth, R. M. (2002). 2-Arylindene metallocenes: conformationally dynamic catalysts to control the structure and properties of polypropylenes. *Acc. Chem. Res.*, *35*, 765–773.
24. Rush, S., Reinmuth, A., Risse, W. (1997). Palladium(II)-catalyzed olefin addition polymerizations of 3,3-dialkyl-substituted cyclopropenes. *Macromolecules*, *30*, 7375–7385.
25. McLain, S. J., Feldman, J., McCord, E. F., Gardner, K. H., Teasley, M. F., Coughlin, E. B., Sweetman, K. J., Johnson, L. K., Brookhart, M. (1998). Addition polymerization of cyclopentene with nickel and palladium catalysts. *Macromolecules*, *31*, 6705–6707.
26. Ittel, S. D., Johnson, L. K., Brookhart, M. (2000). Late-metal catalysts for ethylene homo- and copolymerization. *Chem. Rev.*, *100*, 1169–1203.
27. Wang, W., Nomura, K. (2005). Remarkable effects of aluminum cocatalyst and comonomer in ethylene copolymerizations catalyzed by (Arylimido)(aryloxo)vanadium complexes: efficient synthesis of high molecular weight ethylene/norbornene copolymer. *Macromolecules*, *38*, 5905–5913.
28. Hasan, T., Ikeda, T., Shiono, T. (2004). Ethene-norbornene copolymer with high norbornene content produced by aftsa-fluorenylamidodimethyltitanium complex using a suitable activator. *Macromolecules*, *37*, 8503–8509.
29. Jansen, J. C., Mendichi, R., Sacchi, M. C., Tritto, I. (2003). Kinetic studies of the copolymerization of ethylene with norbornene by ansa-zirconocene/methylaluminoxane catalysts: evidence of a long-lasting “quasi-living” initial period. *Macromol. Chem. Phys.*, *204*, 522–530.
30. Yoshida, Y., Saito, J., Mitani, M., Takagi, Y. S., Matsui, S., Ishii, S.-I., Nakano, T., Kashiwa, N., Fujita, T. (2002). Living ethylene/norbornene copolymerisation catalyzed by titanium complexes having two pyrrolide-imine chelate ligands. *Chem. Commun.*, *1298–1299*.
31. Li, X. F., Dai, K., Ye, W. P., Pan, L., Li, Y. S. (2004). New titanium complexes with two β-enaminoketonato chelate ligands: syntheses, structures, and olefin polymerization activities. *Organometallics*, *23*, 1223–1230.
32. Kiesewetter, J., Kaminsky, W. (2003). Ethene/norbornene copolymerization with palladium(II) α-diimine catalysts: from ligand screening to discrete catalyst species. *Chem. Eur. J.*, *9*, 1750–1758.
33. Yoshida, Y., Mohri, J., Ishii, J., Mitani, M., Saito, J., Matsui, S., Makio, H., Nakano, T., Tanaka, H., Onda, M., Yamamoto, Y., Mizuno, A., Fujita, T. (2004). Living copolymerization of ethylene with norbornene catalyzed by bis(pyrrolide-imine) titanium complexes with MAO. *J. Am. Chem. Soc.*, *126*, 12023–12032.
34. Hong, M., Wang, Y.-X., Mu, H.-L., Li, Y.-S. (2011). Efficient synthesis of hydroxylated polyethylene via copolymerization of ethylene with 5-norbornene-2-methanol using bis(β-enaminoketonato)titanium catalysts. *Organometallics*, *30*, 4678–4686.
35. Arriola, D. J., Carnahan, E. M., Hustad, P. D., Kuhlman, R. L., Wenzel, T. T. (2006). Catalytic production of olefin block copolymers via chain shuttling polymerization. *Science*, *312*, 714–719.
36. Hustad, P. D., Kuhlman, R. L., Arriola, D. J., Carnahan, E. M., Wenzel, T. T. (2007). Continuous production of ethylene-based diblock copolymers using coordinative chain transfer polymerization. *Macromolecules*, *40*, 7061–7064.
37. Zintl, M., Rieger, B. (2007). Novel olefin block copolymers through chain-shuttling polymerization. *Angew. Chem. Int. Ed.*, *46*, 333–335.

38. Chung, T. C., Rhubright, D. (1991). Synthesis of functionalized polypropylene. *Macromolecules*, *24*, 970–972.

39. Chung, T. C. (2002). Synthesis of functional polyolefin copolymers with graft and block structures. *Prog. Polym. Sci.*, *27*, 39–85.

40. Drent, E., Budzelaar, P. H. M. (1996). Palladium-catalyzed alternating copolymerization of alkenes and carbon monoxide. *Chem. Rev.*, *96*, 663–681.

41. Bianchini, C., Meli, A. (2002). Alternating copolymerization of carbon monoxide and olefins by single-site metal catalysis. *Coord. Chem. Rev.*, *225*, 35–66.

42. Belov, G. P., Novikova, E. V. (2004). Polyketones as alternating copolymers of carbon monoxide. *Russ. Chem. Rev.*, *73*, 267–291.

43. Durand, J., Milani, B. (2006). The role of nitrogen-donor ligands in the palladium-catalyzed polyketones synthesis. *Coord. Chem. Rev.*, *250*, 542–560.

44. García Suárez, E. J., Godard, C., Ruiz, A., Claver, C. (2007). Alternating and non-alternating Pd-catalysed co- and terpolymerisation of carbon monoxide and alkenes. *Eur. J. Inorg. Chem.*, *2582–2593*.

45. Drent, E., van Dijk, R., van Ginkel, R., van Oort, B., Pugh, R. I. (2002). The first example of palladium catalysed non-perfectly alternating copolymerisation of ethene and carbon monoxide. *Chem. Commun.*, 964–965.

46. Hearley, A. K., Nowack, R. J., Rieger, B. (2005). New single-site palladium catalysts for the nonalternating copolymerization of ethylene and carbon monoxide. *Organometallics*, *24*, 2755–2763.

47. (a) di Lena, F., Matyjaszewski, K. (2010). Transition metal catalysts for controlled radical polymerization. *Prog. Polym. Sci.*, *35*, 959–1021; (b) Fetzer, L., Toniazzo, V., Ruch, D., di Lena, F. (2012). Transition-metal catalysts for controlled radical polymerization: a first update. *Isr. J. Chem.*, (2012). *52(3-4)*, 221–229.

48. Boffa, L. S., Novak, B. M. (2000). Copolymerization of polar monomers with olefins using transition-metal complexes. *Chem. Rev.*, *100*, 1479–1494.

49. Stockland, R. A., Jordan, R. F. (2000). Reaction of vinyl chloride with a prototypical metallocene catalyst: stoichiometric insertion and β-Cl elimination reactions with rac-(EBI)$ZrMe^+$ and catalytic dechlorination/oligomerization to oligopropylene by rac-(EBI)$ZrMe_2$/MAO. *J. Am. Chem. Soc.*, *122*, 6315–6316.

50. Stockland, R. A., Foley, S. R., Jordan, R. F. (2003). Reaction of vinyl chloride with group 4 metal olefin polymerization catalysts. *J. Am. Chem. Soc.*, *125*, 796–809.

51. Foley, S. R., Shen, H., Qadeer, U. A., Jordan, R. F. (2004). Generation and insertion reactivity of cationic palladium complexes that contain halogenated alkyl ligands. *Organometallics*, *23*, 600–609.

52. Li, W., Zhang, X., Meetsma, A., Hessen, B. (2004). Palladium-catalyzed copolymerization of ethene with acrolein dimethyl acetal: catalyst action and deactivation. *J. Am. Chem. Soc.*, *129*, 12246–12247.

53. Williams, B. S., Leatherman, M D., White, P. S., Brookhart, M. (2005). Reactions of vinyl acetate and vinyl trifluoroacetate with cationic diimine Pd(II) and Ni(II) alkyl complexes: identification of problems connected with copolymerizations of these monomers with ethylene. *J. Am. Chem. Soc.*, *127*, 5132–5146.

54. Wu, F., Foley, S. R., Burns, C. T., Jordan, R. F. (2005). Acrylonitrile insertion reactions of cationic palladium alkyl complexes. *J. Am. Chem. Soc.*, *127*, 1841–1853.

55. Groux, L. F., Weiss, T., Reddy, D. N., Chase, P. A., Piers, W. E., Ziegler, T., Parvez, M., Benet-Buchholz, J. (2005). Insertion of acrylonitrile into palladium methyl bonds in neutral and anionic Pd(II) complexes. *J. Am. Chem. Soc.*, *127*, 1854–1869.
56. Wu, F., Jordan, R. F. (2006). Acrylonitrile insertion reactions of palladium alkyl complexes that contain neutral or anionic bidentate phosphine ligands. *Organometallics*, *25*, 5631–5637.
57. Michalak, A., Ziegler, T. (2001). DFT studies on the copolymerization of α-olefins with polar monomers: comonomer binding by nickel- and palladium-based catalysts with Brookhart and Grubbs ligands. *Organometallics*, *20*, 1521–1532.
58. Philipp, D. M., Muller, R. P., Goddard, W. A. I., Storer, J., McAdon, M., Mullins, M. (2002). Computational insights on the challenges for polymerizing polar monomers. *J. Am. Chem. Soc.*, *124*, 10198–10210.
59. Szabo, M. J., Galea, N. M., Michalak, A., Yang, S.-Y., Groux, L. F., Piers, W. E., Ziegler, T. (2005). Copolymerization of ethylene with polar monomers: chain propagation and side reactions. A DFT theoretical study using zwitterionic Ni(II) and Pd(II) catalysts. *J. Am. Chem. Soc.*, *127*, 14692–14703.
60. von Schenck, H., Strömberg, S., Zetterberg, K., Ludwig, M., Åkermark, B., Svensson, M. (2001). Insertion aptitudes and insertion regiochemistry of various alkenes coordinated to cationic (σ-R)(diimine)palladium(II) (R = $-CH_3$, $-C_6H_5$). A theoretical study. *Organometallics*, *20*, 2813–2819.
61. Zhang, Z., Lu, X., Xu, Z., Zhang, Q., Han, X. (2001). Role of halide ions in divalent palladium-mediated reactions: competition between β-heteroatom elimination and β-hydride elimination of a carbon–palladium bond. *Organometallics*, *20*, 3724–3728.
62. Watson, L. A., Yandulov, D. V., Caulton, K. G. (2001). C–D_0 (D_0 = π-donor, F) cleavage in $H_2CCH(D0)$ by $(Cp_2ZrHCl)n$: mechanism, agostic fluorines, and a carbene of Zr(IV). *J. Am. Chem. Soc.*, *123*, 603–611.
63. Kraft, B. M., Jones, W. D. (2002). Mechanism of vinylic and allylic carbon–fluorine bond activation of non-perfluorinated olefins using $Cp^*_2ZrH_2$. *J. Am. Chem. Soc.*, *124*, 8681–8689.
64. Clot, E., Mégret, C., Kraft, B. M., Eisenstein, O., Jones, W. D. (2004). Defluorination of perfluoropropene using $Cp^*_2ZrH_2$ and Cp^*_2ZrHF: a mechanism investigation from a joint experimental–theoretical perspective. *J. Am. Chem. Soc.*, *126*, 5647–5653.
65. Zhao, H., Ariafard, A., Lin, Z. (2006). β-Heteroatom versus β-hydrogen elimination: a theoretical study. *Organometallics*, *25*, 812–819.
66. Baird, M. C. (2000). Carbocationic alkene polymerizations initiated by organotransition metal complexes: an alternative, unusual role for soluble Ziegler–Natta catalysts. *Chem. Rev.*, *100*, 1471–1478.
67. Chen, C.-L., Chen, Y.-C., Liu, Y.-H., Peng, S.-M., Liu, S.-T. (2002). Initiation steps for the polymerization of vinyl ethers promoted by cationic palladium aqua complexes. *Organometallics*, *21*, 5382–5385.
68. (a) Yamamoto, T., Yamamoto, A., Ikeda, S. (1972). Activation of iron-alkyl bonds in dialkylbis (dipyridyl)iron(II) by interaction with olefins. *Bull. Chem. Soc. Jpn.*, *45*, 1104; (b) Yamamoto, T., Yamamoto, A., Ikeda, S. (1972). The mechanism of vinyl polymerization by dialkylbis(dipyridyl)iron(II). *Bull. Chem. Soc. Jpn.*, *45*, 1111; (c) Ikariya, T., Yamamoto, A. (1974). Preparation and properties of ligand-free methylcopper and of copper alkyls coordinated with 2,2′-bipyridyl and tricyclohexylphosphine. *J. Organomet. Chem.*, *72*, 145; (d) Miyashita, A., Yamamoto, T., Yamamoto,

A. (1977). Thermal stability of alkylcopper(I) complexes coordinated with tertiary phosphines. *Bull. Chem. Soc. Jpn.*, *50*, 1109.

69. Schaper, F., Foley, S. R., Jordan, R. F. (2004). Acrylonitrile polymerization by Cy_3PCuMe and $(Bipy)_2FeEt_2$. *J. Am. Chem. Soc.*, *126*, 2114–2124.
70. Johnson, L. K., Mecking, S., Brookhart, M. (1996). Copolymerization of ethylene and propylene with functionalized vinyl monomers by palladium(II) Catalysts. *J. Am. Chem. Soc.*, *118*, 267–268.
71. Michalak, A, Ziegler, T. (2001). DFT studies on the copolymerization of α-olefins with polar monomers: ethylene–methyl acrylate copolymerization catalyzed by a Pd-based diimine catalyst. *J. Am. Chem. Soc.*, *123*, 12266–12278.
72. Guan, Z. (2003). Control of polymer topology through late-transition-metal catalysis. *J. Polym. Sci. Part A:Polym. Chem.*, *41*, 3680–3692.
73. Sen, A., Bokar, S. (2007). Perspective on metal-mediated polar monomer/alkene copolymerization. *J. Organomet. Chem.*, *692*, 3291–3299.
74. Bokar, S., Yennawar, H., Sen, A. (2007). Methacrylate insertion into cationic diimine palladium(II)–alkyl complexes and the synthesis of poly(alkene-block-alkene/carbon monoxide) copolymers. *Organometallics*, *26*, 4711–4714.
75. Gibson, V. C., Tomov, A. (2001). Functionalised polyolefin synthesis using [P,O]Ni catalysts. *Chem. Commun., 1964–1965.*
76. Rünzi, T., Guironnet, D., Göttker-Schnettmann, I., Mecking, S. (2010). Reactivity of methacrylates in insertion polymerization. *J. Am. Chem. Soc.*, *132*, 16623–16630.
77. Luo, S., Jordan, R. F. (2006). Copolymerization of silyl vinyl ethers with olefins by (α-diimine)PdR^+. *J. Am. Chem. Soc.*, *128*, 12072–12073.
78. Berkefeld, A., Mecking, S. (2008). Coordination copolymerization of polar vinyl monomers $H_2C{=}CHX$. *Angew. Chem. Int. Ed.*, *47*, 2538–2542.
79. Nakamura, A., Ito, S., Nozaki, K. (2009). Coordination-insertion copolymerization of fundamental polar monomers. *Chem. Rev.*,*109*, 5215–5244.
80. Ito, S., Nozaki, K. (2010). Coordination–insertion copolymerization of polar vinyl monomers by palladium catalysts. *Chem. Rec.*, *10*, 315–325.
81. Drent, E., van Dijk, R., van Ginkel, R., van Oort, B., Pugh, R. I. (2002). Palladium catalysed copolymerisation of ethene with alkylacrylates: polar comonomer built into the linear polymer chain. *Chem. Commun., 744–745.*
82. Skupov, K. M., Marella, P. R., Simard, M., Yap, G. P. A., Allen, N., Conner, D., Goodall, B. L., Claverie, J. P. (2007). Palladium aryl sulfonate phosphine catalysts for the copolymerization of acrylates with ethane. *Macromol. Rapid Commun.*, *28*, 2033–2038.
83. Guironnet, D., Roesle, P., Ruenzi, T., Goettker-Schnetmann, I., Mecking, S. (2009). Insertion polymerization of acrylate. *J. Am. Chem. Soc.*, *131*, 422–423.
84. Luo, S., Vela, J., Lief, G. R., Jordan, R. F. (2007). Copolymerization of ethylene and alkyl vinyl ethers by a (phosphine-sulfonate)PdMe catalyst. *J. Am. Chem. Soc.*, *129*, 8946–8947.
85. Weng, W., Shen, Z., Jordan, R. F. (2007). Copolymerization of ethylene and vinyl fluoride by (phosphine-sulfonate)Pd(Me)(py) catalysts. *J. Am. Chem. Soc.*,*129*, 15450–15451.

86. Kochi, T., Noda, S., Yoshimura, K., Nozaki, K. (2007). Formation of linear copolymers of ethylene and acrylonitrile catalyzed by phosphine sulfonate palladium complexes. *J. Am. Chem. Soc.*, *129*, 8948–8949.

87. Ito, S., Munakata, K., Nakamura, A., Nozaki, K. (2009). Copolymerization of vinyl acetate with ethylene by palladium/alkylphosphine-sulfonate catalysts. *J. Am. Chem. Soc.*,*131*, 14606–14607.

88. Ito, S., Kanazawa, M., Munakata, K., Kuroda, J., Okumura, Y., Nozaki, K. (2011). Coordination-insertion copolymerization of allyl monomers with ethylene. *J. Am. Chem. Soc.*,*133*, 1232–1235.

89. Skupov, K. M., Piche, L., Claverie, J. P. (2008). Linear polyethylene with tunable surface properties by catalytic copolymerization of ethylene with N-vinyl-2-pyrrolidinone and N-isopropylacrylamide. *Macromolecules*, *41*, 2309–2310.

90. Runzi, T., Frohlich, D., Mecking, S. (2010). Direct synthesis of ethylene-acrylic acid copolymers by insertion polymerization. *J. Am. Chem. Soc.*, *132*, 17690–17691.

91. Bouilhac, C., Runzi, T., Mecking, S. (2010). Catalytic copolymerization of ethylene with vinyl sulfones. *Macromolecules*, *43*, 3589–3590.

92. Leblanc, A., Grau, E., Broyer, J.-P., Boisson, C., Spitz, R., Monteil, V. (2011). Homo- and copolymerizations of (meth)acrylates with olefins (styrene, ethylene) using neutral nickel complexes: a dual radical/catalytic pathway. *Macromolecules*, *44*, 3293–3301.

93. Chen, E. Y.-X. (2009). Coordination polymerization of polar vinyl monomers by single-site metal catalysts. *Chem. Rev.*, *109*, 5157–5214.

94. Webster, O. W., Hertler, W. R., Sogah, D. Y., Farnham, W. B., Rajan Babu, T. V. (1983). Group-transfer polymerization. 1. A new concept for addition polymerization with organosilicon initiators. *J. Am. Chem. Soc.*, *105*, 5706.

95. Kochi, T., Nakamura, A., Ida, H., Nozaki, K. (2007). Alternating copolymerization of vinyl acetate with carbon monoxide. *J. Am. Chem. Soc.*,*129*, 7770–7771.

96. Nakamura, A., Munakata, K., Kochi, T., Nozaki, K. (2008). Regiocontrolled copolymerization of methyl acrylate with carbon monoxide. *J. Am. Chem. Soc.*, *130*, 8128–8129.

97. Nozaki, K., Kusumoto, S., Noda, S., Kochi, T., Chung, L. W., Morokuma, K. (2010). Why did incorporation of acrylonitrile to a linear polyethylene become possible? Comparison of phosphine-sulfonate ligand with diphosphine and imine-phenolate ligands in the Pd-catalyzed ethylene/acrylonitrile copolymerization. *J. Am. Chem. Soc.*, *132*, 16030–16042.

98. Bettucci, L., Bianchini, C., Claver, C., Garcia Suarez, E. J., Ruiz, A., Meli, A., Oberhauser, W. (2007). Ligand effects in the non-alternating CO–ethylene copolymerization by palladium(II) catalysis. *Dalton Trans.*, 5590–5602.

99. Noda, S., Nakamura, A., Kochi, T., Chung, L. W., Morokuma, K., Nozaki, K. (2009). Mechanistic studies on the formation of linear polyethylene chain catalyzed by palladium phosphine–sulfonate complexes: experiment and theoretical studies. *J. Am. Chem. Soc.*, *131*, 14088–14100.

100. Li, S., Ye, Z. (2010). Synthesis of narrowly distributed ω-telechelic hyperbranched polyethylenes by efficient end-capping of Pd-diimine-catalyzed ethylene "living" polymerization with styrene derivatives. *Macromol. Chem. Phys.*, *211*, 1917–1924.

101. Hong, S. C., Jia, S., Teodorescu, M., Kowalewski, T., Matyjaszewski, K., Gottfried, A. C., Brookhart, M. (2002). Polyolefin graft copolymers via living polymerization

techniques: preparation of poly(n-butyl acrylate)-graft-polyethylene through the combination of Pd-mediated living olefin polymerization and atom transfer radical polymerization. *J. Polym. Sci. A Polym. Chem.*, *40*, 2736–2749.

102. http://www.cas.org/content/chemical-substances. Access date: March 18th 2014, Published year 2014.
103. Kirkpatrick, P., Ellis, C. (2004). Nature insight - chemical space. *Nature*, *432*, 823–865 and references therein.
104. (a) Katritzky, A. R., Kuanar, M., Slavov, S., Hall, C. D. (2010). Quantitative correlation of physical and chemical properties with chemical structure: utility for prediction. *Chem. Rev.*, *110*, 5714–5789; (b) Berhanu, W. M., Pillai, G. G., Oliferenko, A. A., Katritzky, A. R. (2012). Quantitative structure–activity/property relationships: the ubiquitous links between cause and effect. *ChemPlusChem*, *77*, 507–517.
105. Todeschini, R., Consonni, V. (2009). *Molecular Descriptors for Chemoinformatics* (2 volumes). Wiley-VCH.
106. Matyjaszewski, K. (2012). Atom transfer radical polymerization (ATRP): current status and future perspectives. *Macromolecules*, *45*, 4015–4039.
107. di Lena, F., Chai, C. L. L. (2010). Quantitative structure-reactivity modeling of copper-catalyzed atom transfer radical polymerization. *Polym. Chem.*, *1*, 922–930 and references therein.

3

GENERAL ASPECTS OF COPOLYMERIZATION

ALEX VAN HERK

3.1 COPOLYMERIZATION IN CHAIN REACTIONS

In synthesizing copolymers, one is interested in the average chain composition in terms of the average ratio of the two monomers in the copolymer chains, the composition distribution, and the block length distribution within a chain (blocky copolymers, alternating copolymers, etc.). All these characteristics are controlled by the feed ratio of the two monomers and also the copolymer reactivity ratios. The copolymer reactivity ratios are different for different polymerization mechanisms (for the same monomers) comparing for example anionic and radical polymerizations. However, the reactivity ratios, in general, are not very sensitive to changes in temperature, pressure, and solvent. Exceptions to this rule will be discussed in detail in Section 3.3. This chapter will give the general background to copolymerization, composition drift, medium effects on copolymerizations, and methods to extract copolymerization parameters from experimental data [1–7]. Furthermore, short-chain effects (influence of initiation, transfer, and termination) on copolymerizations [8,9] will be discussed in Section 3.4. It is not really possible to produce block copolymers with free radical polymerization in a controlled way. However, the controlled radical polymerization techniques available allow one to produce block copolymers within well-controlled structures. In Section 3.5, the controlled radical polymerization as a way to produce controlled block copolymers will be discussed.

Synthesis and Applications of Copolymers, First Edition. Edited by Anbanandam Parthiban.

3.1.1 Derivation of the Copolymerization Equation

In this section, we discuss the copolymerizations with two monomers that are built in unbranched polymers via chain reactions in general and free radical polymerizations specifically. In this chapter, in most cases we talk about the chain end radical reactivity but it can also be read as the cationic or anionic chain end in ionic polymerizations. This explanation is very common and was adapted from an introduction of free radical copolymerization [1].

If the penultimate unit in the radical chain end does not influence the reactivity of the radical and if Flory's principle for constant reactivity is valid, then only four propagation steps are involved in the copolymerization reaction:

$$\begin{aligned} -\mathrm{M}_1\bullet + \mathrm{M}_1 &\xrightarrow{k_{11}} -\mathrm{M}_1\bullet \\ -\mathrm{M}_1\bullet + \mathrm{M}_2 &\xrightarrow{k_{12}} -\mathrm{M}_2\bullet \end{aligned} \qquad r_1 = \frac{k_{11}}{k_{12}} \tag{3.1}$$

$$\begin{aligned} -\mathrm{M}_2\bullet + \mathrm{M}_2 &\xrightarrow{k_{22}} -\mathrm{M}_2\bullet \\ -\mathrm{M}_2\bullet + \mathrm{M}_1 &\xrightarrow{k_{21}} -\mathrm{M}_2\bullet \end{aligned} \qquad r_2 = \frac{k_{22}}{k_{21}} \tag{3.2}$$

The reactivity ratios represent the preference of the chain end radical $-\mathrm{M}_1\bullet$ for addition of monomer 1 (homopropagation) over that of the addition of monomer 2 (crosspropagation) expressed in r_1 and of chain end radical $-\mathrm{M}_2\bullet$ for addition of monomer 2 over that of monomer 1 (r_2). In this derivation, it is assumed to deal with infinitely long chains, which means that effects of the initiator fragment, transfer, and termination on specific addition of one over the other monomer is neglected. Also the number of transitions from one monomer to the other monomer unit in the chain is high and equal.

We can define the probability that a specific chain end radical reacts with a specific monomer:

$$p_{11} = \frac{k_{11}[\mathrm{M}_1]}{k_{11}[\mathrm{M}_1] + k_{12}[\mathrm{M}_2]} \qquad p_{12} = 1 - p_{11} \tag{3.3}$$

$$p_{22} = \frac{k_{22}[\mathrm{M}_2]}{k_{22}[\mathrm{M}_2] + k_{21}[\mathrm{M}_1]}, \qquad p_{21} = 1 - p_{22} \tag{3.4}$$

The mole fraction of blocks M_1 with length i (mole fraction relative to all other blocks M_1) is

$$x_i = p_{11}^{i-1}\,(1 - p_{11}) \tag{3.5}$$

so that the average block length is

$$\bar{i}_n = \sum x_i i = \frac{1}{1 - p_{11}} = r_1 \frac{[\mathrm{M}_1]}{[\mathrm{M}_2]} + 1 \tag{3.6}$$

and for the M_2 blocks:

$$\bar{j}_n = \frac{1}{1 - p_{22}} = r_2 \frac{[M_2]}{[M_1]} + 1 \tag{3.7}$$

If $\overline{P}_n \rightarrow \infty$ or, more precisely formulated, if there are many $-M_1M_2-$ and $-M_2M_1-$ transitions per copolymer chain, there are as many M_1 blocks as there are M_2 blocks, the following holds for the composition of the instantaneously formed copolymer:

$$\frac{d[M_1]}{d[M_2]} = \frac{\bar{i}_n}{\bar{j}_n} = \frac{r_1[M_1]/[M_2] + 1}{r_2[M_2]/[M_1] + 1} \tag{3.8}$$

This is the *copolymerization equation*, which is often expressed in the fraction of free monomer $f_1 = [M_1]/\{[M_1] + [M_2]\}$ and the fraction monomer built into the instantaneously formed copolymer $F_1 = d[M_1]/\{d[M_1] + d[M_2]\}$:

$$F_1 = \frac{r_1 f_1^2 + f_1 f_2}{r_1 f_1^2 + 2 f_1 f_2 + r_2 f_2^2} \tag{3.9}$$

Obviously the average block length is of importance for the properties of the copolymer and one can calculate the average block length of a 1 : 1 copolymer; for $\bar{j}_n = \bar{i}_n$ it holds that $[M_1]/[M_2] = (r_2/r_1)^{1/2}$, so that

$$\bar{i}_n = \bar{j}_n = 1 + (r_1 r_2)^{1/2} \tag{3.10}$$

So for a copolymer with two reactivity ratios close to zero, one gets an alternating copolymer with a typical blocklength of 1 for both monomers. For a copolymer produced with two reactivity ratios much higher than 1, one would get a blocky copolymer (but not a diblock copolymer because the number of crosspropagation steps cannot be controlled).

It is also possible that one of the reactivity ratios equals zero, in that case homopropagation of that monomer is not possible but crosspropagation can take place (e.g., maleic anhydride).

It is seen that k_i and k_t do not occur in the copolymerization equation, that is, the copolymer composition is independent of the overall reaction rate and the initiator concentration (related to the long chain limit, not valid for very small copolymers, see Section 3.4). The reactivity ratios are generally independent of the type of initiator, inhibitor, retarder, chain transfer agent, but are (slightly) dependent on temperature and pressure. Occasionally, they depend on the type of solvent, especially for polar or charged monomer combinations (see Section 3.3).

In general, $f_1 \neq F_1$, so that both the composition of the monomer feed mixture and the copolymer composition change as the conversion increases. This phenomenon is referred to as *composition drift*. As the conversion increases, the distribution of the chain composition broadens. This can lead to heterogeneous copolymers.

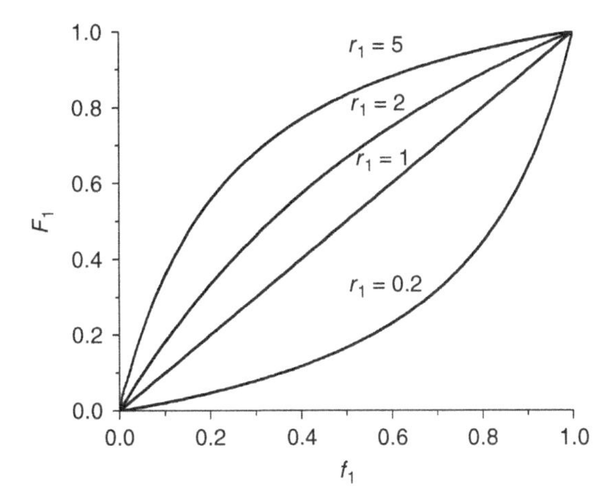

Figure 3.1 Ideal copolymerizations with $r_1r_2 = 1$. (Adapted with permission from [1].)

3.1.2 Types of Copolymers

Depending on the value of the reactivity ratios, we can distinguish between a number of cases.

1. *Ideal copolymerization* (Figure 3.1) occurs if $r_1r_2 = 1$, that is, each radical end has the same preference for one of the monomers. In this case, $d[M_1]/d[M_2] = r_1[M_1]/[M_2]$, and for a 1 : 1 copolymer $\bar{j}_n = \bar{i}_n = 2$.
   The monomers succeed each other in very short blocks of length 1–3 monomer units.
   A special case is where $r_1 = r_2 = 1$ or $k_{11} = k_{12}$ and $k_{22} = k_{21}$.
   The reactivity of both radicals will be in general different, in other words $k_{11} \neq k_{21}$! In this case, $f = F$ always holds, independently of conversion. The copolymerization of (for instance) styrene (M_1) and *p*-methoxystyrene (M_2) is close to an ideal copolymerization with $r_1 = 1.16$, $r_2 = 0.82$, and $r_1r_2 = 0.95$ (60 °C).
2. *Alternating copolymerization* (Figure 3.2) occurs if $r_1 = r_2 = 0$, that is, each radical reacts exclusively with the other monomer, not with its own. In this case, $d[M_1]/d[M_2] = 1$ and for each monomer feed composition $\bar{j}_n = \bar{i}_n = 1$. An example is the copolymerization of vinyl acetate (M_1) with maleic anhydride (M_2), for which $r_1 = 0.055$, $r_2 = 0.003$, and $r_1r_2 = 0.00017$.
3. *Nonideal copolymerization*

   a. $0 < r_1r_2 < 1$, this applies to most cases (Figure 3.3). The copolymerization behavior is in between alternating and ideal. The product r_1r_2 is a measure of the tendency for alternating behavior.
      An interesting case occurs if $r_1 < 1$ and $r_2 < 1$. The F–f curve intersects the diagonal, and we speak of *azeotropic copolymerization*. In the intersection, the copolymer composition does not change as conversion proceeds

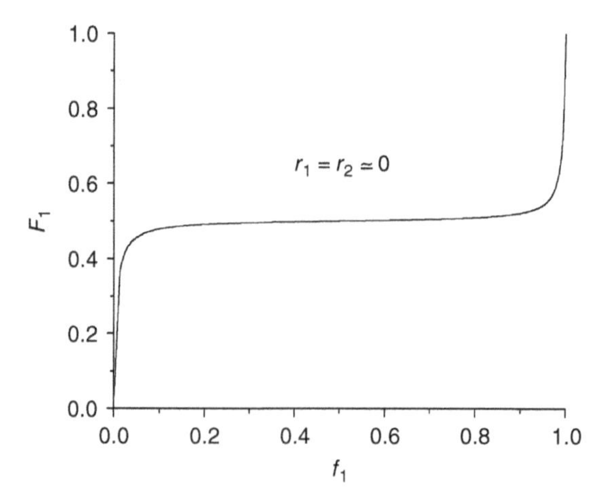

Figure 3.2 Alternating copolymerization. (Adapted with permission from [1].)

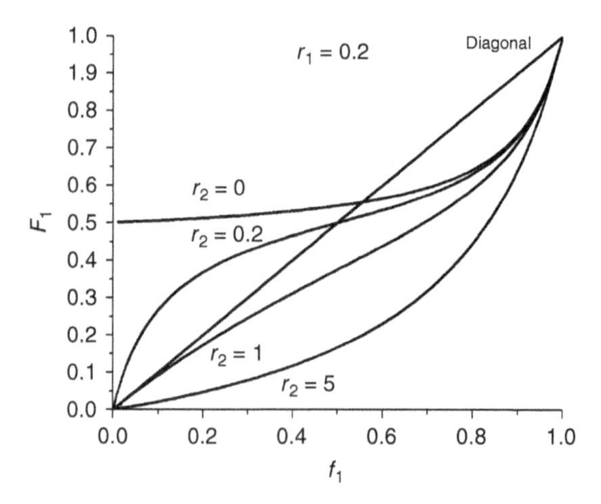

Figure 3.3 Nonideal copolymerization f–F curves. (Adapted with permission from [1].)

because $f_1 = F_1 = (1 - r_2)/(2 - r_1 - r_2)$. However, this is an unstable equilibrium, the direction of the composition drift points away from the azeotrope.

b. $r_1 r_2 > 1$ with $r_1 < 1$ and $r_2 > 1$. For a 1 : 1 copolymer, $\bar{j}_n = \bar{i}_n > 2$. There is a certain tendency to form larger blocks because of the preference of M_2 for homopropagation.

c. $r_1 r_2 > 1$ with $r_1 > 1$ and $r_2 > 1$. This would lead to block copolymerization and in the extreme case of $r_1 \gg 1$ and $r_2 \gg 1$ even to simultaneous homopolymerizations. This behavior is very rare though, and the number of crossovers from monomer 1 to monomer 2 cannot be controlled. So if block copolymers are required, specific techniques must be employed.

3.1.2.1 Homogeneous–Heterogeneous Copolymers

Homogeneous copolymers are formed only if one of the following conditions is satisfied:

- low conversion, this can be achieved in a continuous stirred tank reactor (CSTR) where the steady-state conversion is kept low and the unreacted monomers are recycled;
- maintain a constant monomer feed composition (f_1 = constant) by the addition of monomer(s);
- azeotropic copolymerization behavior ($f_1 = F_1 = f_{az}$)
- $r_1 = r_2 = 1$ ($f_1 = F_1$)
- $r_1 = r_2 = 0$ ($F_1 = 1/2$)
- *Homogeneous copolymers* generally have one glass transition temperature (Figure 3.4).

If a batch polymerization is performed to high conversion under non-azeotropic conditions, then the monomer feed composition changes and the composition of the copolymer (composition drift) also changes. It is then possible that the copolymer formed initially is no longer miscible in the copolymer formed in a later part of the reaction. Phase separation then occurs and one speaks of heterogeneous copolymers. Such a *heterogeneous copolymer* has the properties of the different polymers it is composed of (e.g., may have two glass transition temperatures, Figure 3.4).

3.1.3 Polymerization Rates in Copolymerizations

Besides the chemical composition (distribution) resulting from a copolymerization also the rate of the copolymerization is of importance, for example, to estimate the needed cooling capacity. The rate of a polymerization is proportional to the product of the propagation rate coefficient, monomer concentration, and the propagating

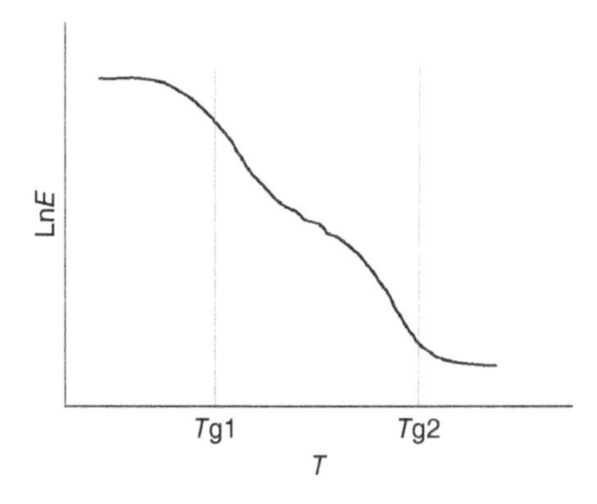

Figure 3.4 ln E versus temperature showing two glass transition temperatures (T_g's) in a heterogeneous copolymer.

radical concentration. In order to calculate the rate in a copolymerization, we can use the averages of each of the three quantities. The average propagation rate coefficient $\langle k_p \rangle$ is given by the general equation:

$$\langle k_p \rangle = \frac{\overline{r_1} f_1^2 + 2 f_1 f_2 + \overline{r_2} f_2^2}{(\overline{r_1} f_1 / \overline{k_{11}}) + (\overline{r_2} f_2 / \overline{k_{22}})} \quad (3.11)$$

where in the terminal model $\overline{r_1} = r_1, \overline{r_2} = r_2, \overline{k_{11}} = k_{11}, \overline{k_{22}} = k_{22}$ with $r_1 = k_{11}/k_{12}$ and $r_2 = k_{22}/k_{21}$.

In the case that the penultimate unit also influences the reactivity, the penultimate model applies. There are eight different propagation steps in the penultimate unit model. The rate coefficients are systematically defined as follows (example):

$$-M_1M_2 \bullet + M_1 \xrightarrow{k_{121}} -M_2M_1 \bullet \quad (3.12)$$

We can further define $\overline{k_{11}}$ and $\overline{k_{22}}$ as

$$\overline{k_{11}} = \frac{k_{111}(f_1 r_1 + f_2)}{f_1 r_1 + f_2/s_1} \text{ and } \overline{k_{22}} = \frac{k_{222}(f_2 r_2 + f_1)}{f_2 r_2 + f_1/s_2} \quad (3.13)$$

$$\overline{r_1} = \frac{r_1'(f_1 r_1 + f_2)}{f_1 r_1' + f_2} \text{ and } \overline{r_2} = \frac{r_2'(f_2 r_2 + f_1)}{f_2 r_2' + f_1}$$

$$r_1 = \frac{k_{111}}{k_{112}}, r_2 = \frac{k_{222}}{k_{221}}, r_1' = \frac{k_{211}}{k_{212}}, r_2' = \frac{k_{122}}{k_{121}}, s_1 = \frac{k_{211}}{k_{111}}, s_2 = \frac{k_{122}}{k_{222}}$$

In general, the compositional data of a copolymerization can be well described by the ultimate model. The k_p data in copolymerizations, for example, obtained by the pulsed laser method [2] are best described by the penultimate unit model. The description of the compositional data is less model sensitive.

3.2 MEASURING COPOLYMERIZATION PARAMETERS

It is clear that the reactivity ratios are the most important parameters in copolymerization. For many monomer pairs, these reactivity ratios are determined experimentally and tabulated, for example, in publications or in the *Polymer Handbook*. For many years, there was no good agreement between values obtained for the same monomer pairs and the values obtained from different labs. In addition to the differences in the experimental methods and accuracy, the main reason was the data treatment and the statistical methods used. At the Eindhoven University of Technology for many years, we tried to contribute to the development of proper statistical methods.

Three basic methods exist to measure the important reactivity ratios:

1. Low conversion experiments measuring the copolymer composition and/or the sequence distribution (e.g., with NMR) versus the feed composition using Equation 3.9.
2. High conversion experiments measuring the monomer feed composition as a function of conversion of monomers, using the integrated copolymerization equation.
3. Low conversion experiments measuring the full chemical composition distribution (CCD) and fitting these data with the theoretical distribution depending on the reactivity ratios.

At this point, it is important to stress that the correct statistical methods should be used to extract reactivity ratios from the experimental data [3]. Especially in the field of determining reactivity ratios with approach 1 (the most commonly used approach), although the correct methods are outlined in literature already a long time ago, we can still find many examples in recent literature where major errors are made. Not only it is important to use correct methods of parameter estimation but also the issue of experimental design and model discrimination is of paramount importance for obtaining accurate values for the parameters [3].

Method 2, where the feed composition is followed as a function of conversion, is not often applied because it is a more complex method in terms of experimental accuracy and the risk of including systematic errors. Furthermore, using only one experiment usually results in low accuracy of the reactivity ratios obtained.

Method 3 is relatively new and can obtain reactivity ratios from a single experiment, but one has to be able to measure the CCD very accurately. Although gradient polymer elution chromatography does give a rough indication of the CCD, only matrix-assisted laser desorption/ionization time-of-flight mass spectrometry (MALDI-ToF MS) is really able to give an accurate CCD [4].

Method 3 has recently been evaluated and it is able to reproduce many literature reactivity ratios and seems to be very powerful [4]. As only a single experiment is needed and the MALDI-ToF MS data can be inherently segregated in different CCDs at different chain lengths, one can obtain information about chain length-dependent reactivity ratios and also influence of feed ratios on reactivity ratios. For this reason, we will elaborate a bit more on this new method.

The method consists of obtaining the reactivity ratios from a single MALDI-ToF-MS spectrum converted to a CCD that is then fitted with either a Monte Carlo approach to numerically simulate a first-order Markov chain or the analytical form of the first-order Markov chain. A single CCD derived from a MALDI-ToF-MS spectrum (Figure 3.5) proved to give very good estimates of the reactivity ratios of comonomers from copolymers synthesized by free radical polymerization of styrene and (meth)acrylates, ring-opening polymerization of lactones and lactides, or ring-opening copolymerization of anhydrides and epoxides [4].

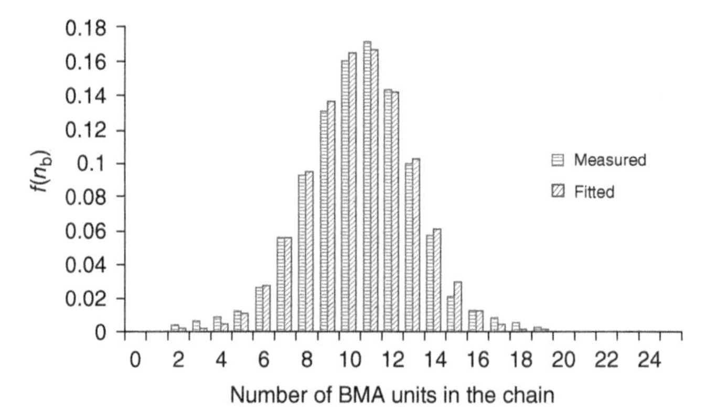

Figure 3.5 Distributions of chemical composition for an MMA/BMA copolymer recorded (▤) and fitted by using the Monte Carlo method followed by minimizing the sum of squares of residuals (▨) for chain length 20. (Reprinted with permission from [4]. Copyright © 2012 American Chemical Society.)

In Figure 3.5, the experimentally obtained CCD is shown for a copolymerization of methyl methacrylate (MMA) and butyl methacrylate (BMA) as well as the fitted CCD, resulting in copolymerization parameters of $r_1 = 0.81$ and $r_2 = 1.28$.

In copolymerization sometimes solvent effects are observed, these solvent effects can even be induced by the change in feed composition (take, e.g., copolymerization in the aqueous phase with acrylic acid as one of the comonomers). It is therefore very useful to have a method to determine the reactivity ratios at one feed composition, making it possible to study the effect of changing the feed composition on the reactivity ratios. A fast and reliable method that can result in reactivity ratios at one feed composition is therefore highly desired.

With MALDI-ToF MS, it is even possible to look at effects of chain length on composition and reactivity ratios because this information is also contained in the spectra (Figure 3.6).

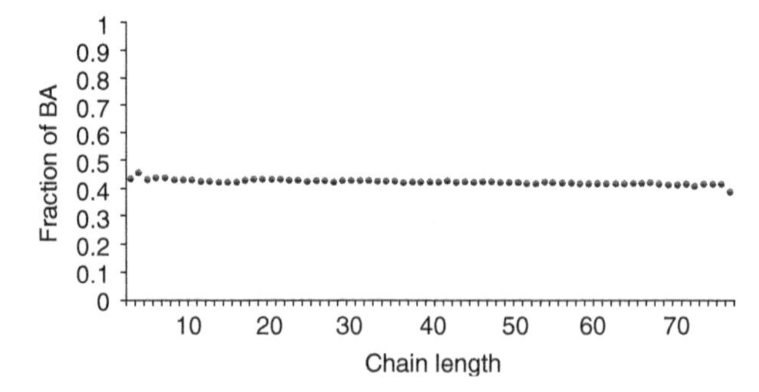

Figure 3.6 Composition versus chain length for the copolymer of styrene and butyl acrylate as obtained from MALDI-ToF MS. (Reprinted with permission from [4]. Copyright © 2012 American Chemical Society.)

The data obtained for the reactivity ratios as a function of chain length were surprising in the sense that it was expected that chain length dependence would be observed. However, the region of chain lengths below five to six monomer units has not been studied, and in that region these effects are visible as shown for somewhat smaller oligomers produced in the aqueous phase of emulsion copolymerizations (see Section 3.4).

3.3 INFLUENCE OF REACTION CONDITIONS

As reactivity ratios are constants obtained under certain conditions such as solvent, temperature, and pressure, it is important to know whether one can use the reactivity ratios under slightly different conditions. Because reactivity ratios are ratios of propagation rate coefficients, we can expect the reactivity ratios not to change as much as the individual rate coefficients. For example, the temperature dependency of reactivity ratios depends on the difference between activation energies of the individual rate parameters. So often using reactivity ratios at a slightly different temperature does not matter. Similar statements can be made for pressure effects. With regard to solvents for many uncharged and not so polar monomers solvent effects are also minor. With regard to charged or highly polar monomers, solvent effects can be expected [10]. In this aspect, bulk copolymerizations and solution copolymerizations can give different reactivity ratios too. In some cases, the monomers can form associates with themselves or with the other monomer or with the created polymer. In the latter case, we can see the effect of the so-called templating (co)polymerization [5].

Overall, we can say that reactivity ratios are relatively insensitive to the reaction conditions. It is even the case that reactivity ratios that are obtained in bulk copolymerizations can also be employed in heterogeneous polymerization techniques such as emulsion copolymerizations. In this case, the monomer ratios that are effective at the locus of polymerization (often not equal to the overall monomer ratios) should be used.

3.4 SHORT-CHAIN EFFECTS IN COPOLYMERIZATION

Generally, producing polymer chains of considerable length i.e. high molecular weight polymers, are preferred. However, in some cases, particularly for making block copolymers which are used as stabilizers, oligomers are produced. Also the increasing use of controlled radical (co)polymerizations leads to the production of relatively short chains initially. As the general copolymerization equations (Eqs. 3.9 and 3.11) are derived on the basis of infinitely long chains and neglecting effects of initiation, termination, and transfer, for very short copolymer chains additional effects can be anticipated.

In principle, reactivity ratios should not depend on chain length of the growing polymer chain. Propagation reactions are chemically controlled (radical reactivity toward a certain monomer), and there are no diffusion limitations (as there are for termination reactions). Some reports on chain length-dependent reactivity ratios [6] and chain length-dependent propagation rate coefficients [7] were based on artifacts in the analytical methods used. However, the first few propagation steps are chain length dependent, mainly because the reactivity of the initiator-derived radical toward the two monomers can be different from that of the long-chain radical. As we have penultimate unit and even pen-penultimate unit effects, the effect of the initiator can stretch to three to four addition steps. It is not easy to study the first few propagation steps in radical copolymerizations. One recent study was focusing on the short oligomers produced in the aqueous phase of emulsion copolymerizations [8,9]. The unique possibilities of MALDI-ToF MS allowed studying the composition of these oligomers as a function of chain length and using different initiators. One way of looking at chain length-dependent reactivity ratios is to look at the average composition (expressed in fraction of a monomer unit in the copolymer, see Figure 3.6) as a function of chain length at low conversion.

From these experiments, indeed it was observed that there are initiator-derived radical preferences for one monomer over the other monomer. Furthermore, there are kinetic effects induced by differences of the propagation rate coefficients. For example, if we form an oligomer of three units of methylacrylate and one unit of styrene, this oligomer is produced in much less time than one with three units of styrene and 1 unit of methylacrylate, simply because the homopropagation rate coefficient of methylacrylate is about 400 times faster than that of styrene. This means that if we study composition as a function of chain length after a certain growth time of the oligomers, we see a deviation from the average. Of course this averages out at longer chain lengths because these chains are produced close to the average composition mainly and the chains deviating from the average composition, in a relative sense, do not take a substantially different time to be formed (three units of a monomer more or less in the same chain length on a total of 1000 monomer units hardly make a difference on the time needed to grow this chain). Overall these different effects result in curvatures in the F versus chain length curves up to 10 monomer units in the oligomer [9].

3.5 SYNTHESIS OF BLOCK COPOLYMERS WITH CONTROLLED CHAIN ARCHITECTURE

In the previous sections, we discussed the different aspects of copolymerization. It is clear that with normal free radical polymerization one cannot produce block copolymers within a controlled chain architecture (e.g., diblock or triblock copolymers or gradient copolymers). The reason is that in free radical polymerization, it takes only a fraction of a second for a chain to initiate, propagate, and terminate. In other words, if we would want to change the monomer feed composition during the growth of one single chain, we need to change it within this fraction of

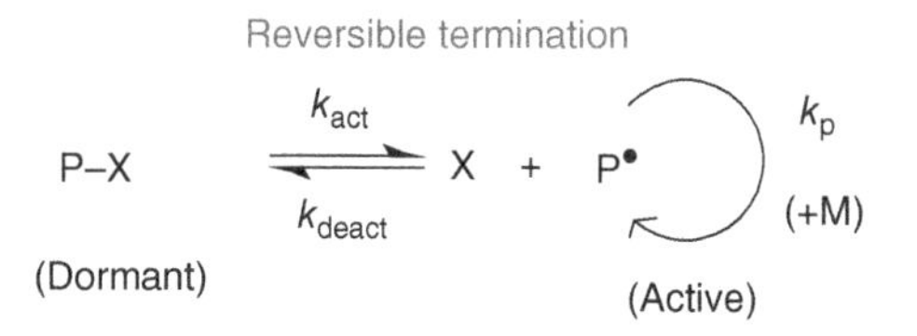

Figure 3.7 General scheme for reversible termination.

a second which is not possible. With the development of several controlled radical polymerization techniques, the growth time of a single chain can be extended over a considerable time span. The way this can be done is suppressing the termination reaction by protecting the free radical with some kind of molecule X (Figure 3.7) in an equilibrium reaction. For example, stable nitroxide radicals can be used as X.

The principle is as follows: a species X is added to a radical polymerization and forms a complex PX, the dormant form of the growing chain. The dormant species is in a dynamic equilibrium with the active species. In the period that the radical is in the active form, it can add monomer units and propagate (and terminate). As the equilibrium lies far to the left (Figure 3.7), the concentration of active radicals is far below that in a free radical polymerization and therefore also the rate of termination which depends on the square of the radical concentration. In this way termination is suppressed, and the growth time of the polymer chains has been extended from seconds to hours. Unfortunately, also the overall rate of polymerization has been reduced substantially. However, in this much longer time of growth of the polymer chain, one can change the monomer feed and produce diblock, triblock, and gradient copolymers easily. If initiation is fast, all the chains have more or less the same growth time and growth of the chains stops when either no more monomer is present or a chain stopper is added to the system. In this way, also the molecular weight distribution can be very narrow (low polydispersity index, PDI).

If one is only after making block copolymers but then accepts higher values for the PDI, a very interesting alternative is to use reversible transfer. In addition to bimolecular termination, another way of terminating a growing chain is by a transfer of the radical to a transfer agent, for example, with a mercaptan. If the transfer reaction is reversible also, these chains have a dormant and an active state and the total growth time of the chains can be extended as to produce block copolymer again. Reversible addition fragmentation transfer (RAFT) reactions contain normal initiator in addition to the RAFT agent, and initiation occurs throughout the reaction. In RAFT polymerizations, normal termination can also occur although one normally operates in the so-called transfer-dominated regime. Because the level of radicals is much higher than in the reversible termination approach, the rate of polymerization is similar to normal free radical polymerizations. As a trade-off, the PDI values are much higher than in the reversible termination approach but one can still produce block copolymers. An excellent review of the possibilities of controlled radical polymerization is given in Reference 11.

REFERENCES

1. Van Herk, A. M., editor. *Chemistry and Technology of Emulsion Polymerization*. Second ed. John Wiley & Sons, Chichester, UK, 2013, pp. 34–41.
2. Van Herk, A. M. (2000). Pulsed initiation polymerization as a means of obtaining propagation rate coefficients in free-radical polymerizations. II. Review up to 2000. *Macromol. Theor. Simul.*, *9*, 433–441.
3. Van Herk, A. M. (1995). Least-squares fitting by visualization of the sum of squares space. *J. Chem. Ed.*, *72*, 138–140.
4. Huijser, S., Mooiweer, G. D., van der Hofstad, R., Staal, B. B. P., Feenstra, J., van Herk, A. M., Koning, C. E., Duchateau, R. (2012). Reactivity ratios of comonomers from a single MALDI-ToF-MS measurement at one feed composition. *Macromolecules*, *45*, 4500–4510.
5. Challa, G., Tan, Y. Y. (1981). Template polymerization. *Pure Appl. Chem.*, *53*, 627–641.
6. Semchikov, Y. D., Smirnova, L. A., Knyazeva, T. Ye., Bulgakova, S. A., Sherstyanykh, V. I. (1990). Dependence of copolymer composition upon molecular weight in homogeneous radical copolymerization. *Eur. Polym. J.*, *26*, 883–887.
7. Willemse, R. X. E., Staal, B. B. P., van Herk, A. M., Pierik, S. C. J., Klumperman, L. (2003). Application of matrix-assisted laser desorption ionization time-of-flight mass spectrometry in pulsed laser polymerization. Chain-length-dependent propagation rate coefficients at high molecular weight: an artifact caused by band broadening in size exclusion chromatography. *Macromolecules*, *36*, 9797–9803.
8. Daswani, P., Rheinhold, F., Ottink, M., Staal, B., van Herk, A. (2012). Method to isolate and characterize oligomers present in the aqueous phase in emulsion copolymerization. *Eur. Polym. J. 48*, 296–308.
9. Daswani, P. (2012). Entry in emulsion copolymerization (Ph.D. thesis Eindhoven University of Technology).
10. O'Driscoll, K. F., Davis, T. P., Klumperman, B., Madruga, E. L. (1995). Solvent effects in copolymerization. *Macromol. Rapid Commun. 16*, 207–210.
11. Matyjaszewski, K. (2005). Macromolecular engineering: from rational design through precise macromolecular synthesis and processing to targeted macroscopic material properties. *Prog. Polym. Sci.*, *30*, 858–875.

4

POLYMERS BEARING REACTIVE, PENDANT CYCLIC CARBONATE (CC) GROUP: SYNTHESES, POST-POLYMERIZATION MODIFICATIONS, AND APPLICATIONS

SATYASANKAR JANA

4.1 INTRODUCTION

In the last few decades, there is a tremendous development in the synthesis and application of polymers with controlled structures and functionalities. Property of a polymer highly depends on the presence of pendant functional groups and dictates its application as a functional material. Very often, polymers with expected functional groups are either difficult to synthesize or cannot be synthesized in a controlled manner by the direct polymerization or copolymerization of related functional monomer because of different issues such as stability, low tolerance of functional monomer during polymerization, side reactions, or simply lower reactivity/inhibition to polymerization. The situation has improved by the rapid development of controlled radical polymerization (CRP) techniques, such as atom transfer radical polymerization (ATRP) or reversible addition fragmentation chain transfer (RAFT)-mediated polymerization, which has better functional group tolerance and, therefore, an ability to use a wide range of functional monomers as compared to that of traditional anionic and cationic polymerization techniques. In spite

Synthesis and Applications of Copolymers, First Edition. Edited by Anbanandam Parthiban.

of many such related developments, there are still difficulties in synthesizing polymers with many pendant functionalities directly and especially in a controlled manner. A viable alternative approach is the polymerization of a monomer bearing reactive functional groups and the subsequent post-polymerization modification of the polymer produced. This post-polymerization strategy is highly attractive because this leads to a combinatorial approach to the synthesis of polymer libraries with different pendant groups having identical average chain length and chain length distributions. This approach to library synthesis does require only one optimization reaction of polymerization of the precursor monomer with a highly reactive functional group and an efficient post-polymerization reaction methodology. Thus, recently there is a rapid development going on for the discovery of new highly efficient reactions including the application of the polymer modification especially after the introduction of the concept of click chemistry or click reactions.

Recently, Klok and coworkers [1] have reviewed the strategy of post-polymerization modification of different pendant functional groups. Chemoselective modification reactions of different important pendant functional groups that draw attention in recent times are briefly presented in Table 4.1. Modification of very common pendant functional groups such as $-CO_2H$, $-COCl$, $-OH$, or $-NH_2$ groups has long been used in polymer chemistry and is excluded from the table. In this chapter, emphasis is being placed on the synthesis, modification, and applications of pendant cyclic carbonate (CC) monomers and polymers therefrom. Polymers obtained by the polyaddition reaction of CC group using multiple CC monomers or by the ring-opening reaction of CC groups will no longer have any pendant CC groups for further modifications, and therefore these polymers are not discussed here. Similarly, polymers having only end-functional CC groups are also avoided although same chemistry as the pendant CC polymers could be applied. Pendant CC polymers are interesting [2] in the following different aspects: (i) the synthesis of CC monomers involves the utilization of CO_2 gas that causes global warming, (ii) ultrafast photopolymerizing nature of CC (meth)acrylate monomers, (iii) highly polar and metal coordinating nature of CC groups of polymer, which leads to the possible application as polymer/gel electrolyte, and (iv) room temperature curing of pendant CC groups without releasing any by product, which leads to the possible application as thermosetting resins for coatings (Figure 4.1).

4.2 CYCLIC CARBONATE (CC) MONOMERS AND POLYMERS

4.2.1 Cyclic Carbonate (CC) Monomers and Their Synthesis

Different CC groups containing (meth)acrylate monomers (Figure 4.2) have been reported to be synthesized by several researchers as presented in Scheme 4.1. Propylene carbonate acrylate, **M1** (PCA, alternative name: glycerol carbonate acrylate, (2-oxo-1,3-dioxolan-4-yl)methyl acrylate), was first reported by D'alelio et al. [13] by the reaction of acryloyl chloride or acrylic anhydride with glycerol carbonate (GC) (also known as *4-(hydroxymethyl)-1,3-dioxolan-2-one*), the later was used as the CC source. At present, GC is commercially available.

TABLE 4.1 Different Pendant Functional Groups and Their Modification Reactions

Pendant Groups (Type)	General Functionalizing Agent and Mechanism	Modified Groups	Related Reviews and References
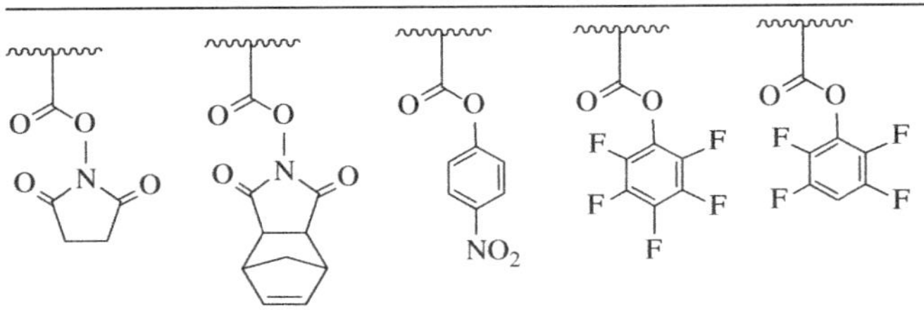 O, O, N, O; O, O, N, O; O, O, NO_2; O, O, F, F, F, F, F; O, O, F, F, F, F **(Activated esters)**	R–XH (X = NH/O) (Nucleophilic substitution)	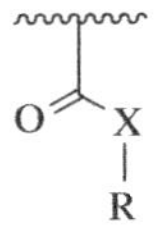 O, X, R (X = NH/O)	[3]
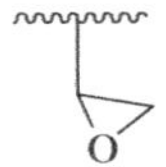 O **(Epoxide/oxirane)**	R–XH (X = NH/ NR′/O/COO/S) (Nucleophilic addition)	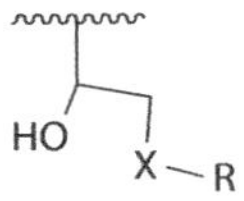 HO, X–R (X = NR′/O/COO/S) (Crosslinked pdt when X = NH)	[4]
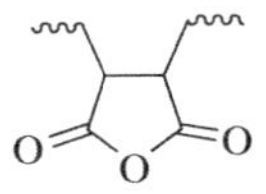 O, O, O **(Anhydride)**	R–XH (X = NH/O) (Nucleophilic addition)	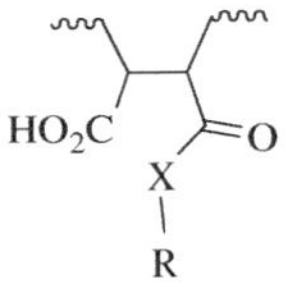 HO_2C, O, X, R (X = NH/O)	[5a–d]

(continued)

TABLE 4.1 ***(Continued)***

Pendant Groups (Type)	General Functionalizing Agent and Mechanism	Modified Groups	Related Reviews and References
	$R'-NH_2$ ($R' = OH/ -(CH_2)_2-SO_3H$) (Nucleophilic addition)	O N O R′	[5e–h]
(Oxazolone)	R–XH (X = NH/NR′/O) (Nucleophilic addition)	O O O NH O X–R (X = NH/NR'/O)	[6]
N=C=O **(Isocyanate)**	RXH (X = O/NH/S) (addition reaction)	HN O X–R (X = O/NH/S)	[7]

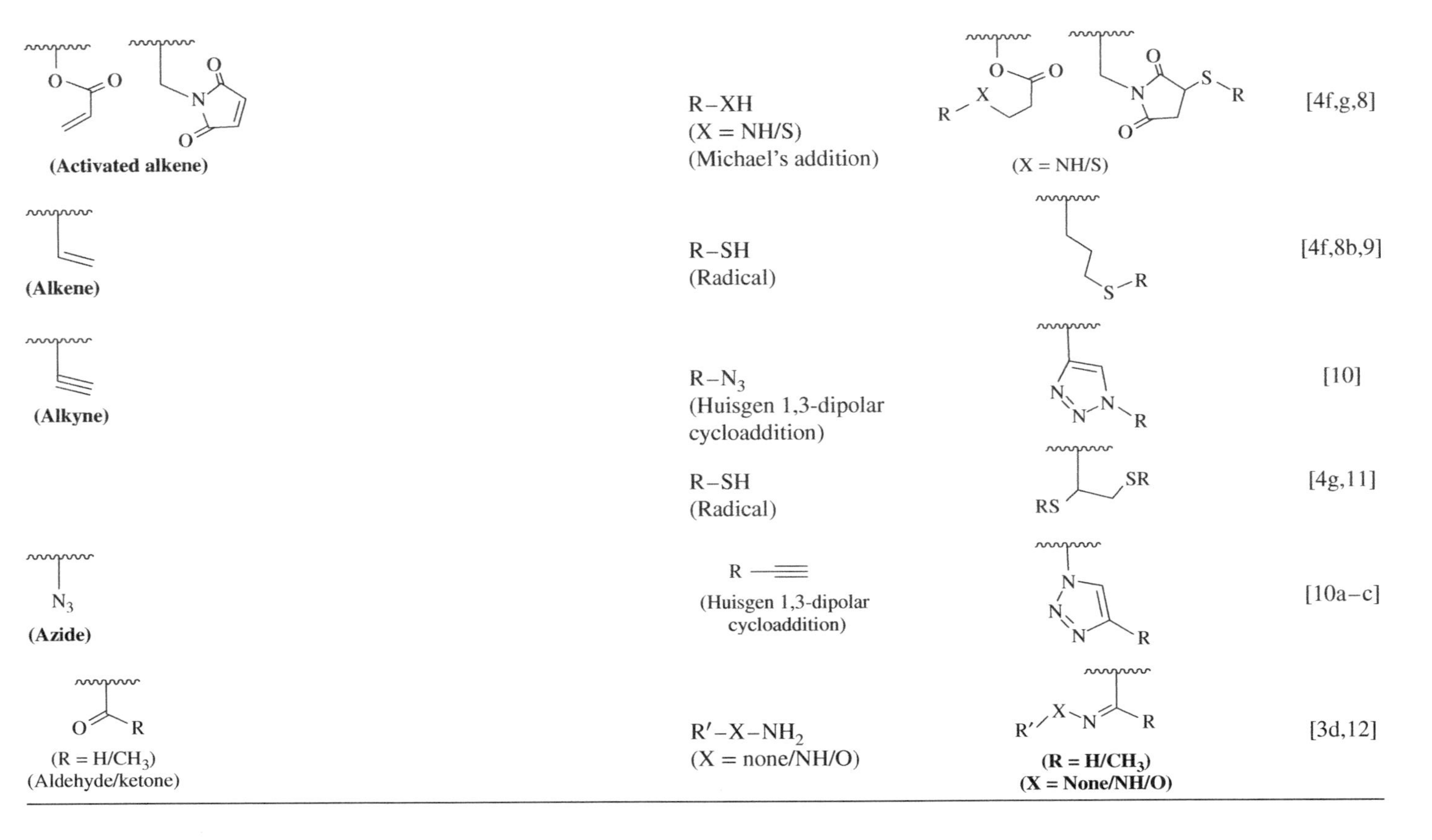

Functional group	Reagent (reaction)	Product	References
(Activated alkene)	R–XH (X = NH/S) (Michael's addition)	(X = NH/S)	[4f,g,8]
(Alkene)	R–SH (Radical)		[4f,8b,9]
(Alkyne)	R–N_3 (Huisgen 1,3-dipolar cycloaddition)		[10]
	R–SH (Radical)		[4g,11]
N_3 (Azide)	R —≡ (Huisgen 1,3-dipolar cycloaddition)		[10a–c]
(R = H/CH_3) (Aldehyde/ketone)	R′–X–NH_2 (X = none/NH/O)	(R = H/CH_3) (X = None/NH/O)	[3d,12]

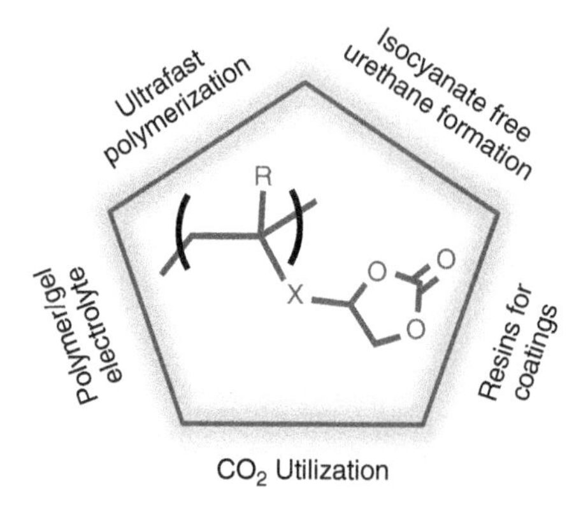

Figure 4.1 Important aspects of monomers and polymers bearing pendant CC group.

Otherwise, GC can be synthesized [14] by following reactions: (i) transesterification of glycerol with a source of carbonate [14b], for example, ethylene carbonate (EC), propylene carbonate (PC), or dimethyl carbonate; (ii) direct carboxylation of glycerol using CO_2 [15]; and (iii) carbonylation of glycerol using CO or urea [16]. While synthesizing PCA, M1, D'alelio et al. experienced polymerization during purification by distillation [15]; however, the acid chloride route was later adopted by many other researchers to synthesize propylene carbonate (meth)acrylates (PC(M)A) [17], **M1** and **M2**, which were purified by column chromatography. Synthesis of **M2** by transesterification of GC with methyl methacrylate (MMA) was reported by Feng et al. [18]. Reaction of glycerol carbonate chloroformate (GCC) with methacrylic acid (MAA) was also reported to produce **M2** [19]. Recently, we have developed a protocol for the *N,N′*-dicyclohexylcarbodiimide (DCC)-mediated coupling of MAA and GC [20] to synthesize **M2**, which was subsequently purified by column chromatography. The other route to synthesize **M2** was by the insertion or addition reaction of CO_2 (gas) with glycidyl methacrylate (GMA) [21]. Both metal salts [22] and organic amine (onium) salts [23] or mixture [24] were employed for this purpose. The recent discovery of addition reaction of CO_2 with GMA using $InBr_3/PPh_3$ catalyst system by Shibata et al. [22a] is very interesting, considering the formation of **M2** with >85% yield at 1 atm pressure and in 2 h at room temperature without the use of any external solvent.

Monomers with longer aliphatic spacer chain between polymerizable (meth)acrylate and functional CC groups, **M3** ((2-oxo-1,3-dioxolan-4-yl)butyl acrylate) and **M4** ((2-oxo-1,3-dioxolan-4-yl)butyl methacrylate) (Figure 4.1), were synthesized by Britz et al. [17d] by the addition reaction of CO_2. Vinyl monomers with both acid and CC functional groups, **M5** (2-oxo-1,3-dioxolan-4-yl)methyl maleate) and **M6** (2-oxo-1,3-dioxolan-4-yl)methyl itaconate), were reported by D'alelio et al. [13] by the reaction of GC with maleic anhydride (MAH) and itaconic anhydride, respectively. Interestingly, in both these cases, monomers were obtained in crystalline and stable forms unlike the monomer **M1**.

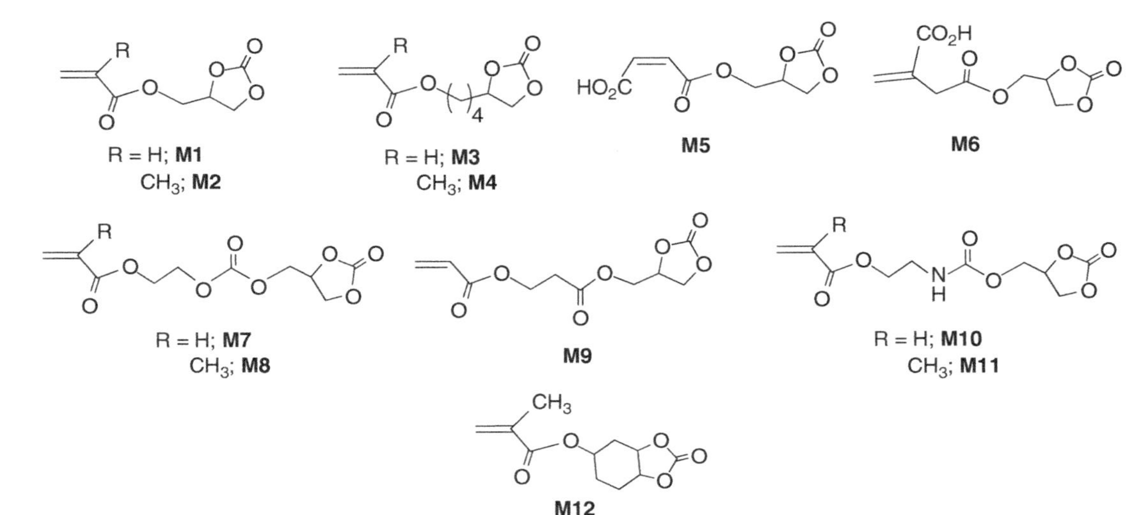

Figure 4.2 Different (meth)acrylate monomers used to synthesize pendant cyclic carbonate (CC) polymers.

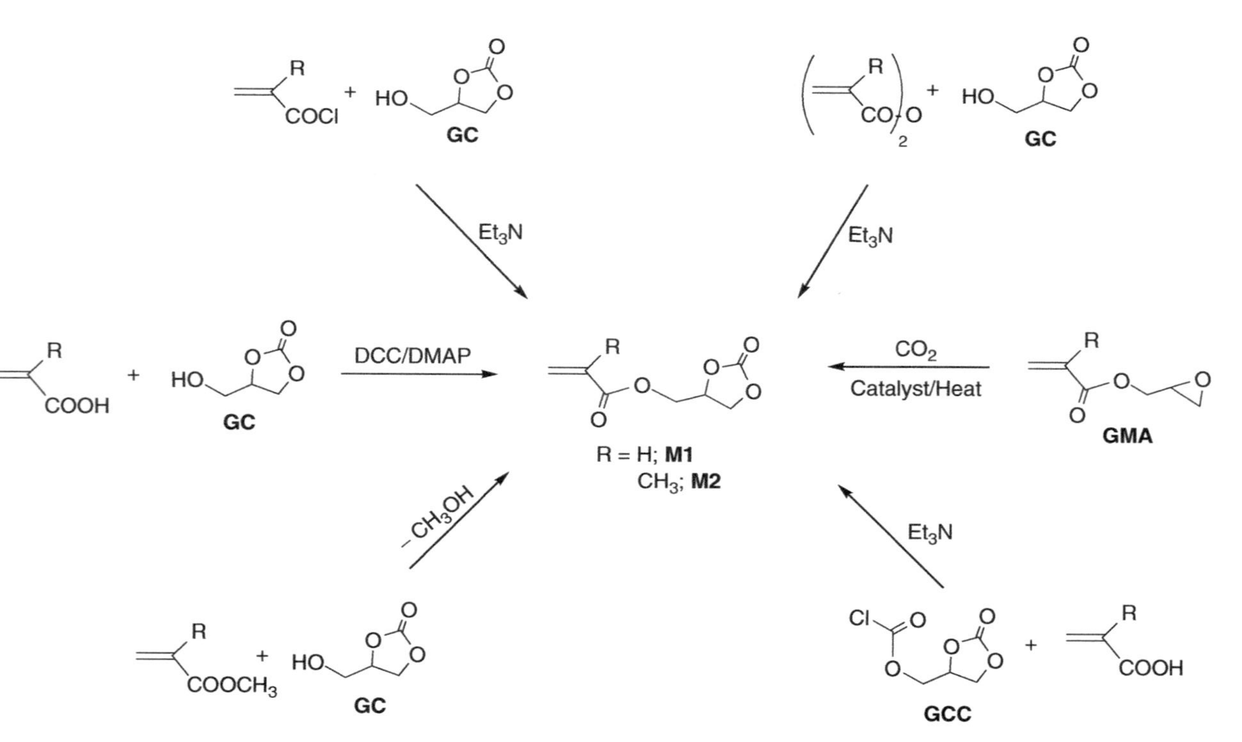

Scheme 4.1 Synthesis of propylene carbonate (meth)acrylate, PC(M)A, **M1** and **M2**, monomers.

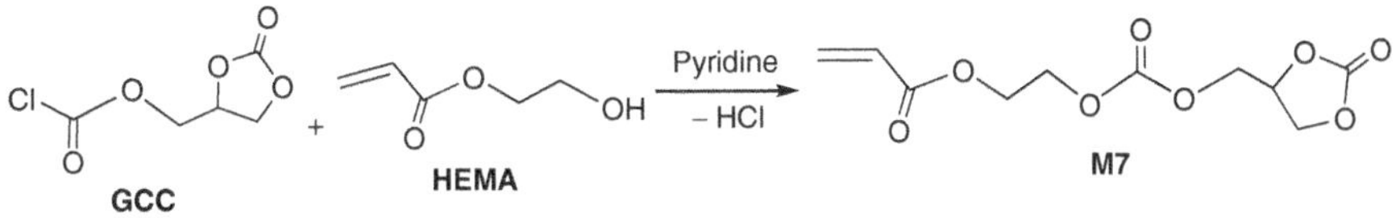

Scheme 4.2 Synthesis of **M7** using glycerol carbonate chloroformate (GCC).

Monomers bearing acyclic carbonate or ester spacer group present in between (meth)acrylate and CC groups **M7–M9** were synthesized from GC methacrylate chloroformate (Scheme 4.2) [19,25]. Monomers bearing flexible urethane spacer group present in between (meth)acrylate and CC groups **M10** and **M11** were synthesized by two different procedures as shown in Scheme 4.3 [17b,26]. Very recently, a sterically hindered CC methacrylate monomer **M12** has been reported by Barkakaty et al. [27] by the addition reaction of CO_2 with the corresponding oxirane compound catalyzed by amidine.

Among other CC vinyl monomers (Figure 4.3), vinyl ethylene carbonate (VEC), **M13** (4-vinyl-1,3-dioxolan-2-one), was first reported by the reaction of 3-butene-1,2-diol with diethyl carbonate by Bissinger et al. [28]. Later it was synthesized by Prichard et al. [29] by the addition of CO_2 to 3,4-epoxy-1-butene. This monomer is now commercially available from different sources. The phenyl-substituted monomer **M14** (4-phenyl-5-vinyl-1,3-dioxolan-2-one) (Figure 4.3) was synthesized by Ochiai et al. [30] by the addition reaction of CO_2. Vinylene carbonate (VC) monomer **M15** (1,3-dioxol-2-one) (Figure 4.3) is at present commercially available. Otherwise, it can be synthesized by chlorination–dehydrochlorination of EC [31].

Propylene carbonate vinyl ether (PCVE), **M16** ((2-oxo-1,3-dioxolan-4-yl)methyl vinyl ether), was synthesized by the reaction of CO_2 with glycidyl vinyl ether (GVE) [32]. Nishikubo et al. [32b] also reported the synthesis of **M16** from GVE using β-butyrolactone and sodium hydrogen carbonate along with the production of by-products. Norborene derivative with CC functional group (**M17**) was reported by Welker et al. [33a] by the Diels–Alder reaction between cyclopentadiene and VEC (**M13**), and the addition reaction of CO_2 with the corresponding epoxide was reported by Endo and coworkers [33b], respectively. Very recently, Miyata et al. [34] have reported the synthesis of styrene-based monomer having five-membered CC structure attached to it by the LiBr-catalyzed addition reaction of CO_2 with its corresponding epoxide.

4.2.2 Polymerization of Cyclic Carbonate (CC) Monomers

In this section, polymers with modifiable pendant CC groups will be discussed. Thus, polycarbonates, poly(ether carbonate)s, or poly(ester carbonate)s obtained by the ring-opening polymerizations [35] of CC groups are outside the scope of this discussion.

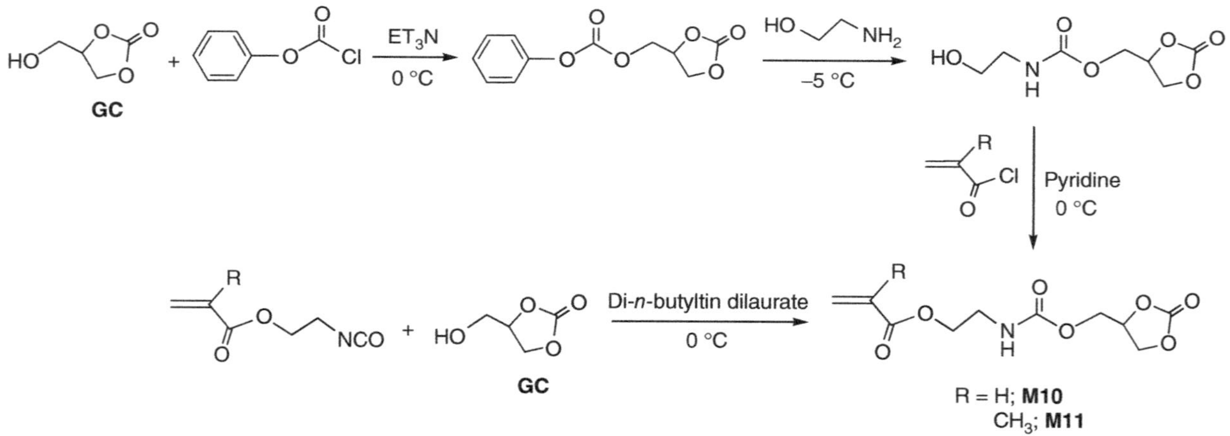

Scheme 4.3 Synthesis of monomers with urethane spacer group, **M10** and **M11**, using glycerol carbonate (GC).

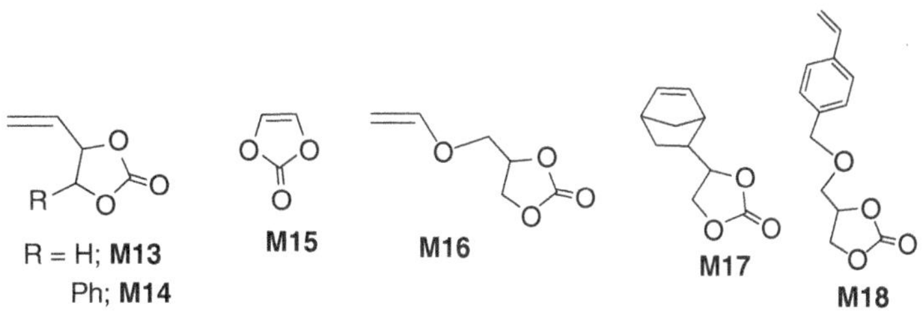

R = H; **M13**
Ph; **M14**
M15
M16
M17
M18

Figure 4.3 Examples of other cyclic carbonate (CC) vinyl monomers.

In general, CC-based methacrylate monomers (Figure 4.2) are highly reactive and susceptible to polymerize easily even while attempting to purify the monomer by distillation [13,21a]. Alternatively, the polymerization of CC monomers can be avoided by purifying it using column chromatography [17d,20,21b]. Other CC vinyl monomers **M13–M17** (Figure 4.3) are not that reactive as their (meth)acrylate analogs. All the polymerization methods of the CC-containing monomers reported to date are discussed in three different categories:

1. thermal free radical polymerization (FRP) of CC monomers
2. photopolymerization of CC monomers, and
3. cationic polymerization, anionic polymerization, controlled radical polymerization (CRP), and ring-opening metathesis polymerization (ROMP) of CC monomers.

4.2.2.1 Thermal Free Radical Polymerization (FRP) of CC Monomers

Although there are a few initial reports of gelation [13,18,19] and difficulties during purification, storage, and homopolymerization of CC (meth)acrylate monomers partly because of the presence of impurities and chain transfer reaction from the CC ring [21b], the CC (meth)acrylate monomers were later successfully reported to be homopolymerized and copolymerized to yield soluble linear polymers by many research groups. Among all CC vinyl monomers, most of the reported results of polymerization were performed using PC(M)A (i.e., **M1** and **M2**). **M1** and **M2** (Figure 4.2) were homopolymerized [13,17a,17d,21b,36] and copolymerized [13,17d,19,21b,36] with styrene (St), MMA, methyl acrylate (MA), butyl methacrylate (BMA), hexyl methacrylate (HMA), vinyl acetate (VAc), and so on by free radical polymerization (FRP) using either 2,2′-azobisisobutyronitrile (AIBN) or benzoyl peroxide (BPO) as the radical source. The recent free radical homo- and copolymerization of **M1** and **M2** with BMA by Britz et al. [17d] was reported to produce (co)polymer of M_n 9–77 k and of polydispersity index (PDI) 1.9–3.6, which is common for a conventional FRP. Polymers produced in dimethylformamide (DMF) were reported to be higher molecular weight than using other solvents, for example, THF, methanol, and toluene, possibly because of the chain transfer reactions and poor solubility of polymer produced in these later mentioned solvents. It was also reported [17d] that upon polymerization, the methacrylate

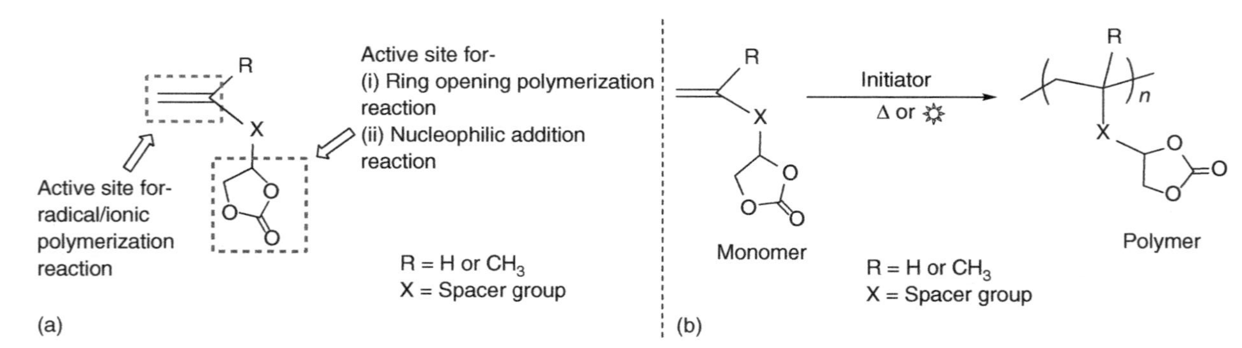

Scheme 4.4 (a) Multifunctional CC monomers and (b) thermal or photoinitiated synthesis of pendant CC polymers.

monomer, **M2** generally produced polymer of higher degree of polymerization (DP) than that of the acrylate analog, that is, monomer **M1**. The reactivity of propylene carbonate methacrylate (PCMA), **M2**, is reported to be similar to that of BMA.

Homopolymerization of **M1** or **M2** by radical initiators is generally fast and produce white brittle solid that are soluble only in highly polar organic solvents, for example, DMF, dimethylsulfoxide (DMSO), and propylene carbonate. The brittleness of the polymer can be reduced by the copolymerization with other suitable flexible vinyl monomer. Copolymerization also improves the solubility of the resultant polymer. Incorporation of **M1** and **M2** into the copolymer could be confirmed by the presence of five-membered CC absorption peak at $\sim$1800 cm^{-1} in the FT-IR spectra of copolymers [21b] and indeed by the use of other characterization methods such as nuclear magnetic resonance (NMR) spectroscopy. In general, copolymerization of these CC (meth)acrylate monomers is an attractive means to incorporate reactive pendant CC functionality into a variety of polymers of commercial interest. However, the enhanced reactivity of these CC (meth)acrylate monomers is a matter of concern for storage and handling of these monomers for large scale industrial synthesis.

Monomers with long-aliphatic spacer unit, **M3** and **M4** (Figure 4.2), were homopolymerized by Britz et al. using AIBN [17d]. Polymerizations proceeded well in DMF than those in other solvents such as toluene and methanol. D'Alelio et al. [13] have reported the free radical homopolymerization and copolymerization of CC monomers with both acid and CC functional groups, that is, **M5** and **M6** (Figure 4.2). Although homopolymerization of **M6** was successful, **M5** was very reluctant to homopolymerize under similar conditions. Copolymerization behavior and reactivity ratios of these monomers with St, MMA, MA, and VAc were reported by D'Alelio et al. [13]. CC monomers M7-M10 were mainly reported to be polymerized by UV using photoinitiators [17b,19,25,26].

Ochiai et al. [26b] have reported free radical homopolymerization and copolymerization of **M11** (Figure 4.2) with butyl (meth)acrylate (B(M)A) and diethylene glycol methyl ether methacrylate (DEGMEM) to produce flexible polymer films. This group has also performed AIBN-induced thermal polymerization of sterically hindered CC methacrylate monomer **M12** [27] to yield a relatively high molecular weight homopolymer. Not much literature is available on the polymerization reaction of VEC, **M13**, which is commercially available at present. The earliest polymerization attempts [2] by two independent research groups, namely, Asahara et al. and Semenova et al., have reported low yield of homopolymer of **M13**. The copolymerization reactions of this monomer with other vinyl monomers such as St, MMA, BMA, EA and maleic anhydride (MAH) were reported to produce copolymers with low incorporation of **M13**. However, **M13** was reported to be readily copolymerized [2] with VAc. Webster et al. [37,38] have also explored the copolymerization behavior of **M13** with different vinyl ester monomers and butyl acrylate (BA) and reported the similar observation. Interestingly, they were able to incorporate **M13** to a varying degree, maximum up to 50% as the copolymer with vinyl esters to obtain polymers that are soluble in a range of organic solvents. They were also able to incorporate **M13** into a series of VAc/BA latexes [37,38]. A unique feature of **M13** is that it can be copolymerized with olefin monomers using transition metal catalysts [39]. The homopolymerization and copolymerization behavior of the phenyl-substituted version of VEC monomer **M14** with VAc and MAH was studied by Ochiai et al., which yielded copolymers with as high as ~66% incorporation of **M14**. The other unsaturated CC monomer, VC, **M15** appears to be one of the highly reactive 1,2-disubstituted vinyl monomers and therefore could be homo- and copolymerized easily [40]. However, Ding et al. [40e] reported gelation during bulk polymerization of **M15**, whereas low molecular weight soluble polymers were obtained by solution polymerization in DMF and DMSO. The first reported polymerization using glycerol carbonate vinyl ether (GCVE), **M16**, was performed by Nishikubo et al. [32b]. It was homopolymerized by cationic polymerization and copolymerized with MA, MAH, acrylonitrile (AN), and *N*-phenylmaleimide (NPM) by FRP. Later Moon et al. [32c] and Ha et al. [41] reported the synthesis of copolymer **M16** with NPM and AN, respectively, for studying the miscibility in polymer blends. There was no report available in the literature yet for the FRP of monomer **M17**. However, **M18** had been homo- and copolymerized [34] with styrene to yield (co)polymers of moderate molecular weight.

4.2.2.2 Photopolymerization of CC Monomers The photopolymerization process that converts the highly reactive liquid monomer to solid polymer is very useful for a variety of applications including dental materials, contact lenses, lithographic processes, adhesives, and fiber optic coatings. After the first discovery of the outstanding reactivity of CC (meth)acrylate monomers by Decker et al. [25a,42], there are ongoing efforts to study the photopolymerization behavior of CC (meth)acrylate monomers especially **M1** and **M2** (Figure 4.2) in detail. Indeed these monomers can be polymerized by UV irradiation in presence of a suitable photointiator such as α,α-dimethoxydeoxybenzoin or 2,2-dimethoxy-2-phenylacetophenone (DMPA). In general, polymerization rates of these mono(meth)acrylates having secondary CC functionalities were reported to be very fast, as a matter of seconds and also dependent on temperature [17,26,43e] (Figure 4.4). Recently, Bowman and coworkers have reported the reactivity of such highly reactive monomers [17,26,43] and demonstrated that the polymerization reaction can even be continued extensively in dark [44] after UV initiation unlike that of the polymerization of traditional acrylates and diacrylates. CC (meth)acrylate monomers (**M1** and **M2**) or monomers with ester and acyclic carbonate spacer groups (**M7–M9**) were used for this study. The mechanistic explanations behind the enhanced reactivity of CC (meth)acrylate monomers have been explained by the presence of intramolecular interactions because of the CC moieties [17,26,43c,e].

4.2.2.3 Cationic Polymerization, Anionic Polymerization, Controlled Radical Polymerization (CRP) and Ring-Opening Metathesis Polymerization (ROMP) of CC Monomers Polymers with reactive functionalities and controlled structures are of significant interest for numerous applications. In this section, reports on controlled/living polymerization results of CC monomers are highlighted. The first of this kind of approach, the *cationic* polymerization of CC monomer GCVE (**M16**) (Figure 4.3), was carried out by Nishikubo et al. in dichloromethane

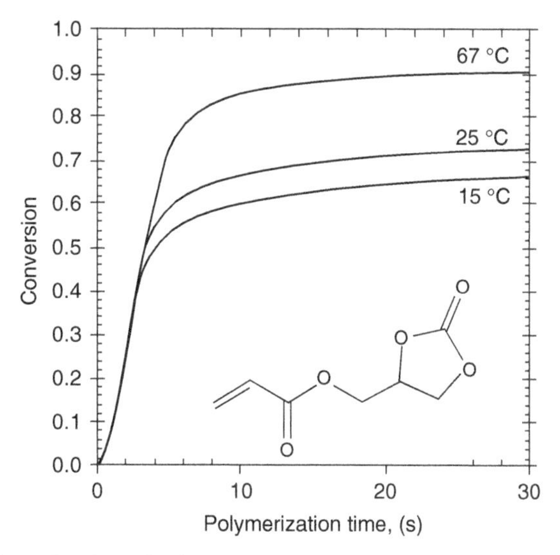

Figure 4.4 Effect of polymerization temperature on the steady-state kinetics of the CC acrylate polymerization (UV light 5 mW cm^{-2}; initiator used was 0.1 wt% of DMPA). (Reprinted with permission from [43c]. Copyright © 2008 American Chemical Society.)

and toluene at 0 °C to −50 °C using various catalysts such as $BF_3 \cdot OEt_2$, I_2, and CF_3SO_3H [32b]. Polymers were obtained in high yield; however, molecular weight and PDI characteristics were not reported.

The first *anionic* homopolymerization of PCMA, **M2**, was reported by Britz et al. [17d] using diphenyl lithium in THF. Similar attempts to anionically polymerize acrylate version of CC monomers **M1** and **M3** failed for different reasons. However, the controllability of the homopolymerization of **M2** was poor as the PDI achieved in this attempt was 2.2. Correlation between theoretical molecular weight and observed molecular weight was not presented as well.

Recently, we have reported copper-mediated ATRP of PCMA (**M2**) (Scheme 4.5) to yield a polymer with controlled molecular structure [20]. The molecular weight of the polymer increased with the conversion, and the PDI achieved was much lower (<1.5) than that of the polymer obtained by the FRP of **M2** using AIBN. Some PEGylated ABA-type triblock and (AC)B(AC) statistical terpolymers with very low PDIs (Figure 4.5) were also synthesized by a similar technique using dibromofunctional poly(ethylene glycol) (PEG) macroinitiators. Owing to the similar reactivity of PCMA or **M2** with MMA, PCMA was introduced statistically within the (AC)B(AC) terpolymers as confirmed by ^{1}H NMR spectroscopy and other characterizations [20].

The other type of (co)polymerization of CC monomers including ROMP and palladium-catalyzed addition copolymerization of monomer **M17** (Figure 4.3) with 5-butyl-2-norborene (BNB) derivative was reported (Scheme 4.6) by Sudo et al. [33b] although PDI of the obtained copolymers was broad (2.3–3.7).

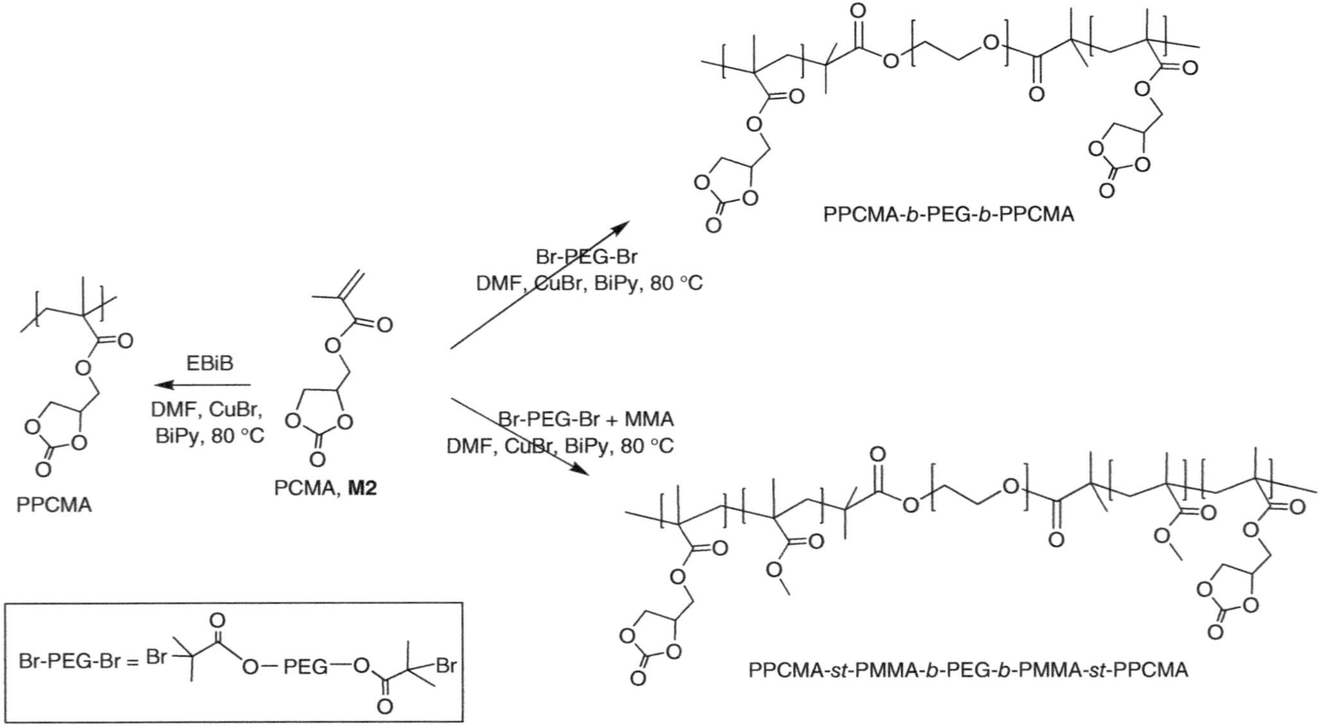

Br-PEG-Br
DMF, CuBr, BiPy, 80 °C

PPCMA-*b*-PEG-*b*-PPCMA

EBiB
DMF, CuBr,
BiPy, 80 °C

PPCMA

PCMA, **M2**

Br-PEG-Br + MMA
DMF, CuBr, BiPy, 80 °C

PPCMA-*st*-PMMA-*b*-PEG-*b*-PMMA-*st*-PPCMA

Br-PEG-Br = Br–C(CH$_3$)$_2$–C(=O)–O–PEG–O–C(=O)–C(CH$_3$)$_2$–Br

Scheme 4.5 Synthesis of PCMA copolymers by ATRP.

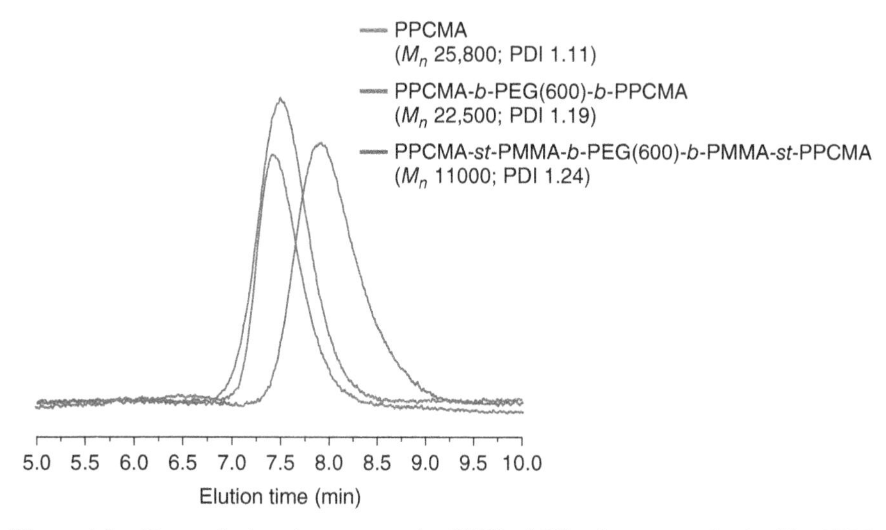

Figure 4.5 Size exclusion chromatography (SEC) of CC polymers synthesized by ATRP. (Reprinted with permission from [20]. Copyright © 2010 Wiley Periodicals Inc.) (*See insert for color representation.*)

4.2.3 Alternative Route to Synthesize Pendant CC (Co)polymers by CO_2 Addition/Fixation Reaction

Owing to the highly reactive and sensitive nature of CC monomers and the difficulties in controlling the polymerization of CC (meth)acrylate monomer, an alternate route to synthesize pendant CC polymers without the need of synthesizing the related CC monomer would be a significant development. In this approach, (co)polymers of glycidyl (meth)acrylate, G(M)A, with controlled structures could be first synthesized and subsequently converted to the pendant CC (co)polymers by the addition reactions of CO_2 (Scheme 4.7) [45–48]. Solution-phase conversion proceeds effectively for these copolymers. Interestingly, the CO_2 fixation/addition reaction was performed in solid state [45c,46a,48], which avoided the additional usage of solvent and purification steps. However, this solid-phase reaction was reported to produce a cross-linked product in the presence of a highly dense oxirane/epoxide group [48b–c]. Cross-linking can be avoided by the use of GMA copolymers with lower GMA concentration [48b–c]. Copolymers containing both oxirane and internal catalytic moieties (e.g., quaternary ammonium salts) (Scheme 4.8) were also reported [45c,48a] to produce copolymers with pendant CC groups by the addition reaction of CO_2 gas both in liquid and solid states.

Recently, Ochiai et al. [49] have also reported one-step, low-energy intensive synthesis of pendant CC (co)polymers by the concurrent radical polymerization and CO_2 fixation of GMA.

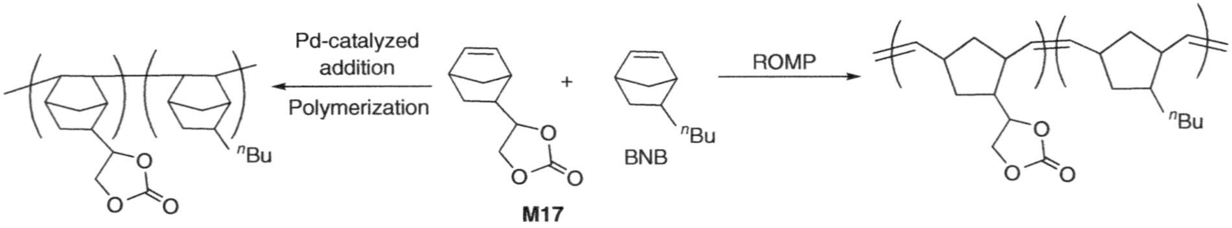

Pd-catalyzed addition Polymerization

^{n}Bu

M17

BNB

ROMP

Scheme 4.6 (Co)polymerization of norborene monomer with CC functional group, **M17**.

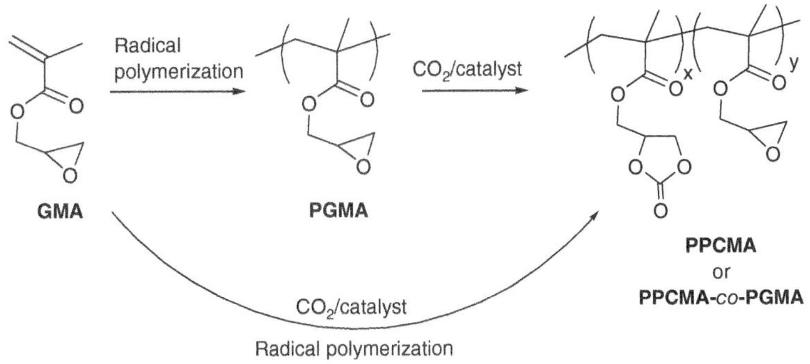

GMA PGMA PPCMA or PPCMA-*co*-PGMA

Radical polymerization; CO_2/catalyst; CO_2/catalyst Radical polymerization

Scheme 4.7 Alternate synthetic routes to synthesize pendant CC polymers.

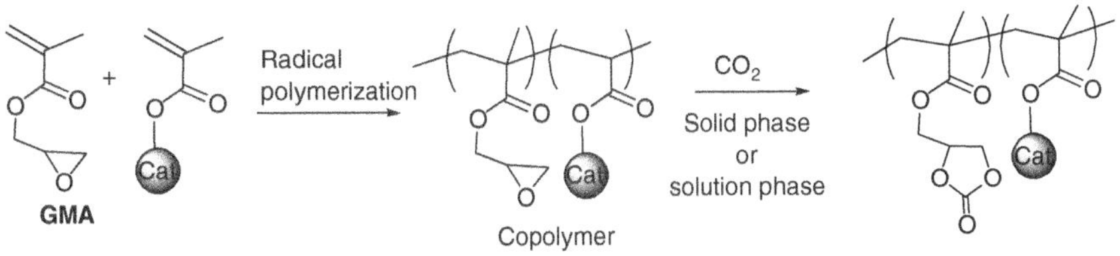

GMA Copolymer

Radical polymerization; CO_2 Solid phase or solution phase

Scheme 4.8 CO_2 fixation of a GMA-based copolymer with internal catalytic system.

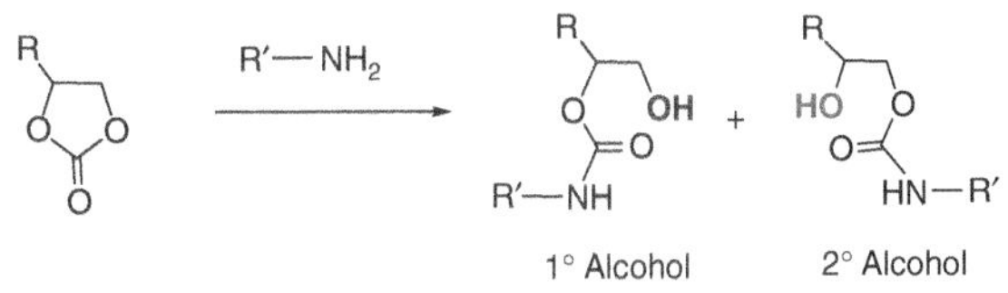

R′—NH$_2$

1° Alcohol 2° Alcohol

Scheme 4.9 Aminolysis of CC functional group.

4.3 CHEMICAL MODIFICATION OF PENDANT CC POLYMERS

CC group undergoes ring-opening reaction with a few nucleophiles. The most widely studied reaction among these is aminolysis, that is, the reaction of CC group with amines (Scheme 4.9). The reaction conditions employed is normally mild and the extent of ring opening by the nucleophilic attack depends on the type of amine, solvent polarity, reaction temperature, and so on. The kinetics of this aminolysis reaction was studied by different researchers [21b,37,50]. The reaction rate is fastest for the primary amine that is attached to a primary carbon atom. This aminolysis reaction with a primary amine produces a mixture of primary and secondary hydroxyurethane groups with about 1 : 4.4 to 1 : 1.2 ratios [50b,51a]. This reaction is also tolerant to many functional groups and impurities present and thus can be performed in a range of solvents including water, alcohol and ionic liquid. The reaction is claimed to be very chemoselective [51].

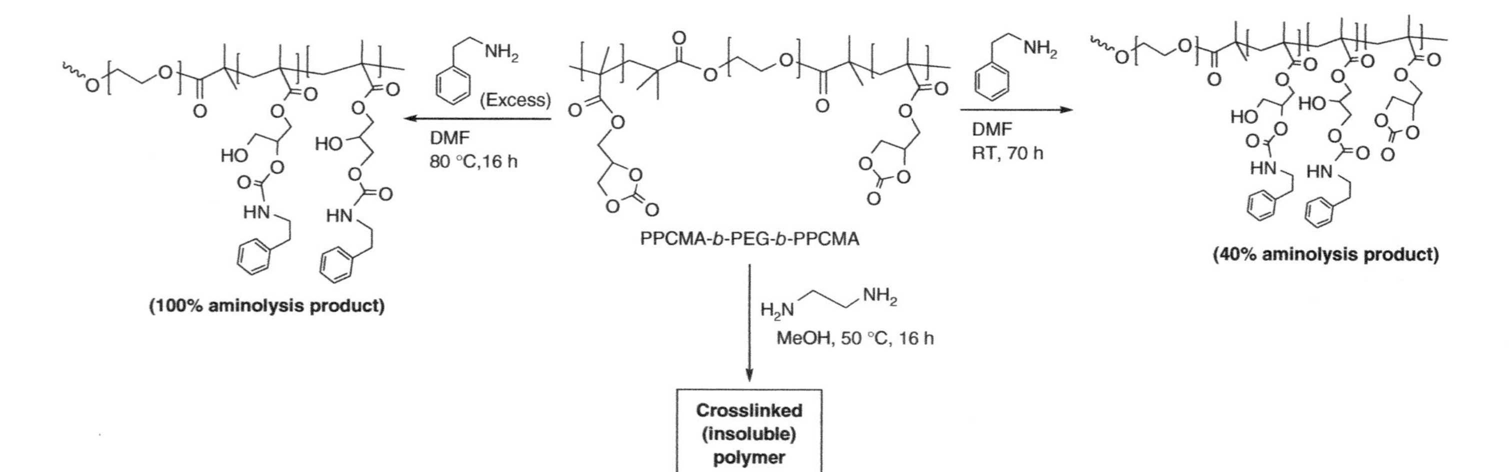

Scheme 4.10 Synthesis of functional polymers by nucleophilic addition reaction of amine with polymer pendant CC groups.

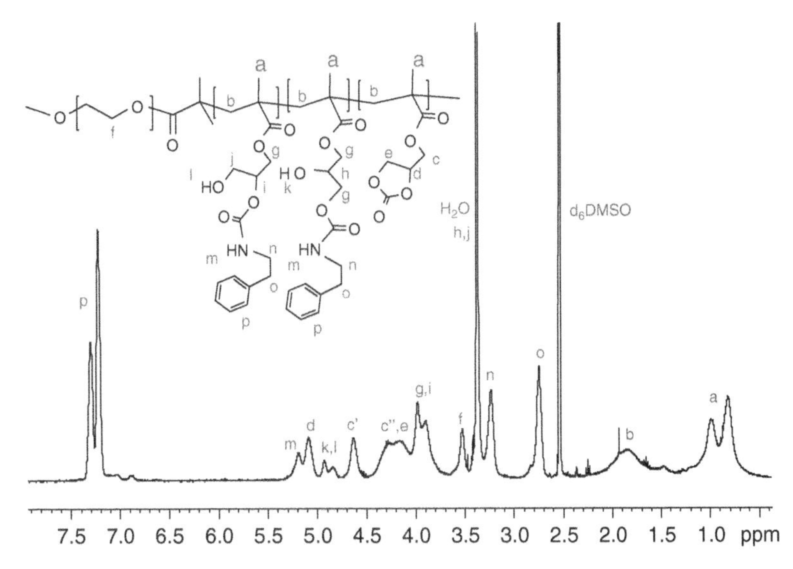

Figure 4.6 ^{1}H NMR spectrum of a multifunctional polymer obtained by the partial (~40%) aminolysis of a pendant CC copolymer. (Reprinted with permission from [20]. Copyright © 2010 Wiley Periodicals Inc.)

Although the facile polyaddition reaction of low molecular weight difunctional CC with diamines was [51,52] exploited to synthesize polyhydroxyurethanes in a nontoxic isocyanate-free route by numerous researchers, it would not be discussed further in this section as the main focus here is to discuss about the reactions of polymers with pendant CC groups.

Similar to that of low molecular weight CC compounds, polymers with pendant CC groups, synthesized from different CC vinyl monomers (e.g., **M1**, **M2**, **M14**, **M16**, **M17**), react very easily with monofunctional primary amines to produce a range of functional polymers [20,22b,31,32b,33b,50a]. This aminolysis reaction yielded pendant hydroxyurethane polymers (Scheme 4.10 and Figure 4.6) that resulted in a change in all the physical properties including solubility, transparency, brittleness, and film formation properties [20] etc. This is a nonsacrificial atom-efficient reaction as the reactive CC group is converted to a group having active primary and secondary hydroxyl functionalities after the aminolysis reaction. Ideally these hydroxyl groups can further be exploited by other modification reactions as well. The extent of aminolysis can be quantified by the elemental analysis where elemental nitrogen increases with the conversion. The reaction can also be monitored by the disappearance of the carbonyl resonances of CC group at ~2000 cm^{-1} in FT-IR spectra and by tracking the changes of proton resonances in ^{1}H NMR spectrum of polymer as shown in Figure 4.6. Reaction of pendant CC polymers with diamines, however, produced cross-linked network products [20] as expected.

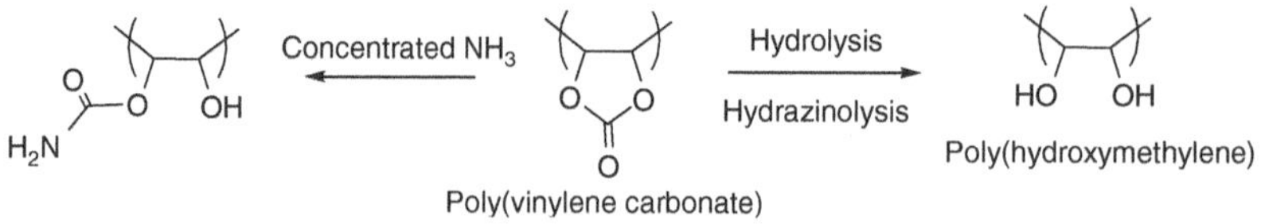

Concentrated NH_3 Hydrolysis Hydrazinolysis

Poly(vinylene carbonate) Poly(hydroxymethylene)

Scheme 4.11 Functionalization reaction of poly(vinylene carbonate).

A few other types of functionalization reactions were reported for the polymers obtained from VC (**M15**) [40a,53], which is not actually a pendant CC polymer but a backbone CC polymer. The completely hydrolyzed polymer of poly(vinylene carbonate), that is, poly(hydroxymethylene) (Scheme 4.11), is reported to be insoluble in common organic solvents presumably because of very strong intermolecular hydrogen bonding. However, poly(hydroxymethylene) is soluble in anhydrous hydrazine and the saturated LiCl solution in water or DMSO [53].

There are some striking differences between the CC and epoxy or oxirane groups. In general, CC group is more tolerant to impurities and solvents than to epoxides. As mentioned before, epoxy or oxirane group is reported to react with a number of different functionalities including primary and secondary amines, alcohol, phenol, acid, acid chloride, and thiol (Table 4.1) [4], whereas CC group mainly react with primary and secondary amines. Reactions with other nucleophiles are not well studied. Reaction of pendant CC polymers with primary amine produces soluble pendant hydroxyurethane polymers. However, similar reaction with epoxy or oxirane compound produced a cross-linked structure [4a] because of the subsequent reaction of epoxy group with the secondary amine, which is formed by the initial reaction of primary amine and epoxy functionality. Although secondary amine reacts with CC group, albeit at a slower rate than primary amine, the product of primary amine with CC group is an unreactive hydroxyurethane group unlike that of the addition product of primary amine with oxirane and therefore does not cross-link. Otherwise, oxirane group can easily be hydrolyzed [49a] using dilute acid or can be reacted with acid chloride [54], which is not so easy for the CC groups (Scheme 4.12).

4.4 APPLICATIONS OF PENDANT CC POLYMERS

4.4.1 Fixing CO_2 into Polymer

Carbon dioxide (CO_2) is a green house gas and an abundant, inexpensive, attractive raw material available as a by-product in various industrial processes as well as biorenewable. Among many studies of chemical fixation of CO_2, the reaction of oxiranes or epoxides with CO_2 to produce a five-membered CC is well established and is practiced industrially [55]. In this regard, polymers with highly polar and reactive pendant CC groups are attractive as a permanent CO_2 storage functional material. CO_2 content in a few representative pendant CC polymers is presented in Figure 4.7. The production of pendant CC (co)polymers is also an alternative

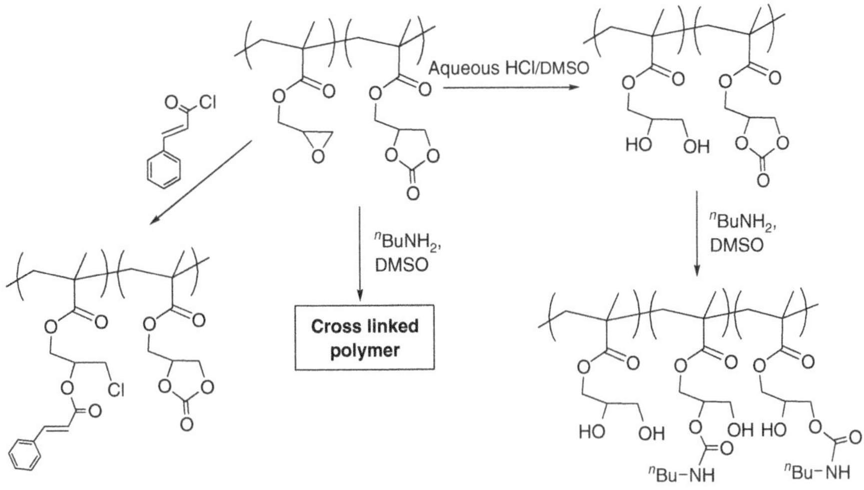

Scheme 4.12 Reactivity differences between epoxy or oxirane and CC groups.

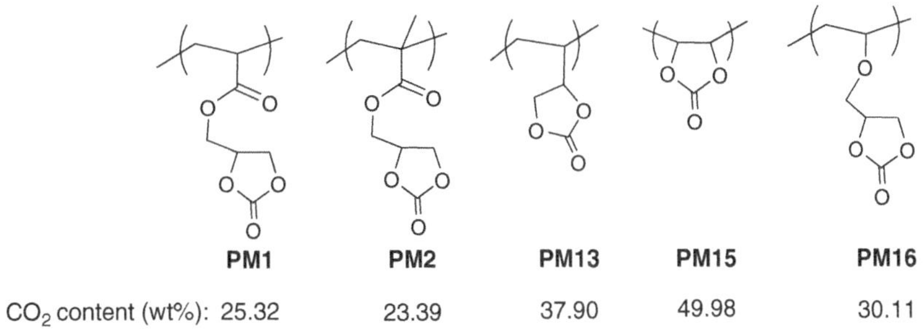

	PM1	PM2	PM13	PM15	PM16
CO_2 content (wt%):	25.32	23.39	37.90	49.98	30.11

Figure 4.7 CO_2 content in different pendant CC polymers (**PM1** is the polymer of monomer **M1**).

approach to produce CO_2-based polymers in reference to the polycarbonates produced by the copolymerization of CO_2 and epoxide. However, there are a few technical challenges for the industrial applications of these polymers as follows: (i) the pendant CC-based vinyl monomers are highly reactive and thus pose a challenge during synthesis, purification, storage, and transport; (ii) pendant CC homopolymers are highly brittle; and (iii) owing to the highly polar nature of the CC groups, pendant CC polymers are mostly soluble in highly polar solvents such as DMF, DMSO, EC, and propylene carbonate (PC). The flexibility and the solubility of CC polymers can be improved by copolymerizing CC monomers with suitable comonomer or through the functionalization of CC groups of polymer by ring-opening reactions. The difficulties in monomer purification can be solved by the simultaneous polymerization and CO_2 fixation reactions of the corresponding oxirane monomer, for example, GMA as reported by Endo et al. [49b]. Finally, it is necessary to broaden the application of pendant CC polymers before the process is to be recognized as an effective CO_2 utilization process.

4.4.2 Surface Coating

Thermosetting coating formulations that cure at ambient or low baked conditions are of enormous interest for paint and coating industries. The intriguing reaction between five-membered CC and amines is of particular interest as the reaction proceeds rapidly at room temperature or at a slightly elevated temperature, particularly without any release of volatile by-product. In addition, the reaction produces flexible urethane group without the use of toxic isocyanate compounds. Many researchers have reported the use of pendant CC polymers for coating formulations. Webster et al. [37,56] used copolymers based on VEC (**M13**) and low molecular weight diamines and triamines to obtain good gloss, hardness, and solvent-resistant coatings. They have also formulated a powder coating composition using a mixture or solid pendant CC polymer and a solid carbamate salt of amines [57]. Coating formulations based on polymers from CC-based (meth)acrylate monomers (i.e., **M1**, **M2**) and pure or protected multifunctional amines (e.g., ketimine-blocked amine) as cross-linking agents were reported by many researchers as well [58]. Water-based coating formulations have also been designed recently using CC copolymers based on VEC (**M13**) monomer [59].

4.4.3 Solid or Gel Polymer Electrolyte for Lithium-Ion Batteries

Highly polar CC compounds are known to dissolve metal salts and to solvate ions, for example, lithium ions in a tetrahedral solvent shell as shown in Figure 4.8a. Thus, ethylene and propylene carbonates (EC and PC) are used for the lithium-ion batteries. However, the use of EC- or PC-based liquid organic electrolyte poses potential safety hazards such as leakage in the case of accident and overheating and ignition in the case of short-circuiting. Solid- or gel-based materials with similar solvating characteristics are potential for these applications. Golden et al. [21c] have first reported the lithium-ion interaction and carried out lithium-ion conductivity measurements using pendant CC-based linear and gel polymers synthesized from PCA (**M1**) and VC (**M15**) (Figure 4.2). Britz et al. [17d] synthesized various pendant CC (meth)acrylate (co)polymers utilizing monomers **M1–M4** to blend with lithium salts to obtain high lithium-ion conductivity. However, blending these pendant CC polymers with propylene carbonate enhances the ionic conductivity further significantly. Recently, we have studied lithium-ion conductivity of PCMA (**M2**)-based homopolymers and different block copolymers having flexible and solvating PEG as a middle block and achieved lithium-ion conductivity of $2–5 \times 10^{-6}$ S cm^{-1} at room temperature [20]. PCMA (**M2**)-based homo- and copolymers were also reported to be soluble in various ionic liquids and the ionic liquid-swelled PCMA network, that is, ion gels produced (Figure 4.8b,c) in the presence or absence of lithium salts were transparent, flexible, and highly ion conductive [60].

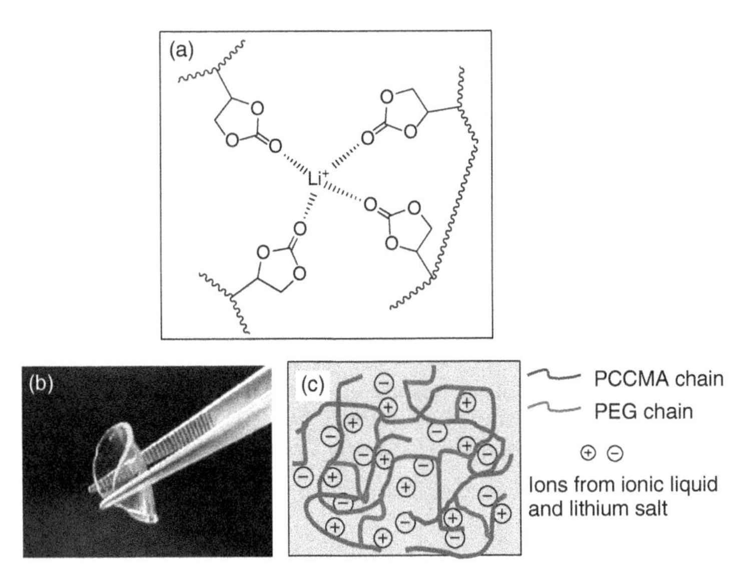

Figure 4.8 (a) Lithium-ion coordination with CC groups of polymer. (Reprinted with permission from [17d]. Copyright © 2007 American Chemical Society.) (b) Photograph and (c) schematic presentation of a transparent, flexible, and highly conductive ion gel based on pendant CC copolymer. (Reprinted with permission from [60]. Copyright © 2011 Royal Society of Chemistry.) (*See insert for color representation.*)

4.4.4 Enzyme Immobilization

The facile ring-opening reaction of pendant CC groups of polymers by the nucleophilic addition of amino groups could potentially be exploited for the immobilization of proteins or enzymes. However, mainly the (co)polymers or the network gel of VC, **M15** (Figure 4.3), is reported to be used till date for this purpose. Chen et al. [61] have used VC (**M15**)-grafted LDPE film or the poly(vinylene carbonate)–Jeffamine hydrogel for the immobilization of alkaline phosphatase. Ding et al. [62] used either poly(vinylene carbonate)–diamine gel or VC-based copolymers (Figure 4.9) for the attachment of trypsin as recyclable catalysts. Recently, Türünc et al. [63] have reported immobilization of α-amylase on inorganic–organic hybrid network with pendant CC groups and reported improved thermal stability and higher activity than the free enzyme. The other example of enzyme immobilization is the work published by Mauz et al. [64] where they have used **M2** and **M16** (Figures 4.2 and 4.3) as comonomers as well as a source of CC.

4.4.5 Photopolymerization

As discussed before, CC (meth)acrylates (especially monomers **M1**, **M2**, and **M7–9** in Figure 4.2) undergo ultrafast polymerization under UV irradiation

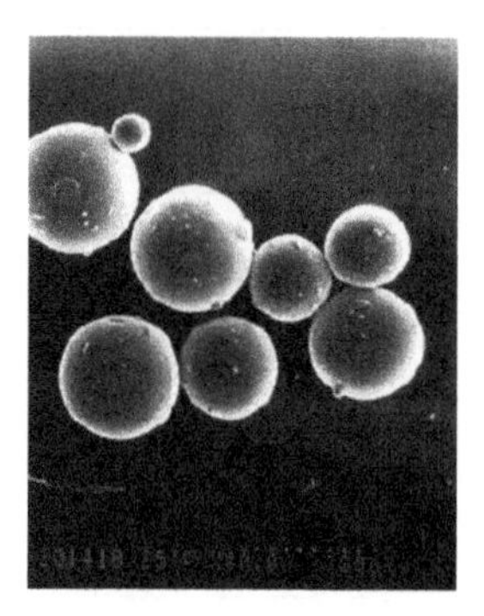

Figure 4.9 Scanning electron micrograph of spherical copolymer beads of vinylene carbonate (VC), **M15** and hydroxyethyl methacrylate used for trypsin immobilization. (Reprinted with permission from [62b]. Copyright © 2002 Wiley Periodicals Inc.)

[17,25a,26,42–44]. The photopolymerization process is widely used in different industrial applications such as coatings, adhesives, lithographic process, and dental restoratives, owing to the rapid conversion of liquid monomer to solid polymer especially under ambient conditions. However, various other criteria are to be satisfied, for example, residual unsaturation, oxygen inhibition, and different mechanical properties before the actual application. Podszun et al. [65] have capitalized the ultrafast polymerization of CC-functionalized mono(meth)acrylates to prepare multifunctional oligomers for dental applications.

4.4.6 Polymer Blends

Copolymer of PCMA (**M2**) and ethyl acrylate (EA) and copolymer of PCMA (**M2**) and St were used by Park et al. [46c,47a] for the miscibility study with PMMA or poly(vinyl chloride) and was reported to be compatible over the whole range of composition. In a separate study, they have used PCMA homopolymer for the miscibility study with the copolymer of MMA and EA [46d].

4.5 CONCLUSION

A range of different CC vinyl monomers are available for the synthesis of pendant CC (co)polymers, some of which are commercially available at present. Polymers bearing pendant CC functionalities are very promising materials for a number of applications including the synthesis of pendant multifunctional polymers by exploiting especially the very simple chemoselective aminophilic reaction of CC group. Intrinsically, syntheses of CC (co)polymers are a matter of interest as the material is derived by utilizing CO_2. Its applications in diverse key areas including polymer electrolytes for lithium batteries, as thermosetting coating resin, synthesis of side-chain urethane polymer, and so on make these polymers most interesting. However, the synthesis, purification, and stability of many CC vinyl monomers are still matter of concern for large scale industrial utilization or applications.

REFERENCES

1. Gauthier, M. A., Gibson, M. I., Klok, H. A. (2009). Synthesis of functional polymers by post-polymerization modification. *Angew. Chem. Int. Ed. Engl.*, *48*, 48–58.
2. Webster, D. C. (2003). Cyclic carbonate functional polymers and their applications. *Prog. Org. Coat.*, *47*, 77–86.
3. (a) Theato, P. (2008). Synthesis of well-defined polymeric activated esters. *J. Polym. Sci. Part A: Polym. Chem.*, *46*, 6677–6687; (b) Reza, A. (1994). Polymer synthesis via activated esters: a new dimension of creativity in macromolecular chemistry. *Adv. Polym. Sci.*, *111*, 1–41; (c) Vosloo, J. J., Tonge, M. P., Fellows, C. M., D'Agosto, F., Sanderson, R. D., Gilbert, R. G. (2004). Synthesis of comblike poly(butyl methacrylate)using reversible addition-fragmentation chain transfer and activated ester. *Macromolecules*, *37*, 2371–2382; (d) Li, R. C., Hwang, J., Maynard, H. D. (2007). Reactive block copolymer scaffolds. *Chem. Commun.*, *44*, 3631–3633.
4. (a) Edmondson, S., Huck, W. T. S. (2004). Controlled growth and subsequent chemical modification of poly(glycidyl methacrylate) brushes on silicon wafer. *J. Mater. Chem.*, *14*, 730–734; (b) Nishikubo, T., Iizawa, T., Takahashi, E., Nono, F. (1985). Study of photopolymers. 26. Novel synthesis of self-sensitized photosensitive polymers by addition reactions of poly(glycidyl methacrylate) with nitroaryl cinnamate. *Macromolecules*, *18*, 2131–2135; (c) Tsyalkovsky, V., Kelp, V., Ramaratnam, K., Lupitskyy, R., Minko, S., Luzinov, I. (2008). Fluorescent reactive core-shell composite nanoparticles with a high surface concentration of epoxy functionalities. *Chem. Mater.*, *20*, 317–325; (d) Barbey, R., Klok, H. A. (2010). Room temperature, aqueous post-polymerization modification of glycidyl methacrylate-containing polymer brushes prepared via surface-initiated atom transfer radical polymerization. *Langmuir*, *26*, 18219–18230; (e) Liu, Y., Klep, V., Zdyrko, B., Luzinov, I. (2004). Polymer grafting via ATRP initiated from macroinitiator synthesized on surface. *Langmuir*, *20*, 6710–6718; (f) Hall, D. J., Berghe, H. M. V. D., Dove, A. P. (2011). Synthesis and post-polymerization modification of maleimide-containing polymer by 'thiol-ene' and Diels-Alder chemistries. *Polym. Int.*, *60*, 1149–1157; (g) Hoyle, C. E., Lowe, A. B., Bowman, C. N. (2010). Thiol-ene click chemistry: a multifaceted toolbox for small molecule and polymer synthesis. *Chem. Soc. Rev.*, *39*, 1355–1387.
5. (a) Rzayev, Z. M. O., Turk, M., Uzgoren, A. (2010). Bioengineering functional copolymers. XV. Synthesis and characterization of poly(N-isopropyl acrylamide-co-3,4.dihydro-2H–pyran-alt-maleic anhydride)s and their PEO branched derivatives. *J. Polym. Sci. Part A: Polym. Chem.*, *48*, 4285–4295; (b) Jiang, X., Cui, L., Wang, H., Xu, K., Zhou, C., Li, J. (2009). Synthesis and electrostatic assembly of an optically active derivative of poly(ethylene-*alt*-maleic anhydride) with L-phenylalanine as chiral material. *Reac. Func. Polym.*, *69*, 619–622; (c) Donati, I., Gamini, A., Vetere, A., Campa, C., Paoletti, S. (2002). Synthesis, characterization and preliminary biological study of glycoconjugates of polt(styrene-*co*-maleic acid). *Biomacromolecules*, *3*, 805–812; (d) Stoilova, O., Ignatova, M., Manolova, N., Godjevargova, T., Mita, D. G., Rashkov, I. (2010). Functionalized electrospun mats from styrene-maleic anhydride copolymers for immobilization of acetylcholinesterase. *Eur. Polym. J.*, *46*, 1966–1974; (e) Parthiban, A. (2009). Water swellable and water soluble polymers and use thereof. *PCT Int. Appl.* WO 2009154568 A1 20091223; (f) Parthiban, A., Yu, H., Chai, C. L. L. (2009). Amphiphilic water and organo soluble grafted copolymer. *Polym. Prepr.*, *50*(*2*), 428–429; (g) Parthiban, A. (2011). Water swellable and water soluble polymers and use thereof, *US pat. pub. no. 2011*0092636; (h) Parthiban, A. (2011). Water swellable and water soluble polymers and use thereof, SG167476.

6. (a) Tully, D. C., Roberts, M. J., Geierstanger, B. H., Grabbs, R. B. (2003). Synthesis of reactive poly(vinyl oxazolones) via nitroxide-mediated living free radical polymerization. *Macromolecules*, *36*, 4302–4308; (b) Heilmann, S. M., Rasmussen, J. K., Krepski, L. R. (2001). Chemistry and technology of 2-alkenyl azlactones, *J. Polym. Sci. Part A: Polym. Chem.*, *39*, 3655–3677; (c) Fournier, D., Pascual, S., Fontaine, L. (2004). Copper-mediated living radical polymerization of 2-vinyl-4,4.dimethyl-5-oxazolone. *Macromolecules*, *37*, 330–335; (d) Fournier, D., Pascual, S., Montembault, V., Fontaine, L. (2006). Elaboration of well-defined Rasta resins and their use as supported catalytic systems for atom transfer radical polymerization. *J. Polym. Sci. Part A: Polym. Chem.*, *44*, 5316–5328.
7. (a) Moraes, J., Maschmeyer, T., Perrier, S. (2011). "Clickable" polymers via combination of RAFT polymerization and isocyanate chemistry. *J. Polym. Sci. Part A: Polym. Chem.*, *49*, 2771–2782; (b) Hattermer, E., Zentel, R., Mecher, E., Meerholz, K. (2000). Synthesis and characterization of novel multifunctional high-T_g photorefractive materials obtained via reactive precursor polymers. *Macromolecules*, *33*, 1972–1977; (c) Dorr, M., Zentel, R., Dietrich, R., Meerholz, K., Brauchle, C., Wichern, J., Zippel, S., Boldt, P. (1998). Reactions on vinyl isocyanate/maleimide copolymers: NLO functionalized polymers with high glass transitions for nonlinear optical applications. *Macromolecules*, *31*, 1454–1465.
8. (a) Rieger, J., Butsele, K. V., Lecomte, P., Detrembleur, C., Jerome, R., Jerome, C. (2005). Versatile functionalization and grafting of poly(ϵ-caprolactone) by Michael-type addition. *Chem. Comm.*, 274–276; (b) Hoyle, C. E., Bowman, C. N. (2010). Thiol-ene click chemistry. *Angew. Chem. Int. Ed.*, *49*, 1540–1573; (c) Pounder, R. J., Stamford, M. J., Brooks, P., Richards, S. P., Dove, A. P. (2008). Metal free thiol-maleimide click reaction as a mild functionalization strategy for degradable polymers, *Chem. Comm.*, 5158–5160.
9. (a) Stranix, B. R., Gao, J. P., Barghi, R., Salha, J., Darling, G. D. (1997). Functional polymers from (vinyl)polystyrene. Short route to binding functional groups to polystyrene resin through a dimethylene spacer: bromine, sulfur, phosphorous, silicon, hydrogen, boron and oxygen. *J. Org. Chem.*, *62*, 8987–8993; (b) Justynska, J., Hordyjewicz, Z., Schlaad, H. (2005). Toward a toolbox of functional block copolymers via free-radical addition of mercaptans. *Polymer*, *46*, 12057–12064.
10. (a) Mansfield, U., Pietsch, C., Hoogenboon, R., Becer, R., Schubert, U. S. (2010). Clickable initiators, monomers and polymers in controlled radical polymerizations-a prospective combination in polymer science. *Polym. Chem.*, *1*, 1560–1598. (b) Harvison, M. A., Lowe, A. B. (2011). Combining RAFT radical polymerization and click/highly efficient coupling chemistries: a powerful strategy for the preparation of novel materials. *Macromol. Rapid Commun.*, *32*, 779–800; (c) Akeroyd, N., Klumperman, B. (2011). The combination of living radical polymerization and click chemistry for the synthesis of advanced macromolecular architectures. *Eur. Polym. J.*, *47*, 1207–1231; (d) Hong, M., Liu, J. Y., Li, B. X., Li, Y. S. (2011). Facile functionalization of polyethylene via click chemistry. *Macromolecules*, *44*, 5659–5665.
11. (a) Hoogenboom, R. (2010). Thiol-yne chemistry: a powerful tool for creating highly functional materials. *Angew. Chem. Int. Ed.*, *49*, 3415–3417; (b) Lowe, A. B., Hoyle, C. E., Bowman, C. N. (2010). Thiol-yne chemistry: a powerful and versatile methodology for materials synthesis. *J. Mater. Chem.*, *20*, 4745–4750.
12. (a) Christman, K. L., Maynard, H. D. (2005). Protein micropatterns using a pH-responsive polymer and light. *Langmuir*, *21*, 8389–8393; (b) Yang, S. K., Weck, M.

(2008). Modular covalent multifunctionalization of copolymers. *Macromolecules*, *41*, 346–351.

13. D'alelio, G. F., Huemmer, T. (1967). Preparation and polymerization of some vinyl monomers containing 2-oxo-1,3-dioxolane group. *J. Polym. Sci. Part A1*, *5*, 307–321.
14. (a) Sonnati, M. O., Amigoni, S., Taffin de Givenchy, E. P. T., Darmanin, T., Choulet, O., Guittard, F. (2013). Glycerol carbonate as a versatile building block for tomorrow: Synthesis, reactivity, properties and applications. *Green Chem.*, *15*, 283–306; (b) Dibenedtto, A., Angelini, A., Aresta, M., Ethiraj, J., Fragale, C., Nocito, F. (2011). Converting wastes into added value products: from glycerol to glycerol carbonate, glycidol and epichlorohydrine using environmentally friendly synthetic routes. *Tetrahedron*, *67(7)*, 1308–1313 and references therein.
15. Climent, M. J., Corma, A., Frutos, P. D., Iborra, S., Noy, M., Veltry, A., Concepcion, P. (2010). Chemicals from biomass: synthesis of glycerol carbonate by transesterification and carbonylation with urea with hydrotalcite catalysts. The role of acid-base pairs. *J. Catal.*, *269*, 140–149 and references therein.
16. (a) Doro, F., Winnertz, P., Leitner, W., Prokofieva, A., Müeller, T. E. (2011). Adopting a Wacker-type catalyst system to the palladium-catalyzed oxidative carbonylation of aliphatic polyols. *Green Chem.*, *13(2)*, 292–295 and references therein; (b) Hu, J., Gu, Y., Guan, Z., Mo, W., Li, T., Li, G. (2011). An efficient palladium catalyst system for the oxidative carbonylation of glycerol to glycerol carbonate. *ChemSusChem.*, *4*, 1767–1771.
17. (a) Katz, H. E. (1987). Preparation of soluble poly(carbonyldioxyglyceryl methacrylate). *Macromolecules*, *20*, 2026–2027; (b) Jansen, J. F. G. A., Dias, A. A., Dorschu, M., Coussens, B. (2003). Fast monomer: factors affecting the inherent reactivity of acrylate monomers in photoinitiated acrylate polymerization. *Macromolecules*, *36*, 3861–3873; (c) Kilambi, H., Beckel, E. R., Berchtold, K. A., Stansbury, J. W., Bowman, C. N. (2005). Influence of molecular dipole on monoacrylate monomer reactivity. *Polymer*, *46(13)*, 4735–4742; (d) Britz, J., Meyer, W. H., Wegner, G. (2007). Blend of poly(methacrylates) with 2-oxo-(1,3)dioxolane side chains and lithium slats as lithium ion conductors. *Macromolecules*, *40*, 7558–7565.
18. Feng, J. C., Hill, S. (1961). Polymerizable esters of acrylic and methacrylic acid and polymers thereof, *US patent*, 2967173.
19. Brosse, J. C., Couvret, D., Chevalier, S., Senet, J. P. (1990). Monomèrs acryliques à function carbonate cyclique, 1 synthèse at polymerization. *Makromol. Chem. Rapid Commun.*, *11*, 123–128.
20. Jana, S., Yu, H., Parthiban, A., Chai, C. L. L. (2010). Controlled synthesis and functionalization of PEGylated methacrylates bearing cyclic carbonate pendant groups. *J. Polym. Sci. Part A: Polym. Chem.*, *48*, 1622–1632.
21. (a) Brindopke, G. (1989). Process for preparation of 2-oxo-1,3-dioxolanes. *US patent*, 4835289; (b) Kihara, N., Endo, T. (1992). Synthesis and reaction of polymethacrylate bearing cyclic carbonate moieties in the side chain. *Makromol. Chem.*, *193*, 1481–1492; (c) Golden, J. H., Chew, B. G. M., Zax, D. B., DiSalvo, F. J., Frechet, J. M. J., Tarascon, J. M. (1995). Preparation of propylene carbonate acrylate and poly(propylene carbonate acrylate) electrolyte elastomer gels. ^{13}C NMR evidence for Li+-cyclic carbonate interaction. *Macromolecules*, *28*, 3468–3470.
22. (a) Shibata, I., Mitani, I., Imakuni, A., Baba, A. (2011). Highly efficient synthesis of cyclic carbonates from epoxides catalyzed by indium tribromide system. *Tetrahedron*

Lett., *52*(*6*), 721–723; (b) Wong, W. L., Cheung, K. C., Chan, P. H., Zhou, Z. Y., Lee, K. H., Wong, K. Y. (2007). A tricarbonyl rhenium(I) complex with a pendant pyrrolidinium moiety as a robust and recyclable catalyst for chemical fixation of carbon dioxide in ionic liquid. *Chem. Commun.*, *21*, 2175–2177; (c) Jin, L., Jing, H., Chang, T., Bu, X., Wang, L., Liu, Z. (2007). Metal porphyrin/phenyltrimethylammonium tribromide: high efficient catalysts for coupling reaction of CO_2 and epoxides. *J. Mol. Catal. A: Chem.*, *261*(2), 262–266; (d) Kossev, K., Koseva, N., Troev, K. (2003). Calcium chloride as co-catalyst of onium halides in the cycloaddition of carbon dioxide to oxiranes. *J. Mol. Catal A: Chem.*, *194*, 29–37; (e) Baba, A., Nozaki, T., Matsuda, H. (1987). Carbonate formation from oxiranes and carbon dioxide catalyzed by organotin halide-tetraalkylphophonium halide complexes, *Bull. Chem. Soc. Japan.*, *60*, 1552–1554.

23. (a) Park, S. W., Choi, B. S., Park, D. W., Oh, K. J., Lee, J. W. (2007). Chemical kinetics of carbon dioxide with glycidyl methacrylate using immobilized tributylamine supported on poly(styrene-co-vinylbenzyl chloride) as catalyst. *Green Chem.*, *9*, 605–610 and references therein; (b) Chang, T., Jing, H., Jin, L., Qiu, W. Quaternary onium tribromide catalyzed cyclic carbonate synthesis from carbon dioxide and epoxides. *J. Mol. Catal. A: Chem.*, *264*, 241–247; (c) Park, D. W., Yu, B. S., Jeong, E. S., Kim, I., Kim, M. I., Oh, K. J., Park, S. W. (2004). Comparative studies on the performance of immobilized quaternary ammonium salt catalyst for the addition of carbon dioxide to glycidyl methacrylate. *Catal. Today*, *98*, 499–504; (d) Caló, V., Nacci, A., Monopoli, A., Fanizzi, A. (2002). Cyclic carbonate formation from carbon dioxide and oxiranes in tetrabutylammonium halides as solvents and catalysts. *Org. Lett.*, *4*(*15*), 2561–2563.
24. Barkakaty, B., Morino, K., Sudo, A., Endo, T. (2010). Amidine-mediated delivery of CO_2 from gas phase to reaction system for highly efficient synthesis of cyclic carbonates from epoxides. *Green Chem.*, *12*, 42-44.
25. (a) Decker, C., Moussa, K. (1990). A new class of highly reactive acrylic monomers, 1 light-induced polymerization. *Makromol. Chemie, Rapid Commun.*, *11*, 159–167; (b) Burgard, M., Rollat, A., Piteau, M., Senet, J. P. (1980). *Eur. Pat. App.*, EP 18259 A1 19801029; (c) Finger, W. J., Lee, K. S., Podszun, W. (1996). Monomers with low oxygen inhibition as enamel/dentin adhesives, *Dent. Mater.*, *12*, 256–261.
26. (a) Beckel, E. R., Stansbury, J. W., Bowman, C. N. (2005). Evaluation of potential ionic contribution to the polymerization of highly reactive (meth)acrylate monomers. *Macromolecules*, *38*, 9474–9481 (see supporting information); (b) Ochiai, B., Ootani, Y., Maruyama, T., Endo, T. (2007). Synthesis and properties of polymethacrylate bearing cyclic carbonate through urethane linkage. *J. Polym. Sci. Part A: Polym. Chem.*, *45*, 5781–5789.
27. Barkakaty, B., Morino, K., Sudo, A., Endo, T. (2011). Synthesis of methacrylic monomer having pendant cyclic carbonate-easy CO_2 fixation and radical polymerization. *J. Polym. Sci. Part A: Polym. Chem.*, *49*, 545–549.
28. Bissinger, W. E., Fredenburg, R. H., Kadesch, R. G., Kung, F., Langston, J. H., Stevens, H. C., Strain, F. (1947). Some reactions of butadiene monochlorohydrine, 1-chloro-3-buten-2-ol. *J. Am. Chem. Soc.*, *69*, 2955–2961.
29. Prichard, W. W. (1950). Vinylethylene carbonate and its preparation. *US patent*, 2511942.
30. Ochiai, B., Matsuki, M., Nagai, D., Miyagawa, T., Endo, T. (2005). Radical polymerization behavior of a vinyl monomer bearing five-membered cyclic carbonate structure and reactions of the obtained polymers with mines. *J. Polym. Sci. Part A: Polym. Chem.*, *43*, 548–592.

31. Newman, M. S., Addor, R. W. (1955). Synthesis and reaction of vinylene carbonate. *J. Am. Chem. Soc.*, *77*, 3789–3793.

32. (a) Nishikubo, T., Kameyama, A., Yamashita, J., Tomoi, M., Fukuda, W. (1993). Insoluble polystyrene-bound quaternary onium salt catalysts for the synthesis of cyclic carbonates by the reaction of oxiranes with carbon dioxide. *J. Polym. Sci. Part A: Polym. Chem.*, *31*, 939–947; (b) Nishikubo, T., Kameyama, A., Sasano, M. (1994). Synthesis of functional polymers bearing cyclic carbonate groups from (2-oxo-1,3-dioxolan-4.yl)methyl vinyl ether. *J. Polym. Sci. Part A: Polym. Chem.*, *32*, 301–308; (c) Moon, J. Y., Jang, H. J., Kim, K. H., Park, D. W., Ha, C. S., Lee, J. K. (2000). Synthesis of poly[(2-oxo-1,3-dioxolan-4.yl) methyl vinyl ether-*co*-*N*-phenylmaleimide and its miscibility in blend with styrene-acrylonitrile or poly(vinyl chloride). *J. App. Polym. Sci.*, *77*, 1809–1815.

33. (a) Welker, M. E., Franks, M. A. (2002). Diels-Alder adducts of epoxybutene and epoxybutene derivatives. *US patent*, 6380447; (b) Sudo, A., Morishita, H., Endo, T. (2010). Synthesis of norborane monomer having cyclic carbonate moiety based on CO_2 fixation and its transition metal-catalyzed polymerizations. *J. Polym. Sci. Part A: Polym. Chem.*, *48*, 3896–3902.

34. Miyata, T., Matsumoto, K., Endo, T., Yonemori, S., Watanabe, S. (2012). Synthesis and radical polymerization of styrene-based monomer having five membered cyclic carbonate structure. *J. Polym. Sci. Part A: Polym. Chem.*, *50*, 3046–3051.

35. (a) Ochiai, B., Endo, T. (2005). Carbon dioxide and carbon disulfide as resources for functional polymers. *Prog. Polym. Sci.*, *30*, 183–215 and references therein; (b) Darensbourg, D. J. (2010). Chemistry of carbon dioxide relevant to its utilization: a personal perspective. *Inorg. Chem.*, *23*, 10765–10780.

36. (a) Wendler, K., Fedtke, M., Pabst, S. (1993). The copolymerization of styrene with the cyclic carbonate of glycidyl methacrylate. *Die Angew. Makromol. Chem.*, *213*, 65–72; (b) Schneider, K., Neerman, H. (1978). *European Patent*, 1088.

37. Webster, D. C., Crain, A. L. (2000). Synthesis and applications of cyclic carbonate functional polymers in thermosetting coatings. *Prog. Org. Coat.*, *40*, 275–282 and references therein.

38. Webster, D. C., Crain A. L. *Functional Polymers: Modern Synthetic Methods and Novel Structures*. In: Patil, A. O., Schulz, D. N., Novak, B. M., editors. American Chemical Society, Washington, DC, 1998, *704*, pp. 303–320.

39. Turner, S. R., Mackenzie, P. B., Jones, A. S., McDevitt, J. P., Killian, C. M., Ponasik Jr., J. A. (2000). Polymers containing functionalized olefin monomers. *US patent*, 6090900.

40. (a) Field, N. D., Schaefgen, J. R. (1962). High molecular weight poly(vinylene carbonate)and derivatives. *J. Polym. Sci.*, *58*, 533–543; (b) Hayashi, H., Smets, G. Copolymerization of vinylene carbonate. *J. Polym. Sci.*, *27*, 275–283; (c) Judge, J. M., Price, C. C. (1959). The copolymerization characteristics of vinylene carbonate, γ-crotonolactone and methyl bicycle(2,2,1)-2-heptene-5-carboxylate. *J. Polym. Sci.*, *41*, 435–443; (d) Wei, X., Shriver, D. F. (1998). Highly conductive polymer electrolytes containing rigid polymer. *Chem. Mater.*, *10*, 2307–2308; (e) Ding, L., Li, Y., Liang, Y., Huang, J. (2001). Polymerization of vinylene carbonate as well as aminolysis and hydrolysis of poly(vinylene carbonate). *Eur. Polym. J.*, *37*, 2453–2459.

41. Ha, C. S., Yoo, G., Park, D. W., Jo, N. J., Lee, J. K., Cho, W. J. (2002). Synthesis of poly{[(2-oxo-1,3-dioxolan-4.yl)methyl vinyl ether]-co-acrylonitrile} and its miscibility with SAN. *Polym. Int.*, *52*, 1023–1030.

42. (a) Decker, C., Moussa, K. (1991). A new class of highly reactive monomers, light-induced copolymerization with difunctional oligomers. *Makromol. Chem.*, *192*, 507–522; (b) Decker, C., Moussa, K. (1991). Photopolymerisation de monomers multifonctionnels-V. Resines polyurethannes-acrylates. *Eur. Polym. J.*, *27*, 881–889; (c) Moussa, K., Decker, C. (1993). Light-induced polymerization of new highly reactive acrylic monomers. *J. Polym. Sci. Part A: Polym. Chem.*, *31(9)*, 2197–2203.
43. (a) Kilambi, H., Stansbury, J. W., Bowman, C. N. (2007). Deconvoluting the impact of intermolecular interactions on the polymerization kinetics of ultrarapid mono(meth)acrylates. *Macromolecules*, *40(1)*, 47–54; (b) Kilambi, H., Reddy, S. K., Beckel, E. R., Stansbury, J. W., Bowman, C. N. (2007). Influence of secondary functionalities on the reaction behavior of monovinyl (meth)acrylates. *Chem. Mater.*, *19(4)*, 641–643; (c) Berchtold, K. A., Nie, J., Stansbury, J. W., Bowman, C. N. (2008). Reactivity of monovinyl (meth)acrylates containing cyclic carbonates. *Macromolecules*, *41(23)*, 9035–9043.
44. Kilambi, H., Reddy, S. K., Schneidewind, L., Stansbury, J. W., Bowman, C. N. (2007). Copolymerization and dark polymerization studies for photopolymerization of novel acrylic monomers. *Polymer*, *48(7)*, 2014–2021.
45. (a) Kihara, N., Endo, T. (1992). Incorporation of carbon dioxide into poly(glycidyl methacrylate). *Macromolecules*, *25*, 4824–4825; (b) Kihara, N., Endo, T. (1994). Solid-state catalytic incorporation of carbon dioxide into oxirane-polymer. Conversion of poly(glycidyl methacrylate) to carbonate-polymer under atmospheric pressure. *J. Chem. Soc. Chem. Commun.*, 937–938; (c) Kihara, N., Endo, T. (1994). Self-catalyzed carbon dioxide incorporation system. The reaction of copolymers bearing epoxide and a quarternary ammonium group with carbon dioxide. *Macromolecules*, *27*, 6239–6244; (d) Sakai, T., Kihara, N., Endo, T. (1995). Polymer reaction of epoxide and carbon dioxide. Incorporation of carbon dioxide into epoxide polymers. *Macromolecules*, *28*, 4701–4706; (e) Yamamoto, S. I., Hayashi, T., Kawabata, K., Moriya, O., Endo, T. (2002). Fixation of carbon dioxide into polysilsesquioxane containing glycidyl groups. *Chem. Lett.*, 816–817; (f) Yamamoto, S., Moriya, O., Endo, T. (2003). Efficient fixation of carbon dioxide into poly(glycidyl methacrylate) containing pendant crown ether. *Macromolecules*, *36*, 1514–1521.
46. (a) Ochiai, B., Iwamoto, T., Miyagawa, T., Nagai, D., Endo, T. (2004). Solid-phase incorporation of gaseous carbon dioxide into oxirane-containing copolymers. *J. Polym. Sci. Part A: Polym. Chem.*, *42*, 3812–3817; (b) Ochiai, B., Iwamoto, T., Miyagawa, T., Nagai, D., Endo, T. (2004). Direct incorporation of gaseous carbon dioxide into solid-state copolymer containing oxirane and quaternary ammonium halide structure as self-catalytic function. *J. Polym. Sci. Part A: Polym. Chem.*, *42*, 4941–4947; (c) Park, S. Y., Park, H. Y., Lee, H. S., Park, S. W., Ha, C. S., Park, D. W. (2001). Synthesis of poly[(2-oxo-1,3-dioxolane-4.yl)methyl methacrylate-co-ethyl acrylate] by incorporation of carbon dioxide into epoxide polymer and the miscibility behavior of its blends with poly(methyl methacrylate) or poly(vinyl chloride). *J. Polym. Sci. Part A: Polym. Chem.*, *39*, 1472–1480; (d) Park, S. Y., Lee, H. S., Ha, C. S., Park, D. W. (2001). Synthesis of poly[(2-oxo-1,3-dioxolane-4.yl)methyl methacrylate by polymer reaction of carbon dioxide and miscibility of its blends with copolymers of methyl methacrylate and ethyl acrylate. *J. Polym. Sci. Part A: Polym. Chem.*, *39*, 2161–2169; (e) Park, S. Y., Park, H. Y., Lee, H. S., Park, S. W., Ha, C. S., Park, D. W. (2002). *J. Macromol. Sci., Pure Appl. Chem.*, *39*, 573–589.
47. (a) Park, S. Y., Park, H. Y., Woo, H. S., Ha, C. S., Park, D. W. (2002). Synthesis of poly[(2-oxo-1,3-dioxolane-4.yl)methyl methacrylate-co-styrene] by addition reaction

of carbon dioxide and its compatibility with poly(methyl methacrylate) or poly(vinyl chloride). *Polym. Adv. Tech.*, *13*, 513–521; (b) Yamamoto, S. I., Moriya, O., Endo, T. (2005). Ring size effect of crown ether on the fixation of carbon dioxide into an oxirane polymer. *Macromolecules*, *38*, 2154–2158; (c) Ochiai, B., Endo, T. (2007). Polymer-supported pyridinium catalysts for synthesis of cyclic carbonate by reaction of carbon dioxide and oxirane. *J. Polym. Sci. Part A: Polym. Chem.*, *45*, 5673–5678.

48. (a) Yamamoto, S. I., Kawabata, K., Moriya, O., Endo, T. (2005). Effective fixation of carbon dioxide into poly(glycidyl methacrylate) in the presence of pyrrolidone polymers. *J. Polym. Sci. Part A: Polym. Chem.*, *43*, 4578–4585; (b) Ochiai, B., Iwamoto, T., Miyazaki, K., Endo, T. (2005). Effective gas-solid phase reaction of atmospheric carbon dioxide into copolymers with pendant oxirane groups: effect of comonomer component and catalyst on incorporation behavior. *Macromolecules*, *38*, 9939–9943; (c) Ochiai, B., Iwamoto, T., Endo, T. (2006). Selective gas-solid phase fixation of carbon dioxide into oxirane-containing polymers: synthesis polymer bearing cyclic carbonate group. *Green Chem.*, *8*, 138–140.
49. (a) Ochiai, B., Hatano, Y., Endo, T. (2008). Fixing carbon dioxide concurrently with radical polymerization for utilizing carbon dioxide by low-energy cost. *Macromolecules*, *41*, 9937–9939; (b) Ochiai, B., Hatano, Y., Endo, T. (2009). Facile synthesis of polymers bearing cyclic carbonate structure through radical solution and precipitation polymerizations accompanied by concurrent carbon dioxide fixation. *J. Polym. Sci. Part A: Polym. Chem.*, *47*, 3170–3176.
50. (a) Couvret, D., Brosse, J. C., Chevalier, S., Senet, J. P. (1990). Monomèrs acryliques à function carbonate cyclique, modification chimique de copolymères à groupements carbonate cyclique lateraux. *Makromol. Chem.*, *191*, 1311–1319; (b) Steblyanko, A., Choi, W., Sanda, F., Endo, T. (2000). Addition of five-membered cyclic carbonate with amine and its application to polymer synthesis. *J. Polym. Sci. Part A: Polym. Chem.*, *38*, 2375–2380.
51. (a) Ochiai, B., Inoue, S., Endo, T. (2005). Salt effect on polyaddition of bifunctional cyclic carbonate and diamine. *J. Polym. Sci. Part A: Polym. Chem.*, *43*, 6282–6286; (b) Ochiai, B., Satoh, Y., Endo, T. (2005). Nucleophilic polyaddition in water based on chemo-selective reaction of cyclic carbonate with amine. *Green Chem.*, *7*, 765–767; (c) Ochiai, B., Satoh, Y., Endo, T. (2009). Polyaddition of bifunctional cyclic carbonate with diamine in ionic liquids: in situ ionic composite formation and simple separation of ionic liquid. *J. Polym. Sci. Part A: Polym. Chem.*, *47*, 4629–4635 and references therein; (d) Kihara, N., Endo, T. (1993). Synthesis and properties of poly(hydroxyurethane)s. *J. Polym. Sci. Part A: Polym. Chem.*, *31*, 2765–2773.
52. (a) Ochiai, B., Nakayama, J. I., Mashiko, M., Kaneko, Y., Nagasawa, T., Endo, T. Synthesis and crosslinking reaction of poly(hydroxyurethane) bearing a secondary amine structure in the main chain. *J. Polym. Sci. Part A: Polym. Chem.*, *43*, 5899–5905; (b) Ochiai, B., Inoue, S., Endo, T. (2005). One-pot non-isocyanate synthesis of polyurethanes from bisepoxide, carbon dioxide and diamine. *J. Polym. Sci. Part A: Polym. Chem.*, *43*, 6613–6618.
53. Akkapeddi, M. K., Reimschuessel, H. K. Stereochemical studies on poly(hydroxymethylene) and poly(phenylvinylene glycol). *Macromolecules*, *11*, 1067–1074.
54. Park, S. Y., Park, H. Y., Lee, H. S., Park, S. W., Park, D. W. (2002). Synthesis and application of terpolymer bearing cyclic carbonate and cinnamoyl groups. *Opt. Mater.*, *21*, 331–335.

55. Sakakura, T., Choi, J. C., Yasuda, H. (2007). Transformation of carbon dioxide. *Chem. Rev.*, *107*, 2365–2387.
56. Webster, D. C., Crain, A. L. (2000). Proceedings of the International Waterborne, High-Solids, and Powder Coatings Symposium, Eastman Chemical Company, Kingsport, TN, University of Southern Mississippi, Department of of Polymer Science, *27*, 240–253, 2000:824765 CAN134:297211 CAPLUS.
57. Webster, D. C. (2002). Powder coatings from cyclic carbonate functional polymers and amine carbamate salts. *US patent*, 6339129.
58. (a) Stanssens, D., Elshout, W. V. D., Tijssen, P. (1997). *Belgium Patent*, 1 009 543; (b) Harui, N., Iwamura, G., Kumada, H. (1990). *Japan Patent*, 02 053 880; (c) Just, C., Dürr, H., Brindopke, G. (1988). Curable mixture and their use. *US patent*, 4772666; (d) Wamprecht, C., Blum, H., Pedain, J. (1991). Moisture-hardening binder compositions containing copolymers and blocked polyamines. *US patent*, 5045602; (e) Iwamura, G., Kinoshita, H., Kometani, A. (1994). Thermosetting resin composition. *US patent*, 5374699; (f) Iwamura, G., Kinoshita, H., Kometani, A. (1989). Thermosetting resin composition, *US patent*, 5393855; (g) Brindopke, G., Hoenel, M. (1989). Polymers containing amino groups, their preparation and their use. *US patent*, 4882391.
59. (a) Rabasco, J. J., Smith, C. D., Bott, R. H. (2003). Water based emulsion copolymers incorporating vinyl ethylene carbonate. *US patent*, 6593412; (b) Rabasco, J. J., Smith, C. D., Bott, R. H. (2004). Water based emulsion copolymers incorporating vinyl ethylene carbonate. *US patent*, 6756438; (c) Ramesh, S., Lessek, P., Bremser, W. (2002). Water-based coating composition having carbonate-amine cross-linking, method of preparing the same and a cured film thereof. *US patent*, 6403709.
60. Jana, S., Parthiban, A., Chai, C. L. L. (2010). Transparent, flexible and highly conductive ion gels from ionic liquid compatible cyclic carbonate network. *Chem. Commun.*, *46*, 1488–1490.
61. (a) Chen, G., Does, L. V. D., Bantjes, A. (1993). Investigations on vinylene carbonate. V. Immobilization of alkaline phosphatase onto LDPE films cografted with vinylene carbonate and *N*-vinyl-*N*-methylacetamide. *J. Polym. Sci. Part A: Polym. Chem.*, *47*, 25–36; (b) Chen, G., Does, L. V. D., Bantjes, A. (1993). Investigations on vinylene carbonate. VI. Immobilization of alkaline phosphatase onto poly(vinylene carbonate)-jeffamine hydrogel beads. *J. Polym. Sci. Part A: Polym. Chem.*, *48*, 1189–1198.
62. (a) Ding, L., Qu, B. (2001). New supports for enzyme immobilization based on the copolymers of poly(vinylene carbonate) and α-(2-aminoethylene amino)-ω-(2-aminoethylene amino)-poly(ethylene oxide). *React. Func. Polym.*,*49*(*1*), 67–76; (b) Ding, L., Li, Y., Jiang, Y., Cao, Z., Huang, J. (2002). New supports for enzyme immobilization based on copolymers of vinylene carbonate and β-hydroxyethylene acrylate. *J. App. Polym. Sci.*, *83* (*1*), 94–102.
63. Türünc, Q., Kahraman, M. V., Akdemir, Z. S., Apohan, N. K., Güngör, A. (2009). Immobilization of α-amylase onto cyclic carbonate bearing hybrid material. *Food Chem.*, *112*(*4*), 992–997.
64. Mauz, O., Noetzel, S., Sauber, K. (1988). Crosslinked polymers with carbonate esters groups, and a process for their preparation. *US patent*, 4767620.
65. Podszun, W., Krüger, J., Finger, W., Heiliger, L., Casser, C. (1998). Urethane (meth)acrylates containing cyclic carbonate groups. *US patent*, 5763622.

5

MONOMERS AND POLYMERS DERIVED FROM RENEWABLE OR PARTIALLY RENEWABLE RESOURCES

ANBANANDAM PARTHIBAN

5.1 BUILDING BLOCKS FROM RENEWABLE RESOURCES

Concerns of depleting fossil fuels and ever increasing emission of "global warming" gases such as CO_2 are two of the major driving forces behind the development of so-called bio-refinery processes and efforts to make, in particular, building blocks from renewable resources. As majority of chemicals, including monomers and polymers, are derived from one or two processes associated with the refining of crude oil, its depletion would naturally affect the continued supply of monomers and polymers as well as their price. Interestingly, until recently, efforts were directed mainly to derive fuels from renewable resources such as vegetable oils. As it emerged clear that such developments lead to the rise in cost of vegetable oils that are also consumed by humans, a conflict arose that shifted the source of renewable resource to nonedible matters such as algae; switch grass; stalks of food grains such as rice and wheat; food waste such as used cooking oil; fruit punch; and nonedible oils such as those extracted from the nuts of jatropha. Later it turned out that the renewable fuel resource required to substitute fossil fuels is so huge that space in terms of aerable land suitable for agriculture, and other resources such as water, nutrients in the form of fertilizer, and so on, required to make this feasible would bring enormous strain in existing

Synthesis and Applications of Copolymers, First Edition. Edited by Anbanandam Parthiban.

infrastructure. As a result, nowadays, more attention is devoted in the development of gas hydrates, the recovery of crude oil, and natural gases from sediments of rocks. Substantial efforts are also being made to enhance the recovery of oil from old or mature wells. Because of this, the attention turned toward other lesser requirements such as deriving chemicals from renewable resources. It may be noted that historically, the research activities on renewables increased on every occasion as the price of crude oil escalated because of disruptions in supply or production. However, of late, the efforts are more sustained and intense as major chemical companies not only have shifted and reoriented their line of business but also have announced major collaboration with biotech companies, which make use of renewable feed stocks for producing chemicals by bio-based processes. There exist many challenges in such ventures because of the contrasting nature of petroleum-based refining process and the bio-based approach. Fundamentally, the conventional refining process follows the "bottom-up" approach, whereby small molecules are converted to bigger and complex moieties through various chemical transformations. On the contrary, typically, in a bio-based process, complex compounds are broken down, predominantly, with the help of enzymes to form small molecules largely referred to as "building blocks" [1]. By nature, bio-based processes are volume inefficient, in particular, enzymatic processes. As a result, the so-called building blocks are present in extremely dilute solutions, and hence its separation is often challenging. Unlike the conventional refining processes that involve predominantly hydrocarbons, bio processes are largely based on compounds carrying large number of oxygen and other hetero atoms, which pose a challenge in subsequent metal-catalyzed chemical transformations [1b]. In spite of these challenges, there have been successful instances of making chemicals by bio-based processes. Renewable chemicals and building blocks are produced by three primary processes: biochemical processes such as fermentation; enzyme catalyzed thermochemical processes such as pyrolysis and gasification; and chemical catalysis [2]. Ethanol, butanol, 1,3-propane diol, lactic acid, and succinic acid [3] are some of the chemicals produced by fermentation. Syngas produced from biomass by thermochemical means can be converted to higher olefins and hydrocarbons. 1,2-Propylene glycol, ethylene glycol, and acrylic acid are some of the examples for raw materials obtained by chemical catalytic processes (Figure 5.1) [2]. Furfural is another example of raw material obtained by chemical-catalyzed process. For example, hydrolysis of biomass using diluted H_2SO_4 produces furfural and levulinic acids. Du Pont has utilized levulinic acid for making pyrrolidones and lactones [4]. Levulinic acid is also potentially useful for making compounds like bisphenol-A with a pendant carboxylic acid group at the β-carbon of the quarternary carbon flanked between two phenols at the *para*-position [4b]. Itaconic acid (IA) with two carboxylic acid groups, one of which is conjugated with an unsaturated bond can be regarded as α-substituted acrylic or methacrylic acid that can also be potentially produced by bioprocesses [5]. Before the advent of petroleum refining, biomass was used as a renewable source for making chemicals such as ethanol, butanol, acetic acid, citric acid, and

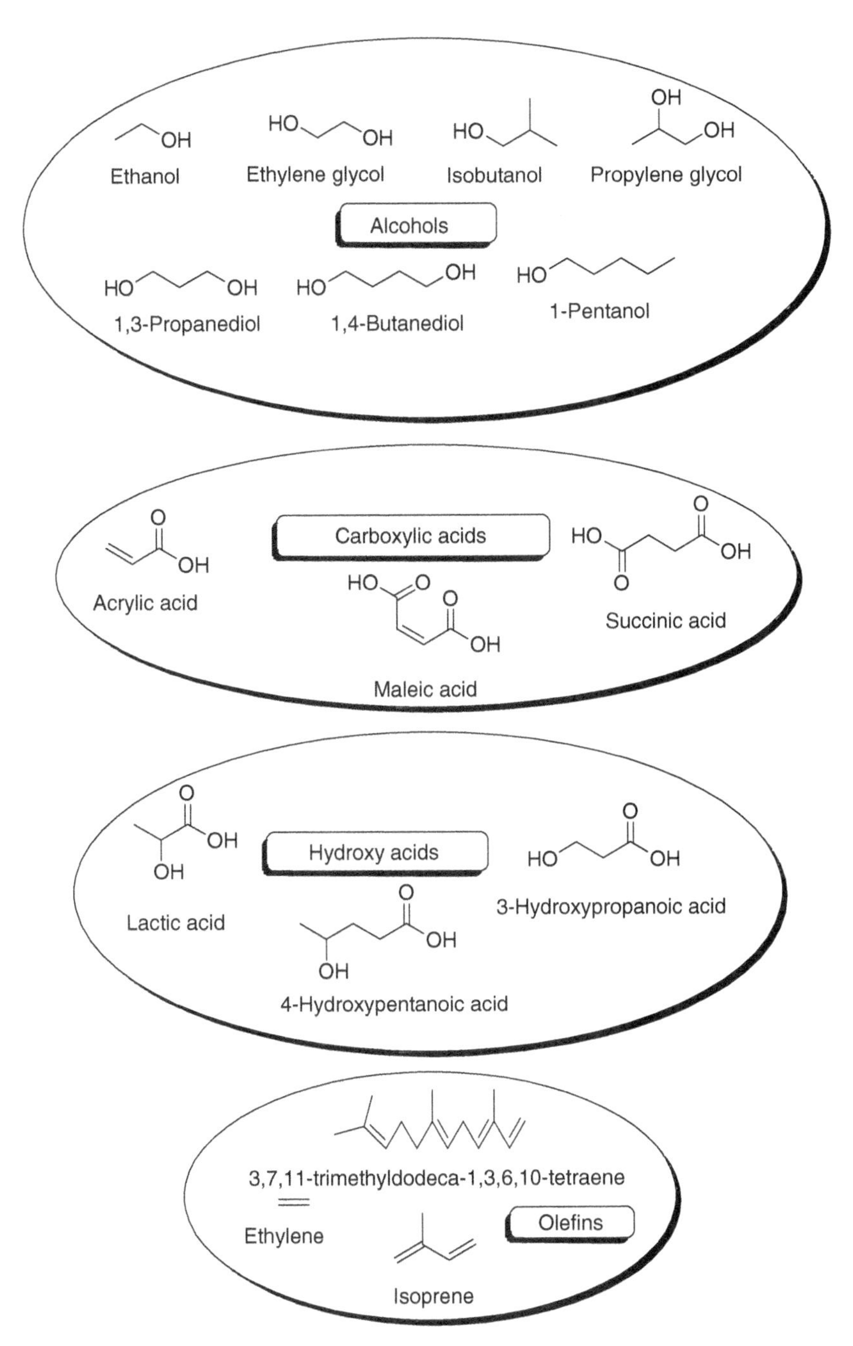

Figure 5.1 Selected examples of chemicals derived from renewable feedstocks.

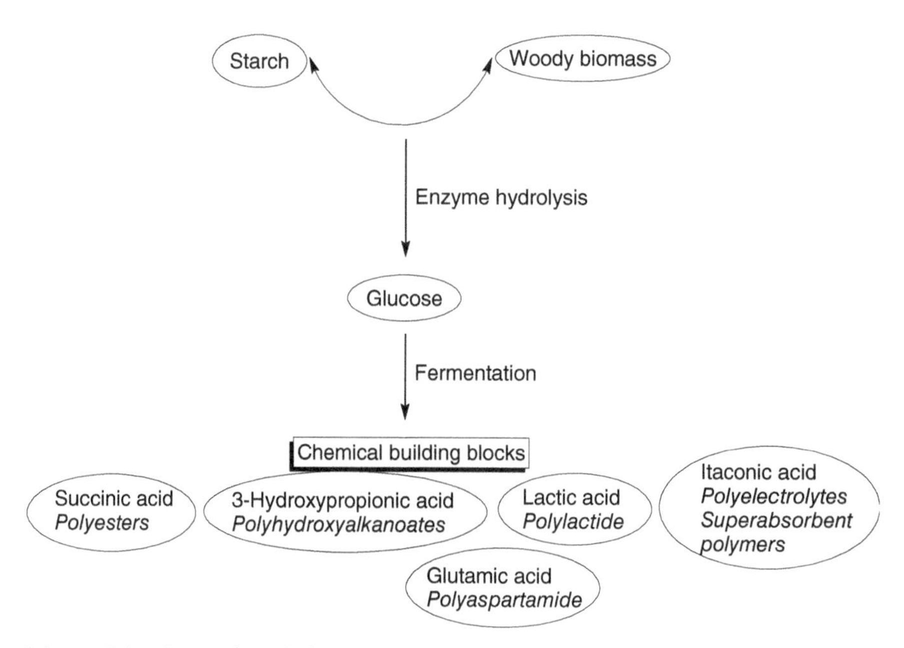

Scheme 5.1 Conversion of biomass to monomers. (Reprinted with permission from [4b]. Copyright © 2007 American Chemical Society.)

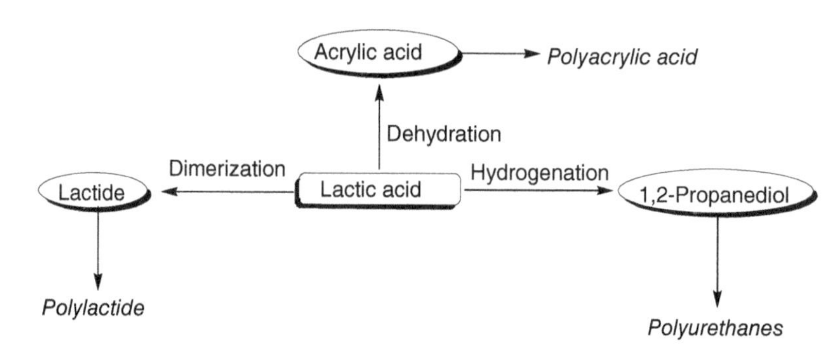

Scheme 5.2 Potential derivatives of lactic acid. (Reprinted with permission from [4b]. Copyright © 2007 American Chemical Society.)

lactic acid [6]. Schemes 5.1–5.4 describe the utility of bio-derived building blocks for further conversion to monomers.

2-Hydroxy isobutyric acid (2-HIBA) is a specialty chemical that is used as a pharmaceutical intermediate and also as a complexing agent for lanthanide- and actinide-based heavy metals. As there is no metabolic pathway existing for 2-HIBA, it has not been considered as a bio-derived building block. Octane-boosting fuel additive such as methyl *tert*-butyl ether (MTBE) could potentially produce 2-HIBA during biodegradation. However, the presence of *tert*-butyl group makes MTBE

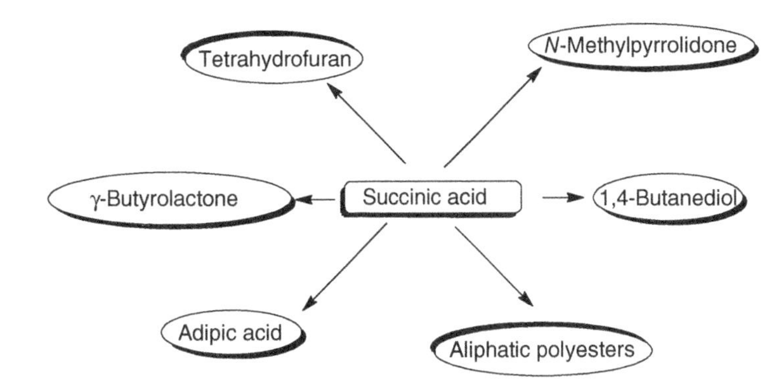

Scheme 5.3 Chemicals that can be derived from succinic acid [6].

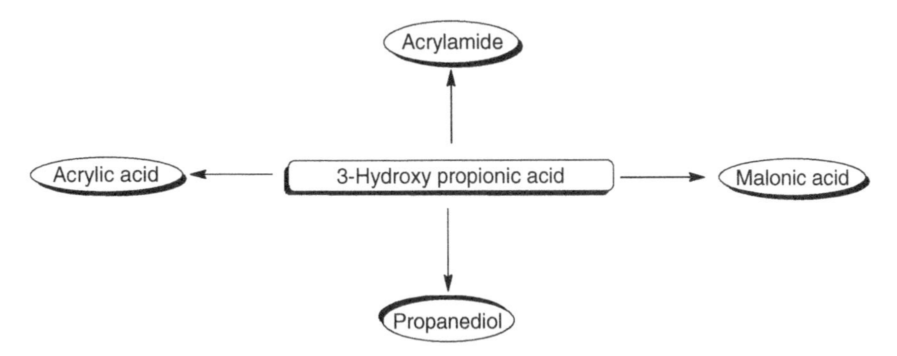

Scheme 5.4 Building block platform of 3-hydroxypropionic acid [6].

poorly biodegradable. Other processes that can potentially yield 2-HIBA are biooxidation of *tert*-butanol and bioisomerization of 3-HIBA to 2-HIBA. 2-HIBA is a useful building block molecule for various monomers as shown in Scheme 5.5 [7]. 2-HIBA has also been prepared by chemical route in four steps from bio-based lactic acid. 2-HIBA has been converted to tetramethylglycolide, which is proposed as a racemization free monomer. Tetramethylglycolide has been reportedly polymerized using *tert*-BuLi at 130 °C for 8 h to yield aliphatic polyester with characteristics such as $\langle M_n \rangle = 22{,}000$, $\langle M_w \rangle = 32{,}000$, glass transition temperature $T_g = 70.8$ °C, and melting transition $T_m = 190.7$ °C [8]. The presence of additional methyl group as compared to polylactide has resulted in increased backbone rigidity, leading to the formation of polymer whose T_g and T_m are well above that of polylactide.

5.2 POLYESTERS INCORPORATED WITH ISOSORBIDE

As isosorbide (1,4 : 3,6-dianhydro-D-glucitol) is produced from starch, it is one of the renewable resource monomers that is also available commercially. The two hydroxyl groups of isosorbide have been used for making numerous

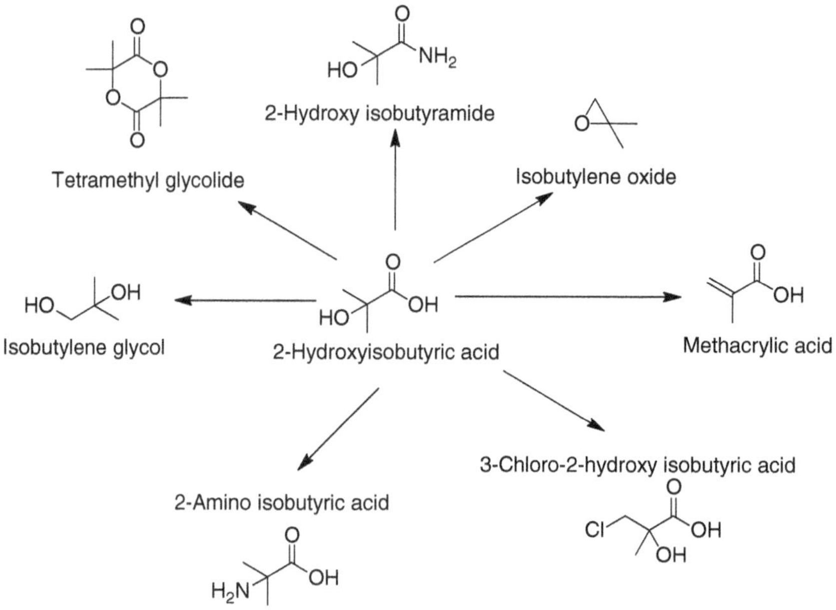

Tetramethyl glycolide
2-Hydroxy isobutyramide
Isobutylene oxide
Isobutylene glycol
2-Hydroxyisobutyric acid
Methacrylic acid
2-Amino isobutyric acid
3-Chloro-2-hydroxy isobutyric acid

Scheme 5.5 Chemicals potentially derived from 2-hydroxyisobutyric acid [7].

polyethers, polyesters, and polycarbonates since latter half of the 1990s [9]. Although isosorbide is useful for increasing the T_g of aromatic polyesters such as poly(butylene terephthalate), its incorporation normally leads to some undesired properties such as yielding colored aromatic polyesters, which are otherwise bright white solids in bulk and are capable of forming highly transparent materials upon processing into films. It also lowers the melting temperature and affects degree of crystallinity. In fact, when incorporated above 30 mol%, the crystallization is suppressed completely [10].

5.2.1 Poly(hydroxy ester)s Derived from Macrolides

Macrolides are lactones with ring size ≥12 atoms. Even though lactones of six- and seven-membered rings undergo ring-opening polymerization (ROP) chemically in a facile manner, macrolides are difficult to polymerize chemically. A catalytic route has recently been reported with poly(ω-pentadecalactone) PPDL of high polydispersity, 2.1–2.8 [11a]. However, macrolides are readily converted to high molecular weight polyesters when catalyzed by lipases [11b]. The polyesters prepared from macrolides such as PPDL have been proposed as suitable biodegradable alternatives for polyethylenes (PEs) of certain grades. PDLs are also a class of naturally occurring macrocyclic musks that are used in fragrance industry. However, in general, the melting point of poly(hydroxy acid)s are lower than that of PE with the exception of polyglycolic acid. Table 5.1 compares the melting point of various poly(hydroxy acid)s or lactones with PE [12].

TABLE 5.1 Comparison of Melting Point, T_m of Polylactones with PE

Polymer	T_m (°C)
Polyethylene	~120
Polyglycolic acid	230
Poly(3-hydroxy propionate)	70
Poly(4-hydroxy butyrate)	50
Poly(ε-caprolactone)	60
Poly(10-hydroxy capric acid)	80
Poly(ω-penta decalactone)	100

In order to replace PE in applications such as tapes and fibers, an attempt was made to prepare PPDL in high molecular weight. For the polymerization catalyzed by Novozyme 435, long reaction time, 72 h was required for polymerizing 34 g of monomer at 85 °C [13].

5.2.2 Semicrystalline Polymers from Fatty Acids

Dicarboxylic acids and diols were derived from fatty acids obtained from plant oils through various chemical transformations involving Pd-catalyzed carbonylation, hydrogenation, and so on (Scheme 5.6). The authors claim that this process is one of "complete feedstock molecule utilization." The produced diols and dicarboxylic acids have long methylene units of 16–22 between the two functional groups and thus yield semicrystalline polymers [14]. As in the case of macrolides, the T_m of semicrystalline polyesters is lower than that of linear PE.

5.2.3 Cyclic Ester Derived from a Natural Precursor

In yet another example of converting renewable compound into monomer through chemical transformations, menthol which is a naturally occurring terpene derivative was converted to a seven-membered lactone, namely, (−) menthide. (−) Menthide was subsequently polymerized to a high molecular weight polyester (Scheme 5.7) [15]. The ring strain of (−) menthide was approximately same as that of (ε-caprolactone). The rate of polymerization was significantly lower than that of lactide under similar conditions and the reaction followed first-order kinetics. As is usually the case for many of the cyclic ester polymerizations, (−) menthide also underwent inter- and intramolecular transesterification reactions.

5.2.4 Polymerization of Dilactone Derived from 12-Hydroxy Stearic Acid

13,26-Dihexyl-1,14-dioxacyclohexacosane-2,15 dione (12HSAD) is a dilactone possessing 12-hydroxy stearate sequence. This was subjected to homo- and copolymerization [16]. It may be noted that 12-hydroxy stearic acid is a long-chain

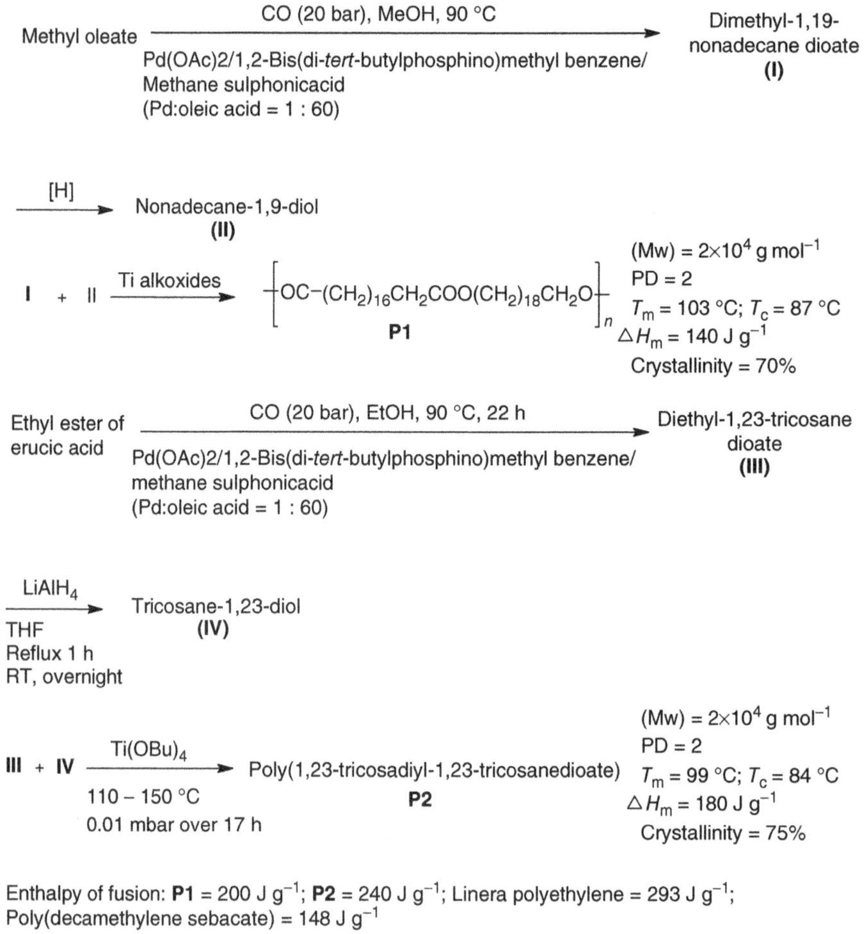

Methyl oleate —CO (20 bar), MeOH, 90 °C; Pd(OAc)2/1,2-Bis(di-*tert*-butylphosphino)methyl benzene/ Methane sulphonicacid (Pd:oleic acid = 1 : 60)→ Dimethyl-1,19-nonadecane dioate **(I)**

—[H]→ Nonadecane-1,9-diol **(II)**

I + **II** —Ti alkoxides→ $\left[OC{-}(CH_2)_{16}CH_2COO(CH_2)_{18}CH_2O \right]_n$ **P1**

(Mw) = $2{\times}10^4$ g mol^{-1}
PD = 2
T_m = 103 °C; T_c = 87 °C
ΔH_m = 140 J g^{-1}
Crystallinity = 70%

Ethyl ester of erucic acid —CO (20 bar), EtOH, 90 °C, 22 h; Pd(OAc)2/1,2-Bis(di-*tert*-butylphosphino)methyl benzene/ methane sulphonicacid (Pd:oleic acid = 1 : 60)→ Diethyl-1,23-tricosane dioate **(III)**

—$LiAlH_4$; THF, Reflux 1 h, RT, overnight→ Tricosane-1,23-diol **(IV)**

III + **IV** —$Ti(OBu)_4$; 110 – 150 °C, 0.01 mbar over 17 h→ Poly(1,23-tricosadiyl-1,23-tricosanedioate) **P2**

(Mw) = $2{\times}10^4$ g mol^{-1}
PD = 2
T_m = 99 °C; T_c = 84 °C
ΔH_m = 180 J g^{-1}
Crystallinity = 75%

Enthalpy of fusion: **P1** = 200 J g^{-1}; **P2** = 240 J g^{-1}; Linera polyethylene = 293 J g^{-1}; Poly(decamethylene sebacate) = 148 J g^{-1}

Scheme 5.6 Chemical conversion of plant-oil-derived unsaturated fatty acids to linear polyesters. (Reprinted with permission from [14]. Copyright © 2010 Wiley Periodicals Inc.)

fatty acid derived from castor oil. The polymerization of dilactone was found to be poor. Even when copolymerized with L-lactide (LL), high molecular weight polymer could be obtained only when 12HSAD content was low. It was also important to prepare the copolymers in a stepwise manner, with the polymerization of 12HSAD first using Et_2Zn followed by LL. The yield of copolymers went down with increasing content of 12HSAD. Also, the 12HSAD composition in the copolymer was lower than the monomer ratio in feed.

5.2.5 Thermoplastic Elastomers Derived from Polylactide and Polymenthide

ABA-type triblock copolymers function as thermoplastic elastomers when block A is made up of hard and or semicrystalline segment and block B is soft as well as amorphous. Due to the immiscible nature of A and B blocks,

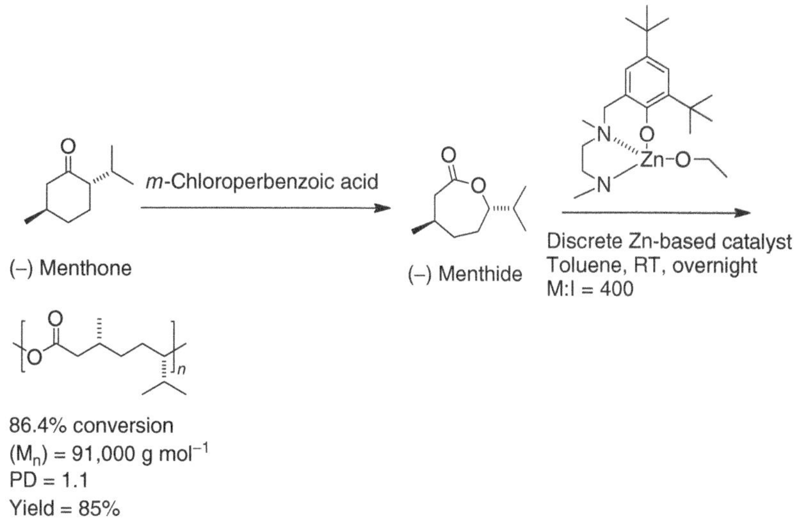

m-Chloroperbenzoic acid

(–) Menthone

(–) Menthide

Discrete Zn-based catalyst
Toluene, RT, overnight
M:I = 400

86.4% conversion
(M_n) = 91,000 g mol^{-1}
PD = 1.1
Yield = 85%

Scheme 5.7 Ring-opening polymerization of substituted lactone derived from a "cool" natural precursor. (Reprinted with permission from [15]. Copyright © 2005 American Chemical Society.)

microphase separation occurs where the soft, rubbery phase of block B is physically cross-linked by the semicrystalline and/or high T_g block. This interaction gives strength and elasticity to the triblock copolymer. Mechanical behavior such as stress–strain properties of these microphase-separated triblock copolymers is similar to that of vulcanized rubber with the added benefit of being processable. Among the renewable monomers, lactide is a good choice for forming semicrystalline polymers thus suitable for constructing A block in a ABA triblock copolymer. However, the elasticity and ductility that are fundamental requirements for a thermoplastic elastomer are poor for polylactide as it tends to fracture at very low strain. Thus, a suitable renewable resource monomer that can function as B block could combine well with polylactide to form completely biorenewable thermoplastic elastomer. In this regard, polymenthide (PM) as a suitable B block for ABA triblock copolymer has been reported [17]. PM is an amorphous polymer with a glass transition temperature of −25 °C. A physical blend of homopolymers of polylactide and PM of similar molecular weight are immiscible as the blend showed two T_g's corresponding to each homopolymer. A triblock copolymer of PL–PM–PL was prepared as shown in Scheme 5.8. PL–PM–PL triblock copolymers showed two T_g's in the differential scanning calorimetric analysis with the T_g of PM block at −22 °C and that of PL varied from 20 °C to 51 °C according to the molecular weight of PL block. The mechanical properties of renewable triblock copolymer were, in general, poorer than the commercial thermoplastic elastomer.

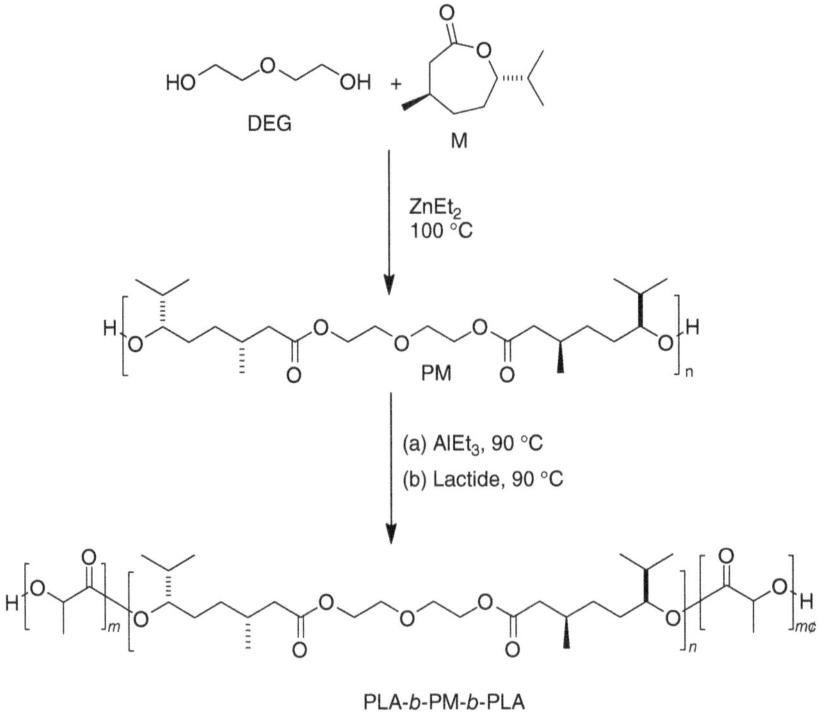

DEG

M

$ZnEt_2$
100 °C

PM

(a) $AlEt_3$, 90 °C
(b) Lactide, 90 °C

PLA-*b*-PM-*b*-PLA

Scheme 5.8 Synthesis of renewable resource thermoplastic elastomer, poly(lactide-*block*-menthide-*block*-lactide). (Reprinted with permission from [17]. Copyright © 2007 American Chemical Society.)

5.3 ROSIN AND DEVELOPMENTS ASSOCIATED WITH ROSIN

Rosin is produced from natural resources such as pine and conifer extrudates by distillation of tall oil produced in the Kraft pulp process. Chemically, rosin is a mixture of predominantly about 90% acidic and about 10% neutral compounds. Acid compounds of rosin, in turn, are a mixture of abietic and pimeric acids (Figure 5.2).

5.3.1 Polyamides and Polyesters Derived from Modified Levopimeric Acid

Levopimeric acid was converted to a diacid chloride in four steps by reactions involving Diels–Alder addition, dehydration, decarboxylation, base hydrolysis, and acid chloride formation as shown in Scheme 5.9 [18]. The diacid chloride was reacted with 1,2-diaminoethane and 1,4-butanediol to form polyamide and polyesters, respectively. Low molecular weight polymers were obtained and the thermal stability of polyester was better than that of polyamide.

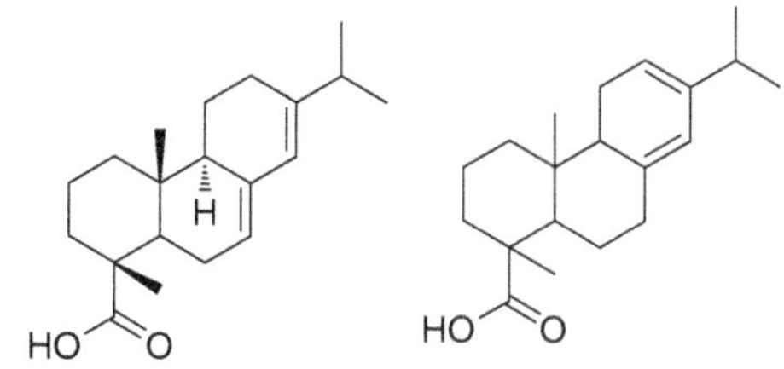
HO
O
H
HO
O

Abietic acid Pimeric acid

Figure 5.2 Rosin derivatives—chemical structure of abietic and pimeric acids.

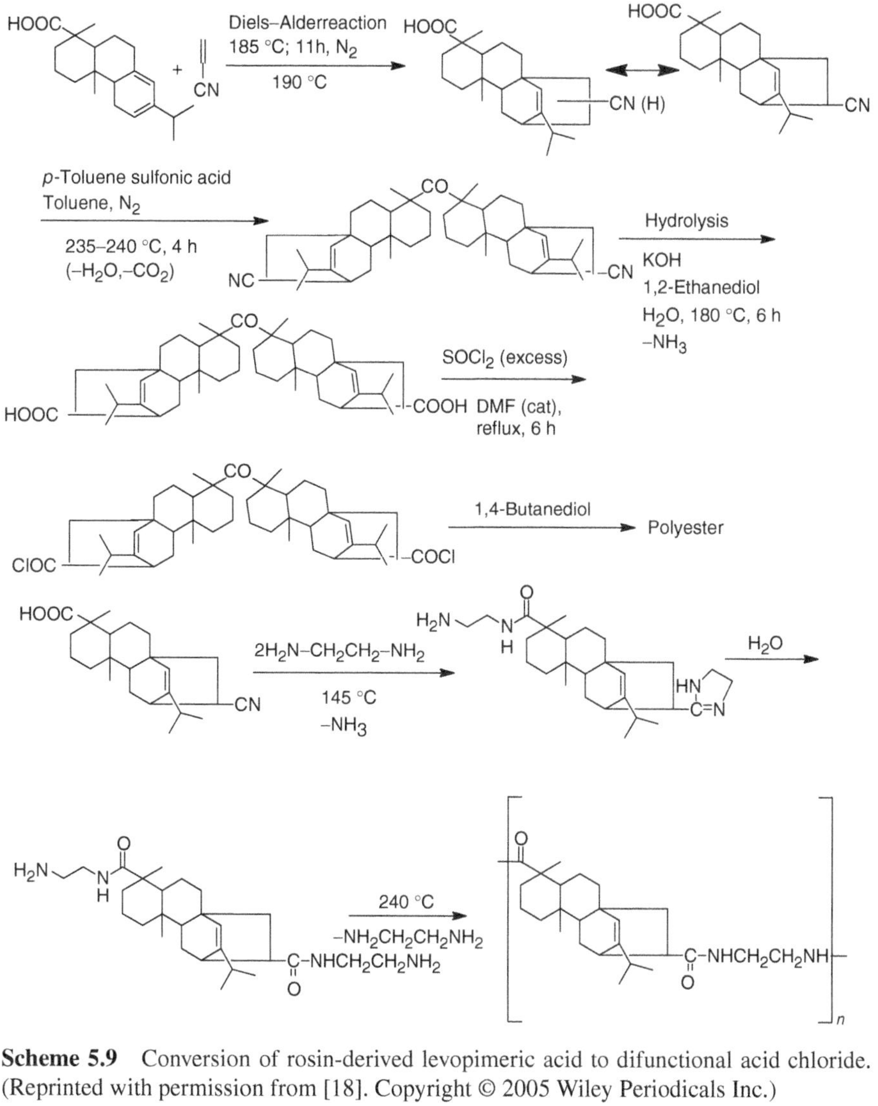
HOOC
CN
Diels–Alderreaction
185 °C; 11h, N2
190 °C
HOOC
CN (H)
HOOC
CN
p-Toluene sulfonic acid
Toluene, N2
235–240 °C, 4 h
(–H2O,–CO2)
NC
CO
CN
Hydrolysis
KOH
1,2-Ethanediol
H2O, 180 °C, 6 h
–NH3
CO
HOOC
COOH
SOCl2 (excess)
DMF (cat),
reflux, 6 h
CO
ClOC
COCl
1,4-Butanediol
Polyester
HOOC
CN
2H2N–CH2CH2–NH2
145 °C
–NH3
H2N
N
H
O
HN
C=N
H2O
H2N
N
H
O
C-NHCH2CH2NH2
O
240 °C
–NH2CH2CH2NH2
O
C-NHCH2CH2NH
O
n

Scheme 5.9 Conversion of rosin-derived levopimeric acid to difunctional acid chloride. (Reprinted with permission from [18]. Copyright © 2005 Wiley Periodicals Inc.)

5.3.2 Radical Polymerization of Modified Dehydroabietic Acid

Dehydroabietic acid was converted to acrylate ester as shown in Scheme 5.10 [19]. Only low molecular weight polymers were obtained. The enthalpy of bulk polymerization was determined to be the same at different temperatures when initiated by AIBN, and the molecular weight was not affected by increasing the polymerization temperature.

5.3.3 ATRP of Vinyl Monomers Derived from Dehydroabietic Acid

The carboxylic acid functionality of dehydroabietic acid was converted to hydroxyl methyl group by reducing the C=O group of ester with $NaBH_4$ and was subsequently converted to an acrylate ester. Acrylate monomers with varying number of methylene spacers were also obtained from dehydroabietic acid (Scheme 5.11) [20]. The acrylate-bearing pendant groups derived from dehydroabietic acid were polymerized under ATRP conditions using Me6Tren ligand and CuBr in tetrahydrofuran (THF) or anisole. Polymerization in THF yielded polymers with polydispersity as high as 3.65, indicating poorly controlled polymerization. Molecular weight as determined by ^{1}H NMR analysis was higher than that of gel permeation chromatography (GPC) analysis, indicating the conformational difference of rosin-derived polyacrylates with polystyrene standards. The polymers exhibited thermoplastic behavior and were amorphous.

5.3.4 Block Copolymers Derived from Dehydroabietic Acid Derivative

2-Acryloxyethyldehydroabietic carboxylate (AOEDAC) was subjected to ATRP using orthogonal initiator such as 2-hydroxyethylbromoisobuyrate. The hydroxyl

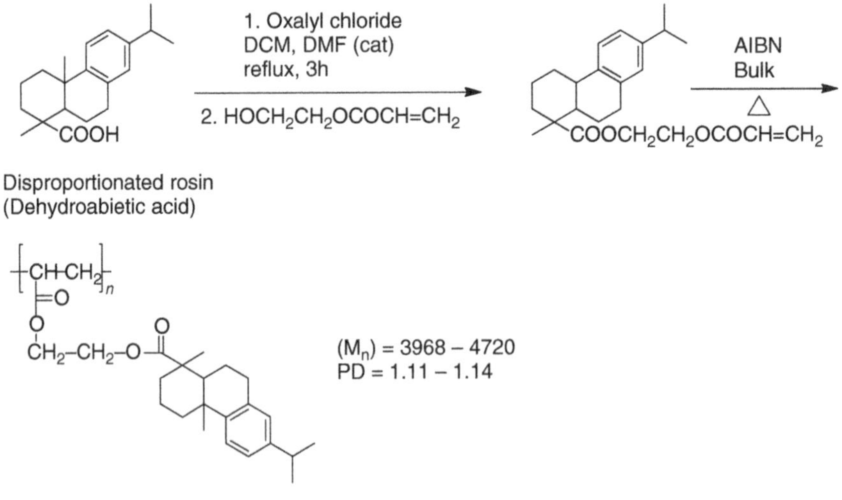

1. Oxalyl chloride
DCM, DMF (cat)
reflux, 3h

2. $HOCH_2CH_2OCOCH{=}CH_2$

COOH

$COOCH_2CH_2OCOCH{=}CH_2$

AIBN
Bulk
Δ

Disproportionated rosin
(Dehydroabietic acid)

$-[CH-CH_2]_n-$

CH_2-CH_2-O

$(M_n) = 3968 - 4720$
PD = 1.11 – 1.14

Scheme 5.10 Preparation of disproportionated rosin-(β-acryloxy ethyl ester). (Reprinted with permission from [19]. Copyright © 2009 Wiley Periodicals Inc.)

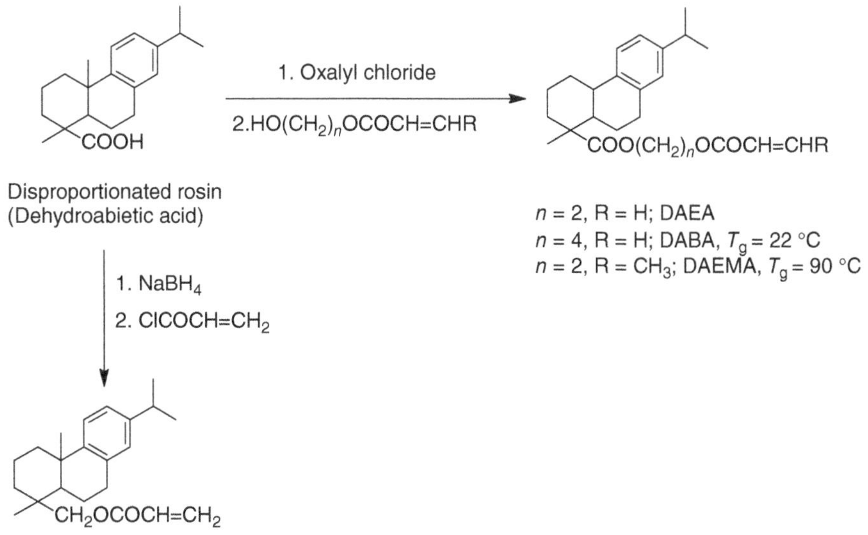

COOH

Disproportionated rosin
(Dehydroabietic acid)

1. Oxalyl chloride
2. $HO(CH_2)_nOCOCH{=}CHR$

$COO(CH_2)_nOCOCH{=}CHR$

$n = 2$, R = H; DAEA
$n = 4$, R = H; DABA, $T_g = 22\ °C$
$n = 2$, R = CH_3; DAEMA, $T_g = 90\ °C$

1. $NaBH_4$
2. $ClCOCH{=}CH_2$

$CH_2OCOCH{=}CH_2$

Scheme 5.11 Synthesis of vinyl monomers from gum Rosin. (Reprinted with permission from [20]. Copyright © 2010 American Chemical Society.)

group was subsequently used for initiating ROP of ε-caprolactone. The diblock copolymers thus prepared underwent complete degradation of ester units in strongly acidic THF like 30% HCl–THF mixture. However, the rosin-acid-derived acrylate polymer was unaffected under these strongly acidic conditions [21].

5.4 POLYURETHANES FROM VEGETABLE OILS

Polyurethanes are one of the most versatile classes of polymers having wide-ranging applications such as adhesives, biomedical materials, cables, coatings, elastomers, fibers, foams, sealants, and so on. Majority of bio-based polyurethanes reported till now are from polyols derived from vegetable oils. As described later, there have been reports that disclose the preparation of diisocyanates derived from triglycerides. Thus, with both components of polyurethane derived from renewable resources, these highly important classes of industrial polymers are fully renewable. One key requirement, though, would be to match the properties of commercially available polyurethanes with the renewable resource polyurethanes.

Polyols derived from vegetable oil such as canola oil were reacted with diphenylmethane diisocyanate to yield polyurethane plastic sheets. The T_g of these plastic sheets depended on the molar ratio of OH/NCO and was inversely proportional to the concentration of OH functionality. Thermogravimetric analysis (TGA) indicated three well-defined degradation profiles for the polyurethane plastic sheet. Degree of cross-linking and stoichiometry of reactants mainly determined the properties of polyurethane plastic sheets [22].

5.4.1 Polyurethanes Derived from Plant Oil Triglycerides

Predominantly polyols are derived from vegetable oils because of their ease of formation. In an alternative approach, isocyanates were derived from plant oil triglycerides (Scheme 5.12) [23]. Plant oil-based triglycerides were converted to allylic bromides. In the next step, the allylic bromides were reacted with AgNCO. Thus, soybean oil was converted to the corresponding isocyanate in two steps in yields of 60–70%. According to the authors, although AgNCO is expensive, the by-product AgBr could potentially be used for regenerating AgNCO. The soybean oil-derived isocyanate was reacted with renewable polyols derived from castor oil and glycerin to form polyurethanes. Similarly, polyureas were formed by the reaction of the isocyanate with triethylenetetramine. Some of the characteristics of polyurethane and polyurea were high elongation, high swelling ratios, and low mechanical strength. These properties make them useful as bio-derived foams.

5.4.2 Long-Chain Unsaturated Diisocyanates Derived from Fatty Acids of Vegetable Origin

1,7-Heptamethylenediisocyanate (HPMDI) was reported to be obtained from azelaic acid via Curtius rearrangement starting from oleic acid [24]. As this reaction involves the formation of diazide, the explosive nature of diazide requires that the ratio of C and O atom numbers ($N_C + N_O$) to nitrogen atom number (N_N), that is, $(N_C + N_O)/(N_N)$ should be well above 2 and be a minimum of 3. For this reason, Curtius rearrangement was applied on long-chain unsaturated dicarboxylic acids obtained by self-metathesis of monounsaturated fatty acids. The unsaturated dicarboxylic acids were converted to diazides that were decomposed to form diisocyanates by refluxing in anhydrous THF [25]. The diisocyanates thus obtained

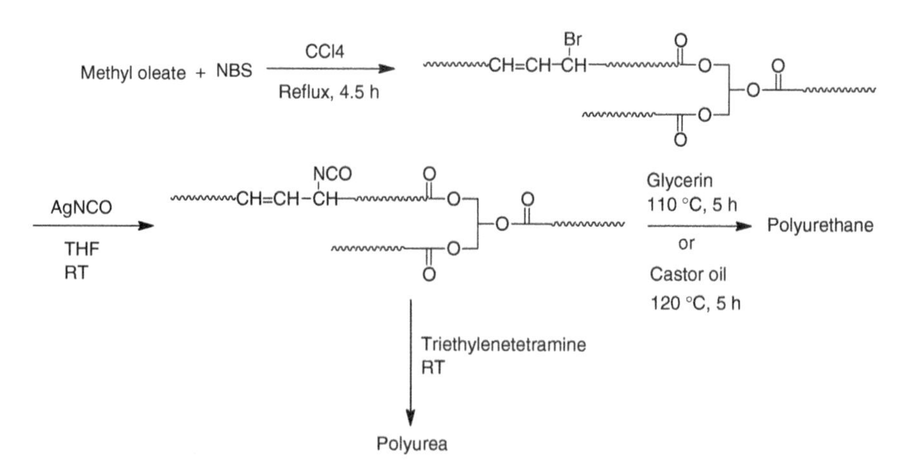

Scheme 5.12 Conversion of brominated triglycerides to isocyanates. (Reprinted with permission from [23]. Copyright © 2008 Wiley Periodicals Inc.)

were reacted with canola diol to yield polymers with $\langle M_n \rangle = 4 \times 10^4$, PD = 1.5, and $T_m = 80\,°C$ and also with canola polyol.

5.5 CO_2 AS RENEWABLE RESOURCE COMONOMER

Polycarbonates are an important class of polymers that are commercially made by reacting diol and phosgene or a derivative of phosgene. Phosgene is not an environment-friendly reagent, and there are active interests to substitute phosgene. A phosgene-free route for producing polycarbonates involves the use of diphenyl carbonate in place of phosgene [26]. Insertion of CO_2 into epoxides to make aliphatic polycarbonates has been known for many decades. Some of the well-known CO_2/epoxide copolymers are formed by using cyclohexene oxide and propylene oxide as comonomers. Major drawbacks of aliphatic polycarbonates are low T_g's and brittle nature. For example, poly(cyclohexene carbonate) has a T_g of 115 °C, whereas the T_g of bisphenol-A polycarbonate is 150 °C. Although poly(propylene carbonate) possesses excellent mechanical properties its T_g is quite low, about 41 °C. Recently, a rigid polycarbonate was reported in which indene oxide was used as epoxide [27].

5.6 RENEWABLE TRIBLOCK COPOLYMER-BASED PRESSURE-SENSITIVE ADHESIVES (PSA)

PSAs are used in labels, stamps, sticky notes, and tapes. The major components of PSAs are tackifiers, plasticizers, and fillers mixed with an elastomeric polymer such as polyacrylate, natural rubber, silicone, or a styrenic block copolymer. Styrenic block copolymers have typically ABA-type arrangement where the mid-segment is made up of soft, rubbery, low T_g polymer, and the end blocks are derived from hard, high T_g polymers. A PSA derived from a completely renewable package consists of triblock copolymer of the type poly(lactide–menthide–lactide) (PL–PM–PL) with a tackifier composed of rosin ester [15,28a]. A two-step polymerization catalyzed by Et_2Zn and Et_3AL was employed for making the triblock copolymer of PL–PM–PL with PL content varying from 20–50 wt% [28b]. These triblock copolymers were microphase separated and exhibited elastomeric properties. It was possible to make PSA formulations based on these triblock copolymers by reducing the PLA content and simultaneously lengthening the PM segment. The renewable rosin ester tackifier was found to be miscible with this triblock copolymer. The tackifier mainly affected the T_g of mid-segment, PM, and the T_g's of PL were unaffected upon varying the concentration of rosin ester from 20–60 wt%.

The three important performance characteristics of a PSA are tack, peel strength, and shear strength. Tack is defined as instant formation of bonding interaction between substrate and adhesive when they come into contact with each other. Peel strength is defined as the force needed to separate the adhesive from substrate. Shear strength is the ability of adhesive to resist flow upon applying a load.

With rosin ester tackifier content of 40 wt% peel adhesion was 3.2 N cm^{-1}, tack was 1.1 N, and shear strength was ≈2500 min. These properties were comparable to commercial duct, paper, and electrical tapes which under identical conditions showed peel adhesion of 1.9–4.2 N cm^{-1}, tack of 0.4–0.6 N, and shear strength of ≈1500 min. However, adhesive failure was observed with the triblock copolymer of PL–PM–PL as indicated by the residue left on stainless steel test plate. Although large amounts of tackifiers are useful for improving the cohesive strength of PSAs because of the tackifier-rich soft phase, it also increases the T_g resulting in lowered tack [28c].

5.7 PHOTOCURABLE RENEWABLE RESOURCE POLYESTER

Photocurable renewable resource polyester was derived from IA or its dimethyl ester (DMI). IA or DMI was polymerized either thermally or by enzymatic process to yield prepolymers in the molecular weight range, $M_n = 940–2200$ g mol^{-1}, with diols such as poly(ethylene glycol), cyclohexanedimethanol, and sorbitol. The linear and branched polyesters thus formed were cross-linked by exposing to UV light, yielding cross-linked materials of varying hydrophilicity as determined by contact angle measurements which varied from 29° to 65°. The photocured polymers also showed different T_g values. These polyester thermosets were soaked in phosphate-buffered solution (PBS) after being sterilized by soaking in ethanol. Ethanol was removed by exposing the thermoset polymer under a germicidal lamp. The cytotoxicity was evaluated using Swiss Albino 3T3 Fibroblasts (SAF) after 24 and 72 h in which polycaprolactone was used as control. After 3 days, majority of thermoset polyesters showed good toxicity profile. A moderate toxicity was observed for the thermoset composed of IA, succinic acid, and sorbitol. A drop in pH as indicated by change in color from red to yellow was also observed [29].

5.8 RENEWABLE RESOURCE-DERIVED WATERBORNE POLYESTERS

With increased awareness about environment combined with tightening regulation of solvents arising from coatings, paint formulators prefer water-soluble or water-dispersible materials. Partially renewable water-soluble unsaturated polyesters were obtained by bulk polycondensation of isosorbide, maleic anhydride, and poly(ethylene glycol) catalyzed by Ti(IV)-*n*-butoxide. The molecular weight characteristics of unsaturated polyester were ⟨M_n = 2830–4450, PD = 1.8–4.5 (PMMA std., hexafluoro isopropanol as eluent). The high polydispersity was proposed to be due to the Michael addition of isosorbide or poly(ethylene glycol) onto the unsaturated bond of maleic acid or due to oxidation processes. The unsaturated polyesters were amorphous materials with thermal resistance above 240 °C which are potential for coating applications [30].

5.8.1 Polyesters Made Up of Isosorbide and Succinic Acid

Bulk polycondensation was employed for making linear and branched polyesters from isosorbide and succinic acids. Terpolyesters were formed by including 2,3-butanediol or 1,3-propanediol as comonomer in the polycondensation. Polyesters with 60–70% isosorbide content are suitable for powder-coating applications. Linear polymers in combination with nonrenewable conventional curing agents showed moderate solvent resistance and mechanical performance. Branched polyesters exhibited good solvent resistance, impact resistance properties, and high hardness [31].

5.8.2 Polyesters Modified with Citric Acid

Citric acid has been proposed as a bio-based alternative to petroleum-derived trimellitic anhydride as functionality enhancing compound in the synthesis of polyesters [32]. A reactive anhydride intermediate is formed from citric acid when it is heated above its melting point. This intermediate seems to react rapidly with hydroxyl compounds leading to the formation of esters.

5.9 POLYMERS FORMED BY COMBINING RENEWABLE RESOURCE MONOMERS WITH THAT DERIVED FROM PETROLEUM FEEDSTOCK

Polylactide or poly(lactic acid) is one of the well-known renewable polymer. The monomer of lactide, a cyclic dimer of lactic acid, is produced exclusively by fermentation of corn starch. The relatively low T_g of polylactide combined with some of the poor mechanical properties prevent it being applied widely. Copolymerization is an often employed strategy to improve polymer properties. Advances like atom transfer radical polymerization (ATRP) allow one to combine monomers to form copolymers, which otherwise polymerize under contrasting conditions. Such a strategy can have two major benefits. First, it is partial substitution of petroleum-derived monomer with a renewable resource monomer. Second, as polylactide is biodegradable, the copolymer will be semidegradable. Even though the degradation can be expected to stop upon consumption or complete degradation of lactide segment of the copolymer, the resulting nondegradable segment derived from vinyl polymer will be lower in molecular weight than the starting copolymer. Unlike the higher molecular weight homopolymer of vinyl monomer, such partially degraded polymer can be expected to undergo faster degradation, and additionally, the partially degraded polymer may not show all the ill-effects of undegraded homovinyl polymer. Such copolymers may also exhibit favorable life cycle analysis as compared to that of completely degradable polymer like polylactide. It is worthwhile to note here that the low molecular weight polymers composed of C–C backbone are far more degradable than the corresponding high molecular weight polymer [33].

Lactide undergoes polymerization by a process termed as *ring-opening polymerization* (ROP) because it proceeds through the opening of six-membered

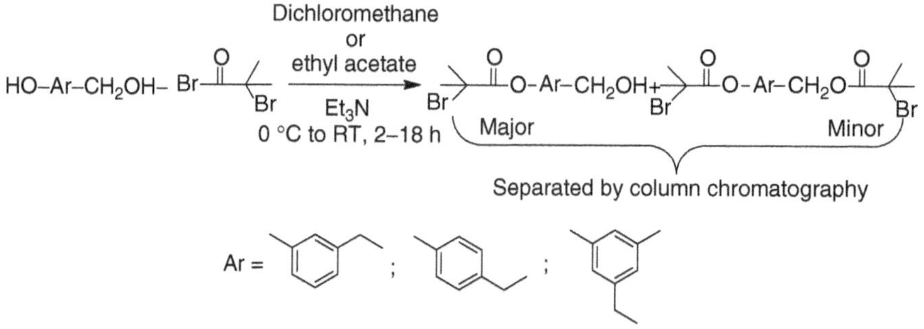

Scheme 5.13 Preparation and chemical structure of bifunctional initiators. (Reprinted with permission from [35]. Copyright © 2008 Wiley Periodicals Inc.)

ring by initiators which are predominantly aliphatic alcohols in the presence of, largely, metal catalysts such as $Sn(oct)_2$. Many of the free radical polymerization techniques developed within the last few decades helps to make polymers with chain end functionalities like –OH. Such terminally functionalized polymers can be used as macroinitiators for ROP of lactides. There are quite a large number of such initiators available in the literature [34]. We reported a variety of bifunctional initiators that were obtained by making use of reactivity difference between phenol and benzyl alcohol in readily available hydroxy benzyl alcohols (Scheme 5.13) [35]. These bifunctional initiators were subjected to ATRP to yield ω-terminal-substituted benzyl alcohol units of vinyl monomers, for example, styrene, which were subsequently employed in the ROP as macroinitiators (Scheme 5.14). Copolymers formed by combining petroleum-derived and

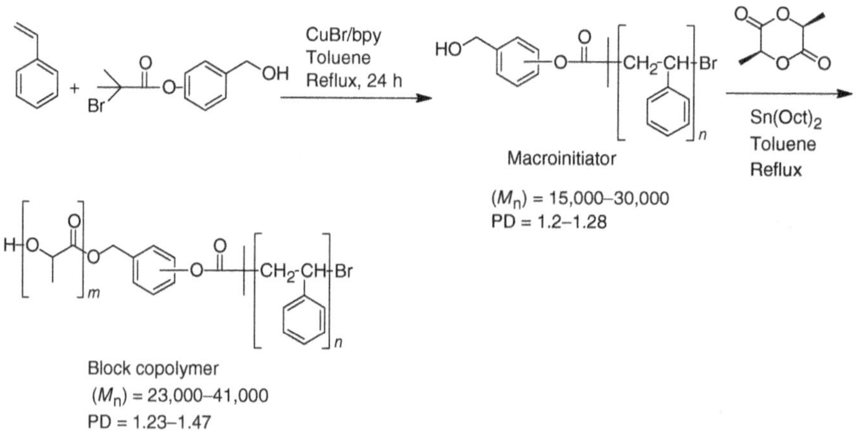

Scheme 5.14 Preparation of macroinitiators and block copolymers. (Reprinted with permission from [35]. Copyright © 2008 Wiley Periodicals Inc.)

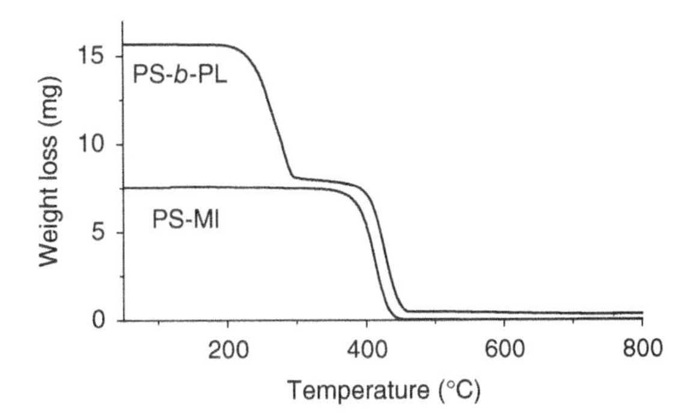

Figure 5.3 Thermogravimetric analysis of copolymer, poly(styrene-*block*-lactide).

renewable resource monomers were obtained in a ratio of 1 : 1 to about 6 : 1. TGA of such copolymers was interesting in that because of the large difference in the thermal stability of these polymeric segments, well-defined stepwise degradation profiles were observed for a copolymer of poly(styrene-*block*-lactide) (Figure 5.3). Both L- and DL-lactides could be polymerized by the macroinitiator. However, the incorporation of lactide depended on the monomer type and mode of addition. When L-lactide alone was used as monomer, lactide content of polymer chain was high. When DL-lactide alone was employed, the lactide content of copolymer was low. However, when a mixture of L- and DL-lactides or a stepwise addition of L-lactide followed by DL-lactide was employed, the lactide content of copolymer increased tremendously (Table 5.2). This strategy was further extended to make ABC-type terpolymers with lactide as one of the segments while varying the A segment from rigid, glassy to soft rubbery. Thus, block copolymers with two semicrystalline segments were obtained in a stepwise process (Scheme 5.15) [36,37].

TABLE 5.2 Ring-Opening Polymerization of Mixture of Lactides with PS-MI[a]

Block Copolymer	M_n	M_w	PD	M_n (PMMA std)	M_n (NMR)	DP_{sty}	DP_{lac}
PS-PL 1[b]	38,017	55,990	1.47	45,827	58,141	283	197
PS-PL 2[c]	40,910	60,198	1.47	47,044	64,195	283	239
PS-PL 3[d]	32,955	43,977	1.33	40,104	36,810	283	49

[a][35a].
[b]One-pot, one-stage process with a mixture of L- and DL-lactides and PS-MI.
[c]One-pot, two-stage process involving the reaction of PS-MI with L-lactide first followed by DL-lactide.
[d]One-pot, two-stage process involving the reaction of PS-MI with DL-lactide first followed by L-lactide.

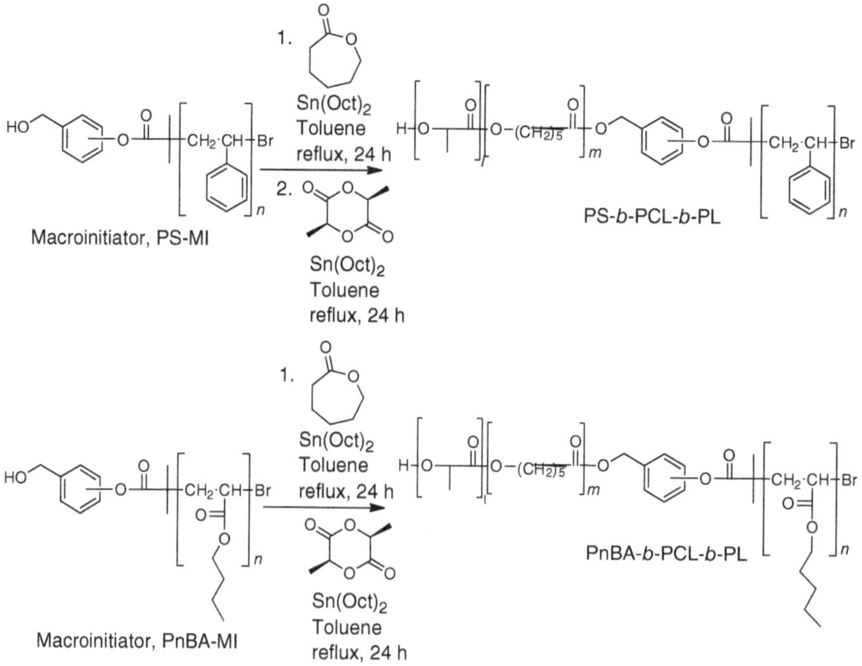

Macroinitiator, PS-MI

1. Sn(Oct)$_2$ Toluene reflux, 24 h

2. Sn(Oct)$_2$ Toluene reflux, 24 h

PS-*b*-PCL-*b*-PL

Macroinitiator, PnBA-MI

1. Sn(Oct)$_2$ Toluene reflux, 24 h

Sn(Oct)$_2$ Toluene reflux, 24 h

PnBA-*b*-PCL-*b*-PL

Scheme 5.15 Preparation of ABC-type triblock copolymers with varying A segment. (Reprinted with permission from [36]. Copyright © 2010 Wiley Periodicals Inc.; reprinted with permission from [37]. Copyright © 2011 Royal Society of Chemistry.)

5.10 CONCLUSION AND OUTLOOK

One of the key advantages often cited about polylactide is its biodegradability that makes it suitable as a solid waste to be disposed of in landfills. However, a recent study questions this assumption [38]. According to this study, the life cycle analysis of biodegradable polymers in comparison to that of nondegradable polymer is unfavorable due to the increased CO_2 emission potential of the former when dumped in a landfill. It has been proposed that the total greenhouse gas emission of such degradable material could be higher when one adds the emission caused by the production of biodegradable material as well.

Although "bioethylene" derived from bioethanol is a potential alternative to ethylene derived from petroleum processes, there is an insurmountable gap between total amount of ethylene and ethanol produced. For example, Brazil and United States are the leading producers of ethanol worldwide with capacities such as 23×10^9 L and 21×10^9 L, respectively, in 2007 [39]. This does not stand anywhere near the amount of total ethylene produced worldwide which stood at 160×10^9 lbs per year in 2012 [40]. It is worth noting here that major part of ethanol produced is diverted for blending with fuels, gasoline, used in automobiles. In addition, though ethanol is produced more efficiently in United States by fermentation of glucose from corn as compared to that of the fermentation of sugar cane in Brazil (energy input:output = 1 : 1.5 vs 1 : 9, respectively), the cost of production is lower in Brazil. Also, corn and its derivatives are a source of food for a variety of animals including humans. On the contrary, the source of bioethanol in Brazil is from a fermentation residue of sugarcane called *bagasse* that is nonedible. Thus, there are inherent contradictions between the two sources of bioethanol. Chemically, glucose is converted to ethanol in the case of corn, whereas constituents of bagasse are cellulose, hemicellulose, and lignin [1b].

The conventional production of often cited renewable resource monomer lactic acid is far from ideal. Currently, every ton of lactic acid produced also generates a ton of $CaSO_4$ as the process involves the separation of lactic acid as Ca salt that is subsequently neutralized with concentrated H_2SO_4 to recover it into acidic form. Thus, the process is not environmentally favorable. Apart from this, the reactor productivities are also low and 50% of production cost is accounted for the separation and purification of lactic acid alone. Such issues may be addressed by development of processes based on membranes, micro- and ultrafiltration, electrodialysis, and so on [4b].

Even though these are challenges to overcome, it should be kept in mind that petroleum feedstock-based conventional processes have evolved over many decades. Considerable efforts and investments would be required to make biorefining processes competitive. It is useful to note that in the early part of twentieth century, alcohols such as ethanol and butanol and organic acids such as acetic acid and citric acid were produced from biomass. With the advent of petroleum refining and its associated processes, such biomass-based processes disappeared. It is also a fact that energy consumption, waste generation, and greenhouse gas emission were not of great concern at that time [6].

Until recently, renewable processes come to the fore with the rise of crude oil prices. Even now, the cost estimation of raw materials derived from biomass could become favorable only when the crude oil prices are very high or beyond a threshold value. Though a complete biorefinery process displacing the petroleum based chemical refineries may be far, far away, a sense of urgency prevails across the world now more than at any other time. Considerable effort is being made and resources are being made available. Such enhanced activities may yield fruitful results in the foreseeable future with sustained interest and investments. As the survival of human race and earth as a whole are at stake, sustainability has taken center stage spurred by the efforts of UN bodies like Intergovernmental Panel on Climate Change (IPCC). In this scenario, it is quite possible that newer business models may emerge which are quite different from the good old petroleum chemistry.

REFERENCES

1. (a) Sanderson, K. (2011). It's not easy being green. *Nature 469*, 18–20; (b) Mathers, R. T. (2012). How well can renewable resources mimic commodity monomers and polymers? *J. Polym. Sci. Part A: Polym. Chem.*, *50*, 1–15.

2. Smith, P. B., Payne, G. F. (2011). Emerging trends in the commercialization of renewable chemicals. *Polym. Prepr., 52(1)*, 84.
3. Bomgardner, M. M. (2011). Biobased chemicals: myriant to build succinic acid plant in Louisiana. *Chem. Eng. News, 89(2)*, 7.
4. (a) Manzer, L. E. Feedstock for the future: using technology development as a guide to product identification. *ACS Symposium Series, 921*. American Chemical Society, Washington, DC, 2006, pp. 40–51; (b) Corma, A., Iborra, S., Velty, A. (2007). Chemical routes for the transformation of biomass into chemicals. *Chem. Rev. 107*, 2411–2502.
5. Willke, Th., Vorlop, K.-D. (2001). Biotechnological production of itaconic acid. *Appl. Microbiol. Biotechnol., 56*, 289–295.
6. Willke, Th., Vorlop, K.-D. (2004). Industrial bioconversion of renewable resources as an alternative to conventional chemistry. *Appl. Microbiol. Biotechnol., 66*, 131–142.
7. Rohwerder, T., Mueller, R. H. (2010). Biosynthesis of 2-hydroxyisobutyric acid (2-HIBA) from renewable carbon. *Microbiol Cell Factories, 9*, 13–22.
8. (a) Watanabe, K., Andou, Y., Shirai, Y., Nishida, H. (2010). Racemization-free monomer: α-hydroxyisobutyric acid from bio-based lactic acid. *Chem. Lett., 39(7)*, 698; (b) Nishida, H., Andou, Y., Watanabe, K., Arazoe, Y., Ide, S., Shirai, Y., (2011). Poly(tetramethyl glycolide) from renewable carbon, a racemization-free and controlled depolymerizable polyester. *Macromolecules, 44*, 12–13.
9. (a) Charbonneau, L. F., Johnson, R. E., Witteler, H. B., Khanarian, G.. (May 16, 2000). Isosorbide containing polyesters and methods for making same. *US patent*, 6,063,464 (assigned to HNA Holdings, Inc., Warren, NJ); (b) Charbonneau, L. F., Johnson, R. E., Witteler, H. B., Khanarian, G.. (Sep. 28, 1999). Polyesters including isosorbide as a comonomer and methods for making same. *US patent* 5,959,066 (assigned to HNA Holdings, Inc., Warren, NJ).
10. (a) Kricheldorf, H. R., Behnken, G. (2007). Influence of isosorbide on glass-transition temperature and crystallinity of poly(butylene terephthalate). *J. Macromol. Sci. Part A-Pure and Appl. Chem., 44(7-9)*, 679–684; (b) Sablong, R., Duchateau, R., Koning, C. E., de Wit, G., van Es, D., Koelewijn, R., van Haveren, J. (2008). Incorporation of isosorbide into poly(butylene terephthalate) via solid-state polymerization. *Biomacromolecules, 9*, 3090–3097.
11. avan der Meulen, I., Gubbels, E., Huijser, S., Sablong, R., Koning, C. E., Heise, A., Duchateau, R. (2011). Catalytic ring-opening polymerization of renewable macrolactones to high molecular weight polyethylene-like polymer. *Macromolecules, 44*, 4301–405; (b) Veld, M. A. J., Fransson, L., Palmans, A. R., Meijer, E. W., Hult, K. (2009). Lactone size dependent reactivity in Candida antartica lipase b: a molecular dynamics and docking study. *ChemBioChem, 10*, 1330–1334.
12. Focarete, M. L., Scandola, M., Kumar, A., Gross, R. A. (2001). Physical characterization of poly(ω-pentadecalactone) synthesized by lipase-catalyzed ring-opening polymerization. *J. Polym. Sci. Part B: Polym. Phys., 39*, 1721–1729.
13. de Geus, M., van der Meulen, I., Goderis, B., van Hecke, K., Dorschu, M., van der Werff, H., Koning, C. E., Heise, A. (2010). Performance polymers from renewable monomers: high molecular weight poly(pentadecalactone) for fiber applications. *Polym. Chem., 1*, 525–533.
14. Quinzler, D., Mecking, S. (2010). Linear semicrystalline polyesters from fatty acids by complete feedstock molecule utilization. *Angew. Chem. Int. Ed., 49*, 4306–4308.

15. Zhang, D., Hillmyer, M. A., Tolman, W. B. (2005). Catalytic polymerization of a cyclic ester derived from a "cool" natural precursor. *Biomacromolecules*, *6*, 2091–2095.
16. Lee, C. W., Masutani, K., Kato, T., Kimura, Y. (2012). Homopolymerization and copolymerization of a dilactone, 13, 26-dihexyl-1,14-dioxa-cyclohexacosane-2,15.dione: synthesis of bio-based polyesters and copolyesters consisting of 12-hydroxystearate sequences. *J. Polym. Sci. Part A: Polym. Chem.*, *50*, 1290–1297.
17. Wanamaker, C. L., O'Leary, L. E., Lynd, N. A., Hillmyer, M. A., Tolman, W. B. (2007). Renewable-resource thermoplastic elastomers based on polylactide and polymenthide. *Biomacromolecules*, *8*, 3634–3640.
18. Bicu, I., Mustata, F. (2005). Polymers from levopimaric acid-acrylonitrile Diels-Alder adduct: synthesis and characterization. *J. Polym. Sci. Part A: Polym. Chem.*, *43*, 6308–6322.
19. Wang, J.-F., Lin, M.-T., Wang, C.-P., Chu, F.-X. (2009). Study on the synthesis, characterization, and kinetic of bulk polymerization of disproportionated rosin (β-acryloy ethyl) ester. *J. Appl. Polym. Sci.*, *113*, 3757–3765.
20. Zheng, Y., Yao, K., Lee, J., Chandler, D., Wang, J., Wang, C., Chu, F., Tang, C. (2010). Well-defined renewable polymers derived from gum rosin. *Macromolecules*, *43*, 5922–5924.
21. Wilbon, P. A., Zheng, Y., Yao, K., Tang, C. (2010). Renewable rosin acid-degradable block copolymers by atom transfer radical polymerization and ring-opening polymerization. *Macromolecules*, *43*, 8747–8754.
22. Kong, X., Narine, S. S. (2007). Physical properties of polyurethane plastic sheets produced from polyols from canola oil. *Biomacromolecules*, *8*, 2203–2209.
23. Cayl, G., Kuesefoglu, S. (2008). Biobased polyisocyanates from plant oil triglycerides: synthesis, polymerization, and characterization. *J. Appl. Polym. Sci.*, *109*, 2948–2955.
24. Hojabri, L., Kong, X., Narine, S. S. (2009). Fatty acid derived diisocyanate and biobased polyurethane produced from vegetable oil: synthesis, polymerization, and characterization. *Biomacromolecules*, *10*, 884–891.
25. Hojabri, L., Kong, X., Narine, S. S. (2010). Novel long chain unsaturated diisocyanate from fatty acid: synthesis, characterization, and application in bio-based polyurethane. *J. Polym. Sci. Part A: Polym. Chem.*, *48*, 3302–3310.
26. Fukuoka, S., Tojo, M., Hachiya, H., Aminaka, M., Hasegawa, K. (2007). Green and sustainable chemistry in practice: development and industrialization of a novel process for polycarbonate produced from CO_2 without using phosgene. *Polym. J.*, *39(2)*, 91–114.
27. Darensbourg, D. J., Wilson, S. J. (2011). Synthesis of poly(indene carbonate) from indene oxide and carbon dioxide-A polycarbonate with a rigid backbone. *J. Am. Chem. Soc.*, *133*, 18610–18613.
28. (a) Wanamker, C. L., O'Leary, L. E., Lynd, N. A., Hillmyer, M. A., Tolman, W. B. (2007). Renewable-resource thermoplastic elastomers based on polylactide and polymenthide. *Biomacromolecules*, *8*, 3634–3640; (b) Shin, J., Martello, M. T., Shreshtha, M., Wissinger, J. E., Tolman, W. B., Hillmyer, M. A. (2011). Pressure-sensitive adhesives from renewable triblock copolymers. *Macromolecules*, *44*, 87–94.
29. Barrett, D. G., Merkel, T. J., Luft, J. C., Yousaf, M. N. (2010). One-step synthesis of photocurable polyesters based on a renewable resource. *Macromolecules*, *43*, 9660–9667.
30. Jasinska, L., Koning, C. R. (2010). Waterborne polyesters partially based on renewable resources. *J. Polym. Sci. Part A: Polym. Chem.*, *48*, 5907–5915.

31. Noordover, B. A. J., van Staalduinen, V. G., Duchateau, R., Koning, C. E., van Benthem, R. A. T. M., Mak, M., Heise, A., Frissen, A. E., van Haveren, J. (2006). Co- and terpolyesters based on isosorbide and succinic acid for coating applications: synthesis and characterization. *Biomacromolecules*, *7*, 3406–3416.
32. Noordover, B. A. J., Duchateau, R., van Benthem, R. A. T. M., Ming, W., Koning, C. E. (2007). Enhancing functionality of biobased polyester coating resins through modification with citric acid. *Biomacromolecules*, *8*, 3860–3870.
33. Swift, G., Baciu, R. (2007). Environmentally degradable polyolefins. *Polym. Prepr.*, *48(1)*, 586–587.
34. Dove, A. P. (2008). Controlled ring opening polymerization of cyclic esters: polymer blocks in self assembled nanostructures. *Chem Commun.*, *48*, 6446–6470.
35. (a) Likhitsup, A., Parthiban, A., Chai, C. L. L. (2008). Combining atom-transfer radical polymerization and ring-opening polymerization through bifunctional initiators derived from hydroxyl benzyl alcohol-preparation and characterization of initiators, macroinitiators, and block copolymers. *J. Polym. Sci. Part A: Polym. Chem.*, *46*, 102–116; (b) Likhitsup, A., Parthiban, A., Chai, C. L. L., Lim, K. S. (2007). Substituted benzyl alcohol based bifunctional initiators suitable for atom transfer radical polymerization and ring opening polymerization. *Polym. Prepr.*, *48(1)*, 504–505; (c) Likhitsup, A., Parthiban, A., Yu, H., Chai, C. L. L. (2007). Block copolymers of styrene and lactide with varying lactide content. *Polym. Prepr.*, *48(2)*, 324–325.
36. Parthiban, A., Likhitsup, A., Yu, H., Chai, C. L. L. (2010). AB- and ABC-type di- and triblock copolymers of poy[styrene-block-(ε-caprolactone)] and poy[styrene-block-(ε-caprolactone)-block-lactide]: synthesis, characterization and thermal studies. *Polym. Int.*, *59*, 145–154.
37. Parthiban, A., Likhitsup, A., Ming Choo, F., Chai, C. L. L. (2010). Triblock copolymers composed of soft and semi-crystalline segments-synthesis and characterization of poy[(n-butyl acrylate)-block-(ε-caprolactone)-block-(L-lactide)]. *Polym. Chem.*, *1*, 333–338.
38. Levis, J. W., Barlaz, M. A. (2011). Is biodegradability a desirable attribute for discarded solid waste? Perspectives from a national landfill greenhouse gas inventory model. *Environ. Sci. Technol.*, *45(13)*, 5470–5476.
39. Morschbacker, A. J. (2009). Bio-ethanol based ethylene. *Polym. Rev.*, *49(2)*, 79–84.
40. Wittcoff, H. A., Reuben, B. G., Plotkin, J. S. *Industrial Organic Chemicals*. Wiley, Hoboken, NJ, 2004.

6

MICROPOROUS ORGANIC POLYMERS: SYNTHESIS, TYPES, AND APPLICATIONS

Shujun Xu
Bien Tan

6.1 INTRODUCTION

Microporous solids, which have pore sizes less than 2 nm in diameter, are an important class of materials that have been studied extensively for several decades [1]. There are several well-known classes of microporous material including activated carbons, silica, and zeolites [2]. More recently, the development of microporous metal-organic frameworks (MOFs) and their exceptional properties has spurred interest in other types of microporous solids. There is significant motivation for the preparation of new porous materials—microporous organic polymers (MOPs), which are composed predominantly of light, nonmetallic elements such as carbon, hydrogen, nitrogen, and boron. According to their structural characteristics, MOPs can be divided into four types: polymers of intrinsic microporosity (PIMs) [3], hypercrosslinked polymers (HCPs) [4], covalent organic frameworks (COFs) [5], and conjugated microporous polymers (CMPs) [6].

There are two classes of MOPs: disordered, amorphous MOPs; and ordered, crystalline MOPs. PIMs, HCPs, and CMPs are amorphous, while COFs are crystalline [7]. The nature of covalent bond-forming chemistry offers many structural possibilities in MOPs. Crystalline MOPs, in general, necessitate a kind of chemistry that is reversible in order to allow the formation of ordered

Synthesis and Applications of Copolymers, First Edition. Edited by Anbanandam Parthiban.

thermodynamic products. Examples of crystalline MOPs include boroxine-ester and boronate-ester formation and nitrile cyclization. Irreversible chemistry, for example, Sonagashira–Hagihara cross-coupling tends to produce amorphous kinetic products in the absence of strong templating units, which could in principle direct such chemistry to form crystalline MOPs.

The main advantage of MOPs is the synthetic diversity that enables the formation of a large variety of structures. A very wide range of polymerization methodologies have been developed to synthesize MOPs. There is enormous scope for post-polymerization modification to introduce specific chemical functionalities. The use of organic building blocks for the preparation of microporous solids presents some exciting possibilities in terms of network design. Synthetic organic chemistry provides access to a diverse array of molecular building blocks that can produce novel network architectures, incorporate functional groups for specific properties, and potentially tune properties through systematic structural modifications. Besides, compared with conventional microporous materials, MOPs have advantages of lighter weight and higher surface area. These materials often exhibit permanent porosity, tunable pore size, and have shown potential as materials for gas storage, separations, and catalysis. Especially the ability to adsorb hydrogen makes them as a novel class of porous materials for hydrogen storage applications.

The chemical and thermal stabilities of MOPs are variable. Generally, main-chain aromatic polymers have very good chemical stability and moderate-to-good thermal stability. For example, HCPs, PIMs, and CMPs all tend to be stable to acids and bases. These materials also exhibit good thermal stability, although less so than inorganic materials such as silica. COFs based on boroxines show good thermal stability, but porosity has been reported to be disturbed by exposure to air. Most MOPs are chemically and thermally stable. They can therefore be handled and derivatized under standard wet chemical conditions without problems such as framework degradation or loss of microporosity.

In addition to the advantage, MOPs present a range of challenges. Some MOPs are relatively expensive. An associated problem is environmental impact, since most organic polymer chemistry is carried out using volatile organic solvents. Likewise, atom economy is poor for some MOP chemistries. The very large surface area of MOPs is a prerequisite to a number of applications. However, large pores of such materials are unstable. In order to support these micropores under dry state, most MOPs are connected by the rigid aromatic rings [8]. From a long-term "life cycle" perspective, the environmental degradability of the aromatic units that comprise most MOPs is poor.

6.2 PREPARATIONS OF MOPS

6.2.1 Polymers of Intrinsic Microporosity

Most polymers have considerable conformational flexibility, and their conformation can be rearranged. Because of rigid and distorted molecular structure of PIMs, the main chain cannot rotate freely, thus macromolecules cannot effectively occupy the various parts of interior space, the rigid part of the contorted molecular structures formed a continuous porous phase. In addition, PIMs can be prepared into insoluble networks or soluble polymers.

6.2.1.1 Insoluble PIM Network McKeown et al. [9] reported the phthalocyanine-containing polymers, which have the strong noncovalent $\pi-\pi$ bond connecting between the various units. These polymers have no porous structure, and their performance in catalysis is poor. In order to improve the catalytic effectiveness of phthalocyanine-based polymers, Budd et al. [3], from the perspective of increasing specific surface area, introduced the microporous structure to these polymers. They used 5,5′,6,6′-tetrahydroxy-3,3,3′,3′-tetramethyl-1,1′-spirobisindane (monomer A1) (Figure 6.1) to connect phthalocyanine unit. In the chemical structure of A1, the spirocenter has a nonlinear shape and a fused ring structure provides rigidity. Bis(phthalonitrile) was prepared using the aromatic nucleophilic substitution reaction between 4,5-dichlorophthalonitrile and monomer A1. Under the template effect of metal ions, it aggregated into the ring composed of tetrameric units resulting in a phthalocyanine-based porous polymer structure (Figure 6.2). The electron spin resonance spectroscopy, UV–visible absorption spectra (UV–vis), and X-ray diffraction (XRD) analysis of the polymer confirm that the rigid spirocyclic linkages prevent close packing of the phthalocyanine components. On the basis of nitrogen (N_2) adsorption studies, surface area of these polymers was between 450 and 950 $m^2\ g^{-1}$. It can be seen from the N_2 adsorption isotherm that significant adsorption occurred at a lower relative pressure ($P/P_0 < 0.01$).

Inspired by the successful synthesis of microporous phthalocyanine-based polymers, Budd et al. [10] further synthesized the porphyrin-based nanoporous network polymers. Metal-containing porphyrin is a very important class of catalyst, which has similar activity such as cytochrome P450 enzymes. The incorporation of the porphyrin unit into porous materials has a great significance to heterogeneous catalysts that are used in various industrial processes. There is a formation of rigid dibenzodioxane units via aromatic nucleophilic substitution reaction between 5,10,15,20-tetrakis(pentafluorophenyl)porphyrin and monomer A1 (Figure 6.3). The spirocenter ensures that porphyrin center and the pentafluorophenyl group are orthogonal to each other.

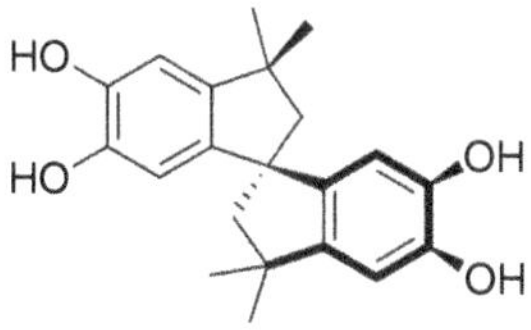

Figure 6.1 Chemical structure of monomer A1. (Reprinted with permission from [3]. Copyright © 2012 Royal Society of Chemistry.)

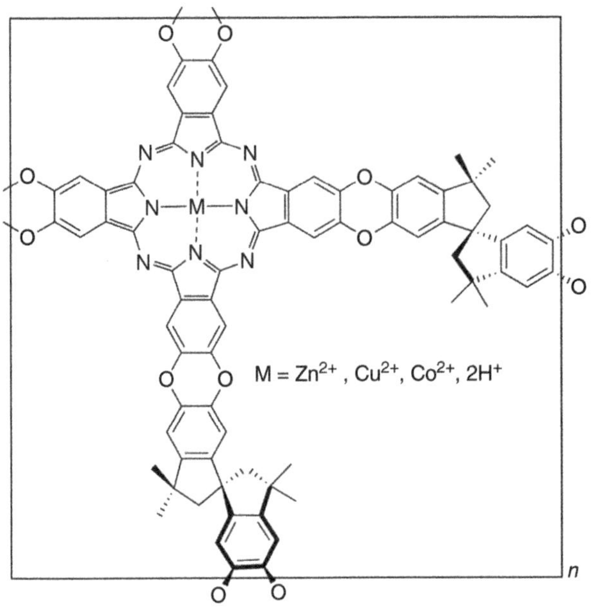

M = Zn^{2+}, Cu^{2+}, Co^{2+}, $2H^+$

Figure 6.2 Chemical structure of phthalocyanin-containing PIM. (Reprinted with permission from [3]. Copyright © 2012 Royal Society of Chemistry.)

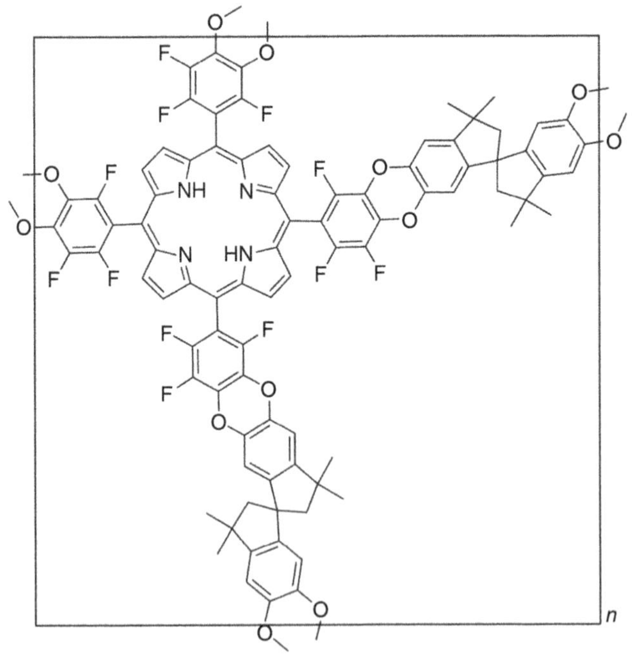

Figure 6.3 Chemical structure of Porph-PIM. (Reprinted with permission from [10]. Copyright © 2002 Royal Society of Chemistry.)

The general method of preparing phthalocyanine- and porphyrin-based PIM involves the use of monomers possessing distorted rigid structures (including the spirocenter or nonplanar structure), which is condensed with another functional monomer by a benzodioxane ring forming reaction. Search for new functional monomers and to develop applications of these porous polymers are of interest at present. Tattershall et al. [11] synthesized 2,3,8,9,14,15-hexachloro-5,6,11,12,17,18-hexaazatrinaphthylene (HATN) from 4,5-dichlorophenylenediamine and hexaketonecyclohexane. This was then condensed with monomer A1 to form porous PIM (Figure 6.4). The surface area of the resulting PIM was in the range 750–850 $m^2 g^{-1}$ as determined by N_2 adsorption studies. It had a large number of active sites to act as a supported catalyst. In addition, these polymers acted as adsorbents for the removal of toxic organic compounds such as phenols from water or gas streams.

Walton et al. [12] selected the cyclotricatechylene (CTC) as functionalities, which has unique C-3 symmetry and rigid electron-rich cavity. CTCs were reacted with tetrafluoroterephthalonitrile to yield microporous polymers (Figure 6.5) with Brunauer–Emmett–Teller (BET) surface area of 830 $m^2 g^{-1}$.

Walton et al. [13] reported PIMs containing triptycene network (Trip-PIM) (Figure 6.6). BET surface area of this material was 1065 $m^2 g^{-1}$. The amount of hydrogen (H_2) adsorption at 1 bar/77 K was about 1.65 wt%, and the amount of H_2 adsorption at 10 bar/77 K was about 2.71 wt%. The bonded support face in the Trip-PIM is vertical to the molecule's growth face, which hinders the support plane

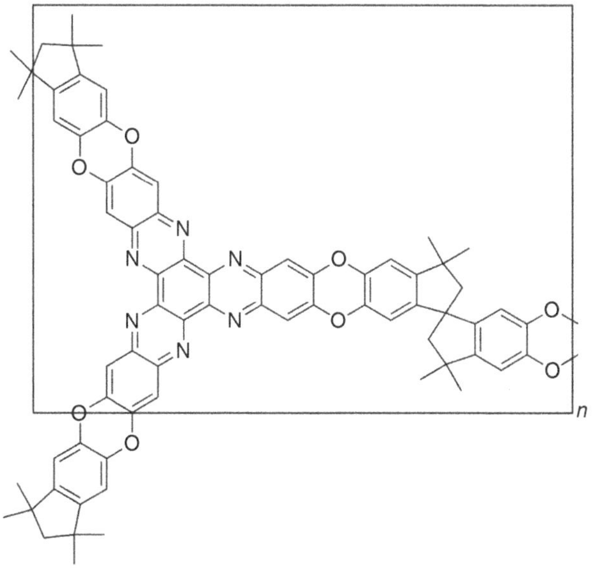

Figure 6.4 Chemical structure of HATN-PIM. (Reprinted with permission from [11]. Copyright © 2003 Royal Society of Chemistry.)

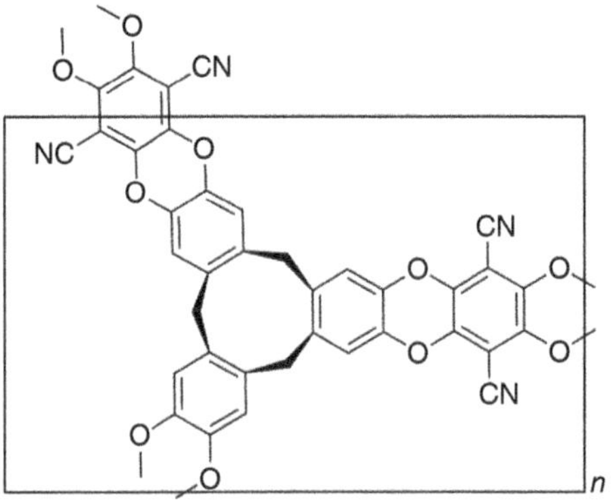

Figure 6.5 Chemical structure of CTC-PIM. (Reprinted with permission from [12]. Copyright © 2006 Wiley Periodicals Inc.)

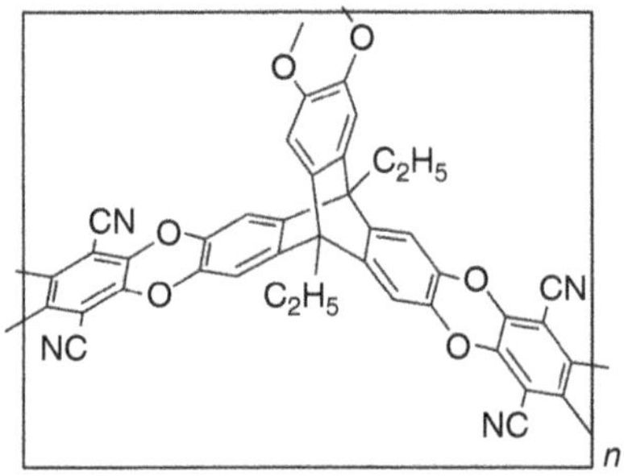

Figure 6.6 Chemical structure of Trip-PIM. (Reprinted with permission from [13]. Copyright © 2006 Royal Society of Chemistry.)

approaching each other. McKeown et al. [14] reported that they synthesized a series of Trip-PIMs where alkyl groups differing in length and degree of branching were attached at the bridge head. The BET surface area varied between 618 and 1760 $m^2\ g^{-1}$ upon varying the nature of pendant groups attached at the bridgehead position. Shorter or branched groups such as the methyl and isopropyl improved the micropore, whose BET surface areas were 1760 and 1601 $m^2\ g^{-1}$, and the H_2 adsorptions were 1.79 and 1.83 wt%, respectively, at 1 bar/77 K. The long-chain alkyl group-substituted PIM, such as octyl Trip-PIM, hindered the formation of rigid porous organic network, resulting in a BET surface area of 618 $m^2\ g^{-1}$, and the H_2 adsorption at 1 bar/77 K was 0.74 wt%. The methyl and isopropyl side chains were unable to change its conformation because they are part of rigid PIM skeleton while the long alkyl side chains have sufficient conformational freedom to occupy the free volume.

In general, insoluble PIM networks showed higher surface area and large hydrogen storage capacity than the linear PIMs. The highly interlinked covalent network structures may have led to the retention of micropore volume to a large extent. Owing to network swelling or restricted access of N_2 molecules to the pore of narrow size, significant adsorption hysteresis was observed most notably in the case of Trip-PIM.

6.2.1.2 Soluble PIMs Inspired by the study of network PIMs, Tattershall et al. [15] realized that PIMs, which are not networked, could also form polymers with stable porosity provided that the molecular structure is rigid and distorted, and it cannot effectively occupy the inner space. Under the guidance of these principles, they synthesized the PIMs 1–6. Monomer A1 and 2,3,5,6-tetrafluoroterephthalonitrile were combined through the formation of benzodioxane to obtain PIM-1 (Figure 6.7). BET surface area of PIM-1 was 850 $m^2\ g^{-1}$ as determined by N_2 adsorption studies. Microanalysis showed that the majority of pore size varied between 0.4 and 0.8 nm and existed as a part of the mesoporous structure. Hysteresis was observed under low pressure.

Thomas et al. [16] extended the dibenzodioxane structure of PIM. They found that from 9,9′-spirobifluorene, they could easily prepare 2,2′-diamino-9,9′-spirobifluorene and 2,2′-dicarboxy-9,9′-spirobifluorene, which are suitable for preparing aromatic polyamide, polyimide (Figure 6.8), and poly(benzimidazole)s. Although spirobifluorene-based polyimide has been reported, microporous properties of these polymers have not been discussed [17]. Unlike that of their linear counterparts, these PIM polyimides exhibited higher solubility and processability.

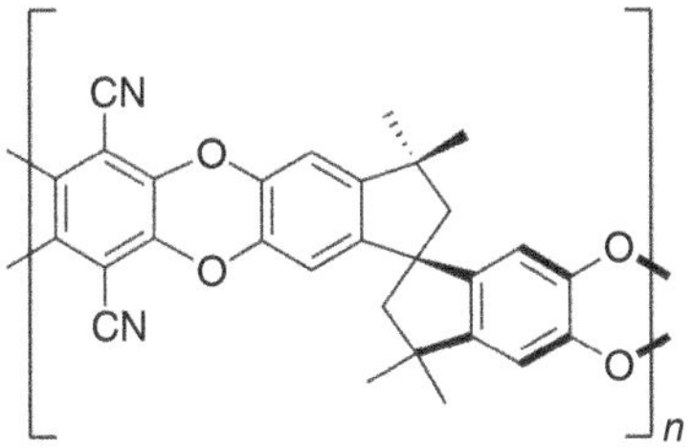

Figure 6.7 Chemical structure of PIM-1. (Reprinted with permission from [15]. Copyright © 2004 Royal Society of Chemistry.)

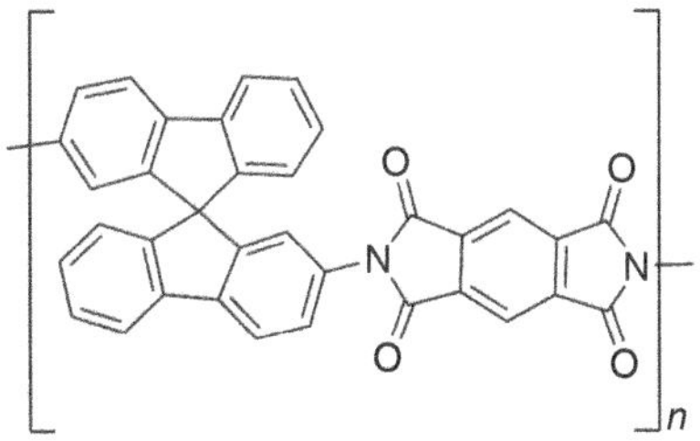

Figure 6.8 Chemical structure of PIM-P4. (Reprinted with permission from [16]. Copyright © 2007 Wiley Periodicals Inc.)

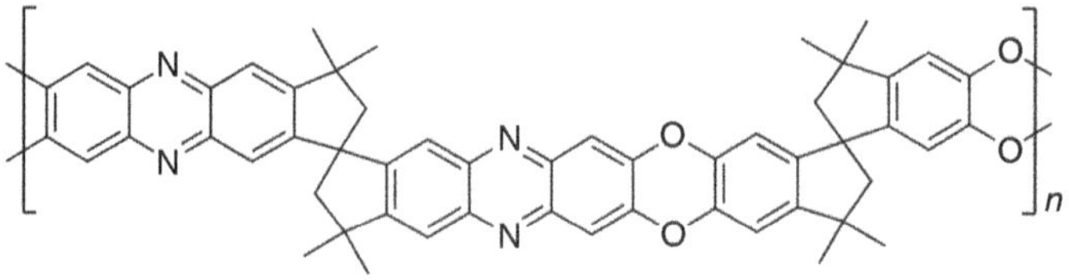

Figure 6.9 Chemical structure of PIM-7. (Reprinted with permission from [18]. Copyright © 2008 American Chemical Society.)

Fritsch et al. [18] introduced pyrazine structure into PIM and prepared a series of new PIMs. At 77 K, N_2 adsorption showed that these polymers had microporosity, and as can be seen from the adsorption isotherm it had significant adsorption at low pressure ($P/P_0 < 0.1$). BET surface area of pyrazine-based PIM was 650 $m^2\ g^{-1}$. At low pressure, N_2 adsorption showed that the pore size distribution of PIM-7 (Figure 6.9) tends to be sub-nanometer and exhibited good film-forming nature, high selection, and permeability to gas mixtures such as O_2/N_2 and CO_2/CH_4, thereby showing its potential as gas separation membranes.

Yampolskii [19] used the dicarboxylic acid anhydride-based spirocyclic structures, which were reacted with different diamines, to synthesize polyimide-based PIMs (PIM-PIs). The chloroform solution of PIM-PIs was used to cast films. The ability of PIM-PIs to separate mixtures of gases was better than the best conventional polyimide.

The rigid and distorted shape of the macromolecular chain is the reason to produce microporosity in the case of PIMs. Thus, PIMs typically contain rigid monomer with spirocyclic structure or nonplanar structure. Compared with that of other MOPs, PIMs deficiency is its disordered molecular structure that makes it difficult to control or regulate the pore size distribution. In addition, the presence of micropores and closed pores makes it difficult to improve the BET surface area significantly. The PIMs' advantage is that, first, functional organic groups and activity point can be introduced into the polymer backbone, which can coordinate with metals used in the field of heterogeneous catalysis; second, linear PIMs can be dissolved in organic solvents, thus facilitating the determination of molecular weight by techniques such as gel permeation chromatography (GPC), thereby enabling to optimize the reaction conditions. In addition, films can be obtained by solution casting technique, and the films thus obtained can be used in applications such as separation of gaseous mixtures and adsorption of gases. Also, PIMs have shown considerable potential for applications such as hydrogen storage and gas adsorption.

6.2.2 Hypercrosslinked Polymer

HCPs are synthesized by irreversible polymerization [4]. It can be prepared by two routes: first, by polymerizing chloromethyl styrene (VBC, vinylbenzyl chloride) to obtain cross-linking precursors, which are subsequently subjected

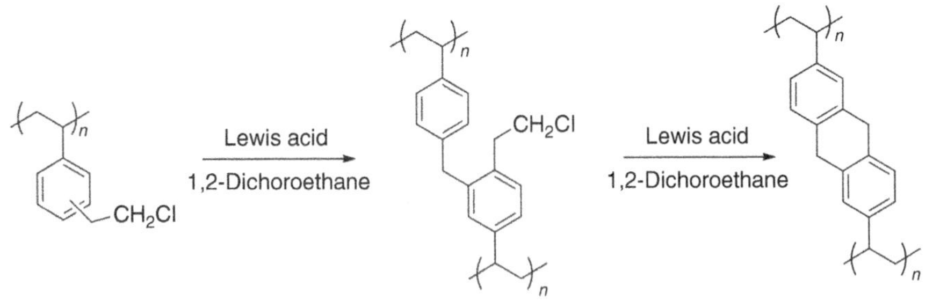

Lewis acid
1,2-Dichoroethane
CH$_2$Cl
Lewis acid
1,2-Dichoroethane

Figure 6.10 HCPs by post-polymerization cross-linking reactions. (Reprinted with permission from [20]. Copyright © 2006 American Chemical Society.)

to Friedel–Crafts (F–C) reaction resulting in polymer network with hypercrosslinked structure (Figure 6.10) [20]; second, through the direct cross-linking reaction of dichloroxylene (DCX), bis(chloromethyl)biphenyl (BCMBP), and bis(chloromethyl) anthracene (BCMA) [21].

Most of these substances have permanent porous structure and exhibit type I gas adsorption isotherms [7]. Hypercrosslinked networks are formed in the solvent. The polymer is isolated by evaporating the solvent. This process accompanies with a significant change in conformation. Network structures tend to shrink, while the rigid structure of the bridge hampers this process at some point. These two factors counterbalance, resulting in very low density porous materials. HCP's adsorption capacity for the polar and nonpolar organic substances, which dissolve in water and organic solvents, is well beyond that of conventional polymer sorbents and activated carbon [22].

Davankov et al. [23] placed styrene–divinylbenzene copolymer in ethylene dichloride (EDC) and cross-linked (post-crosslinked) with solvent-based monochlorodimethyl ether (MCDE) in the presence of $SnCl_4$ after swelling. One molecule of MCDE was linked with phenyl groups of the two different polymer chains, resulting in a rigid diphenyl methyl bridge between the two polymer chains. Through swelling studies, it was found that the cross-linking of styrene and divinyl benzene with MCDE resulted in the formation of HCPs that are more mobile.

Sherrington et al. [20b] found that BET surface of hypercrosslinked p(DVB-VBC) ranged up to 2090 $m^2\ g^{-1}$. These substances can be prepared into blocks, powders, and beads through suspension polymerization, or spherical particles of about 500 nm in diameter through surfactant-free emulsion polymerization. Using N_2 and argon (Ar) as the adsorbent gases to measure adsorption isotherm of HCPs, it was observed that HCPs existed in nonclassical physical states and showed unusual swelling characteristics. This property was mainly utilized to store gases such as methane (CH_4).

Cooper et al. [20a] found the suspension polymerization of VBC and then the intramolecular cross-linking through F–C reaction; and it was possible to obtain spherical porous polymer beads whose diameter varied in the range 50–200 μm. BET surface area was about 1466 m^2 g^{-1}, and H_2 adsorption at 0.1 MPa/77.3 K was 1.28 wt%. Subsequently, Svec et al. [24] placed macroporous and gel-like VBC precursor in dichloromethane for swelling. After swelling through $FeCl_3$-catalyzed F–C reaction, they obtained porous polymer beads with surface areas of 1930 and 1300 m^2 g^{-1}. The polymers stored 1.5 wt% H_2 at 77.3 K/0.12 MPa. Adsorption heat of H_2 in the polymer was 6.6 kJ mol^{-1}.

Cooper et al. [21] employed F–C reaction, using chloromethyl substituted aromatic monomers such as *o*, *m*, *p*-DCX, *p*-DCX and BCMBP, *p*-DCX, and BCMA. A series of porous materials were obtained in one step. N_2 adsorption studies showed that BET surface area of these polymers was above 1904 m^2 g^{-1} at 77 K, and the heat of H_2 adsorption was in the range of 6–7.5 kJ mol^{-1}. Cooper et al. [25] continued to optimize this reaction by differing the ratios of DCX and BCMBP and by cross-linking under the same conditions. When the DCX and BCMBP ratio was 1 : 3, the resulting porous polymer had the largest surface area (1904 m^2 g^{-1}), and this material also adsorbed CH_4 (116 cm^3 g^{-1} at 20 bar/298 K).

Tan et al. [26] used VBC and DVB as comonomers in microemulsion polymerization, which yielded fairly uniform gel-type precursor nanoparticles. By changing the amount of emulsifier, they adjusted the size of these particles in the range of 36–131 nm. The pre-polymer was cross-linked through F–C reaction, and obtained uniform microporous polymer nanoparticles. The BET surface area of these microporous nanaoparticles were around 1500 m^2 g^{-1} with pore volume of 0.56 cm^3 g^{-1} whose H_2 adsorption was 1.59 wt% at 77 K/1.13 bar.

Tan et al. [27] prepared sulfonated and acid-modified hypercrosslinked polymers (SAM-HCPs). It retained the original spherical shape along with porous structure and had sulfonic acid groups as hydrophilic groups as well as active sites. SAM-HCPs possessed a strong ability to absorb toxic metal ions. At 303, 313, and 323 K, the amount of Cu^{2+} adsorbed was 51.45, 54.82, and 57.68 mg g^{-1}, respectively.

HCPs have been prepared by F–C alkylation reaction. The monomers and catalyst are readily available, and the reaction is simple, low cost, and suitable for industrial production. HCPs molecular scaffolds are stable and can tolerate heat or acid effectively. But its disordered structures lead to a wide pore size distribution. Recent studies use microemulsion synthesis techniques to obtain a narrow pore size distribution of HCPs. Thus, HCPs have great application prospect in hydrogen storage. Functional HCPs could be the focus in the near future.

6.2.3 Covalent Organic Frameworks

COFs are porous polymers with ordered structure. Rigid units in their structure arrange neatly to form uniform size pores. Reversible chemistry, for example, boroxine- and boronate-ester formation and nitrile cyclization, tends to result in ordered thermodynamic products. The characteristics of crystalline microporous polymer are precise enough to control their surface chemistry. In addition, the introduction of specific molecular recognition or catalytic point in the polymer structure is conducive to the separation or selective adsorption chemistry and heterogeneous catalysis [28]. The idea of preparation of COFs originated from the synthesis of MOFs (developed by the Yaghi et al. [29]), which had surface area of more than 3000 $m^2\ g^{-1}$. Different from MOFs that contain metals, COFs are entirely constituted by organic elements. During the reaction, bonds are rapidly and reversibly linked under the thermodynamic control and thus form the most stable structure. The ordered network structure is maintained, yielding permanent microporous crystalline polymers.

Yaghi et al. [5] used 1,4-benzene diboronic acid (BDBA) self-condensation to form the first COF-1 (Figure 6.11) in high yields. The planar six-membered boron siloxane ring in the structure is similar to that of the metal clusters in the MOFs. The BET surface area of COF-1 was 711 $m^2\ g^{-1}$ and pore volume was 0.32 $cm^3\ g^{-1}$ with an average pore size of 1.5 nm. The powder XRD showed a highly ordered structure with each mezzanine arranged in a staggered manner. Solvent molecules accumulated within the 1.5 nm pores of COF-1 that can be removed at 200 °C without destroying the crystal structure. COF-5 was synthesized by reacting BDBA and 2,3,6,7,10,11-hexahydroxytriphenylene (HHTP) under similar reaction conditions. The diameter of the pore of COF-5 was 2.7 nm with a narrow pore size distribution, and the BET surface area was 1590 $m^2\ g^{-1}$.

Lavigne et al. [30] changed the monomer's structure and used benzene-1,3,5-triboronic acid (BTBA) and 1,2,4,5-tetrahydroxybenzene (THB) to synthesize COF-18 Å in high yields. COF-18 Å has a lower surface area (1260 $m^2\ g^{-1}$) and smaller pore volume (0.29 $cm^3\ g^{-1}$) than that of COF-5. COF-18 Å degraded

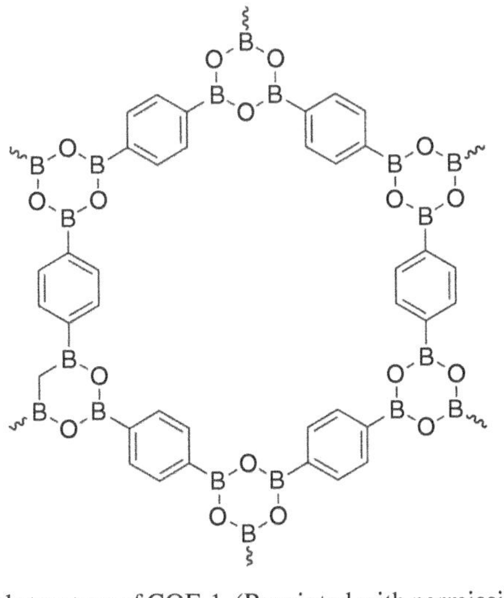

Figure 6.11 Chemical structure of COF-1. (Reprinted with permission from [5]. Copyright © 2006 American Association for the Advancement of Science.)

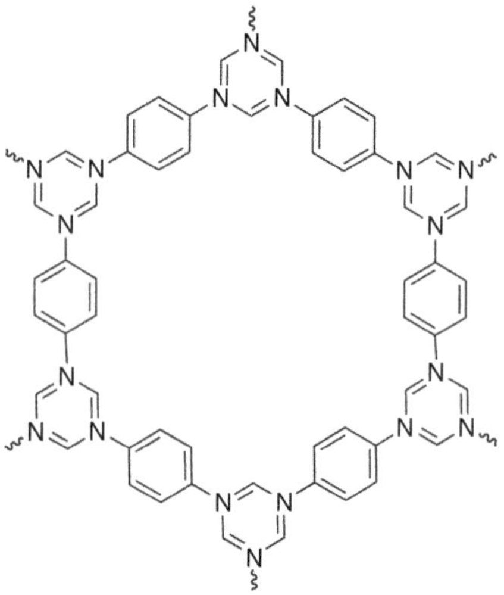

Figure 6.12 Chemical structure of CTF-1. (Reprinted with permission from [30]. Copyright © 2008 Wiley Periodicals Inc.)

in the deuterated water (D_2O) upon adding KOH, and the ^1H NMR analysis of COF-18 Å showed monomer ratio of 3 : 2 (BTBA/THB).

Thomas et al. [31] extended the range of COFs. Terephthalonitrile (DCB) was reacted at 400 °C in molten $ZnCl_2$ as a reaction solvent and as a trimerization catalyst. The crystalline triazine-based organic framework (CTF-1) (Figure 6.12) was formed quantitatively. The BET surface area of CTF-1 was 791 $m^2\ g^{-1}$. The powder XRD analysis showed the presence of ordered six-membered ring structure. Elemental analysis indicated that partial decomposition or carbonization occurred during the reaction. The surface area of COF increased when biphenyl-4,4′-dicarbonitrile (DCBP) instead of DCB was used. The surface area of this polymer was 2475 $m^2\ g^{-1}$ and H_2 adsorption was 1.55 wt% at 1 bar/77 K. Thus, the performance of DCBP network resembled other microporous materials such as most of the MOFs, mesoporous carbon materials, and zeolite. In 2009, Thomas et al. [32] changed the monomer's geometry and function. Condensation was also carried out at different temperatures. They observed that the chemical properties and pore properties can be influenced by reaction conditions, especially by temperature. In 2010, the same authors [33] replaced the monomers with 2,6-naphthalenedinitrile. Under the same reaction conditions, they obtained CTF-2 with well-structured distribution whose surface area was 2255 $m^2\ g^{-1}$ with a pore volume of 1.51 $cm^3\ g^{-1}$.

Yaghi et al. [34] used boric acid as precursors (Figure 6.13). Boric acid was reacted with HHTP to obtain a series of three-dimensional COFs (COF-102, COF-103, COF-105, COF-108), which exhibited excellent physical and chemical properties. Two different network structures, namely, ctn and bor types were formed

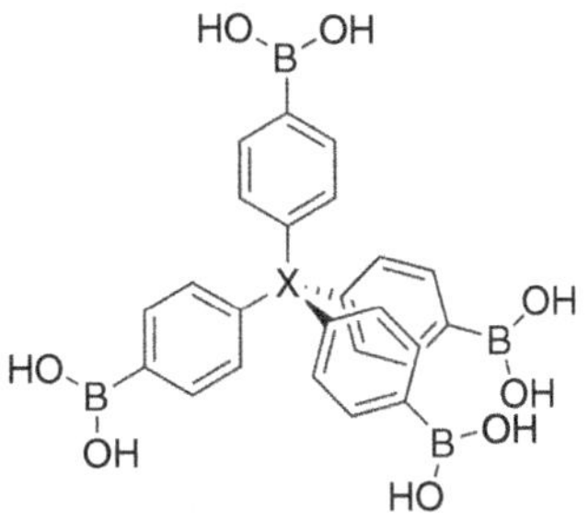

$X = C, Si$

Figure 6.13 Chemical structure of TBPM or TBPS. (Reprinted with permission from [34]. Copyright © 2007 American Association for the Advancement of Science.)

with only COF-108 yielding bor-type structure. The only difference between COF-108 and COF-105 was the central atom of the tetrahedron precursor, which was replaced from Si to C. The other three COFs formed ctn structure. COF-102, COF-103, and COF-105 showed pore diameters as 0.89, 0.96, and 1.83 nm with densities of 0.41, 0.38, and 0.18 g cm^{-3}, respectively. COF-108, which possessed bor-type crystal structure, exhibited two different pores with diameters of 1.52 and 2.96 nm. COF-108 was a crystalline substance of a low density of 0.17 g cm^{-3}. The surface areas of COF-102 and COF-103 were 3472 and 4210 m^2 g^{-1}, respectively, which corresponded to MOF-177 whose surface area was 4500 m^2 g^{-1}.

Yaghi et al. [35] followed different synthetic approach by selecting four tetra-(4-anilyl)methane, which were reacted with terephthaldehyde to form 3D structure of COF-300. The BET surface area of COF-300 was 1360 m^2 g^{-1} with a pore size of 0.72 nm.

Dichtel [36] explored the applications of COFs. They reacted phthalocyanine tetra(acetonide) with 1,4-phenylenebis(boronic acid) catalyzed by Lewis acid to form two-dimensional phthalocyanine-containing COFs. Boric acid connected the phthalocyanine rings to form square grids parallel to the chromophore with a pore size of 2.3 nm. This material had good thermodynamic stability and can be synthesized easily. The substance could absorb a wide range of sunlight through the accumulated phthalocyanine moieties in the COF, which could facilitate the flow of electrons. Thus, phthalocyanine-containing COFs have immense potential in opto-electronic devices.

The biggest feature of COFs is that the skeleton has a long-range order, whose pore size and pore size distribution are very uniform. Compared with other MOPs, they have lower density and higher surface area. In addition, pore size and surface area can be adjusted suitably by changing the interlinking monomer and also by varying the chemical stability. However, the stability of COFs formed by reversible thermodynamic reactions is poor, which limit the application of these materials in certain fields.

6.2.4 Conjugated Microporous Polymers

The $\pi-\pi$ conjugated bonds in CMPs arrange periodically between aromatic groups. The inherent rigidity of the system generates permanent pores. In the dry state, the high surface area can be combined with the system's electronic performance.

Cooper et al. [6] synthesized poly(aryleneethynylene) (PAE) network with permanent porous structure, whose BET surface area was above 834 $m^2\ g^{-1}$. Subsequently, CMPs' BET surface area was increased by 1000 $m^2\ g^{-1}$ [37]. This was the first example of CMPs with a high surface area. Heat treatment of acetylene branch connected to network of the polymer at 350–900 °C yielded porous polymers. Prepolymer network was designed with degradable side chains, and then porous structure was obtained under thermal degradation. Porous polymers were usually obtained by heating at 400 °C, which upon further heating to 900 °C caused carbonation. By employing Sonagashira–Hagihara reaction for cross-linking, a series of CMPs were obtained. BET surface area of PAE networked CMPs (Figure 6.14) was more than 1000 $m^2\ g^{-1}$.

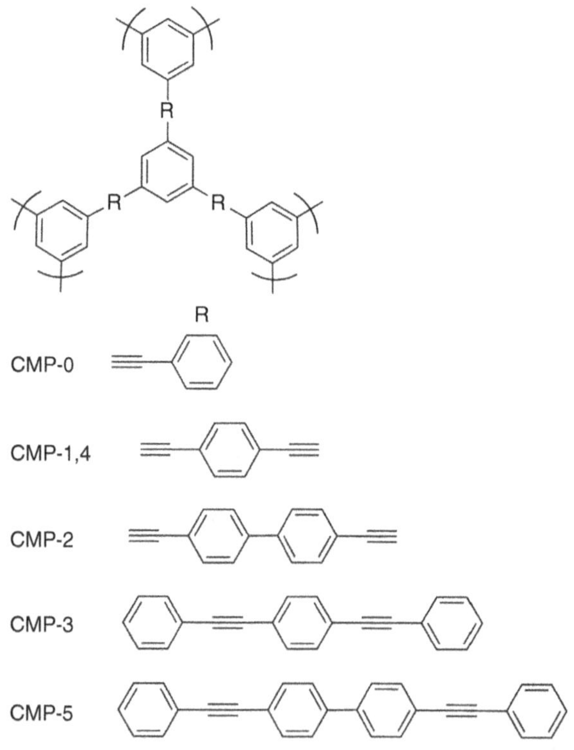

R
R
R
R
CMP-0
CMP-1,4
CMP-2
CMP-3
CMP-5

Figure 6.14 Chemical structure of CMP1-5. (Reprinted with permission from [37]. Copyright © 2008 American Chemical Society.)

Cooper et al. [38] prepared poly(phenylene butadiynylene)s (PPBs) through self-coupling with BET surface area of 842 m^2 g^{-1}. As compared with PAEs, the pore size distribution was very wide and possessed more complex structure. Many kinds of chemical reactions occurred during the polymerization, and the side reactions can be avoided by selecting appropriate reaction conditions [39].

Although PAE polymers and their copolymers were formed with disordered structure, it is possible to fine-tune the pore size, pore volume, and surface area of PAE network by changing the length of the rigid connection. These networks were nonconductive. The structure of these networks can be modified by doping [7].

The monomers used to synthesize porous PAE networks were usually 1,4- or 1,3,5-substituted benzenes. Typically, polymers generated by these monomers exhibited planar conformation. However, PAE networks can also form 3D structures because of the ease of rotating the acetylene bond. In 2008, Thomas et al. [40] used 2,2′,7,7′-tetrabromo-9,9′-spirobifluorene (TBSBF) to generate CMP-P1 with 3D geometry (Figure 6.15). This monomer was also used to prepare porous polyphenylene network and was used in organic light-emitting diode (OLED). These substances also possessed microporosity.

Zhu et al. [2], inspired by the diamond structure, used one or three benzene rings to replace the C–C covalent bond to form aromatic porous network (porous aromatic framework, PAF 1–3) with the 3D structure. Different from the CMPs' amorphous structure, these CMPs possessed a similar configuration as that of diamond. The benzene rings connected each other and stick to tetrahedron vertex. The benzene plane supported mutually, which greatly increased the specific surface area. Among these CMPs, the BET-specific surface area of PAF-1 (5640 m^2 g^{-1}) is the highest of all MOPs. H_2 adsorption at 77 K/48 bar was 10.7 wt%. Carbon dioxide (CO_2) adsorption was 1300 mg g^{-1} at 298 K/40 bar. PAFs possessed very high surface area. The hydrophobic nature of PAFs made them to adsorb benzene and toluene vapors at room temperature.

CMPs can not only be used for the adsorption but can also be applied in the optical area because of its conjugated structure. In 2008, Cooper et al. [41] coupled 1,2,4,5-tetrakis(bromomethyl)benzene (TBMB) to form mesoporous supported poly(*p*-phenylene vinylene) (PPV) network (Figure 6.16). Solid-state NMR and IR data showed that this network had highly branched forms and the network contained a considerable number of end groups. The PPV networks absorbed wavelengths of light in the range 250–400 nm and emitted 500–525 nm

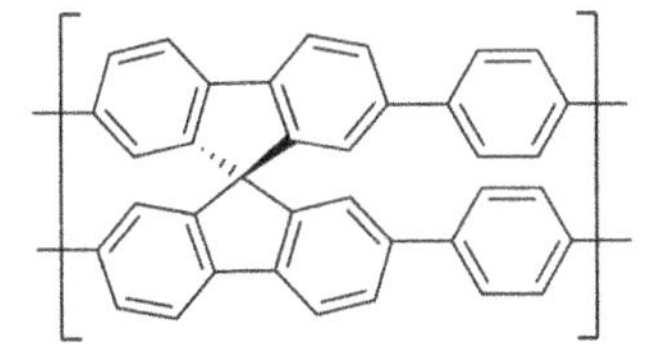

Figure 6.15 Chemical structure of CMP-P1. (Reprinted with permission from [40]. Copyright © 2008 American Chemical Society.)

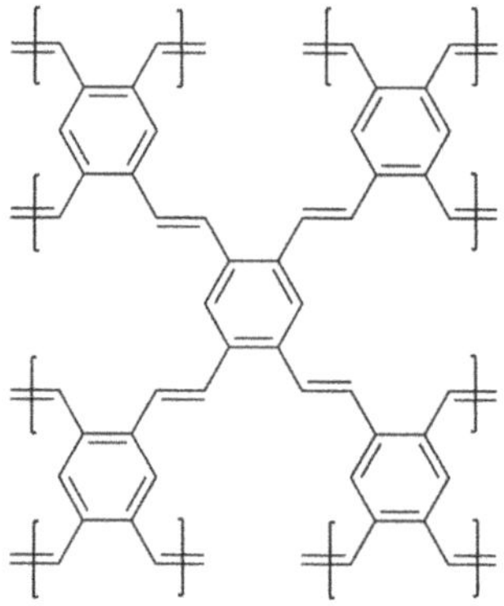

Figure 6.16 Chemical structure of PPV. (Reprinted with permission from [41]. Copyright © 2008 American Chemical Society.)

fluorescence. The electronic effects can be influenced by filling the mesopores with organic or inorganic donor and acceptor compounds. When 1,3,5-tribromobenzene was reacted under the same condition, soluble porous network was produced.

Cooper et al. [42] used poly(triethynylphenyl) amine cross-linked with a series of iodine monomer and obtained NCMP 0–4. NCMPs were chemically and thermodynamically stable. BET surface area of NCMPs was in the range 546–1108 $m^2\ g^{-1}$. The presence of N in the network structure introduced high binding energy sites for metal halides or transition metal complexes, which enabled H_2 storage. The presence of electron-donating triaminobenzene in polymer makes these CMPs suitable for applications such as hybrid photoelectric materials and charge transport materials.

Cooper et al. [43] used spiro-bis(2,5-dibromopropylenedioxythiophene) with different acetylene–benzene cross-linkers to form SPT-CMP1-3 with BET surface areas of 1631, 1601, and 1334 $m^2\ g^{-1}$. SPT-CMP1 showed the highest H_2 adsorption capacity (1.7 wt% at 1.13 bar/77 K). SPT-CMP not only possessed high microporosity but also contained thiophene in the main chain, which introduced low band gap into the polymer, thereby making them potential for producing hybrid optical and conductive porous materials.

Recently, in addition to the potential application of CMPs as photoelectric materials, applications in catalysis had also been reported. In 2010, Wang et al. [44] used the 1,3,5-triethynylbenzene as a linker and Tröger's base as a functional monomer to form CMPs suitable for catalysis. The BET-specific surface area of CMPs was 750 $m^2\ g^{-1}$. Similar to that of the PIMs, the porosity was generated by the rigid and contorted structure of Tröger's base. Considerable catalytic activity was observed for the addition of diethylzinc to 4-chloro-benzaldehyde, and even after three catalytic cycles, activity did not decrease. In the same year, Jiang et al. [45] used ferric porphyrin derivatives as the catalytic monomer for preparing CMP with large specific surface area. In the network of Fe-CMP, there were high density catalytic sites, which are beneficial for exposure to reactants. The large specific surface area was conducive for high conversions. Experiments showed that, under mild conditions, catalytic oxidation of sulfur to the corresponding sulfoxide occurred with high level of activity and selectivity by Fe-CMP. For various sulfur atoms, the Fe-CMP had good catalytic effect and was easy to recover, making them suitable for large-scale industrial applications.

Moreover, Deng et al. [46] doped Li into CMPs and applied them for reversible H_2 storage. High electron affinity of sp^2 carbon network separated electrons from the Li center, which provided strong stability for the H_2 molecule. The structure can notably improve the uptake of H_2 when compared with that of the undoped system. Hydrogen storage was 6.1 wt% at 1 bar/77 K. The CMPs were prepared from 1,3,5-triethynylbenzene. Active sites (carbon–carbon triple bond) in CMP can potentially combine with metal ions. When the Li concentration was higher than 0.5 wt%, H_2 uptake decreased significantly probably because of the high agglomeration of Li.

CMPs have conjugate structure, precisely controlled pore size, large specific surface area, high degree of stability, and can be prepared by a variety of reactions. CMPs are prepared by kinetically controlled reactions. CMPs lack long-range order in their structure. However, its pore size and surface area can be continuously adjusted by changing the alkyne bond length between the aromatic rings. CMPs can be easily modified to introduce functionalities that would make them suitable for catalytic applications. In addition, CMPs have great potential in gas adsorption, electrical conductivity, and other areas. However, CMPs are prepared with expensive transition metal catalysts and monomers. After the reaction, the transition metal catalyst residues in CMPs cannot be removed completely, which restricts their applications.

6.3 HYDROGEN ADSORPTION

The continuous depletion of fossil fuels and possible global warming highlights the serious and urgent need to look at sustainable alternatives, which could replace traditional fossil fuels used extensively in a variety of vehicles and many other systems. Thanks to its high density of energy, pollution-free advantage, and availability in abundance from various sources, hydrogen could undeniably be a perfect candidate to replace fossil fuels [47]. Recently, an inspiring breakthrough in hydrogen production [48] was reported. However, finding a practical way to store sufficient quantities of H_2 in the limited space available in an automobile is still a formidable challenge. A large number of materials have been investigated as physisorptive H_2 adsorbents including carbon [49], zeolites [50], MOFs [51], and COFs [52]. None of these materials meets the current criteria of size, recharge kinetics, cost, and safety required for use in transportation systems.

Organic porous materials a [20,21,37,53] have become favorite candidates for reversible hydrogen storage during the last few years because of obvious advantages over other porous materials: they are composed solely of light elements, can easily be designed and produced in large scale, and are moldable in "monolithic" form [25] required for use in transportation systems.

6.3.1 HCPs for Hydrogen Adsorption

Germain and coworkers tested hydrogen adsorption of both commercial and custom-made cross-linked poly(styrene-*co*-divinylbenzenes), derivatives, and hypercrosslinked poly(styrene-*o*-divinylbenzenes) for hydrogen storage under cryogenic conditions (Table 6.1). These HCPs reversibly adsorbed 1.5 wt % of H_2 under 0.12 MPa of pressure at 77 K. It was found that hypercrosslinked polystyrenes such as Hypersol–Macronet MN200 adsorb more hydrogen than poly(styrene-*co*-divinylbenzene) networks such as Amberlite XAD4, even when the latter exhibits higher surface area [24]. Later, they reported excess and total hydrogen storage capacities of 3.8 and 5.4 wt% of hydrogen adsorption under 4.5 and 8.0 MPa of pressure, respectively [54]. Cooper's group-tested hypercrosslinked poly(divinylbenzene-*co*-vinylbenzyl chloride) (HCP-DVB-VBC) [20a] and hypercrosslinked self-condensation of bischloromethyl monomers (HCP-DCX-BCMBP) [21] were also tested for hydrogen adsorption. HCP-DVB-VBC adsorbed 1.3 wt% under 0.12 MPa at 77 K. Access to high pressure adsorption instrumentation measured a hydrogen storage capacity of 3.0 wt% at 77 K under 3 MPa of pressure. The enthalpy of hydrogen adsorption of this material was 6.6 kJ mol^{-1} at 0.17 wt%. Uniform microporous polymer nanoparticles of HCP-DVB-VBC present higher hydrogen adsorption capacity (1.59 wt%, 77.3 K, 1.13 bar) and faster adsorption rate as compared to that of polydisperse microsize analogs previously reported because of smaller micropore and shorter hydrogen diffusion rate [26]. HCP-DCX-BCMBP demonstrated hydrogen storage capacities of up to 3.7 wt% at 77 K under 1.5 MPa of pressure. The highest adsorption enthalpy exhibited by these materials is 7.5 kJ mol^{-1}, which is higher than 6.6 kJ mol^{-1} measured earlier but still in the cryogenic range. Analyses of the total hydrogen storage capacity and nanopore volume suggest that the nanopores of this material are filled with hydrogen at a density close to that of liquid hydrogen under these saturation conditions.

Studies of hypercrosslinked polystyrenes have demonstrated that they can store significant amounts of hydrogen at cryogenic temperatures. Unfortunately, like other hydrogen storage materials based on physical adsorption, such as MOF, active carbons, and zeolite, they have little hydrogen storage capacity at room temperature. Two different approaches, addition of electron-donating groups and the introduction of smaller pores, have been explored in order to bring the enthalpy of hydrogen adsorption closer to 15 kJ mol^{-1}, which is the limit proposed for reversible hydrogen storage at room temperature. Several groups have shown that aromatic rings are the primary adsorption site in materials such as those that lack metal coordination sites [57] and suggested that adding electron-donating groups to the aromatic ring may increase H_2–polymer interaction [58]. As a partial test, hypercrosslinked polystyrenes were modified with a series of electron-withdrawing groups such as nitro and bromo groups. The resulting materials adsorbed less hydrogen per aromatic ring than an unmodified precursor [54]. It is important to note that this approach is valid only when the added group does not contain moieties, such as ions, that might themselves act as separate hydrogen adsorption sites. Germain and coworkers [59] synthesized a series of

TABLE 6.1 Surface Areas and Hydrogen Adsorption of Commercial Cross-Linked Polymers and Those Synthesized via Suspension Polymerization

		Surface Area, m^2 g^{-1}		
Trade Name	Polymer Type	BET[b]	Langmuir[c]	H_2, wt%[a]
Amberlite XAD4	Poly(styrene-*co*-divinylbenzene)	1060	425	0.8
Locally synthesized [55]	Poly(glycidyl methacrylate-*co*-ethylene glycol dimethacrylate), ammonia treated	192	131	0.2
Locally synthesized using a reported method [56]	Poly(chloromethylstyrene-*co*-divinylbenzene)	359	161	0.4
Locally synthesized using a reported method [56]	Poly(4-vinylpyridine-*co*-divinylbenzene)	551	361	0.6
Amberlite XAD4	Poly(styrene-*co*-divinylbenzene)	1060	425	0.8
Amberlite XAD16	Poly(styrene-*co*-divinylbenzene)	770	336	0.6
Hayesep N	Poly(divinylbenzene-*co*-ethylenedimethacrylate)	460	247	0.5
Hayesep B	Poly(divinylbenzene) modified with poly(ethyleneimine)	570	290	0.5
Hayesep S	Poly(divinylbenzene-*co*-4-vinylpyridine)	510	254	0.5
Wofatit Y77	Poly(styrene-*co*-divinylbenzene)	940	573	1.2
Lewatit EP63	Poly(styrene-*co*-divinylbenzene)	1206	664	1.3
Lewatit VP OC 1064	Poly(styrene-*co*-divinylbenzene)	810	377	0.7
Hypersol–Macronet MN200	Hypercrosslinked polystyrene	840	576	1.3
Hypersol–Macronet MN100 600 477 1.1	Amine-functionalized hypercrosslinked polystyrene	600	477	1.1
Hypersol–Macronet MN500	Sulfonated hypercrosslinked polystyrene	370	266	0.7

[a]Hydrogen storage capacity at 77 K and 0.12 MPa. Portions reproduced with permission from [24]. Copyright 2005, American Chemical Society.
[b]Calculated from nitrogen adsorption isotherms using the BET equation.
[c]Calculated from hydrogen adsorption using the Langmuir equation.

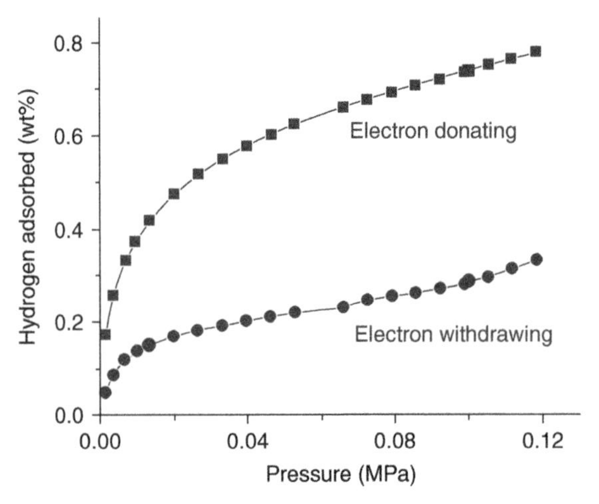

Figure 6.17 Hydrogen uptake of a hypercrosslinked polyaniline before and after converting the electron-donating amine bases to electron-withdrawing groups via protonation. (Reproduced with permission from [60]. Copyright © 2007 Royal Society of Chemistry.)

hypercrosslinked polyanilines with electron-donating amino groups inherent in the HCP. The resulting materials had smaller surface areas than the hypercrosslinked polystyrenes and a maximum hydrogen storage capacity of 2.2 wt% under 3 MPa at 77 K. However, they also exhibited higher adsorption enthalpies, up to 9.3 kJ mol^{-1}, than the porous polystyrenes. It is important to note that the unprotonated amine in polyaniline is a strong electron-donating group; however, the polymer loses nearly all abilities to adsorb hydrogen when this amine becomes electron withdrawing via protonation (Figure 6.17). Later, Germain and coworkers have coupled aryl amines with aryl halides to form networks of aromatic rings linked with nitrogen as a single-atom, trivalent cross-linker. The nitrogen atom acts as an electron-donating group while reducing the cross-link size to that of a single atom and minimizing pore size. The resulting materials had very small pores, in some cases even including pores that are too small for nitrogen adsorption but large enough for hydrogen adsorption. At low surface coverages, these small pore sizes result in hydrogen adsorption enthalpies as high as 18 kJ mol^{-1}, the highest observed to date for a polymer sorbent [59].

6.3.2 PIMs for Hydrogen Adsorption

The first PIMs with surface area of 760 m^2 g^{-1} to 830 m^2 g^{-1} as hydrogen storage materials possessed a hydrogen uptake of up to 0.7 wt% under 0.1 MPa and 1.7 wt% under 1 MPa at 77 K [53a]. Further work by Ghanem et al. [13] reported a PIM with a surface area of 1065 m^2 g^{-1} at 77 K storing 1.7 wt% of hydrogen under 0.1 MPa and 2.7 wt% under 1.0 MPa. Interestingly, this material relied on a triptycene moiety instead of the more common spirocenter as a source of microporosity. Lower hydrogen storage capacities of 0.8–1.3 wt% have been observed at 77 K under 0.1 MPa using fluoropolymers with intrinsic microporosity [61].

6.3.3 COFs for Hydrogen Adsorption

Lavigne et al. synthesized a series of COFs with alkyl chains of varying size. COF-18 Å exhibited rigid, stable pores with high surface (1260 m^2 g^{-1}) and showed the highest uptake of H_2 of 1.5 wt% at 77 K under 0.1 MPa. The hydrogen storage capacity under some conditions decreased as the size of alkyl chains increased [62]. Cooper et al. [63] developed a microwave-based synthetic method that produced COF-5 and COF-102 with BET surface areas up to 2930 m^2 g^{-1} and hydrogen uptakes of 1.7 wt% at 77 K under 0.1 MPa pressure. The most promising COF thus far is a triazine-based framework with a BET surface area of 2480 m^2 g^{-1} with hydrogen uptake of 1.55 wt% under 0.1 MPa at 77 K.

6.3.4 CMPs for Hydrogen Adsorption

Recently, a porous aromatic framework (PAF) with an unprecedented high surface area has been synthesized successfully (BET surface area is 5600 m^2 g^{-1}). PAF-1 possesses local diamond-like tetrahedral bonding of tetraphenylene methane building units to produce exceptional thermal and hydrolytic stabilities. The hydrogen uptake capacity of PAF-1 under 48 bar pressure at 77 K reached 7.0 wt%, which corresponds to an absolute uptake of 10.7 wt%. The H_2 absorption capability of PAF-1 is comparable to the best performing high surface area porous MOFs and COFs, [64] and is the highest among porous organic polymers [8]. From the high pressure hydrogen sorption isotherm data of PAF-1, the isosteric heats of adsorption were calculated using the method described by Zhou et al. [65]. The zero-loading Q_{st} was approximately 4.6 kJ mol^{-1} and shows little loading dependence. This observation is consistent with sorption on a homogeneous porous material [66] and further supports the role of surface area on the total H_2 uptake. A series of tetrahedrally linked CMPs were prepared using a variety of bond-forming chemistries including Sonogashira–Hagihara coupling, Yamamoto coupling, thermal alkyne condensation, and "click" chemistry. These thermally stable polymers exhibit high surface areas (up to 3200 m^2 g^{-1}) and adsorb up to 2.34 wt% hydrogen by mass at 1.13 bar/77 K [67]. Deng and coworkers have demonstrated a strategy to enhance the hydrogen storage capacity of CMP by doping with Li^+ ions. The capacity of these Li-CMP structures for hydrogen under 1 bar at 77 K was 6.1 wt%, which is one of the highest reported to date for physisorption materials such as carbon nanotube-based MOFs under the same conditions [46].

6.4 CARBON DIOXIDE CAPTURE

The concern of global warming caused by the emission of greenhouse gases has drawn unprecedented public attention. Carbon dioxide (CO_2) is the major anthropogenic greenhouse gas, and it accounts for over 60% of the greenhouse effect [68]. The main sources of CO_2 emissions are from the burning of fossil fuels in automobiles and power plants. It has been reported that coal-fired power plants generate about one-third of the CO_2 released to the atmosphere, which result in increased CO_2 emissions in the atmosphere [69]. Since fossil fuels will still be the dominant source of energy in the next few decades and an immediate CO_2-emission halt is impossible, carbon dioxide capture and storage (CCS) technologies thus are one of the plausible way to reduce CO_2 emissions to the atmosphere [70]. Capture and separation of CO_2 are a major road block for the safe storage of carbon into deep geological formations. Conventional CO_2 capture processes involving the chemisorption of CO_2 by alkylamine-containing liquids are costly and inefficient. The major drawbacks include low CO_2 uptake, degradation of the solvent, and the high temperature required to regenerate the adsorbed gas [71]. The development of a low cost means of capturing CO_2 is therefore crucial to stabilize CO_2 levels in the atmosphere.

Adsorption separation by highly porous solids is considered to be one of more promising technologies for CO_2 capture because of their large surface area, low cost, high CO_2 selectivity, and adsorption capacity [72]. Several classes of porous materials such as silicates [73], active carbons [74], zeolites [75], MOFs [76], and zeolitic imidazole frameworks [77] have been investigated as potential adsorbents for CO_2 capture. Porous polymers, which are composed solely of light elements (H, B, C, N, O, Si), add new merits to the family of adsorbents because of their low cost, ease of processing, and high thermal and chemical stability [7]. Porous organic polymers possess a number of potential advantages than that of its analogs. For example, the great choice of monomers available makes it easy to introduce various functional groups in the pore walls [78]. Porous polymer networks (PPNs) can be constructed using a plethora of organic reactions and building blocks, which provides flexibility for the material design to achieve desirable pore properties [79]. Furthermore, polymers are scalable and there are already examples of systems that are produced commercially on a large scale. All of these make porous organic polymers attractive as adsorbents for CO_2 adsorption.

The first porous organic polymers studied for CO_2 adsorption were porous COFs reported by Yaghi and coworkers [52]. By using reversible boronic acid condensation, they prepared a series of COFs with different structural dimensions and pore sizes. Measurements of CO_2 adsorption properties showed that the uptake of excess mass at the surface and the saturation pressures were sensitive to the structure of COFs. COF102 comprised of tetra[4-(dihydroxy)borylphenyl]methane units has a cubic structure with a lattice constant of 27.18 Å. It exhibits the highest CO_2 uptake in this class (1200 mg g^{-1} under 55 bar at 298 K), which exceed the capacities of MOF-5 (970 mg g^{-1}) [80], carbon materials (420 mg g^{-1}), and 370 mg g^{-1} for Norit RB2 and BPL carbon, respectively), and zeolites (220–350 mg g^{-1}) (Table 6.2). According to molecular simulations, the crystal density of COF-102 is 0.41 g cm^{-3} [81]. Owing to such compact atomic packing, COF102 has high capacity even at considerably low pressures, which is advantageous for operation from a safety point of view.

TABLE 6.2 Summary of Porosity Measurement for COFs and the Carbon Dioxide Adsorption at 55 bar and 298 K[a]

Material	Composition	S_{Lang} (m^2 g^{-1})	S_{BET} (m^2 g^{-1})	Pore Size (Å)	CO_2 Uptake (mg g^{-1})
COF-1	C_3H_2BO	970	750	9	230
COF-5	$C_9H_4BO_2$	1990 (3300)	1670 (2050)	27	870
COF-6	$C_9H_4BO_2$	980	750	9	310
COF-8	$C_{14}H_7BO_2$	1400 (2110)	1350 (1710)	16	630
COF-10	$C_{12}H_6BO_2$	2080 (4620)	1760 (1980)	32	1010
COF-102	$C_{25}H_{24}B_4O_8$	4650	3620	12	1200
COF-103	$C_{24}H_{24}B_4O_8Si$	4630	3530	12	1190
BPL carbon		1500	1250		370
Zeolites			260–590		220–350
Norit RB2					420 (40 bar)

[a][52].

Cooper et al. [8] pioneered the CMPs, in which Sonogashira–Hagihara coupling was adopted to generate polymeric frameworks with high microporosity and chemical resistance. Recently, Zhu et al. [2] made a breakthrough. They developed a new class of porous materials and PAFs, with diamond-like structure in which the single bond between two sp^3 carbon atoms is replaced by single or multiple phenyl groups. One such PAF (named as PAF-1) exhibited unprecedented surface area (BET surface area, 5640 m^2 g^{-1}). To date, this exceeds surface areas measured for other materials, either crystalline or amorphous [82]. Owing to its very high porosity and surface area, PAF-1 exhibits excellent carbon dioxide (29.5 mmol g^{-1} at 298 K under 40 bar) storage capacities. Although this CO_2 uptake capacity is slightly lower than post-treated MIL-101 [83], yet it is one of the highest reported for porous materials. Inspired by this work, Lu et al. [84] and Holst et al. [67] synthesized PPNs and CMPs by coupling of tetrahedral monomers, respectively. Lu et al. observed that the gas uptake capacities are directly proportional to the surface areas of materials. PPN-3, with the BET surface area of 2840 m^2 g^{-1}, stored 25.3 mmol g^{-1} of CO_2, which is comparable with that of PAF-1 (29.5 mmol g^{-1}). Network 3 prepared by Lu et al. exhibit high surface area (up to 3200 m^2 g^{-1}) and adsorb up to 7.59 wt% (1.81 mmol g^{-1}) CO_2 by mass under 1.13 bar at 298 K; in fact, this is higher than that reported for PAF-1, despite having a lower surface area. For CO_2 capture at an ambient pressure, materials with very high surface areas may not be optimal [85]. Increasing the heat of adsorption could have more potential to increase the amount of gas adsorbed. Owing to their favorable affinity toward CO_2, amine functionality b [71,86] or accessible nitrogen sites [87] were usually introduced into the pore wall of porous solids. In a recent work, Choi et al. [88] incorporated tris(4-iodophenyl)amine into CMP materials, and a CO_2 capture capacity of 507.7 mg g^{-1} (under 1.06 bar at 195 K) was obtained. Rabbani et al. [89] reported that porous benzimidazole-linked polymer (BILP-1) could adsorb around 188 mg g^{-1} of CO_2 at 273 K under 1 bar. At zero coverage, the Q_{st} was 26.5 kJ mol^{-1}. The CO_2 uptake and Q_{st} were higher than that reported for COFs [52], imine-linked organic cages [53b], and diimide polymers [90]. In addition, BILP-1 shows a high CO_2/N_2 selectivity of 70. According to the authors, the high CO_2 capture capacity and selectivity for CO_2 over N_2 stem from the fact that the imidazole moieties of BILP-1 interact more favorably with the polarizable CO_2 molecules through hydrogen bonding and/or dipole–quadrupole interactions involving the protonated- and proton-free nitrogen sites, respectively. Babarao et al. computationally designed new PAFs by introducing polar organic groups ($-NH_2$, $-OCH_3$, and $-CH_2OCH_2-$ as in THF) to the biphenyl unit and then investigated their separating power toward CO_2 by using grand canonical Monte Carlo (GCMC) simulations [91]. Among these functional PAFs, they found that tetrahydrofuran-like ether-functionalized PAF-1 shows higher adsorption capacity for CO_2 under 1 bar at 298 K (10 mol per kilogram of adsorbent) when compared with that of the amine functionality. The authors ascribed the high CO_2 capacity to the strongest electrostatic interactions between CO_2 and THF. Recently, Torrisi et al. calculated the isosteric heats of sorption for CO_2 in functionalized MOFs and predicted that incorporation of a carboxylic acid group would lead to the highest isosteric heat [92]. Both results challenged the current research emphasis in the literature regarding amino groups for CO_2 captures [86,93]. By choosing proper comonomers, Cooper et al. [94] synthesized a series of CMP materials with different chemical functionalities such as carboxylic acids, amines, hydroxyl groups, and methyl groups. The experimental isosteric heats showed the following order in terms of appended functional groups (Figure 6.18): COOH > $(OH)_2$ > NH_2 > H > $(CH_3)_2$ for adsorbed gas quantities greater than 0.2 mmol g^{-1}. At low absorption, CMP-1-COOH showed a heat of adsorption reaching 33 kJ mol^{-1}, which is higher than the less polar CMP materials in this series. These data corroborate the fundamental computational observation that carboxylic acid functionalities are a good target for CO_2 capture materials.

Recently, the interactions between CO_2- and N-containing organic heterocyclic compounds such as pyridine, imidazole, and tetrazole have been studied [95], revealing that Lewis acid– base interactions are most important, as is hydrogen bonding with the negatively charged oxygen atoms of CO_2. The N-containing heterocylic ring was incorporated within the pores either by the use of N-containing monomer or via post synthetic modifications. Weber's [96] research group carried out systematic work on the intrinsically microporous poly(imide)s. On a binaphthalene-based polyimide, a CO_2 capture capacity of 1.58 mmol g^{-1} at 273 K under 1 bar was observed [96a]. As these polymers are soluble in common organic solvents, they can be cast directly as microporous films and used for gas separation studies. They also synthesized a chiral porous poly(imide) for the first time, which can adsorb significant amounts of CO_2 (8.3 wt% at 273 K under 1 bar) [96c]. In addition, the overall porosity of the chiral polymer is marginally higher than its racemic counterpart. In a recent work by Du et al. [97], tetrazole groups were incorporated into PIM-1 by post-polymerization modification. This approach has the advantage that the molecular structure of the polymer backbone is unaffected,

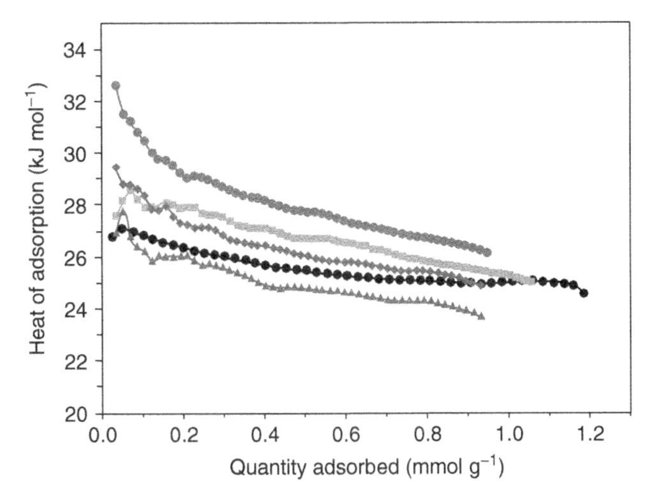

Figure 6.18 Measured isosteric heats of CO_2 for CMP networks. Color coding is as follows: unsubstituted networks (black); $-(CH_3)_2$ (green); $-(OH)_2$ (orange); $-NH_2$ (blue) and $-COOH$ (red). (Reproduced with permission from [94]. Copyright © 2011 Royal Society of Chemistry.) (*See insert for color representation.*)

while the amount of the tetrazole groups throughout the polymer can be adjusted by monitoring the post-polymerization reaction, namely, [2+3] cycloaddition between nitriles and azides. Owing to the strong interactions between CO_2 and the tetrazole, the amount of CO_2 sorption at 273 K in TZPIM-3 (100% conversion of nitrile to tetrazole) is higher than that in PIM-1 in the low pressure range (Figure 6.19), although the BET surface area in TZPIM-3 (30 $m^2\ g^{-1}$) is markedly smaller than that of PIM-1 (700 $m^2\ g^{-1}$). More importantly, the TZPIM membranes demonstrate exceptional gas separation performance, surpassing the most recent upper bounds [98] of conventional and state-of-the-art polymeric membranes for the important gas pairs, such as the CO_2/N_2 separation shown in Figure 6.20.

To date, only few HCPs have been examined for CO_2 capture. Martin et al. [101] synthesized a series of HCPs by copolymerization of *p*-dichloroxylene (*p*-DCX) and 4,4′-bis(chloromethyl)-1,1′-biphenyl (BCMBP). At higher pressure (30 bar), the polymers showed CO_2 uptakes of up to 13.4 mmol g^{-1} (59 wt%), superior to zeolite-based materials (zeolite 13X, zeolite NaX), and activated carbons (BPL, Norit R) available commercially (Table 6.3). Li et al. reported that the HCP prepared by 1,3,5-triphenylbenzene adsorbed 15.9 wt% of CO_2 under 1.13 bar at 273 K.

6.5 SEPARATIONS

MOPs are hot candidates for separation technology because of their high surface areas, chemical and physical stabilities, and micropore sizes, which have molecular

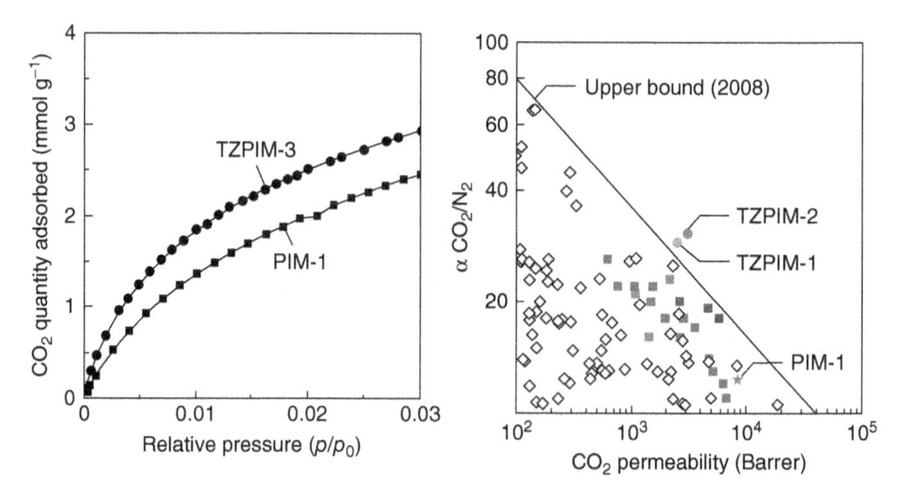

Figure 6.19 (a) CO_2 adsorption isotherms at 0 °C for PIM-1 and TZPIM-3 at a low ratio of gas pressure to saturation pressure. (b) Relationship between CO_2 permeability and CO_2/N_2 selectivity of TZPIMs and PIM-1. Measurements are for single gases and the upper bound is from [41]. TZPIM-1 (55% conversion of nitrile groups to tetrazole groups); TZPIM-2 (70.5% conversion). The squares correspond to the literature data for various PIMs. (Reproduced with permission from [97]. Copyright © 2011, Nature Publishing Group.) (*See insert for color representation.*)

dimensions. And the broad synthetic versatility is another major advantage in terms of designing MOPs with specific functional groups. Various classes of HCPs and PIM have been used in practical separation applications.

6.5.1 HCPs for Separations

Davankov resins, that is, HCPs synthesized via F–C reaction [102], were one of the earliest porous organic polymers which found their way to industry by the end of the 1990s [22]. Since then, various improved HCPs have been used in broad applications. HCPs in particular are quite well developed in terms of chromatographic separations and for the adsorption of specific components from solution mixtures. To give just a few examples, hypercrosslinked polystyrene has been used as a stationary phase for HPLC [103] and as a sorbent for solid-phase extraction of polar compounds from water [104]. The ability to produce bimodal pore size distributions (e.g., large "flow" macropores interconnected with micropores for surface area) [20b] is a significant advantage here, as is the relatively low cost and high stability of these materials. A generic issue for chromatographic separation media is particle size control. HCPs may be produced by suspension polymerization to yield relatively well-controlled polymer beads. In addition, HCPs have been used to adsorb and thus remove toxic organic [105] and inorganic [27,106] contaminations from solution, and also to remove toxic gases [107] to purify air.

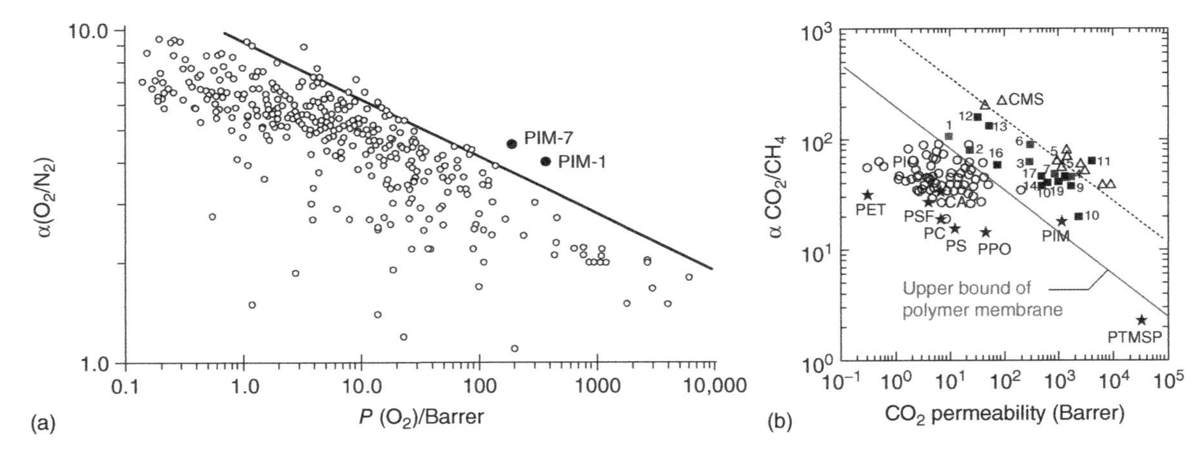

Figure 6.20 (a) Robeson plot showing high O_2/N_2 separation selectivities for PIM-1 and PIM-7. (Reprinted with permission from [99]. Copyright © 1991 Elsevier B. V.) (b) Relation between CO_2 permeability and CO_2/CH_4 selectivity of thermally rearranged polyimides. (Reprinted with permission from [100]. Copyright © 2007 American Association for the Advancement of Science.)

TABLE 6.3 CO_2 Capture Performance of Adsorbent Materials[a]

Materials	CO_2 Heat of Adsorption, kJ mol^{-1}	CO_2 Adsorption Capacity, mmol g^{-1}	Conditions (P, bar; T, K)
HCP-1	23.5	1.7	1, 298
		13.3	30, 298
HCP-2	21.2	1.7	1, 298
		12.6	30, 298
HCP-3	22.1	1.6	1, 298
		11.6	30, 298
HCP-4	21.6	1.6	1, 298
		10.6	30, 298
$Ni_2(BCD)_2$ Dabco[b]	20.4	12.5	15, 298
		14.8	24, 298
$Zn_2(BCD)_2$ Dabco[b]	21.6	2.1	1, 298
		13.7	15, 298
ZIF-70		2.2	1, 298
USO-2-Ni		13.6	25, 298
USO-2-Ni-A[c]		3.2	1, 298
Zeolite 13X	37.2	7.4	32, 298
		3.3	1, 298
Zeolite NaX	49.0	7.8	30, 302
AC-MAXSORB	16.2	25.0	35, 298
		2.1	1, 301
AC-Norit R1	22.0	10	30, 298
AC-Norit RB2		9.5	40, 298
BPL carbon	23.3	8.4	55, 298
		1.9	1, 298
		1.9	1, 301
COF-102		27.3	55, 298
MOF-177	35.0	32.5	30, 298
		20.0	15, 298
MOF-5(IRMOF-1)	34.1	1.9	1, 298
		21.7	35, 298
MCM-48[d]		0.8	1, 298
MIL-53(Al. Cr)[e]	36.0	10	25, 302
MIL-47(V)[e]		11	20, 302
PAF-1[f]		29.5	40, 298

[a][27].
[b]Three-dimensional pillared layer metal-organic framework.
[c]Amine-functionalized MOF.
[d]A type of amine-attached silica.
[e]MIL, materials of Institut Lavoisier (mesoporous MOFs).
[f]Microporous polyphenylene network.

6.5.2 PIMs for Separations

Practical gas separations tend to require relatively thin films to allow high gas fluxes. PIMs have a unique advantage here in that they can be dissolved in organic solvents and cast directly as microporous films [108]. Solution-cast PIM membranes have been used for both gas separation [109] and pervaporation of aqueous phenol solutions [110]. A practical trade-off here is the balance between gas permeability and selectivity: in general, high gas permeabilities tend to be accompanied by low separation selectivities and vice versa, as illustrated by the "Robeson plot". As shown in Figure 5.14, PIM-1 and PIM-7 have been found to exhibit substantially higher O_2/N_2 selectivities [$a(O_2/N_2) > 3.0$] than other polymers of similar permeability [109]. Other "thermally rearranged" [100] polyimides show excellent CO_2/CH_4 separation selectivities. These materials were also shown to function as fuel cell membranes when doped with H_3PO_4, and proton conductivities of 0.15 S cm^{-1} were observed at 130 °C [100] which is higher than polybenzimidazole membranes. In addition to gas separation, PIM have a particular affinity for organic species. Thus, network PIMs show uptake of phenol up to 5 mmol g^{-1} [11] and selective uptake of small dye molecules, but rejecting the large dye naphthol green B [111] from aqueous solution.

6.5.3 CMPs for Separations

The fine control over pore size and pore size distribution in CMPs [37,112] suggests that these materials might also be promising for gas separation applications, although it would first be necessary to solve the nontrivial problem of synthesizing or processing these materials as defect-free thin films or as a molded "sandwiched" layer.

6.6 CATALYSIS

On porous organic polymers as catalysts, a recent review paper [113] has provided a very good insight into the progress. Here, we will describe the catalytic properties of MOPs. Different from the gas absorption properties of the porous materials, specific surface areas are not necessarily good indicators of how these materials will perform in catalysis reactions involving substrates dissolved in solvent. A more important parameter is the pore size, as this dictates the upper size limit for substrates entering the pores. Chen et al. [45] recently reported the synthesis of a CMP-type iron porphyrin network via Suzuki–Miyaura cross-coupling of a [tetrakis(40-bromophenyl)porphyrin]Fe(III) derivative ([*p*-Br]$_4$PFe) and 1,4-phenyldiboronic acid (PDBA). This polymer, FeP-CMP, contains only nanometer-scale pore (0.47 and 2.69 nm) and has good catalytic activity and selectivity for the oxidation of sulfides to sulfones using O_2 as the oxidant, with the larger of its pores allowing for the oxidation of substrates as large as diphenylsulfide. In addition, the constrained spacing of iron porphyrins inside the network likely inhibits oxidative decomposition of these catalysts (via porphyrin attack by a neighboring catalyst) and precludes detrimental formation of μ-oxo dimers. As a consequence, turnover numbers (TONs) as high as 97,000 can be obtained and FeP-CMP can be recycled with no loss of surface area and good retention of catalytic activity. Du et al. [44] reported the synthesis of a porous organic polymer containing Lewis basic nitrogen atoms. Both micropores (0.6 and 1.3 nm in width) and mesopores exist, and these materials are stable up to 350 °C. Given that the Troger's base moiety is a part of this POP, the material was tested for the catalytic alkylation of an aldehyde using diethylzinc, and found to have activity that is slightly less than the homogeneous analog. In 2011, Xie et al. [114] described the synthesis of bpy-ligated, Ru- and Ir-incorporating porous cross-linked polymers (PCPs). Both materials (Ir-PCP and Ru-PCP) possess a relatively broad range of micropores (pore widths = 7.5, 11.0, 13.5, and 16.5 Å) and mesopores. The polymer-immobilized complexes are efficient catalysts for light-driven reactions, such as the aza-Henry reaction, the R-arylation of bromomalonate, and the oxyamination of aldehyde, where yields are comparable to those of the homogeneous analogs. The catalytic activity of both PCPs remained robust after several cycles; analyses of reaction mixtures revealed no leaching of metal ions, underscoring the robust nature of this platform. Recently, Cooper and coworkers [115] extended their cross-coupling strategy to synthesize amorphous metal-organic CMPs (MO-CMPs). Interestingly, among the three types of CMP-CpIr-3 materials tested as catalysts for the reductive amination of several ketones, the one having the lowest surface area was the most active with comparable activities to that of the corresponding homogeneous Ir catalyst.

McKeown and coworkers [3,116] reported the synthesis of microporous, amorphous phthalocyanine-based polymer networks (CoPc-PIM-A and CoPc-PIM-B). Both materials have been used to catalyze the decomposition of hydrogen peroxide, and both show higher activity than powders of CoPc, the molecular cobalt phthalocyanine analog. Catalysis by these porous materials was well behaved over a wide range of conditions, and the corresponding rate expressions were similar to that of the molecular analog, suggesting that the catalytic species are the same in the three cases. The activity of CoPc-PIM-A is 20 times that of insoluble molecular CoPc (powder), presumably because of the exclusive presence of pores in the former, and the consequent enhanced accessibility of reactants to catalytic metal sites. Recently, Makhseed et al. [117] used CoPc-PIM-A to oxidize sulfide to elemental sulfur using O_2 (in air) as the oxidant. Although catalytic activity increased with temperature, it decreased over time because of the accumulation of the sulfur in the pores, which blocks access by new sulfide ions. This interpretation was supported by nitrogen adsorption experiments that showed a large drop in the surface area of CoPc-PIM-A (from 650 to 3 $m^2\ g^{-1}$) after ~13 turnovers. The catalyst can be repeatedly reactivated by extracting the product using hot ethanol. Metalloporphyrins, well-known analogs of the heme cofactor and many metalloenzymes, have been extensively used in both homogeneous and heterogeneous catalyses. McKeown et al. [116] have extended their contorted tetrahedral comonomer strategy to the synthesis of a microporous porphyrin network polymer (FeP-PIM). As one might anticipate, the surface area of FeP-PIM (900–1000 m^2 g^{-1}) is much larger than that of CoPc-PIM-A. Consistent with its enhanced porosity and substantial surface area, FeP-PIM catalyzes the oxidation of hydroquinone at a rate ($\upsilon = 64$ mol-cat^{-1} h^{-1}) that is competitive with (marginally faster, in fact) the rate obtained with a soluble homogeneous analog [$(PhF_5)_4$porphyrin]FeCl ($\upsilon = 43$ mol-cat^{-1} h^{-1}).

Microporous HCPs usually possess a mixture of micro-, meso-, and macropores. HCPs can be swollen by many solvents, making them attractive matrices for the loading of various catalytic metal complexes or precursors. Bronstein and coworkers loaded tetrachloroplatinic acid into a microporous HCPs (surface area = 833 m^2 g^{-1}, mean pore diameter = 2 nm) [118] to form HCPs-encapsulated nanoclusters of platinum (diameter = 1.3 ± 0.3 nm) [119]. These clusters were subsequently identified as consisting of a mixture of $Pt^0/Pt^{II}/Pt^{IV}$ atoms/ions. The fraction present as Pt^0 was found to increase when the clusters were used as catalysts for the oxidation of L-sorbose to 2-keto-L-gulonic acid by O_2, suggesting that L-sorbose acts as a reducing agent at the beginning of the reaction. While the rate of catalysis was comparatively good (0.54 mol mol^{-1} Pt-s), subsequent use of a commercially available HPS that possesses both macropores and micropores (surface area = 738 m^2 g^{-1}, total pore volume = 0.59 cm^3 g^{-1}, 46% of the pore population in the 20–100 nm range, 13% with size <6 nm) yielded a 4.6-fold increase in activity. The lower TON for the purely microporous HCP was attributed to the restricted transport of the substrate through the nanocluster-congested micropores. It was also proposed that the HCP containing both micropores and macropores allows for better access of the reaction medium to the nanoparticles as well as better transport of the L-sorbose substrate to the encapsulated catalyst. These observations clearly point to the importance of transport dynamics in catalysis by porous materials, just as in the case of catalysts on traditional supports.

6.7 PROSPECT

Synthesis and application of MOPs is a new area cutting across disciplines such as chemistry and materials science. Large surface area, low skeletal density, high chemical stability, and diversity of synthesis make MOPs immensely potential for applications in the energy and gas storage, and for making specialty materials. However, the transition metal catalysts or noble metal catalysts used for synthesis of CMPs, PAFs, and some other MOPs are expensive and only produced in the lab. Sustainable mass production of MOPs is an unanswered challenge. Future research work can be carried out on the following aspects: first, to develop the green, low cost synthetic methods for the preparation MOPs. Low cost route to make structure controllable MOPs is particularly important for the industrial and practical applications and for hydrogen storage. Second aspect is to expand into new application areas such as water photolysis, molecular imprinting, energy storage, and sensing.

REFERENCES

1. Maly, K. E. (2009). Assembly of nanoporous organic materials from molecular building blocks. *J. Mater. Chem., 19(13)*, 1781–1787.
2. Ben, T., Ren, H., Ma, S. Q., Cao, D. P., Lan, J. H., Jing, X. F., Wang, W. C., Xu, J., Deng, F., Simmons, J. M., Qiu, S. L., Zhu, G. S. (2009). Targeted synthesis of a porous aromatic framework with high stability and exceptionally high surface area. *Angew. Chem. Int. Ed., 48(50)*, 9457–9460.
3. McKeown, N. B., Makhseed, S., Budd, P. M. (2002). Phthalocyanine-based nanoporous network polymers. *Chem. Commun., 23*, 2780–2781.
4. Tsyurupa, M. P., Davankov, V. A. (2002). Hypercrosslinked polymers: basic principle of preparing the new class of polymeric materials. *React. Funct. Polym., 53 (2–3)*, 193–203.
5. Cote, A. P., Benin, A. I., Ockwig, N. W., O'Keeffe, M., Matzger, A. J., Yaghi, O. M. (2005). Porous, crystalline, covalent organic frameworks. *Science, 310 (5751)*, 1166–1170.
6. Jiang, J. X., Su, F., Trewin, A., Wood, C. D., Campbell, N. L., Niu, H., Dickinson, C., Ganin, A. Y., Rosseinsky, M. J., Khimyak, Y. Z., Cooper, A. I. (2007). Conjugated microporous poly (aryleneethynylene) networks. *Angew. Chem. Int. Ed., 46 (45)*, 8574-8578.
7. Jiang, J. X., Cooper, A. I. Functional metal-organic frameworks: gas storage, separation and catalysis. *Microporous Organic Polymers: Design, Synthesis, and Function*. Springer-Verlag Berlin, Berlin, 2010, Vol. 293, pp. 1–33.
8. Cooper, A. I. (2009). Conjugated microporous polymers. *Adv. Mater., 21(12)*, 1291–1295.
9. McKeown, N. B. (2000). Phthalocyanine-containing polymers. *J. Mater. Chem., 10(9)*, 1979–1995.
10. McKeown, N. B., Hanif, S., Msayib, K., Tattershall, C. E., Budd, P. M. (2002). Porphyrin-based nanoporous network polymers. *Chem. Commun., 23*, 2782–2783.
11. Budd, P. M., Ghanem, B., Msayib, K., McKeown, N. B., Tattershall, C. (2003). A nanoporous network polymer derived from hexaazatrinaphthylene with potential as an adsorbent and catalyst support. *J. Mater. Chem., 13(11)*, 2721–2726.
12. McKeown, N. B., Gahnem, B., Msayib, K. J., Budd, P. M., Tattershall, C. E., Mahmood, K., Tan, S., Book, D., Langmi, H. W., Walton, A. (2006). Towards polymer-based hydrogen storage materials: engineering ultramicroporous cavities within polymers of intrinsic microporosity. *Angew. Chem. Int. Ed., 45(11)*, 1804–1807.
13. Ghanem, B. S., Msayib, K. J., McKeown, N. B., Harris, K. D. M., Pan, Z., Budd, P. M., Butler, A., Selbie, J., Book, D., Walton, A. (2007). A triptycene-based polymer of intrinsic microposity that displays enhanced surface area and hydrogen adsorption. *Chem. Commun., 1*, 67–69.
14. Ghanem, B. S., Hashem, M., Harris, K. D. M., Msayib, K. J., Xu, M. C., Budd, P. M., Chaukura, N., Book, D., Tedds, S., Walton, A., McKeown, N. B. (2010). Triptycene-based polymers of intrinsic microporosity: organic materials that can be tailored for gas adsorption. *Macromolecules, 43(12)*, 5287–5294.
15. Budd, P. M., Ghanem, B. S., Makhseed, S., McKeown, N. B., Msayib, K. J., Tattershall, C. E. (2004). Polymers of intrinsic microporosity (PIMs): robust, solution-processable, organic nanoporous materials. *Chem. Commun., 2*, 230–231.

16. Weber, J., Su, O., Antonietti, M., Thomas, A. (2007). Exploring polymers of intrinsic microporosity-microporous, soluble polyamide and Polyimide. *Macromol. Rapid Commun., 28(18–19)*, 1871–1876.
17. (a) Chou, C. H., Reddy, D. S., Shu, C. F. (2002). Synthesis and characterization of spirobifluorene-based polyimides. *J. Polym. Sci. Part A: Polym. Chem., 40(21)*, 3615–3621; (b) Kim, Y. H., Kim, H. S., Kwon, S. K. (2005). Synthesis and characterization of highly soluble and oxygen permeable new polyimides based on twisted biphenyl dianhydride and spirobifluorene diamine. *Macromolecules, 38(19)*, 7950–7956.
18. Ghanem, B. S., McKeown, N. B., Budd, P. M., Fritsch, D. (2008). Polymers of intrinsic microporosity derived from bis(phenazyl) monomers. *Macromolecules, 41(5)*, 1640–1646.
19. Ghanem, B. S., McKeown, N. B., Budd, P. M., Al-Harbi, N. M., Fritsch, D., Heinrich, K., Starannikova, L., Tokarev, A., Yampolskii, Y. (2009). Synthesis, characterization, and gas permeation properties of a novel group of polymers with intrinsic microporosity: PIM-polyimides. *Macromolecules, 42(20)*, 7881–7888.
20. (a) Lee, J. Y., Wood, C. D., Bradshaw, D., Rosseinsky, M. J., Cooper, A. I. (2006). Hydrogen adsorption in microporous hypercrosslinked polymers. *Chem. Commun., (25)*, 2670–2672; (b) Ahn, J. H., Jang, J. E., Oh, C. G., Ihm, S. K., Cortez, J., Sherrington, D. C. (2006). Rapid generation and control of microporosity, bimodal pore size distribution, and surface area in Davankov-type hyper-cross-linked resins. *Macromolecules, 39(2)*, 627–632.
21. Wood, C. D., Tan, B., Trewin, A., Niu, H. J., Bradshaw, D., Rosseinsky, M. J., Khimyak, Y. Z., Campbell, N. L., Kirk, R., Stockel, E., Cooper, A. I. (2007). Hydrogen storage in microporous hypercrosslinked organic polymer networks. *Chem. Mater., 19(8)*, 2034–2048.
22. Tsyurupa, M. P., Davankov, V. A. (2006). Porous structure of hypercrosslinked polystyrene: state-of-the-art mini-review. *React. Funct. Polym., 66(7)*, 768–779.
23. Davankov, V. A., Pastukhov, A. V., Tsyurupa, M. P. (2000). Unusual mobility of hypercrosslinked polystyrene networks: swelling and dilatometric studies. *J. Polym. Sci. Part B: Polym. Phys., 38(11)*, 1553–1563.
24. Germain, J., Hradil, J., Frechet, J. M. J., Svec, F. (2006). High surface area nanoporous polymers for reversible hydrogen storage. *Chem. Mater., 18(18)*, 4430–4435.
25. Wood, C. D., Tan, B., Trewin, A., Su, F., Rosseinsky, M. J., Bradshaw, D., Sun, Y., Zhou, L., Cooper, A. I. (2008). Microporous organic polymers for methane storage. *Adv. Mater., 20(10)*, 1916–1920.
26. Li, B. Y., Huang, X., Liang, L. Y., Tan, B. (2010). Synthesis of uniform microporous polymer nanoparticles and their applications for hydrogen storage. *J. Mater. Chem., 20(35)*, 7444–7450.
27. Li, B. Y., Su, F. B., Luo, H. K., Liang, L. Y., Tan, B. E. (2011). Hypercrosslinked microporous polymer networks for effective removal of toxic metal ions from water. *Micropor. Mesopor. Mater., 138(1–3)*, 207–214.
28. (a) Kitagawa, S., Kitaura, R., Noro, S. (2004). Functional porous coordination polymers. *Angew. Chem. Int. Ed., 43(18)*, 2334–2375; (b) Yaghi, O. M., O'Keeffe, M., Ockwig, N. W., Chae, H. K., Eddaoudi, M., Kim, J. (2003). Reticular synthesis and the design of new materials. *Nature, 423(6941)*, 705–714; (c) Eddaoudi, M., Kim, J., Rosi, N., Vodak, D., Wachter, J., O'Keeffe, M., Yaghi, O. M. (2002). Systematic design

of pore size and functionality in isoreticular MOFs and their application in methane storage. *Science*, *295(5554)*, 469–472.

29. Li, H., Eddaoudi, M., O'Keeffe, M., Yaghi, O. M. (1999). Design and synthesis of an exceptionally stable and highly porous metal-organic framework. *Nature*, *402(6759)*, 276–279.
30. Tilford, R. W., Gemmill, W. R., zur Loye, H. C., Lavigne, J. J. (2006). Facile synthesis of a highly crystalline, covalently linked porous boronate network. *Chem. Mater.*, *18(22)*, 5296–5301.
31. Kuhn, P., Antonietti, M., Thomas, A. (2008). Porous, covalent triazine-based frameworks prepared by ionothermal synthesis. *Angew. Chem.-Int. Edit.*, *47(18)*, 3450–3453.
32. Kuhn, P., Thomas, A., Antonietti, M. (2009). Toward tailorable porous organic polymer networks: a high-temperature dynamic polymerization scheme based on aromatic nitriles. *Macromolecules*, *42(1)*, 319–326.
33. Bojdys, M. J., Jeromenok, J., Thomas, A., Antonietti, M. (2010). Rational extension of the family of layered, covalent, triazine-based frameworks with regular porosity. *Adv. Mater.*, *22(19)*, 2202–2206.
34. El-Kaderi, H. M., Hunt, J. R., Mendoza-Cortes, J. L., Cote, A. P., Taylor, R. E., O'Keeffe, M., Yaghi, O. M. (2007). Designed synthesis of 3D covalent organic frameworks. *Science*, *316(5822)*, 268–272.
35. Uribe-Romo, F. J., Hunt, J. R., Furukawa, H., Klock, C., O'Keeffe, M., Yaghi, O. M. (2009). A crystalline imine-linked 3-D porous covalent organic framework. *J. Am. Chem. Soc.*, *131(13)*, 4570–4571.
36. Spitler, E. L., Dichtel, W. R. (2010). Lewis acid-catalysed formation of two-dimensional phthalocyanine covalent organic frameworks. *Nat. Chem.*, *2(8)*, 672–677.
37. Jiang, J. X., Su, F., Trewin, A., Wood, C. D., Niu, H., Jones, J. T. A., Khimyak, Y. Z., Cooper, A. I. (2008). Synthetic control of the pore dimension and surface area in conjugated microporous polymer and copolymer networks. *J. Am. Chem. Soc.*, *130(24)*, 7710–7720.
38. Jiang, J. X., Su, F., Niu, H., Wood, C. D., Campbell, N. L., Khimyak, Y. Z., Cooper, A. I. (2008). Conjugated microporous poly(phenylene butadiynylene)s. *Chem. Commun.*, *(4)*, 486–488.
39. Fairlamb, I. J. S., Bauerlein, P. S., Marrison, L. R., Dickinson, J. M. (2003). Pd-catalysed cross coupling of terminal alkynes to diynes in the absence of a stoichiometric additive. *Chem. Commun.*, *(5)*, 632–633.
40. Weber, J., Thomas, A. (2008). Toward stable interfaces in conjugated polymers: microporous poly(p-phenylene) and poly(phenyleneethynylene) based on a spirobifluorene building block. *J. Am. Chem. Soc.*, *130(20)*, 6334–6343.
41. Dawson, R., Su, F. B., Niu, H. J., Wood, C. D., Jones, J. T. A., Khimyak, Y. Z., Cooper, A. I. (2008). Mesoporous poly(phenylenevinylene) networks. *Macromolecules*, *41(5)*, 1591–1593.
42. Jiang, J. X., Trewin, A., Su, F. B., Wood, C. D., Niu, H. J., Jones, J. T. A., Khimyak, Y. Z., Cooper, A. I. (2009). Microporous poly(tri(4-ethynylphenyl)amine) networks: synthesis, properties, and atomistic simulation. *Macromolecules*, *42(7)*, 2658–2666.

43. Jiang, J. X., Laybourn, A., Clowes, R., Khimyak, Y. Z., Bacsa, J., Higgins, S. J., Adams, D. J., Cooper, A. I. (2010). High surface area contorted conjugated microporous polymers based on spiro-bipropylenedioxythiophene. *Macromolecules, 43(18)*, 7577–7582.

44. Du, X., Sun, Y. L., Tan, B. E., Teng, Q. F., Yao, X. J., Su, C. Y., Wang, W. (2010). Troger's base-functionalised organic nanoporous polymer for heterogeneous catalysis. *Chem. Commun., 46(6)*, 970–972.

45. Chen, L., Yang, Y., Jiang, D. L. (2010). CMPs as scaffolds for constructing porous catalytic frameworks: a built-in heterogeneous catalyst with high activity and selectivity based on nanoporous metalloporphyrin polymers. *J. Am. Chem. Soc., 132(26)*, 9138–9143.

46. Li, A., Lu, R. F., Wang, Y., Wang, X., Han, K. L., Deng, W. Q. (2010). Lithium-doped conjugated microporous polymers for reversible hydrogen storage. *Angew. Chem. Int. Ed., 49(19)*, 3330–3333.

47. (a) Grant, P. M. (2003). Hydrogen lifts off—with a heavy load. *Nature, 424(6945)*, 129–130; (b) Schlapbach, L., Zuttel, A. (2001). Hydrogen-storage materials for mobile applications. *Nature, 414(6861)*, 353–358.

48. Kanan, M. W., Nocera, D. G. (2008). In situ formation of an oxygen-evolving catalyst in neutral water containing phosphate and Co2+. *Science, 321(5892)*, 1072–1075.

49. Wang, H. L., Gao, Q. M., Hu, J. (2009). High hydrogen storage capacity of porous carbons prepared by using activated carbon. *J. Am. Chem. Soc., 131(20)*, 7016–7022.

50. Zecchina, A., Bordiga, S., Vitillo, J. G., Ricchiardi, G., Lamberti, C., Spoto, G., Bjorgen, M., Lillerud, K. P. (2005). Liquid hydrogen in protonic chabazite. *J. Am. Chem. Soc., 127(17)*, 6361–6366.

51. Rosi, N. L., Eckert, J., Eddaoudi, M., Vodak, D. T., Kim, J., O'Keeffe, M., Yaghi, O. M. (2003). Hydrogen storage in microporous metal-organic frameworks. *Science, 300(5622)*, 1127–1129.

52. Furukawa, H., Yaghi, O. M. (2009). Storage of hydrogen, methane, and carbon dioxide in highly porous covalent organic frameworks for clean energy applications. *J. Am. Chem. Soc., 131(25)*, 8875–8883.

53. (a) McKeown, N. B., Gahnem, B., Msayib, K. J., Budd, P. M., Tattershall, C. E., Mahmood, K., Tan, S., Book, D., Langmi, H. W., Walton, A. (2006). Towards polymer-based hydrogen storage materials: engineering ultramicroporous cavities within polymers of intrinsic microporosity. *Angew. Chem., Int. Ed. Engl., 118(11)*, 1836–1839; (b) Tozawa, T., Jones, J. T. A., Swamy, S. I., Jiang, S., Adams, D. J., Shakespeare, S., Clowes, R., Bradshaw, D., Hasell, T., Chong, S. Y., Tang, C., Thompson, S., Parker, J., Trewin, A., Bacsa, J., Slawin, A. M. Z., Steiner, A., Cooper, A. I. (2009). Porous organic cages. *Nat. Mater., 8*, 973–978.

54. Germain, J., Svec, F., Fréchet, J. M. J. (2007). *PMSE Prepr., 97*, 272–273.

55. Smigol, V., Svec, F. (1993). Nanoporous polymers for hydrogen storage. *J. Appl. Polym. Sci., 48*, 2033–2039.

56. Fontanals, N., Marce, R. M., Galia, M. (2003). *J. Polym. Sci. Part A: Polym. Chem., 41*, 1927–1933.

57. Buda, C., Dunietz, B. D. (2006). Hydrogen physisorption on the organic linker in metal organic frameworks: ab initio computational study. *J. Phys. Chem. B, 110(21)*, 10479–10484.

58. Lochan, R. C., Head-Gordon, M. (2006). Computational studies of molecular hydrogen binding affinities: The role of dispersion forces, electrostatics, and orbital interactions. *Phys. Chem. Chem. Phys.*, *8(12)*, 1357–1370.
59. Germain, J., Svec, F., Frechet, J. M. J. (2008). Preparation of size-selective nanoporous polymer networks of aromatic rings: potential adsorbents for hydrogen storage. *Chem. Mater.*, *20(22)*, 7069–7076.
60. Germain, J., Frechet, J. M. J., Svec, F. (2007). Hypercrosslinked polyanilines with nanoporous structure and high surface area: potential adsorbents for hydrogen storage. *J. Mater. Chem.*, *17(47)*, 4989–4997.
61. Makhseed, S., Samuel, J., Bumajdad, A., Hassan, M. (2008). Synthesis and characterization of fluoropolymers with intrinsic microporosity and their hydrogen adsorption studies. *J. Appl. Polym. Sci.*, *109(4)*, 2591–2597.
62. Tilford, R. W., Mugavero, S. J., Pellechia, P. J., Lavigne, J. J. (2008). Tailoring microporosity in covalent organic frameworks. *Adv. Mater.*, *20(14)*, 2741–2746.
63. Campbell, N. L., Clowes, R., Ritchie, L. K., Cooper, A. I. (2009). Rapid microwave synthesis and purification of porous covalent organic frameworks. *Chem. Mater.*, *21(2)*, 204–206.
64. (a) Latroche, M., Surblé, S., Serre, C., Mellot-Draznieks, C., Llewellyn, P. L., Lee, J.-H., Chang, J.-S., Jhung, S. H., Férey, G. (2006). Hydrogen storage in the giant-pore metal–organic frameworks MIL-100 and MIL-101. *Angew. Chem.*, *118(48)*, 8407–8411; (b) Latroche, M., Surblé, S., Serre, C., Mellot-Draznieks, C., Llewellyn, P. L., Lee, J.-H., Chang, J.-S., Jhung, S. H., Férey, G. (2006). Hydrogen storage in the giant-pore metal–organic frameworks MIL-100 and MIL-101. *Angew. Chem. Int. Ed.*, *45(48)*, 8227–8231.
65. Zhou, W., Wu, H., Hartman, M. R., Yildirim, T. (2007). Hydrogen and methane adsorption in metal–organic frameworks: a high-pressure volumetric study. *J. Phys. Chem. C*, *111(44)*, 16131–16137.
66. Hübner, O., Glöss, A., Fichtner, M., Klopper, W. (2004). On the interaction of dihydrogen with aromatic systems. *J. Phys. Chem. A*, *108(15)*, 3019–3023.
67. Holst, J. R., Stöckel, E., Adams, D. J., Cooper, A. I. (2010). High surface area networks from tetrahedral monomers: metal-catalyzed coupling, thermal polymerization, and "click" chemistry. *Macromolecules*, *43(20)*, 8531–8538.
68. Bera, P. P., Francisco, J. S., Lee, T. J. (2009). Identifying the molecular origin of global warming. *J. Phys. Chem. A*, *113(45)*, 12694–12699.
69. Kintisch, E. (2007). Preparation and properties of uniform beads based on macroporous glycidyl methacrylate–ethylene dimethacrylate copolymer: Use of chain transfer agent for control of pore-size distribution. Power generation – making dirty coal plants cleaner. *Science*, *317(5835)*, 184–186.
70. Bert Metz, O. D., de Coninck, H., Loos, M., Meyer, L. Preparation and characterization of highly polar polymeric sorbents from styrene–divinylbenzene and vinylpyridine–divinylbenzene for the solid-phase extraction of polar organic pollutants. Carbon dioxide capture and storage. IPCC [M]. Cambridge University Press, Cambridge, UK, 2005, 431 pp.
71. (a) Thallapally, P. K., Tian, J., Radha Kishan, M., Fernandez, C. A., Dalgarno, S. J., McGrail, P. B., Warren, J. E., Atwood, J. L. (2008). Flexible (breathing) interpenetrated metal-organic frameworks for CO_2 separation applications. *J. Am. Chem. Soc.*, *130(50)*, 16842–16843; (b) Demessence, A., D'Alessandro, D. M., Foo, M. L.,

Long, J. R. (2009). Strong CO_2 binding in a water-stable, triazolate-bridged metal-oganic framework functionalized with ethylenediamine. *J. Am. Chem. Soc., 131(25)*, 8784–8786; (c) D'Alessandro, D. M., Smit, B., Long, J. R. (2010). Carbon dioxide capture: prospects for new materials. *Angew. Chem. Int. Ed., 49(35)*, 6058–6082.

72. Wang, Q. A., Luo, J. Z., Zhong, Z. Y., Borgna, A. (2011). CO_2 capture by solid adsorbents and their applications: current status and new trends. *Energy Environ. Sci., 4(1)*, 42–55.
73. Bhagiyalakshmi, M., Anuradha, R., Do Park, S., Jang, H. T. Octa(aminophenyl) silsesquioxane fabrication on chlorofunctionalized mesoporous SBA-15 for CO2 adsorption. *Micropor. Mesopor. Mater. 131(1–3)*, 265–273.
74. Potapov, S. V., Fomkin, A. A., Sinitsyn, V. A. (2009). Adsorption of carbon dioxide on microporous carbon adsorbents. *Russ. Chem. Bull., 58(4)*, 733–736.
75. Chen, C., You, K. S., Ahn, J. W., Ahn, W. S. Synthesis of mesoporous silica from bottom ash and its application for CO_2 sorption. *Korean J. Chem. Eng. 27(3)*, 1010–1014.
76. Ma, S. Q., Zhou, H. C. (2010). Gas storage in porous metal-organic frameworks for clean energy applications. *Chem. Commun., 46(1)*, 44–53.
77. Banerjee, R., Furukawa, H., Britt, D., Knobler, C., O'Keeffe, M., Yaghi, O. M. (2009). Control of pore size and functionality in isoreticular zeolitic imidazolate frameworks and their carbon dioxide selective capture properties. *J. Am. Chem. Soc., 131(11)*, 3875–3877.
78. Weber, J., Antonietti, M., Thomas, A. (2008). Microporous networks of high-performance polymers: elastic deformations and gas sorption properties. *Macromolecules, 41(8)*, 2880–2885.
79. Pandey, P., Katsoulidis, A. P., Eryazici, I., Wu, Y., Kanatzidis, M. G., Nguyen, S. T. (2010). Imine-linked microporous polymer organic frameworks. *Chem. Mater., 22(17)*, 4974–4979.
80. Millward, A. R., Yaghi, O. M. (2005). Metal-organic frameworks with exceptionally high capacity for storage of carbon dioxide at room temperature. *J. Am. Chem. Soc., 127(51)*, 17998–17999.
81. Babarao, R., Jiang, J. (2008). Molecular screening of metal–organic frameworks for CO_2 storage. *Langmuir, 24(12)*, 6270–6278.
82. Abbie, T., Andrew, I. C. (2010). Porous organic polymers: distinction from disorder? *Angew. Chem. Int. Ed., 49(9)*, 1533–1535.
83. Llewellyn, P. L., Bourrelly, S., Serre, C., Vimont, A., Daturi, M., Hamon, L., De Weireld, G., Chang, J.-S., Hong, D.-Y., Kyu Hwang, Y., Hwa Jhung, S., Férey, G. r. (2008). High uptakes of CO_2 and CH_4 in mesoporous metal organic frameworks MIL-100 and MIL-101. *Langmuir, 24(14)*, 7245–7250.
84. Lu, W. G., Yuan, D. Q., Zhao, D., Schilling, C. I., Plietzsch, O., Muller, T., Brase, S., Guenther, J., Blumel, J., Krishna, R., Li, Z., Zhou, H. C. (2010). Porous polymer networks: synthesis, porosity, and applications in gas storage/separation. *Chem. Mater., 22(21)*, 5964–5972.
85. Holst, J. R., Cooper, A. I. (2010). Ultrahigh surface area in porous solids. *Adv. Mater., 22(45)*, 5212–5216.
86. Couck, S., Denayer, J. F. M., Baron, G. V., Remy, T., Gascon, J., Kapteijn, F. (2009). An amine-functionalized MIL-53 metal-organic framework with large separation power for CO_2 and CH_4. *J. Am. Chem. Soc., 131(18)*, 6326–6327.

87. An, J., Geib, S. J., Rosi, N. L. (2009). High and selective CO_2 uptake in a cobalt adeninate metal-oganic framework exhibiting pyrimidine- and amino-decorated pores. *J. Am. Chem. Soc., 132(1)*, 38–39.
88. Choi, J. H., Choi, K. M., Jeon, H. J., Choi, Y. J., Lee, Y., Kang, J. K. (2010). Acetylene gas mediated conjugated microporous polymers (ACMPs): first use of acetylene gas as a building unit. *Macromolecules, 43(13)*, 5508–5511.
89. Rabbani, M. G., El-Kaderi, H. M. (2011). Template-free synthesis of a highly porous benzimidazole-linked polymer for CO_2 capture and H_2 storage. *Chem. Mater., 23(7)*, 1650–1653.
90. Farha, O. K., Spokoyny, A. M., Hauser, B. G., Bae, Y.-S., Brown, S. E., Snurr, R. Q., Mirkin, C. A., Hupp, J. T. (2009). Synthesis, properties, and gas separation studies of a robust diimide-based microporous organic polymer. *Chem. Mater., 21(14)*, 3033–3035.
91. Babarao, R., Dai, S., Jiang, D.-e. (2011). Functionalizing porous aromatic frameworks with polar organic groups for high-capacity and selective CO_2 separation: a molecular simulation study. *Langmuir, 27(7)*, 3451–3460.
92. Torrisi, A., Bell, R. G., Mellot-Draznieks, C. (2010). Functionalized MOFs for enhanced CO_2 capture. *Cryst. Growth Des., 10(7)*, 2839–2841.
93. Vaidhyanathan, R., Iremonger, S. S., Shimizu, G. K. H., Boyd, P. G., Alavi, S., Woo, T. K. (2010). Direct observation and quantification of CO_2 binding within an amine-functionalized nanoporous solid. *Science, 330(6004)*, 650–653.
94. Dawson, R., Adams, D. J., Cooper, A. I., (2011). Chemical tuning of CO_2 sorption in robust nanoporous organic polymers. *Chem. Sci, 2*, 1173–1177.
95. Vogiatzis, K. D., Mavrandonakis, A., Klopper, W., Froudakis, G. E. (2009). Ab initio study of the interactions between CO_2 and N-containing organic heterocycles. *ChemPhysChem, 10(2)*, 374–383.
96. (a) Ritter, N., Antonietti, M., Thomas, A., Senkovska, I., Kaskel, S., Weber, J. (2009). Binaphthalene-based, soluble polyimides: the limits of intrinsic microporosity. *Macromolecules, 42(21)*, 8017–8020; (b) Ritter, N., Senkovska, I., Kaskel, S., Weber, J. (2011). Intrinsically microporous poly(imide)s: structure–porosity relationship studied by gas sorption and X-ray scattering. *Macromolecules, 44(7)*, 2025–2033; (c) Ritter, N., Senkovska, I., Kaskel, S., Weber, J. (2011). Towards chiral microporous soluble polymers—binaphthalene-based polyimides. *Macromol. Rapid Commun., 32(5)*, 438–443.
97. Du, N., Park, H. B., Robertson, G. P., Dal-Cin, M. M., Visser, T., Scoles, L., Guiver, M. D. (2011). Polymer nanosieve membranes for CO_2-capture applications. *Nat. Mater., 10(5)*, 372–375.
98. Robeson, L. M. (2008). The upper bound revisited. *J. Membr. Sci., 320(1–2)*, 390–400.
99. Robeson, L. M. (1991). Correlation of separation factor versus permeability for polymeric membranes. *J. Membr. Sci., 62(2)*, 165–185.
100. Park, H. B., Jung, C. H., Lee, Y. M., Hill, A. J., Pas, S. J., Mudie, S. T., Van Wagner, E., Freeman, B. D., Cookson, D. J. (2007). Polymers with cavities tuned for fast selective transport of small molecules and ions. *Science, 318(5848)*, 254–258.
101. Martin, C. F., Stockel, E., Clowes, R., Adams, D. J., Cooper, A. I., Pis, J. J., Rubiera, F., Pevida, C. (2011). Hypercrosslinked organic polymer networks as potential adsorbents for pre-combustion CO_2 capture. *J. Mater. Chem., 21*, 5475–5483.

102. Davankov, V. A., Rogozhin, S. V., Tsyurupa, M. P. (1971). Macronet polystyrene structures for ionites and method of producing same. 3,729,457.

103. (a) Davankov, V., Tsyurupa, M., Ilyin, M., Pavlova, L. (2002). Hypercross-linked polystyrene and its potentials for liquid chromatography: a mini-review. *J. Chromatogr. A, 965 (1–2)*, 65–73; (b) Sychov, C. S., Ilyin, M. M., Davankov, V. A., Sochilina, K. O. (2004). Elucidation of retention mechanisms on hypercrosslinked polystyrene used as column packing material for high-performance liquid chromatography. *J. Chromatogr. A, 1030(1–2)*, 17–24.

104. Fontanals, N., Galia, M., Cormack, P., Marce, R., Sherrington, D., Borrull, F. (2005). Evaluation of a new hypercrosslinked polymer as a sorbent for solid-phase extraction of polar compounds. *J. Chromatogr. A, 1075(1–2)*, 51–56.

105. (a) Chang, C.-F., Chang, C.-Y., Hsu, K.-E., Lee, S.-C., H??ll, W. (2008). Adsorptive removal of the pesticide methomyl using hypercrosslinked polymers. *J. Hazard. Mater., 155(1–2)*, 295–304; (b) Davankov, V., Pavlova, L., Tsyurupa, M., Brady, J., Balsamo, M., Yousha, E. (2000). Polymeric adsorbent for removing toxic proteins from blood of patients with kidney failure. *J. Chromatogr. B, 739(1)*, 73–80.

106. (a) Tsyurupa, M. P., Tarabaeva, O. G., Pastukhov, A. V., Davankov, V. A. Sorption of ions of heavy metals by neutral hypercrosslinked polystyrene. *Int. J. Polym. Mater., 52(5)*, 403–414;

107. Long, C., Li, Q., Li, Y., Liu, Y., Li, A., Zhang, Q. (2010). Adsorption characteristics of benzene-chlorobenzene vapor on hypercrosslinked polystyrene adsorbent and a pilot-scale application study. *Chem. Eng. J., 160(2)*, 723–728.

108. McKeown, N. B., Budd, P. M. (2006). Polymers of intrinsic microporosity (PIMs): organic materials for membrane separations, heterogeneous catalysis and hydrogen storage. *Chem. Soc. Rev., 35*, 675–683.

109. McKeown, N. B., Budd, P. M., Msayib, K. J., Ghanem, B. S., Kingston, H. J., Tattershall, C. E., Makhseed, S., Reynolds, K. J., Fritsch, D. (2005). Polymers of intrinsic microporosity (PIMs): bridging the void between microporous and polymeric materials. *Chem.-Eur. J., 11 (9)*, 2610–2620.

110. Budd, P. M., Elabas, E. S., Ghanem, B. S., Makhseed, S., McKeown, N. B., Msayib, K. J., Tattershall, C. E., Wang, D. (2004). Solution-processed, organophilic membrane derived from a polymer of intrinsic microporosity. *Adv. Mater., 16(5)*, 456–459.

111. Maffei, A. V., Budd, P. M., McKeown, N. B. (2006). Adsorption studies of a microporous phthalocyanine network polymer. *Langmuir, 22(9)*, 4225–4229.

112. Jiang, J.-X., Su, F., Trewin, A., Wood, C. D., Campbell, N. L., Niu, H., Dickinson, C., Ganin, A. Y., Rosseinsky, M. J., Khimyak, Y. Z., Cooper, A. I. (2008). Conjugated microporous poly(aryleneethynylene) networks. *Angew. Chem. Int. Ed., 47(7)*, 1167.

113. Kaur, P., Hupp, J. T., Nguyen, S. T. (2011). Porous organic polymers in catalysis: opportunities and challenges. *ACS Catal., 1(7)*, 819–835.

114. Xie, Z., Wang, C., deKrafft, K. E., Lin, W. (2011). Highly stable and porous cross-linked polymers for efficient photocatalysis. *J. Am. Chem. Soc., 133(7)*, 2056–2059.

115. Jiang, J.-X., Wang, C., Laybourn, A., Hasell, T., Clowes, R., Khimyak, Y. Z., Xiao, J., Higgins, S. J., Adams, D. J., Cooper, A. I. (2011). Metal–organic conjugated microporous polymers. *Angew. Chem. Int. Ed., 50(5)*, 1072–1075.

116. Mackintosh, H. J., Budd, P. M., McKeown, N. B. (2008). Catalysis by microporous phthalocyanine and porphyrin network polymers. *J. Mater. Chem., 18(5)*, 573–578.

117. Makhseed, S., Al-Kharafi, F., Samuel, J., Ateya, B. (2009). Catalytic oxidation of sulphide ions using a novel microporous cobalt phthalocyanine network polymer in aqueous solution. *Catal. Commun.*, *10(9)*, 1284–1287.

118. Davankov, V. A., Tsyurupa, M. P. (1990). Structure and properties of hypercrosslinked polystyrene--the first representative of a new class of polymer networks. *React. Polym.*, *13(1–2)*, 27–42.

119. Bronstein, L. M., Goerigk, G., Kostylev, M., Pink, M., Khotina, I. A., Valetsky, P. M., Matveeva, V. G., Sulman, E. M., Sulman, M. G., Bykov, A. V., Lakina, N. V., Spontak, R. J. (2004). Structure and catalytic properties of Pt-modified hyper-cross-linked polystyrene exhibiting hierarchical porosity. *J. Phys. Chem. B*, *108(47)*, 18234–18242.

7

DENDRITIC COPOLYMERS

SRINIVASA RAO VINUKONDA

7.1 INTRODUCTION

Although the concept of three-dimensional polymers was first reported by Stockmeyer [1] and Flory [2] in the 1940s and 1950s, this area of research received considerable attention only in the 1990s. Further theoretical investigation on this subject came from Gordon and coworkers [3] and Burchard et al. [4]. More recent works by Müller and coworkers [5], Frey and coworkers [6], Möller and coworkers [7], Dušek et al. [8], Fawcett and coworkers [9], and Galina et al. [10] initiated detailed discussion on the theory. The term *hyperbranched* was coined by Kim and Webster [11] in the late 1980s. Fréchet and coworkers [12] and Kim and Webster [13] first introduced the definition of degree of branching (DB) for hyperbranched polymers.

It is well known that there are a number of similarities and differences between dendrimers and hyperbranched polymers. The similarities arise because the starting materials used for their synthesis are same, that is, AB_n monomers, and the differences are primarily aroused from their variation in the resultant architecture, which is due to the difference in synthetic approaches followed. The hyperbranched polymers attracted attention from academia and industry, as structural perfection may not be a strict prerequisite for many applications as well as for the ease of synthesis, whereas synthesis of dendrimers involves tedious multistep procedures.

The continuous growth of interest in the field of hyperbranched polymers is mainly because of their fairly good solubility, low solution, and melt viscosities, in contrast to dendrimers that are well defined with controlled size, shape, and branching [14,15]. Many groups including Flory [16–19], Kricheldorf et al. [20,21], Kim

Synthesis and Applications of Copolymers, First Edition. Edited by Anbanandam Parthiban.

et al. [22,23], Fréchet et al. [24–27], Voit and coworkers [28], Frey and coworkers [29,30] and Gao et al. [31–34] paid considerable attention to the development of theoretical treatments and synthetic routes for hyperbranched polymers. Among various synthetic strategies that are possible to prepare hyperbranched polymers, "$A_2 + B_3$" is the simplest approach because of greater commercial availability of A_2 and B_3 monomers over AB_n monomers.

Kakimoto and coworkers [35], Hawker and Chu [36], Voit and coworkers [37], and so forth made attempts to investigate the effect of structural parameter on the properties of dendritic polymers. Kakimoto and coworkers [38,34] a studied and compared the hyperbranched polymers synthesized by ideal (self-polymerization of AB_2 monomers) and nonideal ($A_2 + B_3$) polymerization methods. They found that physical properties such as solution viscosity, glass transition temperature, and thermal stability were different from each other because of different packing density, topology, chain entanglement, and intermolecular interaction. They also studied the effect of monomer multiplicity on the DB by using AB_n-type dendron, where $n =$ 2, 4, and 8, and concluded that DB increases with an increase in monomer multiplicity of the starting materials, that is, from 0.32 to 0.72 [34,38b]. It was also suggested that control of DB is possible using this approach. Hawker and Chu [36] also found similar observation in their investigation by the polymerization of AB_2, AB_3, and AB_4 monomers.

Initially, a divergent synthetic approach was used to prepare these polymers. Later, a convergent synthetic approach [39,24], "doublestage" convergent approach [40], and the combination of TERMINI and metal-catalyzed living radical polymerizations [41] were proposed. On the other hand, the uncontrolled chain-growth propagation approach [42,43] was also used for synthesizing highly branched polymers. Polycondensation, self-condensation of vinyl polymerization, and ring-opening multiple branching are the main synthetic approaches for the preparation of highly branched polymers. Significant conformational change occurs when dendrimer reaches specific generation, particularly for generations greater than 4, the structure assumes a densely packed globular shape, which decreases chain entanglements and molecular aspect ratio [44]. This feature imparts attractive solution and bulk properties to dendrimers.

Since the initial report on this class of polymers by Vögtle and coworkers [45] in 1978, many different structural classes of dendritic macromolecules have been reported, which include dendrimers [46], hyperbranched polymers [47], linear–dendritic polymers [48], linear–dendritic copolymers [49], star dendritic copolymers [50], multiarm star polymers [51], and main chain polymers with dendritic side groups [52].

7.2 SYNTHESIS APPROACHES OR STRATEGIES

7.2.1 $AB_2 + A_2$ Approach

In this approach, hyperbranched copolymers are produced in one-pot synthesis by using two different multifunctional monomers such as AB_2 and A_2. Jin et al.

[53] performed a synthesis of hyperbranched copolymer by using activated methylene monomers 4-(4′-chloro-methylbenzy1oxy)phenylacetonitrile and 1,4-bis(chloromethyl)benzene monomers (Scheme 7.1a and b). In this AB_2 monomer, the A ($ClCH_2$) and 2B (CNCH′) sites are attached on the separated aromatic rings by a flexible ether bridge, respectively.

7.2.2 AB_2 + AB Approach

Hyperbranched copolymers can also be synthesized from the AB_2 + AB system (Scheme 7.2). The AB_2 + AB system is equivalent to AB_2 except that AB_2 units are separated from each other by AB units. Hyperbranched aromatic polyamide copolymers were prepared by direct polycondensation of 3-(4-aminophenoxy)benzoic acid (AB monomer) and 3,5-bis(4-aminophenoxy)benzoic acid (AB_2 monomer) in the presence of triphenyl phosphite and pyridine as condensation agents (Scheme 7.3) [54].

Frey et al. [55] prepared hyperbranched random copolymers by using a combination of latent AB_2 glycidol monomer and AB allyl or phenyl glycidyl ethers. In addition, the respective multiarm star block copolymers (BCs) were prepared by the sequential addition of latent AB_2 glycidol monomer and ABR comonomers such as allyl glycidyl ether (AGE) monomer and phenyl glycidyl ether (PGE) (Scheme 7.4). The substituent R of ABR comonomer is inert under the polymerization conditions but permits post-synthetic modification or functionalization of the hyperbranched structure.

Similarly the reaction between AB_2 and B_3 systems also leads to hyperbranched copolymer. In this system, B_3 acts as a central core from which polymerization radiates and offers greater control of molecular shape.

7.2.3 $B_3 + A_2 + B_2$ Approach (Biocatalyst)

In this approach, also hyperbranched copolymers are produced in one-pot synthesis by condensation of three different multifunctional monomers such as $B_3 + A_2 + B_2$. Lipase-catalyzed terpolymerizations were performed with the monomers trimethylolpropane (B_3), 1,8-octanediol (B_2), and adipic acid (A_2) (Scheme 7.5) [56]. Polymerizations were performed in bulk, at 70 °C, for 42 h, using immobilized lipase B from *Candida antarctica* (Novozyme-435) as a catalyst. Variation of trimethylolpropane in the monomer feed gave copolymers with degrees of branching (DB) from 20% to 67%. It is useful for synthesizing hyperbranched polymers only when cross-linking is minimized by limiting conversion and/or diluting the reactants with solvent.

7.2.4 Macromonomers Approach

In this route, instead of simple monomers two or more macromonomers are used to form dendritic BCs. Hedrick et al. [57] synthesized dendritic block copolyesters by using a mixture of two different AB_2 macromonomers of

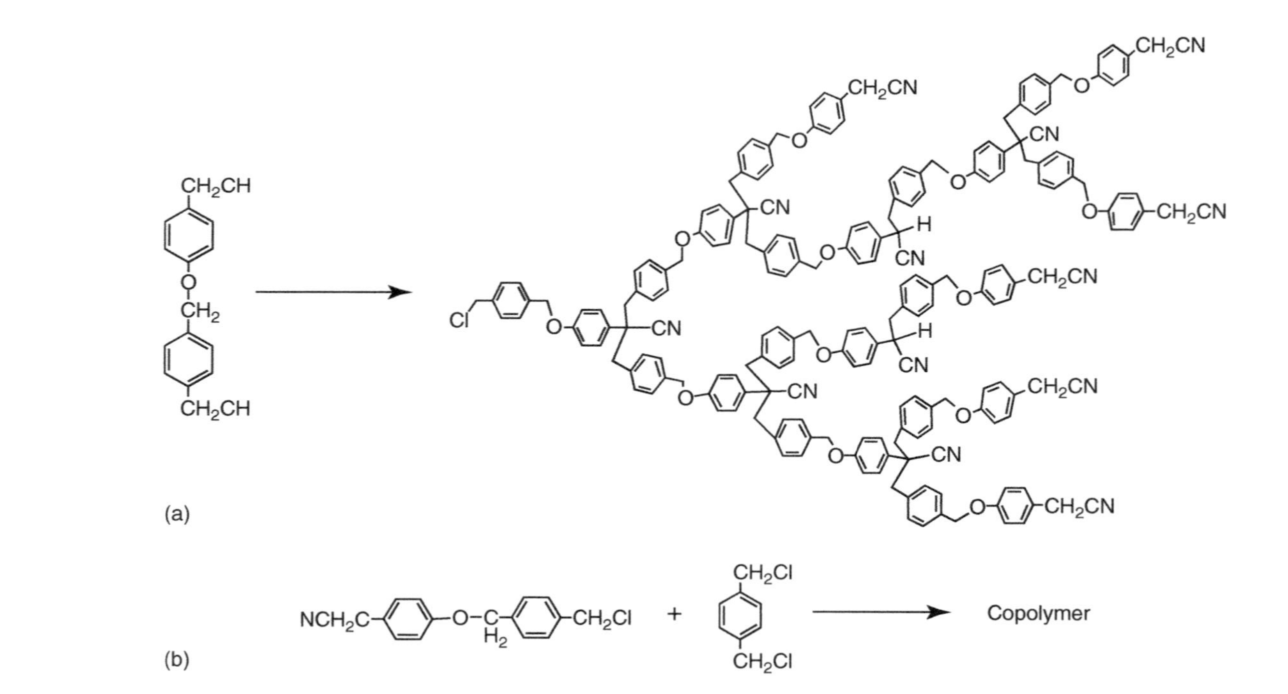

Scheme 7.1 (a) Polymerizations of activated methylene monomers such as AB_2. (Reprinted with permission from [53]. Copyright © 1998 Wiley Periodicals Inc.) (b) Copolymerizations of activated methylene monomers AB_2 and A_2. (Reprinted with permission from [53]. Copyright © 1998 Wiley Periodicals Inc.)

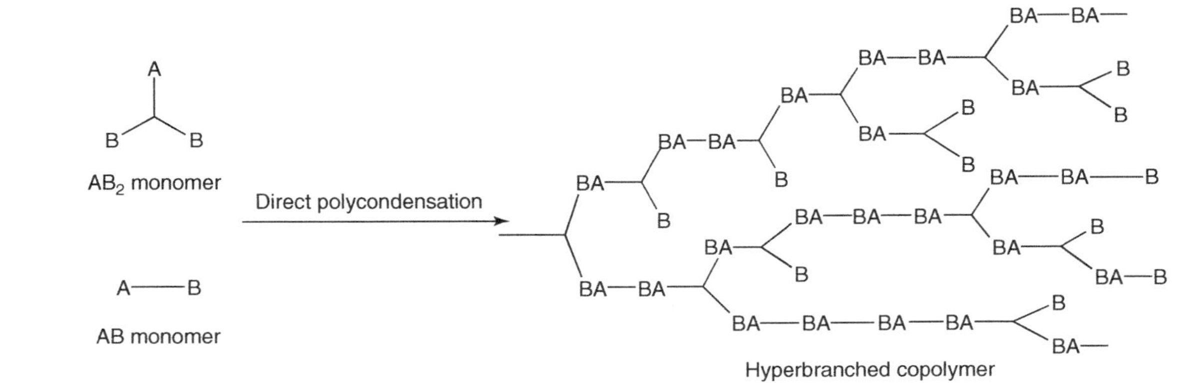

Scheme 7.2 Copolymerization of AB and AB_2 monomers. (Reprinted with permission from [54a]. Copyright © 2000 American Chemical Society.)

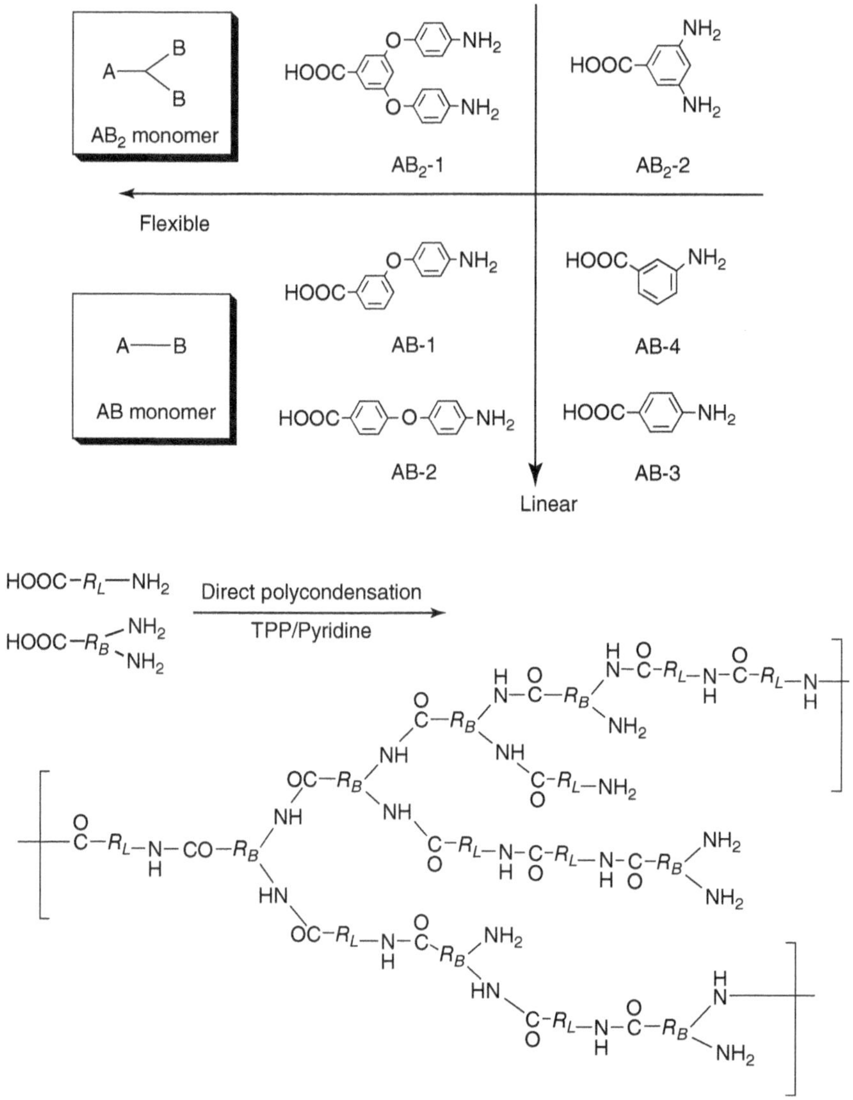

A—<(B)(B) AB$_2$ monomer; AB$_2$-1; AB$_2$-2; Flexible; A—B AB monomer; AB-1; AB-4; AB-2; AB-3; Linear

$HOOC-R_L-NH_2$, $HOOC-R_B(NH_2)_2$ → Direct polycondensation, TPP/Pyridine

Scheme 7.3 Direct polycondensation of AB- and AB$_2$-type monomers. (Reprinted with permission from [54b]. Copyright © 2003 Wiley Periodicals Inc.)

substituted caprolactones by co-condensation (Scheme 7.6). This group also proposed a second route for the synthesis of dendrimer-like star BCs by using a trifunctional 2,2′-bis(hydroxymethyl)propionic acid as initiator followed by generating different blocks for different generations with controlled branching molecular architecture (Scheme 7.7). They were succeeded in controlling the mechanical properties and crystallinity by choosing the copolymer composition of biodegradable dendritic block copolyesters.

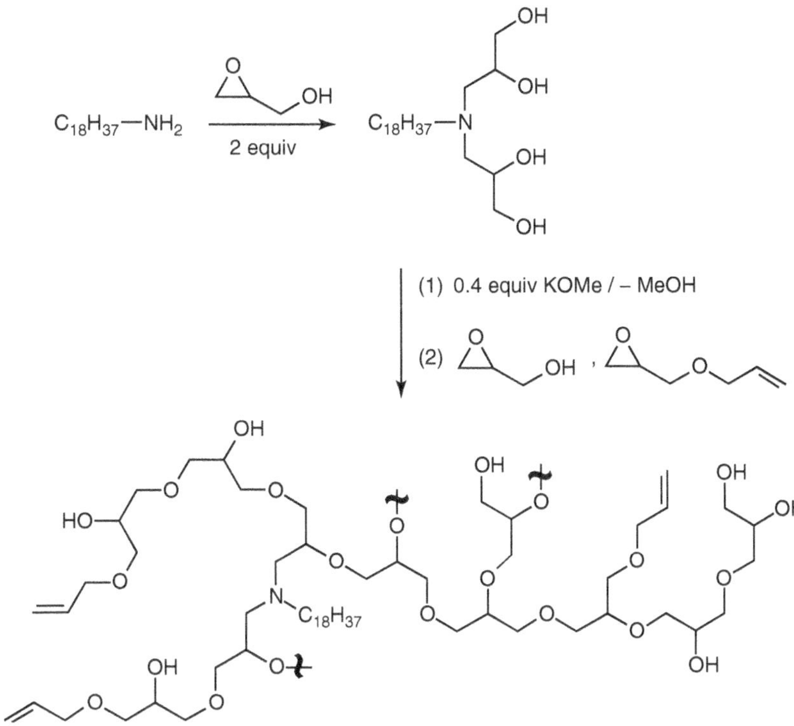

Scheme 7.4 Synthesis of the initiator bis(2,3-dihydroxypropyl)octadecylamine and base-catalyzed random copolymerization of glycidol and allyl glycidyl ether (AGE). (Reprinted with permission from [55]. Copyright © 2000 American Chemical Society.)

7.2.5 Dendrigraft Approach

In this approach, two strategies are used to synthesize dendritic copolymers that are grafting onto and grafting from strategies [58]. Several architectures are possible in dendrigraft polymers as shown in Figure 7.1. In grafting onto approach, the polymeric chains (rather than small molecules) serve as building blocks whereby side chains obtained in a separate step were reacted with substrates bearing suitable coupling sites (Scheme 7.8). It should be noted that the side chain units are different from the backbone units. Each side chain of the substrate contains multiple coupling sites. These coupling sites allow further grafting reaction cycles leading to higher generation (G2, G3) dendritic polymers with additional branching levels, for example, BC side chains through sequential monomer additions to the living polymer. The branching multiplicity (number of side chains grafted per substrate chain) is typically around 10–15, as compared with 2–3 for dendrimers and 1–2 for the hyperbranched systems. These features lead to a very rapid (typically 10-fold) increase in molecular weight and branching functionality for successive grafting reactions (generations). Depending on their structural characteristics (side chain length, branching density, and generation number), polymers with weight-average

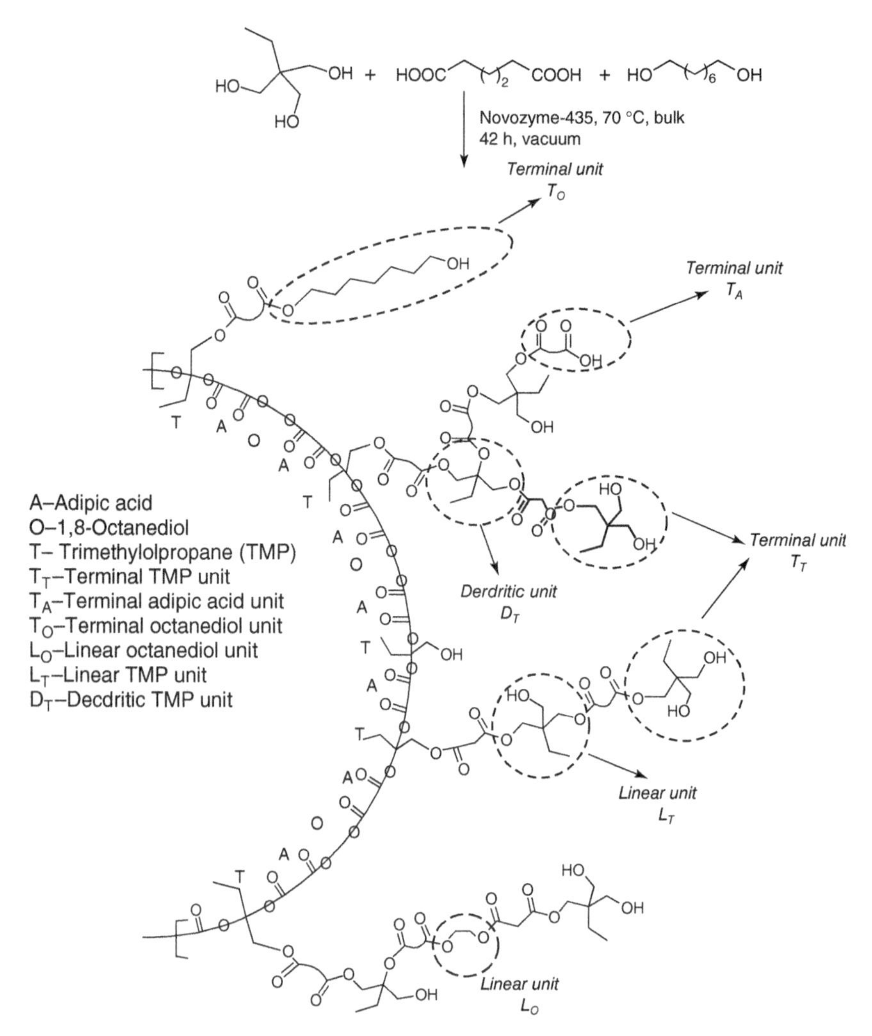

Scheme 7.5 Novozyme-435-catalyzed one-pot polymerization of trimethylolpropane to form terpolyesters. (Reprinted with permission from [56]. Copyright © 2007 American Chemical Society.)

molecular weights (M_w) ranging from 10^4 to 10^8 and branching functionalities (f_w) of $10-10^4$ have been obtained, while maintaining polydispersity index (M_w/M_n) below 1.1. The size distribution of dendritic graft polymers is thus narrower than for hyperbranched polymers, but the structure of polymers is not as strictly controlled as in dendrimers. The dimensions of dendrigraft polymers can be 10–1000 times larger than for dendrimers, even after only two to four grafting cycles.

The use of ionic polymerization techniques is beneficial because it provides extensive control over the structure of these polymers, characterized by side chains of uniform size distributed randomly along the grafting substrate. Several important

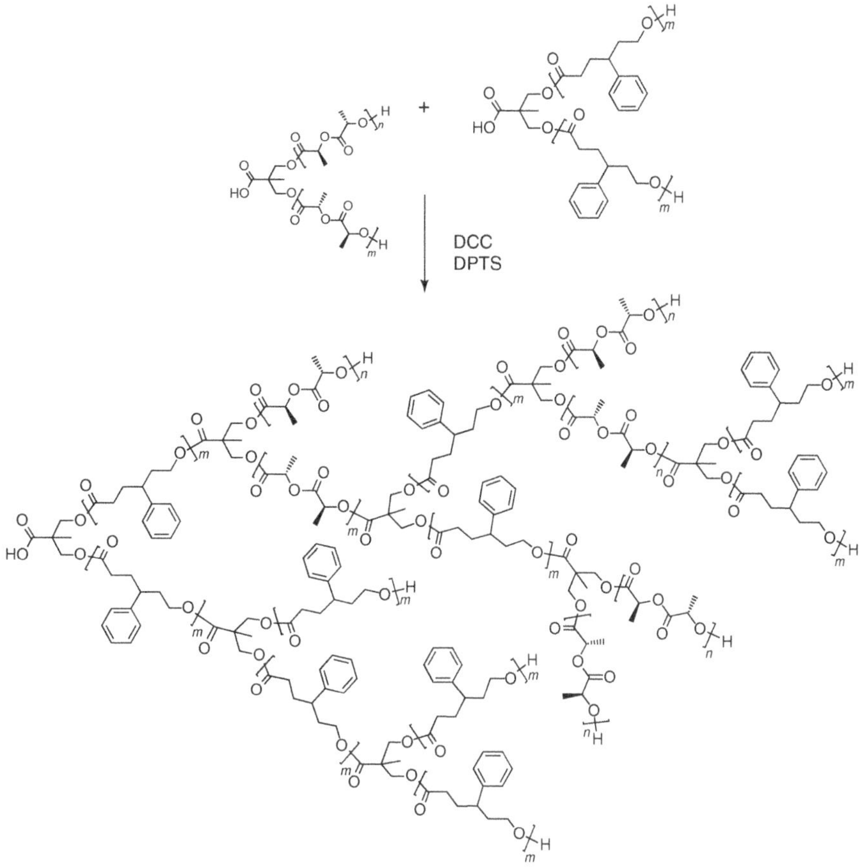

Scheme 7.6 Co-condensation of different AB_2 macromonomers. (Reprinted with permission from [57]. Copyright © 1999 American Chemical Society.)

characteristics distinguish dendritic graft polymers from dendrimers and hyperbranched polymers, the two other dendritic polymer families.

The unusual physical properties of arborescent polystyrenes (PSts) were highlighted in a number of investigations, demonstrating that the polymers displayed increasingly hard sphere-like behavior for higher branching functionalities and for shorter side chains [59].

7.2.6 Linear–Dendritic Copolymers

The building blocks of the linear–dendritic copolymers are only two, but they could be positioned in several distinct configurations because of the presence of multiple anchoring points in both of them [60]. The first general group contains a single monodendron or dendrimer (D) and one (3A), two (3B), or multiple (3C) linear

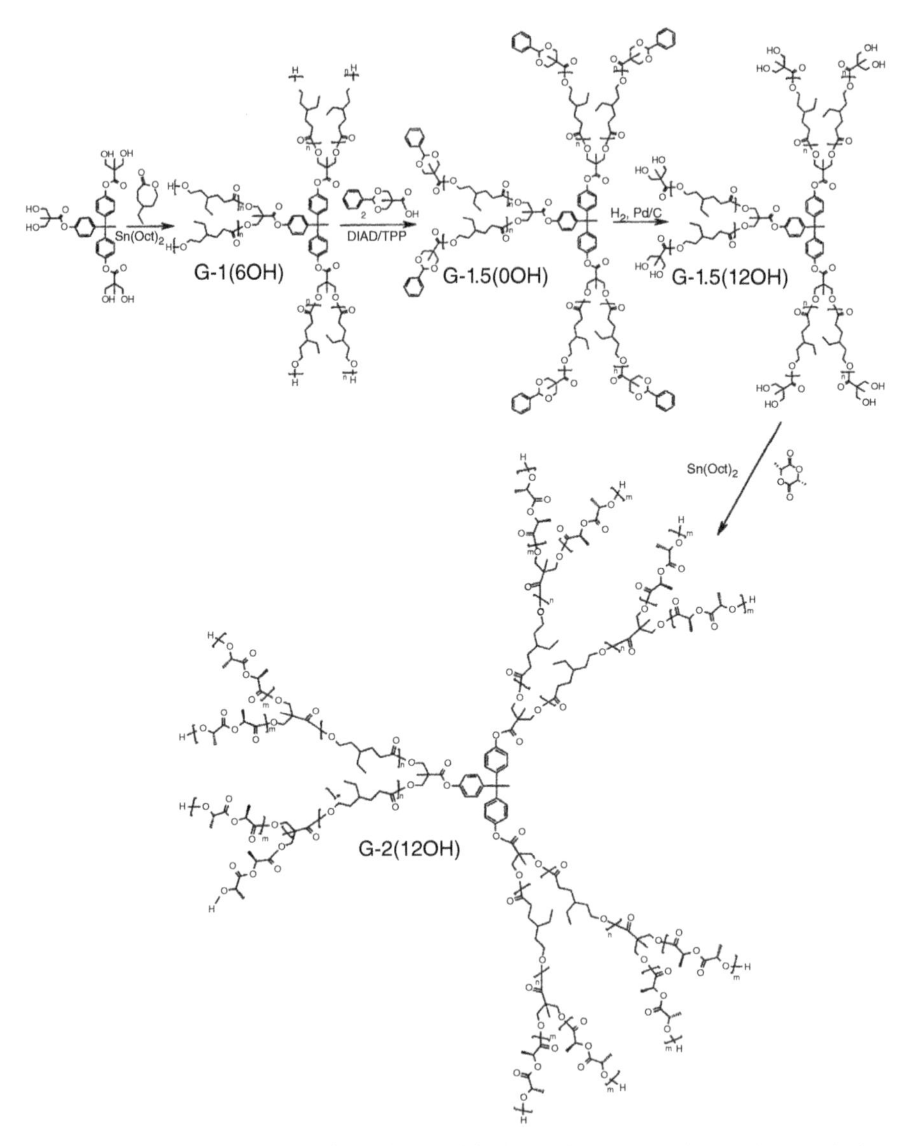

Scheme 7.7 Synthesis of dendrimer-like star block copolymer. (Reprinted with permission from [57]. Copyright © 1999 American Chemical Society.)

segments (L) attached at the "focal" point or at the peripheral functional groups in the D fragments (Figure 7.2).

The size (length and generation) and the polarity of the blocks would determine the relative hydrophobic/hydrophilic balance and ultimately affect the solution and solid-state properties of the hybrids.

The characteristic feature of the second group is the attachment of two monodendrons to the extremities of a single linear chain (4A) or the incorporation

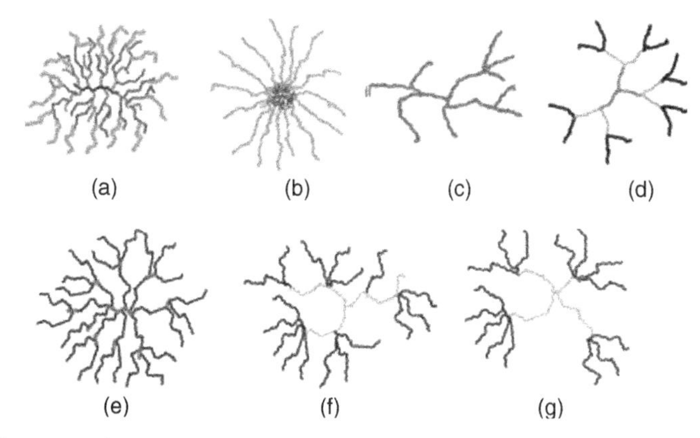

Figure 7.1 Examples or dendrigraft polymer architectures: arborescent copolymers with (a) short or (b) long corona chains; (c) from convergent cationic inimer one-pot synthesis; dendrimer-like polymers with (d) strictly terminal grafting, and (e) branching point grouping; from convergent anionic one-pot synthesis with (f) random branching points, and (g) star-branched structure [58]. (*See insert for color representation.*)

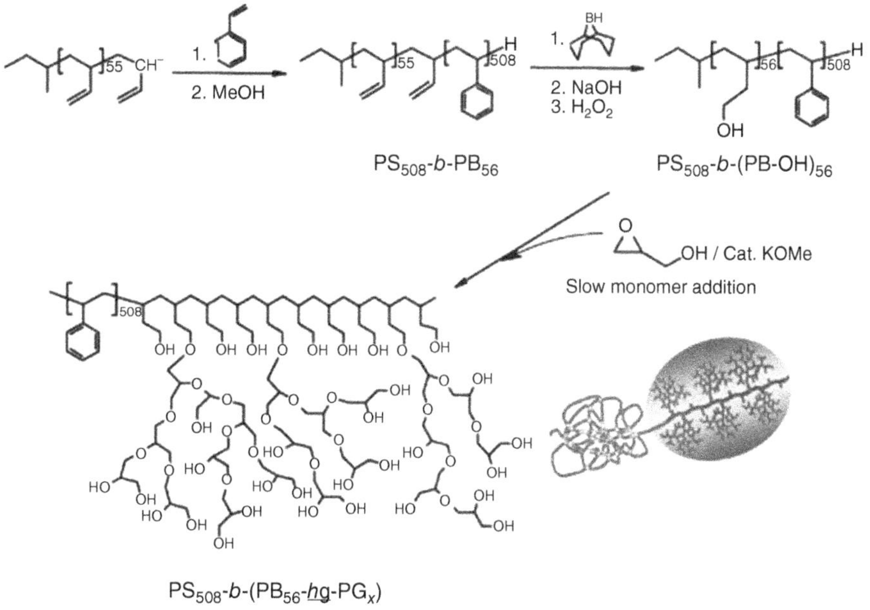

1. (styrene) 2. MeOH

PS_{508}-*b*-PB_{56}

1. (9-BBN) 2. NaOH 3. H_2O_2

PS_{508}-*b*-$(PB\text{-}OH)_{56}$

(glycidol) OH / Cat. KOMe

Slow monomer addition

PS_{508}-*b*-$(PB_{56}$-*hg*-$PG_x)$

Scheme 7.8 Synthetic strategy employed for the preparation of PS-b-PB-hg-PG. (Reprinted with permission from [58b]. Copyright © 2005 Wiley Periodicals Inc.)

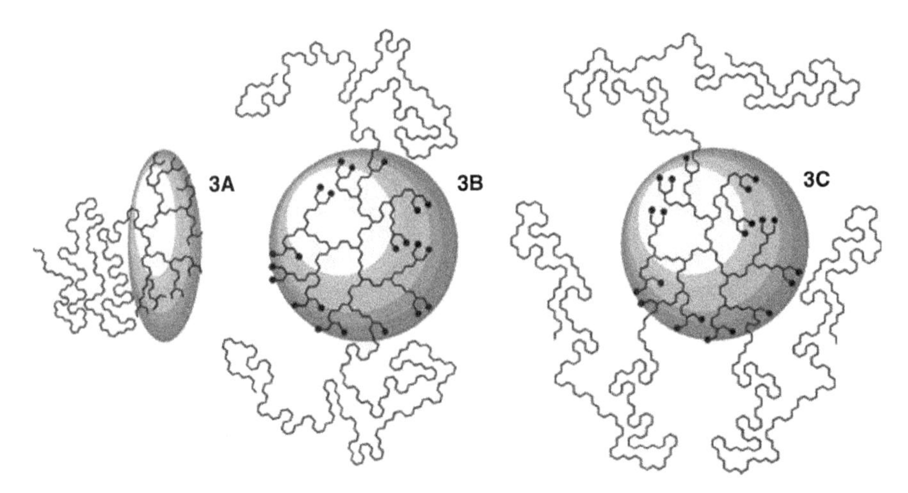

Figure 7.2 Linear–dendritic architecture: 3A, LD copolymer; 3B, LDL copolymer; 3C, L3D copolymer. (Reprinted with permission from [60]. Copyright © 2008 Wiley Periodicals Inc.)

of dendrimers into the main linear chain (4B) (Figure 7.3). The variation of chain length and dendrimer size in this group also offers interesting macromolecular geometries.

The third group unifies the structures, where monodendrons are attached like "pendants" to a main linear chain through short (5A) or long spacers (5B). When the linear chain is "shrunk", 5B is transformed into a star-like macromolecule with the monodendrons anchored at the extremities of the star arms (5C) (Figure 7.4). A special case of 5C arises when the core of the star is not a small multifunctional unit, but a dendrimer. The ratio of the spacer and main chain length to the size (generation) of monodendrons would lead to various interesting geometries from nanocylinders to layered nanospheres.

One strategy involves coupling of preformed end-functionalized linear polymers with reactive dendritic architectures having a complementary functional group at their focal point. The first reported use of preformed fragment coupling was based on the interaction of "living" poly(styrene) dianion, formed by naphthalene–potassium and modified with 1,1-diphenylethylene, and Fréchet-type monodendron with benzyl bromide "focal" point (Scheme 7.9). The reaction proceeded smoothly under slight monodendron excess and afforded quantitatively 4A-type copolymers regardless of the monodendron size (generations 2, 3, and 4) and poly(styrene) chain length (from 474 Da up to 800,000 Da) [48b].

A good example of the "dendrimer-first" strategy could be found in a study of the "living" anionic polymerization of ethylene oxide (EO), initiated by the same dendritic alcoholates (Scheme 7.10) [61]. The resulting monodisperse linear–dendritic copolymers were further transformed into asymmetric dendritic–linear–dendritic copolymers by modification with monodendrons of different generation or with monodendrons having peripheral cyano groups. It was shown that some of these

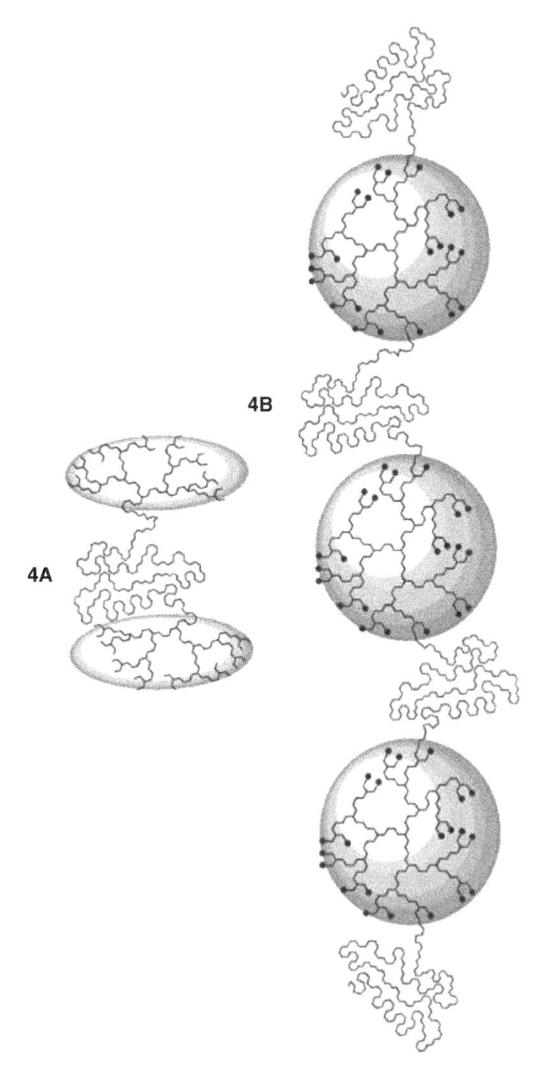

Figure 7.3 Linear–dendritic architecture: 4A, DLD copolymer; 4B, (LD)4L copolymer. (Reprinted with permission from [60]. Copyright © 2008 Wiley Periodicals Inc.)

constructs can self-assemble into two distinct supermolecules depending on the size of the second monodendron.

The 5A-type amphiphilic linear–dendritic copolymers can be formed as shown in Scheme 7.11. This serves as a good example of the layered polarity arrangement [49,62a].

Another approach to dendritic–linear BCs involves a divergent synthetic strategy of the dendron segment from an appropriately end-functionalized linear polymer (Scheme 7.12) [49,62b,c].

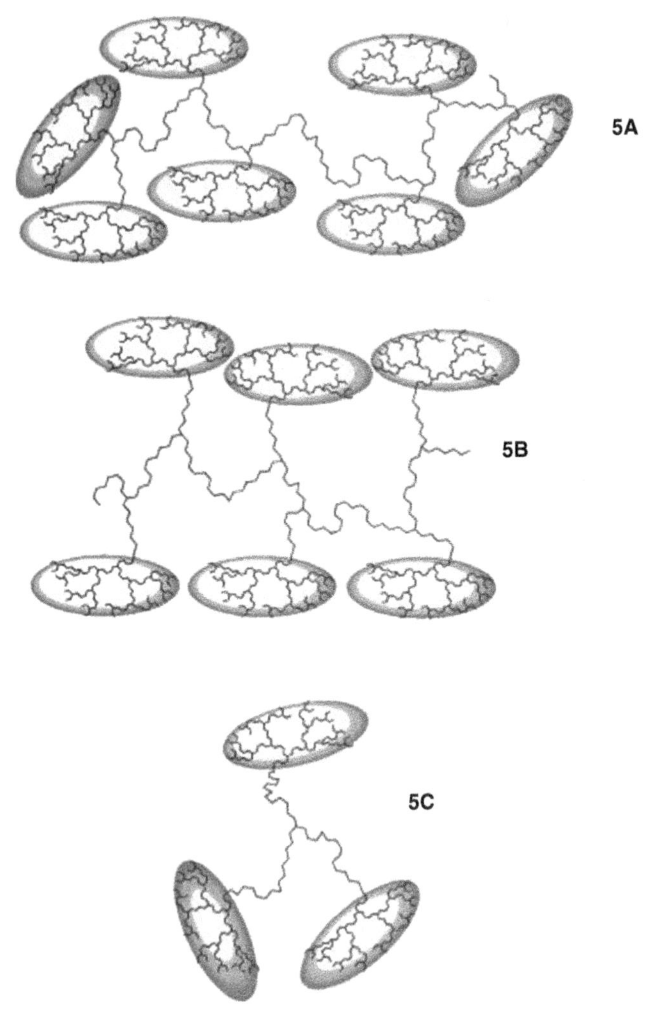

Figure 7.4 Linear–dendritic architecture: 5A, L(D)7 copolymer; 5B, L(L0D)6 copolymer; 5C, (LD)3 star copolymer. (Reprinted with permission from [60]. Copyright © 2008 Wiley Periodicals Inc.)

7.2.7 Living Anionic Polymerization

In this approach, dendrimer-like star-branched polymers and BCs are synthesized by two methods: iterative divergent strategy based on "core-first" initiation method and iterative convergent strategy based on "arm-first" termination method [63]. The divergent "core-first" initiation method involves two elementary reaction steps: controlled/living polymerization from multifunctional precursors and chain-end functionalization to create at least two initiating sites per arm, and repeats the two reaction steps [64–67]. With this methodology, Hedrick and coworkers and

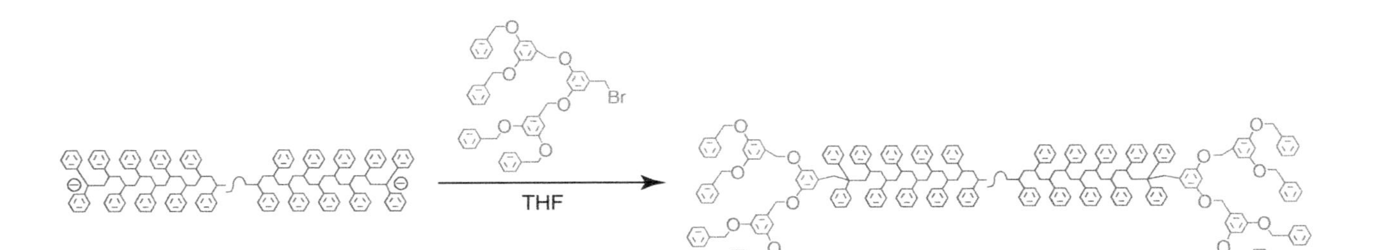

Scheme 7.9 Synthesis of DLD copolymer by coupling of reactive monodendrons to “living” poly(styrene). (Reprinted with permission from [60]. Copyright © 2008 Wiley Periodicals Inc.)

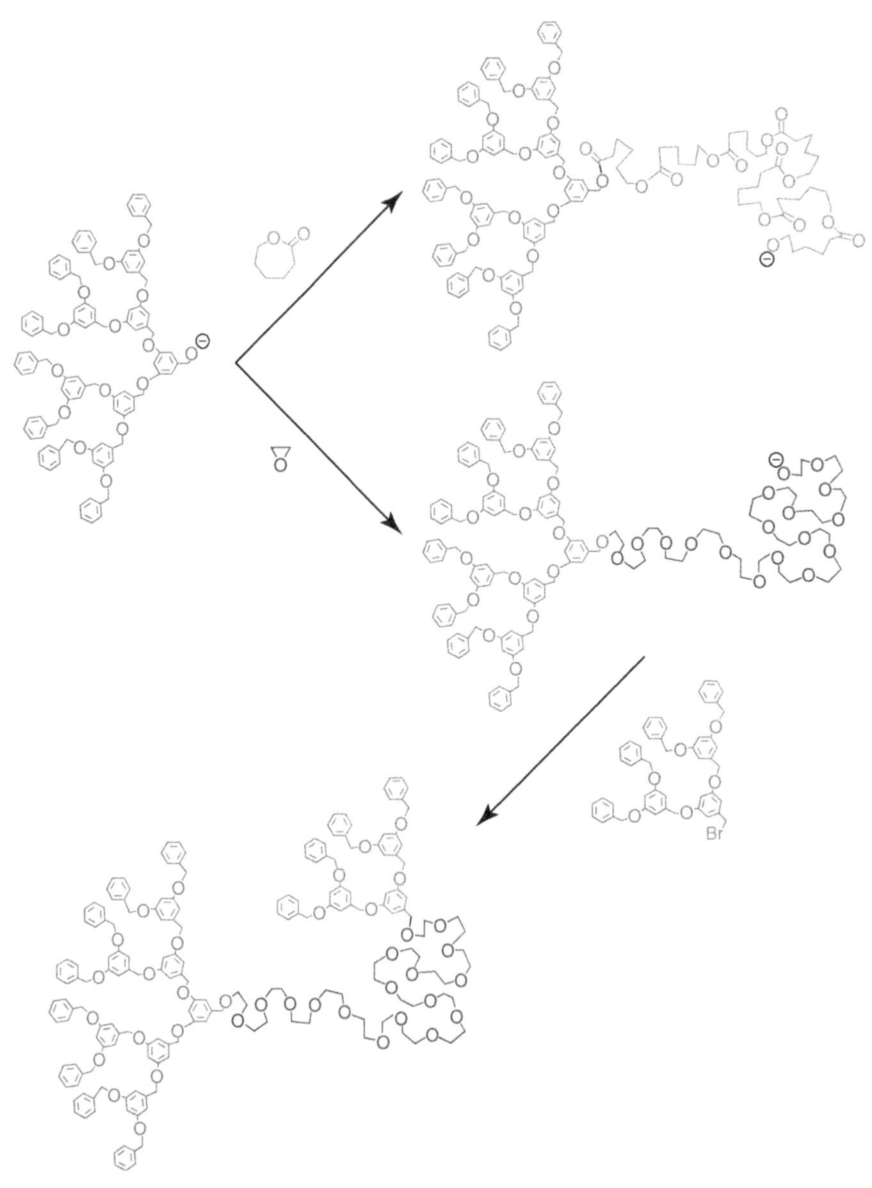

Scheme 7.10 Synthesis of LD and DLD copolymers by activated monodendrons and "living" anionic ring-opening polymerization. (Reprinted with permission from [60]. Copyright © 2008 Wiley Periodicals Inc.)

Gnanou and coworkers were successful in synthesizing various amphiphilic and water-soluble dendrimer-like star-branched BCs. Scheme 7.13 shows the synthetic route for such a BC composed of poly(ethylene oxide) (PEO) and poly(acrylic acid) segments by a combination of the living anionic polymerization of EO and the

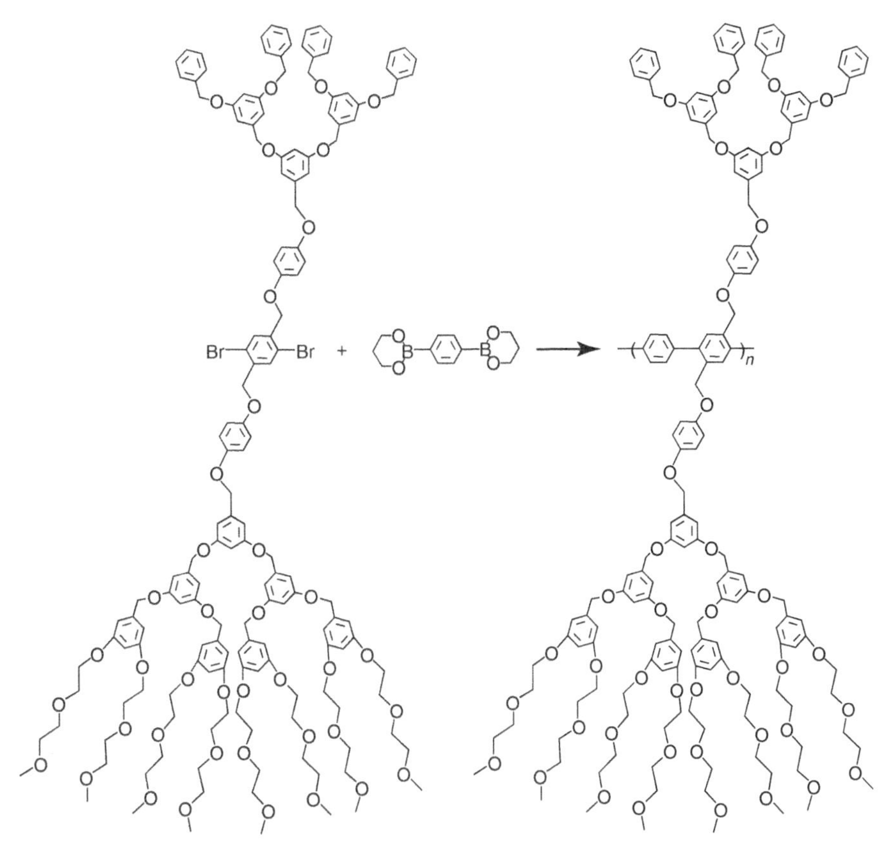

Scheme 7.11 Synthesis of "dendronized" copolymer by polycoupling reaction. (Reprinted with permission from [60]. Copyright © 2008 Wiley Periodicals Inc.)

subsequent atom transfer controlled radical polymerization of *tert*-butyl acrylate (*t*BA), followed by hydrolysis.

In the convergent "arm-first" termination method, dendrimer-like star-branched BCs can be synthesized by the linking reaction of either CH_3SiCl_3 or $SiCl_4$ with living anionic off-center graft copolymers prepared by the addition reaction of macromonomers to living anionic polymers, followed by the anionic polymerization of other monomers [68]. The synthetic outline is illustrated for the dendrimer-like star-branched BC of styrene (St) and isoprene in Scheme 7.14.

As illustrated in Scheme 7.15, a fourth-generation dendrimer-like star-branched BC composed of 16 poly(methyl methacrylate) (PMMA) and 14 poly(2-hydroxyethyl methacrylate) (PHEMA) segments, G-4-OH, was successfully synthesized by reacting the third-generation brominated PMMA (four core arms and two branches at every junction) with a living anionic polymer of the silyl-protected HEMA, followed by deprotection to regenerate the original hydroxyl group [69]. The hydroxyl protection of HEMA was achieved by treatment of

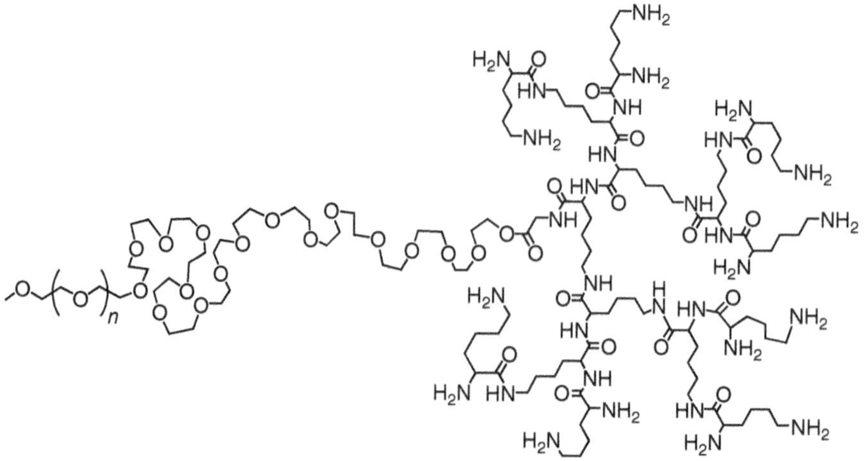

Scheme 7.12 Linear–dendritic copolymer built by divergent dendritic growth of L-lysine on to the hydroxyl terminus of methoxy-poly(ethylene oxide). (Reprinted with permission from [60]. Copyright © 2008 Wiley Periodicals Inc.)

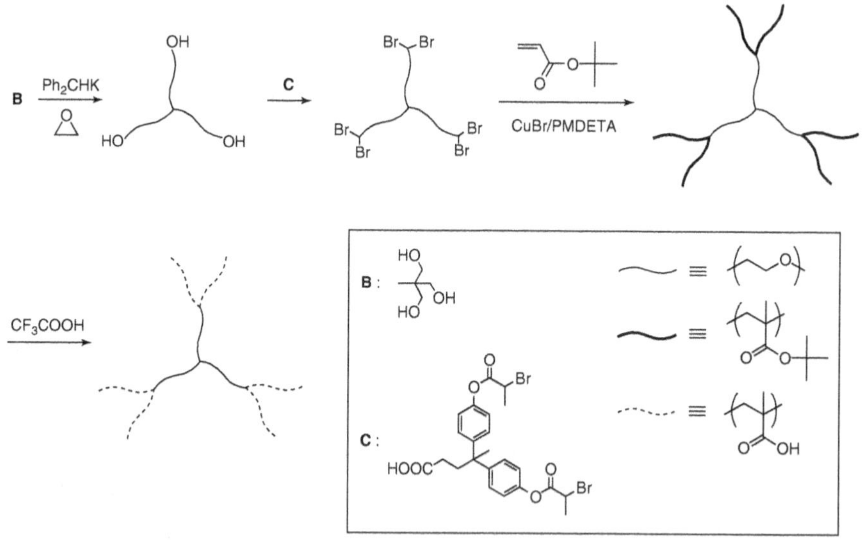

Scheme 7.13 Synthetic scheme of dendrimer-like star-branched block copolymer composed of PEO and poly(acrylic acid) segments by a combination of two living/controlled polymerizations and chain-end functionalization. (Reprinted with permission from [63]. Copyright © 2006 Wiley Periodicals Inc.)

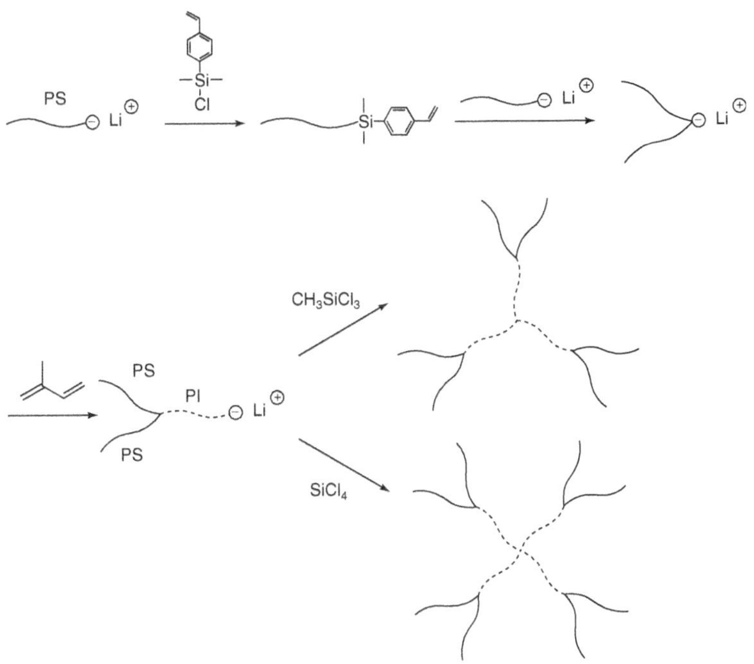

Scheme 7.14 Synthetic scheme of the second-generation star-branched polymers by the reaction of living anionic off-center branched polymer with either $MeSiCl_3$ or $SiCl_4$. (Reprinted with permission from [63]. Copyright © 2006 Wiley Periodicals Inc.)

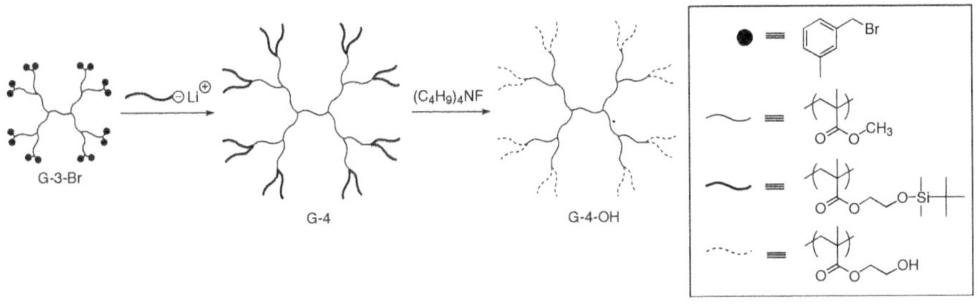

Scheme 7.15 Synthetic scheme of the fourth-generation dendrimer-like star-branched block copolymer composed of PMMA and PHEMA segments. (Reprinted with permission from [63]. Copyright © 2006 Wiley Periodicals Inc.)

HEMA with *tert*-butyldimethylsilyl chloride, and the deprotection was conducted with $(C_4H_9)_4NF$ under neutral conditions [70].

Similarly, a variety of dendrimer-like star-branched BCs have been synthesized by the reaction of the third-generation brominated PMMA with living anionic polymers of the following functionalized methacrylate monomers as shown in

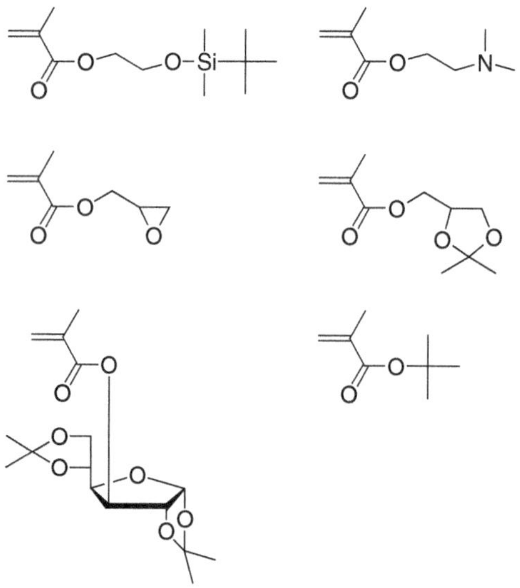

Figure 7.5 Living anionic polymerization-4 Methacrylate monomers with functional groups. (Reprinted with permission from [63]. Copyright © 2006 Wiley Periodicals Inc.)

Figure 7.5. Since the first four PMMA segments radially emerged from a central core in the brominated PMMA used in the linking reaction, all of the resulting BCs were composed of 28 PMMA and 32 functionalized polymer segments as illustrated in Scheme 7.16. *N,N*-Dimethylamino- and epoxy-functionalized polymer chains were introduced as repeating units at the fourth generation by reacting living anionic polymers of 2-(*N,N*-dimethylamino)ethyl [71a] and glycidyl methacrylates [71b], respectively. The quantitative introduction of diol- and glucose (tetraol)-functionalized segments could be achieved via the following two reaction steps: (i) the linking reaction of living anionic polymers of methacrylates substituted with 2,2-dimethyldioxolane [71c] and 1,2 : 5,6-diisopropylidene-α-D-glucofranose

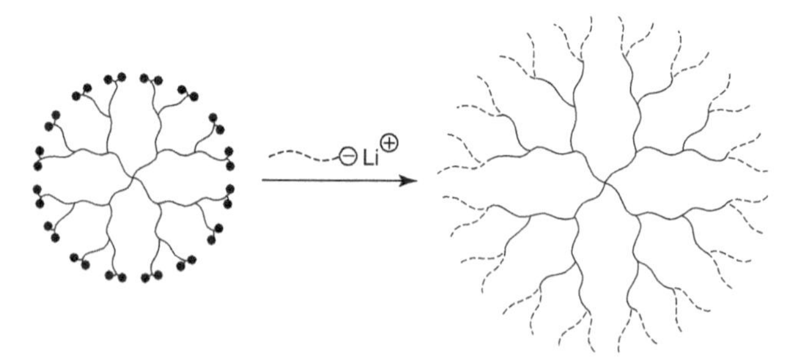

Scheme 7.16 Synthetic scheme of the fourth-generation dendrimer-like star-branched block copolymer composed of PMMA and functionalized polymer segments. (Reprinted with permission from [63]. Copyright © 2006 Wiley Periodicals Inc.)

moieties [71d]; (ii) deprotection of their acetal-protected functionalities under weak acidic conditions. Carboxyl-functionalized segment was also introduced quantitatively by reacting with living poly(*tert*-butyl methacrylate), followed by hydrolysis of the *tert*-butyl ester moiety. Interestingly, *N*,*N*-dimethylamino- and carboxyl-functionalized segments could be converted to water-soluble cationic and anionic segments by treatment with CH_3I and NaOH. Thus, introduction of these functionalized segments endows with interesting and useful characters such as hydrophilicity, water solubility, acid, base, ion, chirality, and high reactivity.

7.2.8 Controlled Living Radical Polymerization

The lack of control in classical radical polymerization procedures has limited the application of this technique for the preparation of well-defined polymer architectures. This situation has changed dramatically because of the progress in controlled radical polymerization (CRP) techniques. Various methods for CRP have been proposed; however, the most successful methods include atom transfer radical polymerization (ATRP), stable free radical polymerization (SFRP), and reversible addition–fragmentation transfer (RAFT) polymerization. These advances provide a simple pathway for the preparation of a wide variety of interesting polymers with well-defined structures.

7.2.8.1 Atom Transfer Radical Polymerization (ATRP) The control over the radical polymerization is based on two principles. First, initiation should be fast, providing a constant concentration of growing polymer chains. Second, because of the persistent radical effect, the majority of growing polymer chains is dormant species that still preserve the ability to grow because of dynamic equilibrium between dormant species and growing radicals established. By keeping the concentration of active species or propagating radicals sufficiently low throughout the polymerization, termination is suppressed (Figure 7.6).

ATRP can be used to derive dendritic copolymers by using dendritic multiarm multifunctional initiators (Scheme 7.17a and b). The macroinitiator can be synthesized via using functionalized initiator, chain transfer reaction, or end-capping method. The synthesis of novel PMMAs with star-like architectures by controlled radical polymerization starting from dendritic 2-, 4-, 6-, and 12-arm multifunctional initiators is described. The more highly functionalized initiators

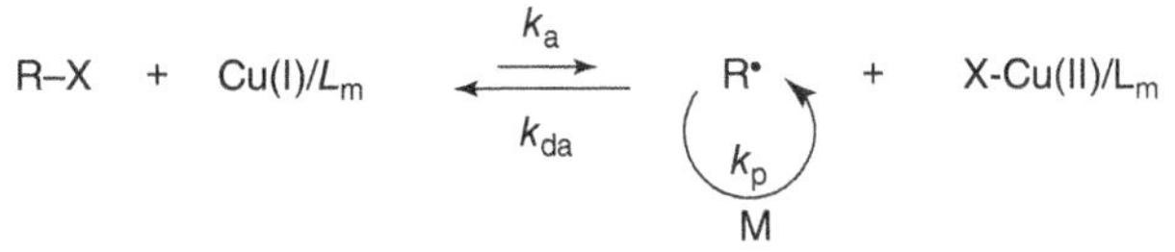

$$\text{R–X} + \text{Cu(I)}/L_m \underset{k_{da}}{\overset{k_a}{\rightleftharpoons}} \text{R}^\bullet \,(k_p,\ \text{M}) + \text{X-Cu(II)}/L_m$$

Figure 7.6 ATRP polymerization where k_a, k_{da}, and k_p are the rate constants of activation, deactivation, and propagation, respectively. Lm is the ligand. (Reprinted with permission from [72a]. Copyright © 2008 Wiley Periodicals Inc.)

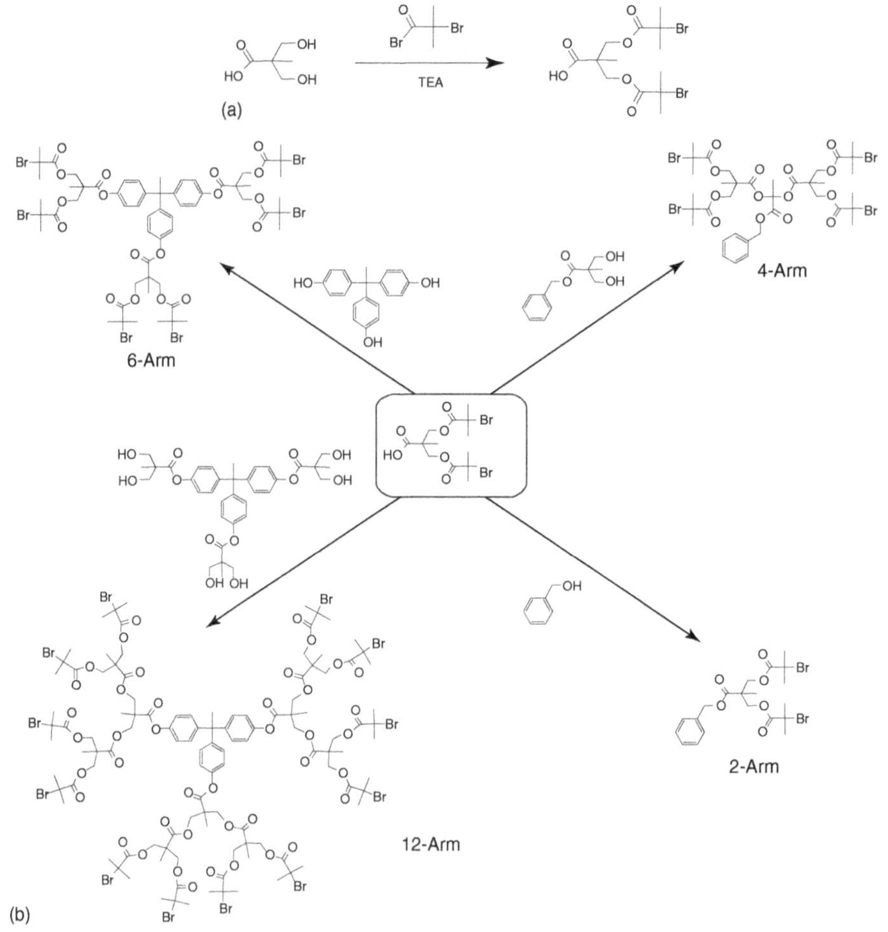

Scheme 7.17 (a) and (b) Synthesis of multiarm dendritic initiators [72].

were obtained by coupling a bromo-functionalized bis-(hydroxymethyl)propionic acid (bis-MPA) first-generation dendron to hydroxyl-functionalized precursors. The functionalized initiator bore a functional group for initiating controlled/living polymerization. In this approach, star-like copolymers of methyl methacrylate (MMA) containing varying amounts of hydroxyethyl methacrylate with slightly higher polydispersities have been synthesized by ATRP (Scheme 7.18) [72].

In another approach, hyperbranched copolymers can be synthesized in one step by using ATRP inimer that contains an acrylate group at one end and an active bromide at the other end. 2-(2-bromoisobutyryloxy) ethyl acrylate (BIEA) is a typical "ATRP inimer" that is acting as both initiator and monomer in ATRP. The incorporability of the acrylate group in chain walking polymerization (CWP) and the catalyst's tolerance toward the α-bromoester functionality allow the successful

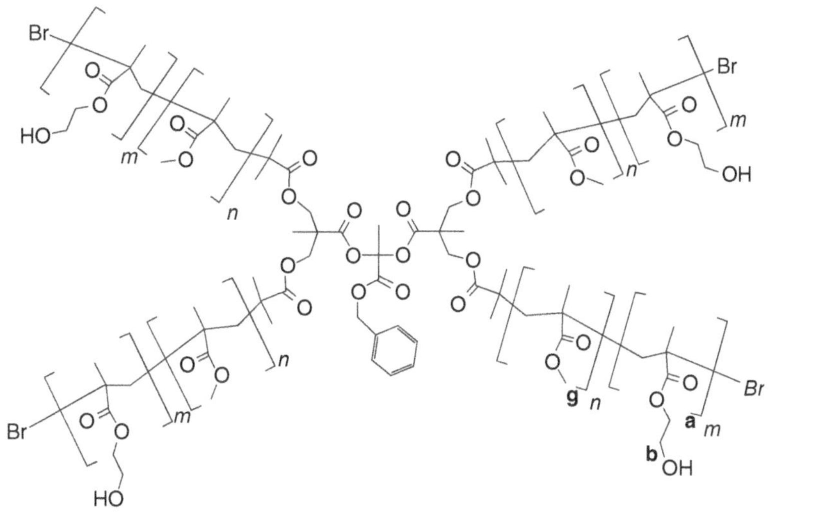

Scheme 7.18 Dendritic copolymer of MMA and hydroxyethyl methacrylate (HEMA) [72].

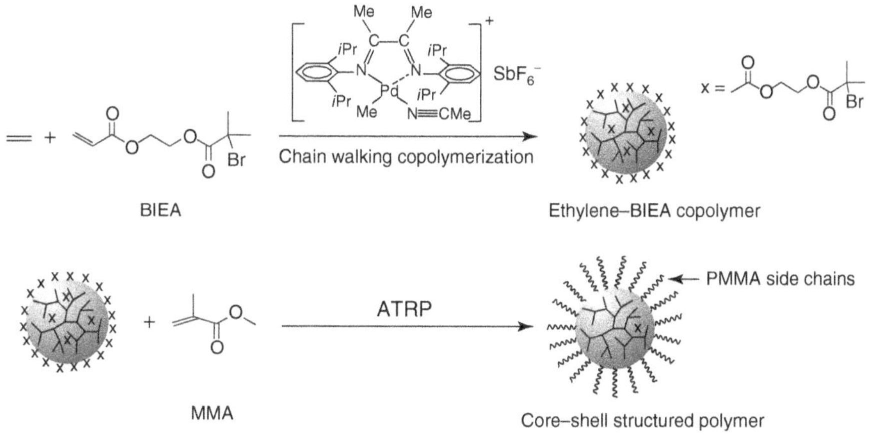

Scheme 7.19 Schematic synthetic procedure for chain walking copolymerization of poly(methyl methacrylate) arms grafted on the hyperbranched polyethylene core. (Reprinted with permission from [73]. Copyright © 2007 Wiley Periodicals Inc.)

synthesis of the hyperbranched copolymers. These hyperbranched polyethylenes (PEs) that contain ATRP initiating sites have been directly used as macroinitiators without further transformation for the ATRP of MMA to synthesize functionalized hyperbranched PEs with PMMA arms grafted on the hyperbranched PE core (Scheme 7.19) [73].

BC can also be synthesized by catalytic olefin polymerization followed by controlled/living polymerization, such as ATRP [74–76]. It required a

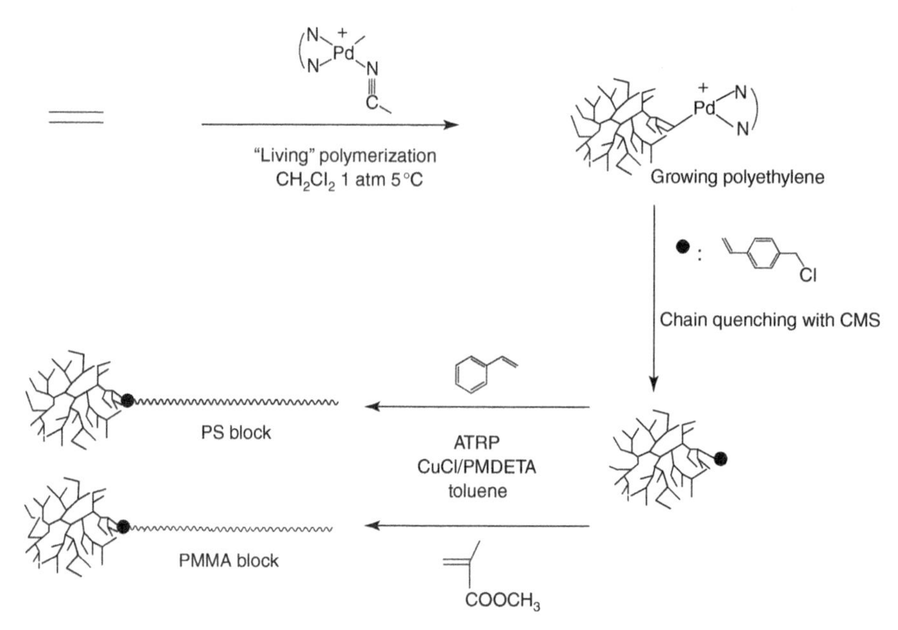

Scheme 7.20 Schematic synthesis of polyethylene end-capped with chloromethyl styrene (CMS-PE) and polyethylene-*b*-polystyrene (PE-*b*-PS) or polyethylene-*b*-poly (methyl methacrylate) (PE-*b*-PMMA) block copolymers via atom transfer radical polymerization (ATRP) of CMS-PE with styrene or methyl methacrylate. (Reprinted with permission from [80]. Copyright © 2010 Wiley Periodicals Inc.)

polyolefin precursor bearing terminal functional group that acts as a macroinitiator to initiate successive controlled/living polymerization. The PE was synthesized by "living" polymerization of ethylene with Pd-diimine catalyst $[(ArN{=}C(Me){-}(Me)C{=}NAr)Pd(CH_3)(N{\equiv}CMe)]^+SbF_6^-$ (Ar = 2,6-$(i\mathrm{Pr})_2C_6H_3$). Chloromethyl styrene (CMS) was used as a chain quenching agent (CQA) to end-cap PE (CMS-PE). The Pd-diimine catalyst featured with chain walking mechanism and polymerized ethylene in a "living" manner to form hyperbranched PE with controllable structure and chain topology [77–79]. The CMS-PE was used as a macroinitiator for successive ATRP of St or MMA to achieve linear-hyperbranched BCs, polyethylene-*block*-polystyrene (PE-*b*-PSt), or polyethylene-*block*-poly(methyl methacrylate) (PE-*b*-PMMA), respectively. The synthetic route is shown in Scheme 7.20 [80].

7.2.8.2 Stable-Free Radical Polymerization (SFRP) SFRP, similar to ATRP, is also based on the reversible termination mechanism (Figure 7.7). However, SFRP uses stable radicals as controlling mediators that react reversibly with propagating radicals to yield dormant chains. Hence, the identity of stable persistent radicals ($T^\bullet$) is crucial for successful SFRP. In order to effectively mediate polymerization, 2,2,6,6-tetramethyl-piperidin-1-yl)oxyl (TEMPO) (and other stable free radicals) should react neither with itself nor with monomer to initiate the growth of new

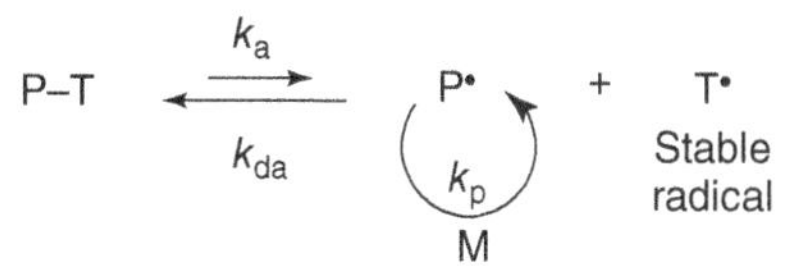

$$\text{P–T} \underset{k_{da}}{\overset{k_a}{\rightleftharpoons}} \text{P}^\bullet (k_p, \text{M}) + \text{T}^\bullet$$

Stable radical

Figure 7.7 SFRP polymerization where k_a, k_{da}, and k_p are the rate constants of activation, deactivation, and propagation, respectively. (Reprinted with permission from [72a]. Copyright © 2008 Wiley Periodicals Inc.)

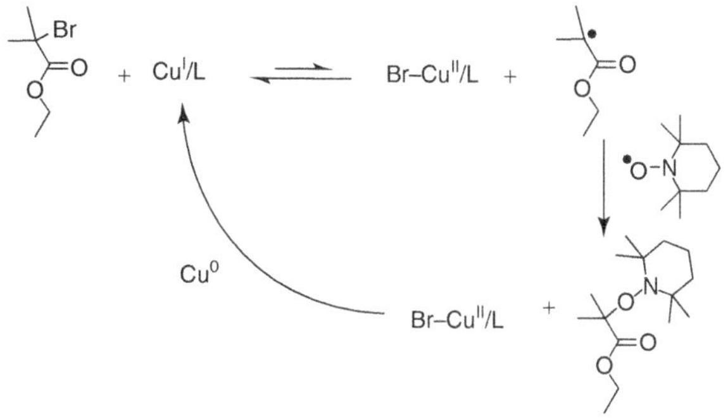

Scheme 7.21 Method of controlled generation and trapping of stable radicals. (Reprinted with permission from [82b]. Copyright © 2007 Elsevier Ltd.)

chains, and it should not participate in side reactions such as the abstraction of β–H atoms. These persistent radicals should also be relatively stable, although their slow decomposition may in some cases help maintain appropriate polymerization rates. Although TEMPO is most efficient in polymerization of St at high temperatures, other nitroxides and alkoxyaminederivatives allow for low temperature polymerization of St as well as methacrylates.

SFRP systems can be initiated in two different ways. Conventional radical initiators can be used in the presence of persistent radicals, as discussed earlier. Alternatively, dormant species can be prepared in advance and used as initiators (so-called unimolecular initiators) [81,82] or macroinitiators for block copolymerization (Scheme 7.21).

SFRP can be used to derive dendritic copolymers by using multifunctional initiators as a core in a two-step or three-step sequence of polymerization. The synthesis of PSt star copolymer is shown in Scheme 7.22 [83]. A trifunctional initiator, 2-phenyl-2-[(2,2,6,6-tetramethyl)-1-piperidinyloxy]ethyl-2,2-bis[methyl(2-bromopropionato)]propionate, was synthesized and used for the synthesis of miktoarm star AB_2 and miktoarm star block AB_2C_2 copolymers via a combination of SFRP and ATRP. In the first step, a PSt macroinitiator with dual ω-bromo functionality was obtained by SFRP of St in bulk at 125 °C. Next, this PSt precursor was used as a macroinitiator for ATRP of *t*BA in the presence of Cu(I)Br and pentamethyldiethylenetriamine at 80 °C, affording miktoarm star (PSt)$(\text{P}t\text{BA})_2$ [where P*t*BA is poly(*tert*-butyl acrylate)]. In the third step, the obtained St$(t\text{BA})_2$

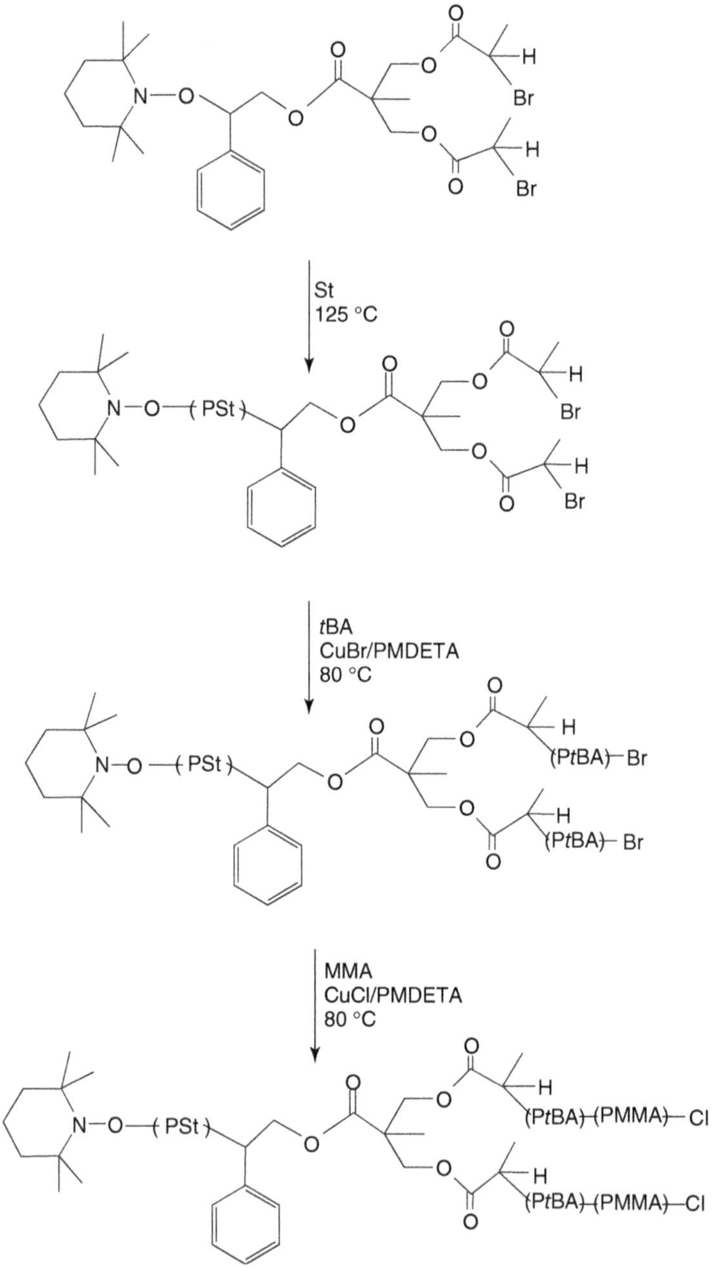
St
125 °C
tBA
CuBr/PMDETA
80 °C
MMA
CuCl/PMDETA
80 °C
PSt
PtBA
PMMA

Scheme 7.22 Synthesis of miktoarm star block copolymer by using multifunctional initiators. (Reprinted with permission from [83]. Copyright © 2003 Wiley Periodicals Inc.)

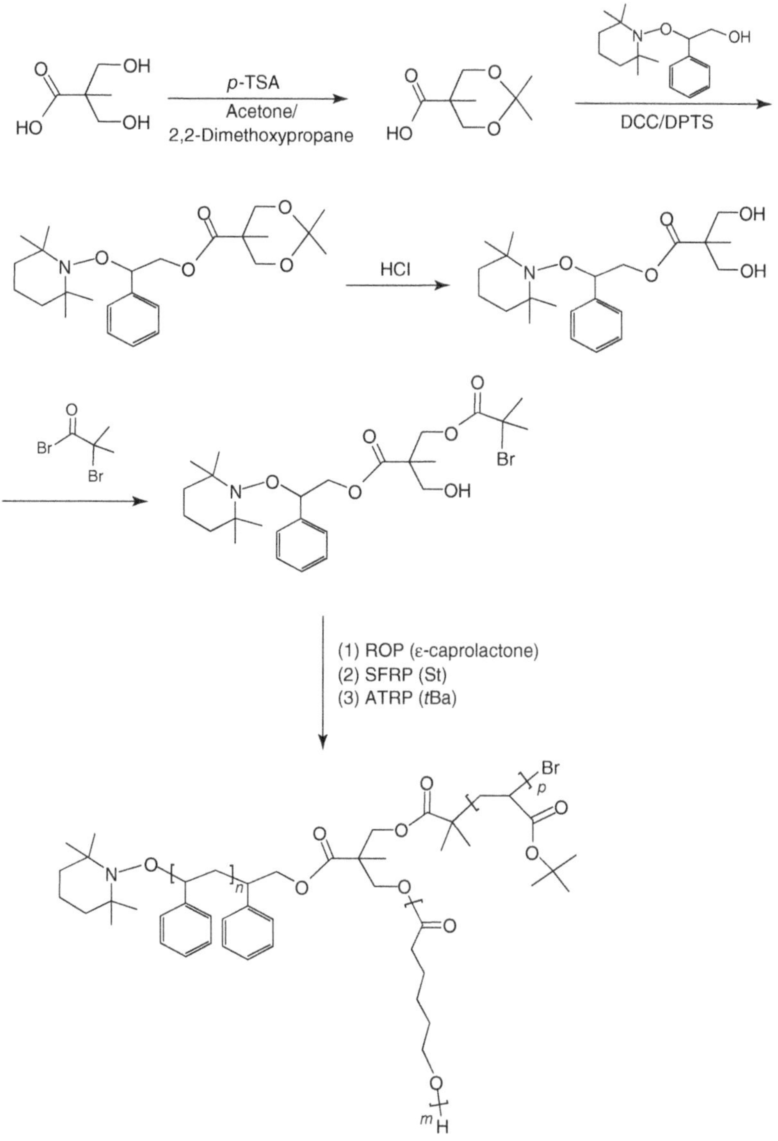

p-TSA
Acetone/
2,2-Dimethoxypropane
DCC/DPTS
HCl
(1) ROP (ε-caprolactone)
(2) SFRP (St)
(3) ATRP (tBa)

Scheme 7.23 Synthesis of caprolactone–styrene star copolymer. (Reprinted with permission from [84]. Copyright © 2004 Wiley Periodicals Inc.)

macroinitiator with two terminal bromine atoms was further polymerized with MMA by ATRP, and this resulted in (PSt)(P*t*BA)$_2$(PMMA)$_2$-type miktoarm star BC.

Similarly the synthesis of caprolactone–St star copolymer is shown in Scheme 7.23 [84]. The polymer was prepared with a core-out method via a combination of ring-opening polymerization (ROP), SFRP, and ATRP. First, ROP of ε-caprolactone was carried out with a miktofunctional initiator, 2-(2-bromo-2-methyl-propionyloxymethyl)-3-hydroxy-2-methyl-propionic acid in 2-phenyl-2-(2,2,6,6-tetramethylpiperidin-1-yl oxy)-ethyl ester, at 110 °C. Second, previously obtained poly(ε-caprolactone) (PCL) was used as a macroinitiator for SFRP of St at 125 °C. As a third step, this PCL–PSt precursor with a bromine functionality in the core was used as a macroinitiator for ATRP of *t*BA in the presence of Cu(I)Br and pentamethyldiethylenetriamine at 100 °C.

7.2.8.3 ***RAFT Polymerization*** The RAFT process has an identical setup as conventional free radical polymerization, in which a conventional chain transfer agent (CTA) is substituted with the RAFT agent. Thus, only monomer, RAFT agent and initiator need to be mixed and freed of oxygen. The mechanism of RAFT polymerization is a reversible transfer process based on reversible exchange of labile end group between a dormant species and an active radical (Figure 7.8). In this process, polymeric propagating radicals produced by free radical initiators are added to CTAs. The resulting intermediate radicals are fragmented by releasing another polymeric radical species to which monomers are added. There exists equilibrium between growing and dormant polymer chains (Figure 7.8). A similar set of reactions is operating in the equilibrium, in which a propagating macroradical reacts with the polymeric RAFT agent. Recurring RAFT events establish the equilibrium between dormant and living chains, by which living/controlled characteristics are induced in the polymerization.

The RAFT process employs effective CTAs to control molecular weight in conventional free radical polymerization with conventional initiators. The typical CTAs include dithiocarbamates [85–87], thiocarbonylthio compounds including dithiobenzoates [88–91], trithiocarbonates [92–94], and dithiocarbonates or MADIX (macromolecular design via interexchange of xanthates) agents [95–97].

However, the structural diversity of RAFT reagents is considerably larger, which ultimately allows for greater control over a wider range of monomers. Both the R and Z groups of a RAFT agent should be carefully selected to provide appropriate control [98]. Generally, R* should be more stable than P_n* in order to efficiently fragment and initiate polymerization. The selection of the R group should take into account the stability of the dormant species and rate of addition of R* to a given monomer. The order of R group leaving ability

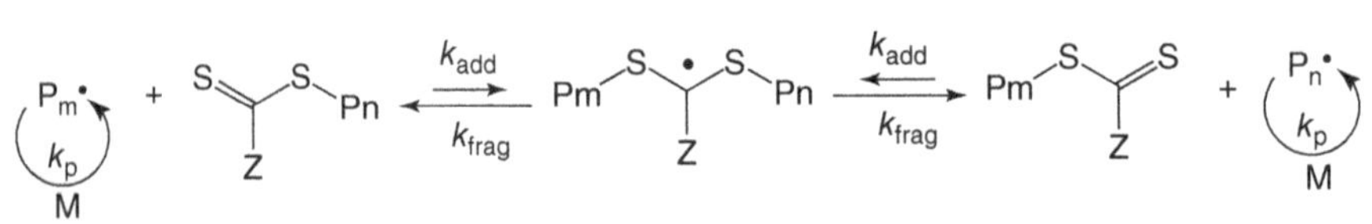

$$P_m^\bullet (M, k_p) + S{=}C(Z){-}S{-}Pn \underset{k_{frag}}{\overset{k_{add}}{\rightleftharpoons}} Pm{-}S{-}\dot{C}(Z){-}S{-}Pn \underset{k_{frag}}{\overset{k_{add}}{\rightleftharpoons}} Pm{-}S{-}C(Z){=}S + P_n^\bullet (M, k_p)$$

Figure 7.8 RAFT polymerization, where k_a, k_{da}, k_p, k_{add}, and k_{frag} are the rate constants of activation, deactivation, propagation, addition, and fragmentation, respectively. (Reprinted with permission from [72a]. Copyright © 2008 Wiley Periodicals Inc.)

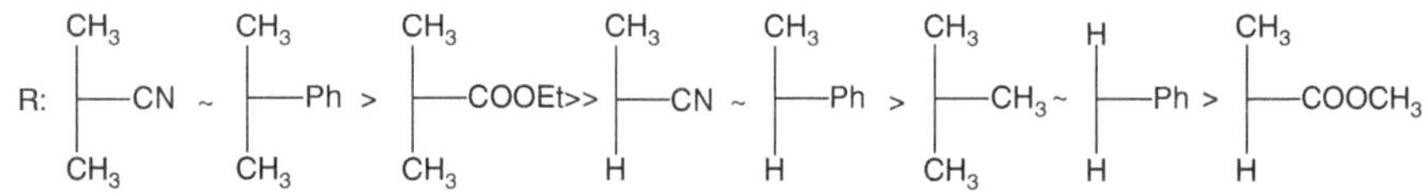

R: $C(CH_3)_2$–CN ~ $C(CH_3)_2$–Ph > $C(CH_3)_2$–COOEt >> CH(CH_3)–CN ~ CH(CH_3)–Ph > $C(CH_3)_2$–CH_3 ~ CH_2–Ph > CH(CH_3)–$COOCH_3$

Figure 7.9 Order of R group leaving ability in RAFT. (Reprinted with permission from [82a]. Copyright © 1996 American Chemical Society.)

(illustrated in Figure 7.9) reflects the importance of both steric and electronic effects. Steric effects in RAFT are much more important than in ATRP. For example, the reactivity of secondary 2-bromopropionitrile in ATRP is higher than that of the tertiary 2-bromoisobutyrate. However, the opposite trend in reactivity is observed in RAFT. Similarly, *t*-butyl halides are inactive in ATRP but are more active than benzyl derivatives in RAFT. In addition, acrylate derivatives are not very active in RAFT in contrast to ATRP. In the RAFT polymerization of MMA with dithiobenzoates (SQC(Ph)SR), the leaving group effectiveness decreased in the order $C(Alkyl)_2CN$ ~ $C(Alkyl)_2Ph$ > $C(Alkyl)_2COOEt$ > $C(CH_3)_2C(=O)NH(Alkyl)$ > $C(CH_3)_2CH_2C(CH_3)_3$ ~ $CH(CH_3)Ph$ > $C(CH_3)_3$ ~ CH_2Ph > $CH(CH_3)COOEt$. In fact, only RAFT reagents with the first two groups were successful in preparing well-defined PMMA [99].

The structure of the Z group is equally important. Stabilizing Z groups such as –Ph is efficient in St and methacrylate polymerization, but they retard polymerization of acrylates and inhibit polymerization of vinyl esters. On the other hand, very weakly stabilizing groups, such as $-NR_2$ in dithiocarbamates or –OR in xanthates, are good for vinyl esters but inefficient for St. Pyrrole and lactam derivatives occupy an intermediate position. Additional fine-tuning is possible with electron-withdrawing or electron-donating substituents. For example, dithio-4-methoxybenzoate is less efficient than dithio-2,5-bis(trifluoromethyl)benzoate. A combination of resonance stabilization and polar effects contribute to the delocalization of charge and spin and stability of the intermediate. Chain transfer constants in a St polymerization were found to decrease in the series where Z is aryl (Ph) > > alkyl (CH_3) ~ alkylthio (SCH_2Ph, SCH_3) ~ N-pyrrolo > > N-lactam > aryloxy (OC_6H_5) > alkoxy > > dialkylamino (Figure 7.10) [100].

Dendritic–linear BC can be synthesized by the combination of dendritic focal point CTAs and RAFT polymerization. The inherent stability and chemical versatility of CTAs utilized in the RAFT technique allow the preparation of well-defined polymers with specific polymer architectures and end-group

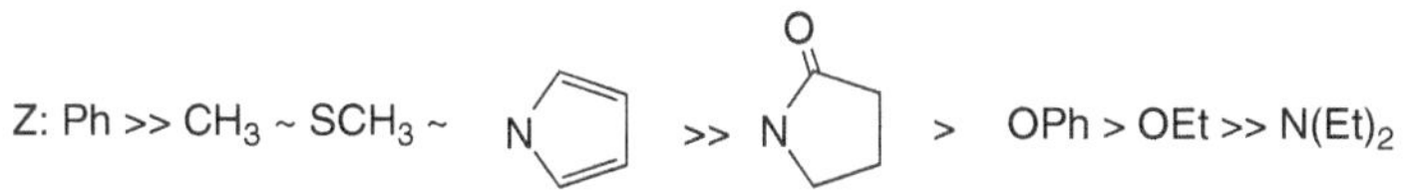

Z: Ph >> CH_3 ~ SCH_3 ~ N-pyrrolo >> N-lactam > OPh > OEt >> $N(Et)_2$

Figure 7.10 Rates of addition decrease and fragmentation increase from left to right for RAFT agents with these Z groups [100].

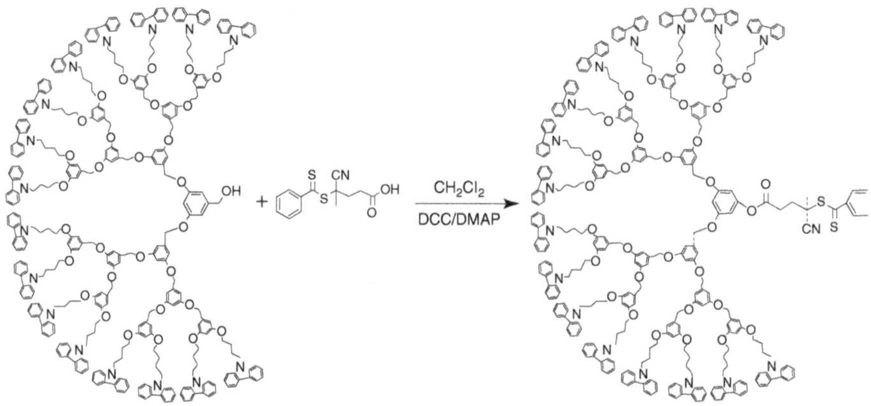

Scheme 7.24 Synthesis of dendritic focal point chain transfer agent. (Reprinted with permission from [100b]. Copyright © 2008 American Chemical Society.)

functionalities. Using dendritic CTAs possessing a single dithioester moiety at the focal point (Scheme 7.24), RAFT polymerization was carried out to attach PSt and PMMA chains of controlled lengths by kinetic control.

In another approach, dendritic CTAs possessing a multiple dithioester moieties at the periphery of dendrimer can be used as a RAFT agent. Dendritic polyester with 16 dithiobenzoate terminal groups was prepared and used in the RAFT polymerization of St to produce star PSt with a dendrimer core (Scheme 7.25). It was found that this polymerization was of living character, the molecular weight of the dendrimer-star polymers could be controlled and the polydispersities were narrow. The dendrimer-like star BCs of St and methyl acrylate (MA) were also prepared by the successive RAFT polymerization using the dendrimer-star PSt as a macro CTA (Scheme 7.26) [101]. Similarly polydisperse hyperbranched polyesters were modified for use as novel multifunctional RAFT agents. The polyester-core-based RAFT agents were subsequently employed to synthesize star polymers of *n*-butyl acrylate and St with low polydispersity (polydispersity index <1.3) in a living free-radical process. The resulting star polymers were subsequently used as platforms for the preparation of star BCs of St and *n*-butyl acrylate with a polyester core with low polydispersities (polydispersity index <1.25) [102].

The rate of polymerization in the steady state remains unchanged in comparison with a conventional polymerization system. It should be noted that the rate of polymerization (R_p) in RAFT polymerizations using dithiobenzoates as the mediating agents was retarded, that is, R_p decreased with an increase in initial RAFT agent concentrations.

7.2.9 Click Chemistry

In 2001, Sharpless et al. [103] described a new concept for conducting organic reactions, which was based on highly selective, simple orthogonal reactions that

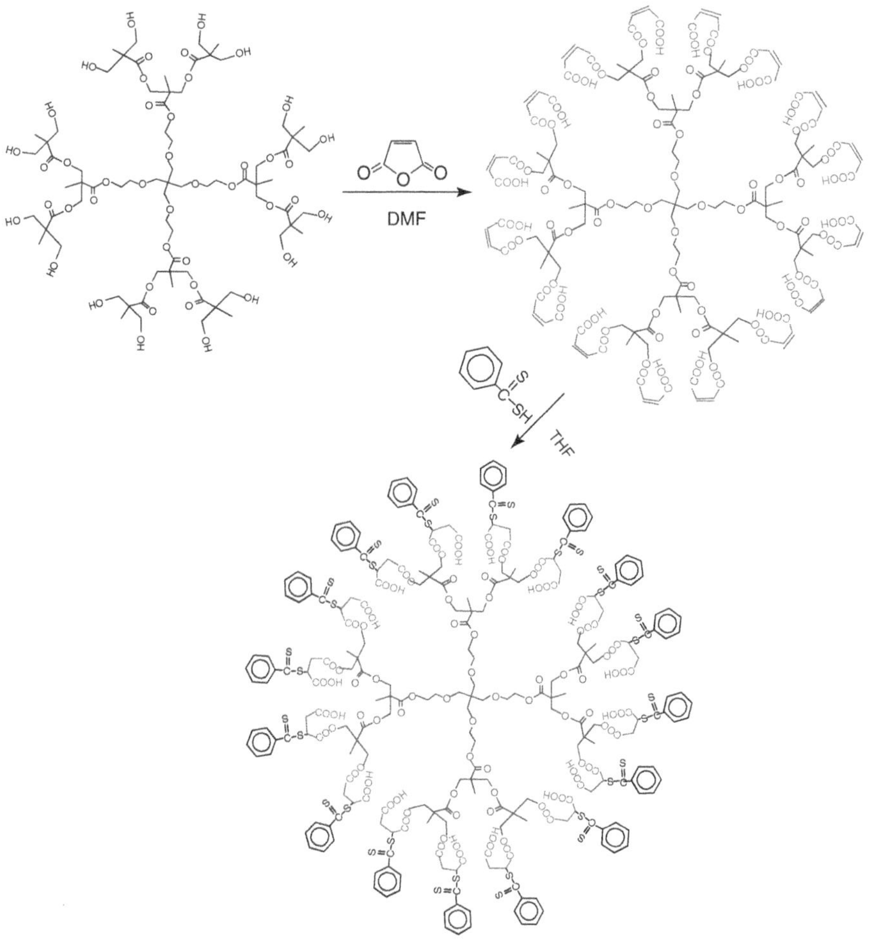

Scheme 7.25 Synthesis of dithiobenzoate-terminated dendrimer. (Reprinted with permission from [101]. Copyright © 2005 Wiley Periodicals Inc.)

do not yield side products and that give heteroatom-linked molecular systems with high efficiency under a variety of mild reaction conditions. Several efficient reactions, which are capable of producing a wide catalog of functional synthetic molecules and organic materials, have been grouped accordingly under the term *click reactions*. Characteristics of modular click reactions include (i) high yields with byproducts (if any) that are removable by nonchromatographic processes, (ii) regiospecificity and stereospecificity, (iii) insensitivity to oxygen or water, (iv) mild, solventless (or aqueous) reaction conditions, (v) orthogonality with other common organic synthesis reactions, and (vi) amenability to a wide variety of readily available starting compounds. Molecular processes considered to fit all or most of these criteria include certain cycloaddition reactions, such as the copper-catalyzed azides/alkynes cycloaddition (CuAAC), thiol-ene and

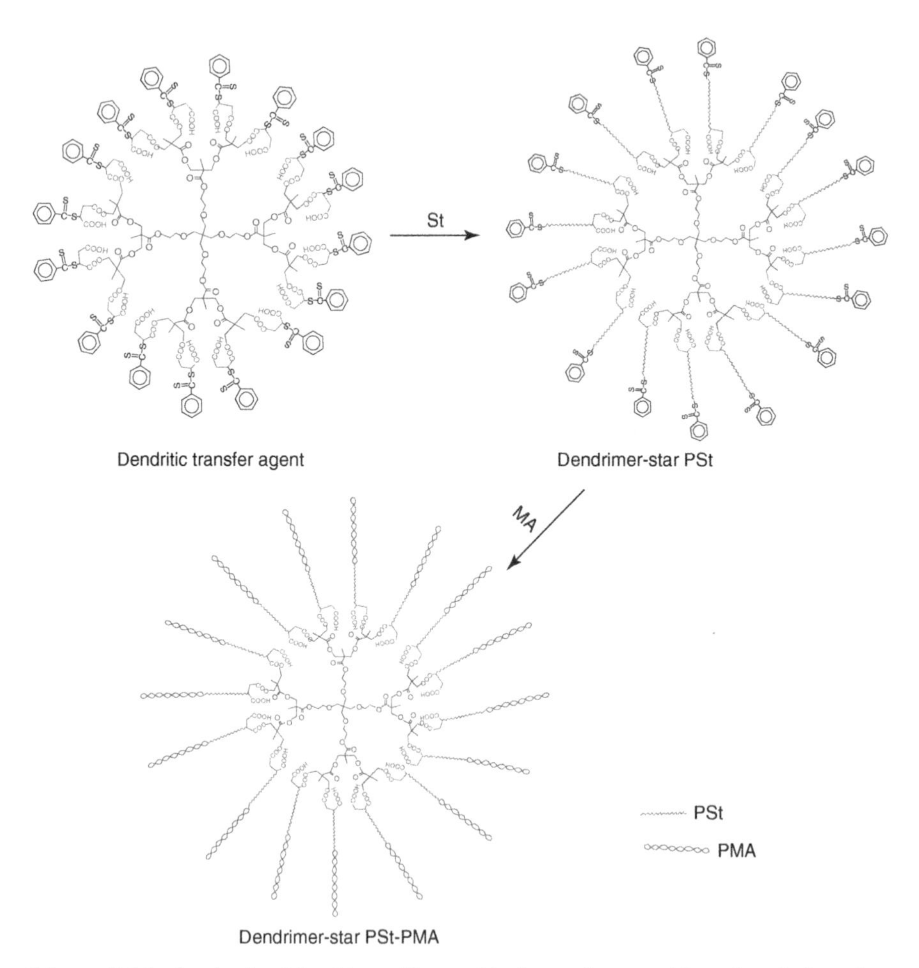

Scheme 7.26 Synthesis of dendrimer-like star block copolymers of St and methyl acrylate by using RAFT polymerization. (Reprinted with permission from [101]. Copyright © 2005 Wiley Periodicals Inc.)

nucleophilic ring-opening reactions. Copper-catalyzed azide/alkyne click reactions in particular have received the most attention, with applications extending to the synthesis of biomedical libraries, dendrimer preparation, synthesis of functional BCs, cross-linking of adhesives for metal substrates (copper/zinc), synthesis of uniformly structured hydrogels, derivatization of cellular surfaces, the *in situ* preparation of enzyme inhibitors, and many others [104–107].

The most versatile synthetic strategy in the preparation of dendritic BCs involves direct coupling of preformed end-functionalized polymers with reactive dendrons that bear a complementary functional group at their focal point. A range of BCs consisting of a linear PMMA block linked to an aliphatic polyester dendron functionalized with azobenzene moieties have been synthesized by sequential ATRP and click chemistry (Scheme 7.27). Two alkyne-functionalized PMMA

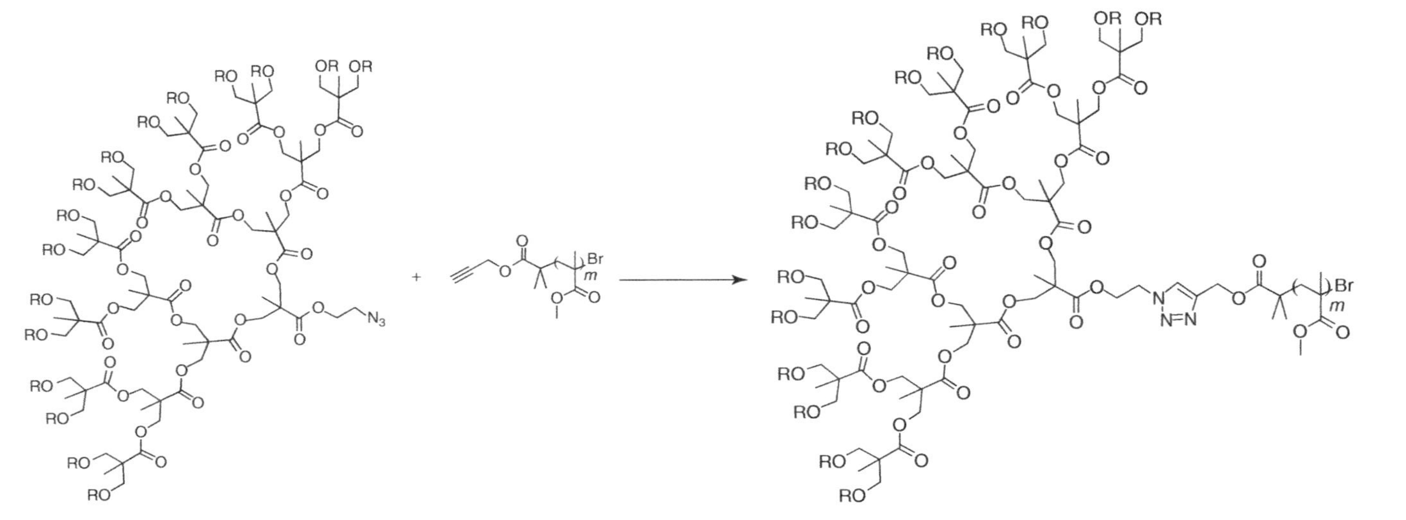

Scheme 7.27 Synthesis of linear–dendritic block copolymer by sequential atom transfer radical polymerization (ATRP) and click chemistry. (Reprinted with permission from [108]. Copyright © 2010 Wiley Periodicals Inc.)

homopolymers with different molecular weights were obtained by ATRP and coupled to generations 2 to 4 of azodendrons bearing an azide group at the focal points [108].

Another approach involves direct coupling of reactive functional groups at the periphery of dendritic polymer to another polymer containing complementary functional group. In this approach, a star graft copolymer composed of hyperbranched polyglycerol (HPG) as core and well-defined asymmetric mixed "V-shaped" identical PSt and poly(*tert*-butyl acrylate) as side chains were synthesized. The V-shaped side chain bearing a "clickable" alkyne group at the conjunction point of two blocks was first prepared through the combination of anionic polymerization of St and atom transfer radical polymerization of *t*BA monomer, and then "click" chemistry was conducted between the alkyne groups on the side chains and azide groups on HPG core (Scheme 7.28) [109].

Similar to CuAAC-based click chemistry, Diels–Alder reaction can also be used to synthesize dendritic copolymers. Dendritic two- and four-arm PMMA-based star polymers with furan-protected maleimide at their focal point $(PMMA)_{2n}$-MI and $(PMMA)_{4n}$-MI were efficiently clicked with the peripheral anthracene-functionalized multiarm star polymer (α-anthryl-functionalized polystyrene)$_m$-poly(divinyl benzene) [(α-anthryl-PS)$_m$-polyDVB] through the Diels–Alder reaction resulting in corresponding multiarm star BCs: $(PMMA)_{2n}$-$(PS)_m$-polyDVB and $(PMMA)_{4n}$-$(PS)_m$-polyDVB, respectively [110].

7.3 PROPERTIES OF DENDRITIC COPOLYMERS

It is well known that the copolymerization is meant for the combined properties of the co-components. Since the copolymers are dendritic in nature, the resultant copolymers show the physical properties such as fairly good solubility, low solution and melt viscosities, and reduced crystallinity. The physical and mechanical properties of dendritic copolymers are strongly influenced not only by the structure and the molar ratio of monomers but also by the size, shape, molecular weight, and DB. DB increases with increased monomer multiplicity of the starting materials AB_n. The packing density, topology, chain entanglement, and intermolecular interaction also influence the properties of resulting copolymers.

7.3.1 Molecular Weight and Molecular Weight Distribution

The polymers were characterized by various techniques including gel permeation chromatography (GPC). However, conventional GPC with linear polymers used for calibration is not an ideal tool to determine the molecular weights of hyperbranched copolymers. This is due to both the highly branched nature of the dendritic blocks and also the tendency of amphiphilic polymers to form micellar aggregates in different solvents. These polymers with their unique chemical and topological features behave quite differently from normal linear polymers. An additional degree of difficulty is introduced in the analysis of these hybrid polymers by the fact that reactions

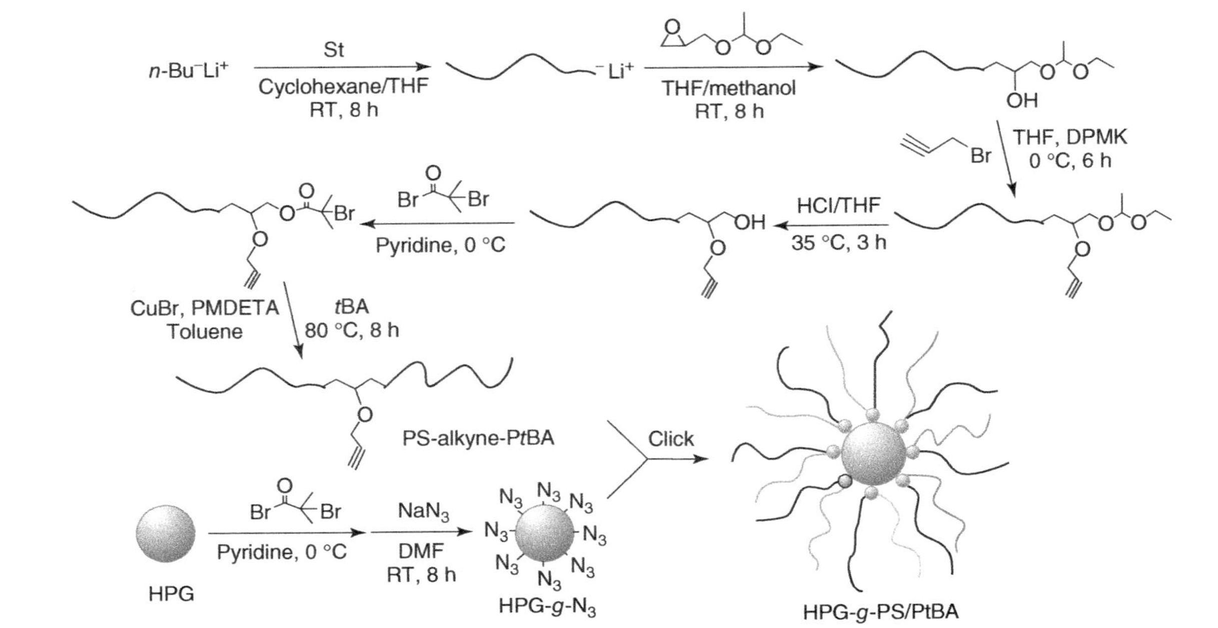

Scheme 7.28 Synthesis of star graft copolymers composed of hyperbranched polyglycerol (HPG) as core and well-defined asymmetric mixed "V-shaped" side chains by combination of atom transfer radical polymerization (ATRP) and click chemistry. (Reprinted with permission from [109]. Copyright © 2009 Wiley Periodicals Inc.)

on polymer chain ends are frequently hard to monitor and quantify. This applies to the functionalization of PEG chain ends with dendrimers, as GPC may not distinguish the difference between a fully functionalized triblock copolymer and a mixture of triblock and monosubstituted diblock copolymers [111]. This difficult analytical task along with the accurate monitoring of polymer molecular weight changes that accompany BC formation, which can be tackled using matrix-assisted laser desorption/ionization-time of flight (MALDI-TOF) mass spectrometry. After ionization, the singly charged and structurally intact ions of polymer detected by MALDI-TOF may provide a bell-shaped distribution of oligomers. Thus, the mass of each oligomer can be identified, and average molecular weights as well as polydispersities of polymers can also be determined. The oligomers are equally spaced in a way that normally reflects the mass of polymer repeat unit. Molecular weights up to almost 40,000 can be accurately analyzed. As the molecular weight and polydispersity of samples increase, MALDI-TOF analysis of molecular weight (MW) distribution becomes less reliable as it cannot be ascertained that the entire molecular weight distribution travels equally through.

For the amphiphilic copolymers with branched dendritic structures, MALDI-TOF spectrometry affords more accurate molecular weight data than the conventional GPC. End-group analysis using MALDI-TOF mass spectrometry proved very useful for the analysis of polymers with relatively low molecular weights. Such end-group analysis can differentiate between the AB dendritic–linear diblocks, ABA triblocks, and linear PEG.

MALDI-TOF mass spectrometry is becoming a technique of choice for the analysis of synthetic polymers. This powerful technique allows the direct analysis of polymers without the complication of fragmentation because of the combination of a soft ionization process with the use of a variety of MALDI matrices [112]. MALDI has been used most extensively in the analysis of biopolymers such as proteins, while other innovative applications such as the evaluation of peptide combinatorial libraries have also emerged [113]. It is only recently that synthetic polymers as well as dendrimers have been successfully analyzed by this technique [114,115].

7.3.2 Degree of Branching (DB)

By definition, a linear polymer has no dendritic segments and DB equals zero. In contrast, dendrimers exhibiting perfect branching have no linear segments and therefore DB equals unity. Because hyperbranched polymers have both linear and dendritic units, a reliable analytical method must be sought to measure the relative concentrations of these units to calculate their DBs that fall in the architectural continuum between linear polymers and dendrimers. Frechet [12] described the DB for AB_2-type homopolymers in Equation 7.1.

$$\mathrm{DB(AB_2)} = \frac{T_{\mathrm{AB_2}} + D_{\mathrm{AB_2}}}{T_{\mathrm{AB_2}} + D_{\mathrm{AB_2}} + L_{\mathrm{AB_2}}} \tag{7.1}$$

where D, T, and L represent the three possible connectivities (dendritic, terminal, and linear) from an AB_2 random polymerization. If conversion of A is high, $D \approx T$, the formula is reduced to the more widely used Equation 7.2 put forth by Frey [116]:

$$\mathrm{DB(AB_2)} = \frac{2D_{\mathrm{AB_2}}}{2D_{\mathrm{AB_2}} + L_{\mathrm{AB_2}}} \tag{7.2}$$

Recently, Frey [117] also discussed the theoretical degrees of branching of AB_m/AB copolymers. Equal reactivity of B functional groups, the absence of intramolecular cyclization, and the absence of secondary reactions must be assumed in this treatment. The theoretical DB for AB/AB_2 copolymerize as a function of the starting mole fraction AB monomer (x_{AB}) and varies between 0.50 and 0.00 for fully hyperbranched and fully linear polymers, respectively. Owing to the additional linear unit found in AB/AB_2 copolymers, the equation for calculating DB (at $p_{\mathrm{A}} = 1$) expands to

$$\mathrm{DB}\left(\frac{\mathrm{AB}}{\mathrm{AB_2}}\right) = \frac{2D_{\mathrm{AB_2}}}{2D_{\mathrm{AB_2}} + L_{\mathrm{AB_2}} + L_{\mathrm{AB}}} \tag{7.3}$$

Given the assumptions present in the theoretical treatment, experimental observations often do not adhere to theory, necessitating experimental approaches to determine DB. Some of these approaches include analysis by NMR spectroscopy (^{1}H or ^{13}C) of the resultant polymer or degradation of the polymer followed by analysis of the resulting dendritic, terminal, and linear fragments [118]. Typically, ^{1}H NMR is used to measure the dendritic and linear values of the polymer after synthesizing suitable model compounds representing dendritic and linear units in order to determine the chemical shifts of key protons for each unit [119]. A basic requirement is that the model compounds and the branched polymer to be measured must be soluble in the same deuterated solvent for comparison. In addition, there must be adequate ^{1}H dispersion and resolution of peaks of the polymer for accurate integration of the dendritic and linear contributions. However, in the case where solubility, dispersion, and resolution are problematic, another method for DB determination must be sought. Particularly the DB of copolymers with $0.25 < x_{\mathrm{AB}} < 1.0$ were not measurable by standard ^{1}H NMR because of severe aggregation and lack of dispersion. In such cases, model reactions of nonpolymerizable small molecules with independent A, B, and B_2 functionalities [(A + B)/(A + B_2) model reaction] whose products represent the linear and dendritic connectivities in the AB/AB_2 copolymer products, was devised. By control of the stoichiometry and utilization of reaction conditions that mimic the polymerizations, the products representing hyperbranched polymer units were produced and their ratios were monitored by ^{1}H NMR and/or HPLC to indirectly determine the DB. The relative concentrations of the linear and dendritic small molecules can then be monitored as a function of time. At reaction times long enough for complete conversion of the A group ($p\mathrm{A} = 1.0$), the DB can then be accurately determined from the relative concentration of the linear and dendritic products. Excellent agreement in DB was observed for copolymer compositions that could be measured and for the corresponding model reactions, validating this indirect method for measuring DB in AB/AB_2 copolymers.

Because the number of functional groups in the monomers was different (2 and 3), some interesting phenomena were observed for the copolymerization of 50/50 of AB_2/AB. The copolymers on different propagating stages were investigated by taking the samples from the polymerization mixture. It was found that the growing molecule contained a higher ratio of AB_2 monomer in the early stage of the polymerization, and dendritic units formed slowly, mainly in the later stage of the polymerization. The stepwise copolymerization of AB_2 and AB monomers was also carried out by direct polycondensation. Polymer II was composed of more dendritic units and less terminal units, and polymer III had long AB linear chains. The solubility of all stepwise copolymers was low and the inherent viscosity was high in comparison with the one-step copolymer. This was caused by the difference in architecture of the copolymers.

The feed ratio of the monomers affected the glass transition temperatures (T_g) and the softening points (T_s). A minimum T_g was observed at a 50% of the AB_2 monomer, whereas T_ss gradually decreased with an increase in the AB_2 monomer and became constant over 60%. Young's modulus determined by the tensile test decreased from 2.4 to 1.6 GPa with an increase in the amount of the AB_2 monomer in the range 0–60%. The decrease might be caused by the less entangled nature of hyperbranched polymers prepared from AB_2 monomers.

7.3.3 Intrinsic Viscosity

Intrinsic viscosity measurements are useful in finding the conformation and solubility parameter of the hyperbranched copolymer in different solvents. The intrinsic viscosity of narrow molecular mass distribution PSt standards, poly(benzyl)ether dendrimers, and their hybrid BCs in a variety of solvents are compared in order to understand the behavior of hybrid copolymers [120]. The relative change in the intrinsic viscosity for all these polymers is almost equivalent, indicating that poly(benzyl ether) dendrimers expand and contract quite readily with the change in solvent. When the solubility parameter of the solvent is same, all the polymers show maximum intrinsic viscosity demonstrating that the thermodynamic interaction with the selected range of solvents was also equivalent. The intrinsic viscosity for hybrid copolymers in a good solvent for both components is found to fall well below the intrinsic viscosity for neat PSt at intermediate molecular masses. Below and above this mass range, the intrinsic viscosity is quite similar to that of PSt. Yet, the transition for intermediate to high molecular mass behavior is quite sharp as seen through a rapid intrinsic viscosity rise at 60–70 kDa. A poorer solvent for the dendrimer, one where the dendrimer intrinsic viscosity is found independent of generation number, produced the same intrinsic viscosity for the hybrid copolymer and virgin PSt. This unique behavior is due to novel interactions between the linear PSt block and the highly branched dendritic block that is a function of the solvent environment.

7.4 APPLICATIONS OF DENDRITIC COPOLYMERS

Dendritic copolymers are intriguing macromolecules, which offer challenge and fascination as purely synthetic objects at the crossroad of organic and polymer chemistry and as promising materials for diverse advanced applications. The ambivalent character of the linear–dendritic architecture opens numerous avenues toward emerging and potential applications. These materials have several applications in various fields depending on the nature of repeating units and end groups. Particularly in the medical field, these materials have many potential applications such as controlled drug delivery, environmental-specific nanoreactors, "semi-artificial" enzymes, and biodevices with blood compatible surface treatment.

Water-soluble dendritic copolymers, consisting of biocompatible building blocks and having both thermoresponsive and photocrosslinkable properties have attracted attention for tissue engineering applications. These materials enable the formation of gels with desired mechanical properties *in situ* to encapsulate bioactive agents and cells for the controlled drug release and to support cell growth. Such polymers are also easy to work clinically because the polymer solutions can be localized within targeted sites after administration due to thermal gelation and then form gels with the desired mechanical properties by photocrosslinking.

For many medical devices, excellent blood compatibility is of paramount importance. Poly(2-methoxy ethyl acrylate) (PMEA) has excellent blood compatibility and PMEA-modified surfaces have recently been demonstrated to promote improved blood compatibility when compared with that of other polymer surfaces. This delicate balance plays a decisive role for surface characteristics of films prepared from the BCs. Mixing of BCs based on PMEA in a polymer matrix is expected to result in segregation of the PMEA block at the surface. That could enhance surface hydrophilicity and improve the biocompatibility. This aspect is of great importance for surface treatment of biodevices with these BCs, in particular, devices made of all polymer materials. Hydrogen bonding and the freezing-bound water on PMEA surfaces, in particular, seem to play a vital role for the PMEA blood compatibility. Furthermore, the differences observed in water contact angles of the PMEA, PMMA, and the BCs hereof are the results of subtle hydrophobic–hydrophilic balances in the materials [121].

Amphiphilic star polymers offer substantial promise for a range of drug delivery applications owing to their ability to encapsulate guest molecules. One appealing but under explored application is transdermal drug delivery using star BC reverse micelles as an alternative to the more common oral and intravenous routes. Amphiphilic star copolymers of polar oligo(ethylene glycol)methacrylate and nonpolar lauryl methacrylate demonstrate the ability to encapsulate polar dyes such as rhodamine B and FITC-BSA in nonpolar media via UV–vis spectroscopic studies and exhibit substantially improved encapsulation efficiencies, relative to self-assembled "1-arm" linear BC analogs [122]. Furthermore, their transdermal carrier capabilities were demonstrated in multiple dye diffusion studies using porcine skin, verifying penetration of the carriers into the stratum corneum.

Over the last decades, PEO (synonym with PEG) has attracted special attention because of its unique properties such as hydrophilicity but uncharged nature of the polymer, good protein repellent, low interfacial energy, high mobility of the polyether chains, better biodistribution, lower toxicity, and immunogenicity. These properties find its applications in the biomedical field, particularly in dendritic and star architectures of PEO and its copolymers. Its applications are very wide in the biomedical field from controlled drug delivery to tissue engineering (substrates for cell culture) [123]. Similar to poly(ethylene glycol), the multiarm star BCs of polyglycerols such as PG-*b*-PHEMA, PG-*b*-poly-(*tert*-butyl acrylate) [PG-*b*-poly(*t*BA)], and PG-*b*-poly(dimethylaminoethyl methacrylate) [PG-*b*-poly(DMAEMA)] have several potential applications such as nanocapsules, soluble support for catalysts, biomineralization, and biomedical application. The design of catalytically active dendritic materials has been achieved either by fixation of catalytically active complexes at structurally perfect dendrimer surfaces, within the interior "core" of the dendrimer, or by the encapsulation of metal nanoparticles. Catalytically active transition-metal complexes have special advantages such as catalyst recovery [124].

Apart from the biological applications, these dendritic copolymers are also playing an important role in the field of electronics such as holographic data storage, photomechanical actuators, and all-optical switches. The incorporation of liquid crystalline blocks into such BCs offers the possibility of altering their self-assembly behavior while allowing the introduction of molecules with optical, electronic, electrooptic, or photoresponsive functionalities. In particular, BCs containing azobenzene units represent an attractive area of research. Azobenzene units undergo isomerization between the trans and cis states when they are irradiated in their absorption bands. Photoinduced anisotropy generated using linearly polarized light has been thoroughly investigated in azobenzene-containing polymers. The confinement of these photoresponsive units in nanosized BC domains gives these materials unique properties that make them interesting in different applications such as volume holographic storage or in the preparation of photoresponsive nanoscopic objects [125].

REFERENCES

1. (a) Stockmeyer, W. H. (1943). Theory of molecular size distribution and gel formation in branched-chain polymers. *J. Chem. Phys.*, *11*, 45–55; (b) Stockmeyer, W. H. (1944). Theory of molecular size distribution and gel formation in branched polymers II. General cross linking. *J. Chem. Phys.*, *12*, 125–131; (c) Stockmeyer, W. H. (1950). Light scattering in multi-component systems. *J. Chem. Phys.*, *18*, 58–61.
2. Flory, P. J. (1952). Molecular size distribution in three-dimensional polymers. IV. Branched polymers containing A–R–Bf-1 type units. *J. Am. Chem. Soc.*, *74*, 2718–2723.
3. (a) Gordon, M. (1962). Good's theory of cascade processes applied to the statistics of polymer distributions. *Proc. R. Soc. Lond.*, *A268*, 240–259; (b) Gordon, M., Scantlebury, G. R. (1964). Non-random polycondensation: statistical theory of the substitution effect. *Trans. Faraday Soc.*, *60*, 604–621.

4. Burchard, W., Schmidt, M., Stockmeyer, W. H. (1980). Information on polydispersity and branching from combined quasi-elastic and integrated scattering. *Macromolecules*, *13*, 1265–1272.
5. (a) Müller, A. H. E., Yan, D., Wulkow, M. (1997). Molecular parameters of hyperbranched polymers made by self-condensing vinyl polymerization. 1. Molecular weight distribution. *Macromolecules*, *30*, 7015–7023; (b) Yan, D., Müller, A. H. E., Matyjaszewski, K. (1997). Molecular parameters of hyperbranched polymers made by self condensing vinyl polymerization. 2. Degree of branching. *Macromolecules*, *30*, 7024–7033; (c) Radke, W., Litvinenko, G., Müller, A. H. E. (1998). Effect of core-forming molecules on molecular weight distribution and degree of branching in the synthesis of hyperbranched polymers. *Macromolecules*, *31*, 239–248; (d) Litvinenko, G. I., Simon, P. F. W., Müller, A. H. E. (2001). Molecular parameters of hyperbranched copolymers obtained by self-condensing vinyl copolymerization, 2. Non-equal rate constants. *Macromolecules*, *34*, 2418–2426.
6. (a) Hölter, D., Burgath, A., Frey, H. (1997). Degree of branching in hyperbranched polymers. *Acta. Polym.*, *48*, 30–35; (b) Frey, H. (1997). Degree of branching in hyperbranched polymers. 2. Enhancement of the db: scope and limitations. *Acta. Polym.*, *48*, 298–309; (c) Hanselmann, R., Hölter, D., Frey, H. (1998). Hyperbranched polymers prepared via the core-dilution/slow addition technique: computer simulation of molecular weight distribution and degree of branching. *Macromolecules*, *31*, 3790–3801.
7. Beginn, U., Drohmann, C., Möller, M. (1997). Conversion dependence of the branching density for the polycondensation of AB_n monomers. *Macromolecules*, *30*, 4112–4116.
8. Dušek, K., Sÿomvársky, J., Smrcková, M., Simonsick, W. J., Wilczek, L. (1999). Role of cyclization in the degree-of-polymerization distribution of hyperbranched polymers. Modelling and experiment. *Polym. Bull. (Berlin)*, *42*, 489–496.
9. (a) Cameron, C., Fawcett, A. H., Hetherington, C. R., Mee, R., McBride, F. V. (1997). Cycles frustrating fractal formation in an AB_2 step growth polymerization. *Chem. Commun.*, 1801–1802; (b) Cameron, C., Fawcett, A. H., Hetherington, C. R., Mee, R., McBride F. V. (1998). Step growth of an AB_2 monomer, with cycle formation. *J. Chem. Phys.*, *108*, 8235–8251.
10. Galina, H., Lechowicz, J. B., Kaczmarski, K. (2001). Kinetic models of the polymerization of an AB_2 monomer. *Macromol. Theory. Simul.*, *10*, 174–178.
11. Kim, Y. H., Webster, O. W. (1988). Hyperbranched polyphenylenes. *Polym. Prepr.*, *29*, 310–311.
12. Hawker, C. J., Lee, R., Fréchet, J. M. J. (1991). The one-step synthesis of hyperbranched dendritic polyesters. *J. Am. Chem. Soc.*, *113*, 4583–4588.
13. Kim, Y. H., Webster, O. W. (1992). Hyperbranched polyphenylenes. *Macromolecules*, *25*, 5561–5572.
14. Inoue, K. (2000). Functional dendrimers, hyperbranched and star polymers. *Prog. Polym. Sci.*, *25*, 453–571.
15. Gao, C., Yan, D. (2004). Hyperbranched polymers: from synthesis to applications. *Prog. Polym. Sci.*, *29*, 183–275.
16. Flory, P. J. (1941). Molecular size distribution in three dimensional polymers. I. Gelation. *J. Am. Chem. Soc.*, *63*, 3083–3090.
17. Flory, P. J. (1941). Molecular size distribution in three dimensional polymers. II. Trifunctional branching units. *J. Am. Chem. Soc.*, *63*, 3091–3096.

18. Flory, P. J. (1941). Molecular size distribution in three dimensional polymers. III. Tetrafunctional branching units. *J. Am. Chem. Soc.*, *63*, 3096–3100.
19. Flory, P. J. (1952). Molecular size distribution in three-dimensional polymers. IV. Branched polymers containing A–R–Bf-1 type units. *J. Am. Chem. Soc.*, *74*, 2718–2723.
20. Kricheldorf, H. R., Zang, Q. Z., Schwarx, G. (1982). New polymer syntheses: 6. Linear and branched poly(3-hydroxy-benzoates). *Polymer*, *23*, 1821–1829.
21. Kricheldorf, H. R., Stukenbrock, T. (1998). New polymer syntheses XCIII. Hyperbranched homo- and copolyesters derived from gallic acid and β-(4-hydroxyphenyl)-propionic acid. *J. Polym. Sci. Part A: Polym. Chem.*, *36*, 2347–2357.
22. Kim, Y. H., Webster, O. W. (1990). Water soluble hyperbranched polyphenylene: "a unimolecular micelle?" *J. Am. Chem. Soc.*, *112*, 4592–4593.
23. Kim, Y. H., Beckerbauer, R. (1994). Role of end groups on the glass transition of hyperbranched polyphenylene and triphenylbenzene derivatives. *Macromolecules*, *27*, 1968–1971.
24. Fréchet, J. M. J. (1994). Functional polymers and dendrimers: reactivity, molecular architecture and interfacial energy. *Science*, *263*, 1710–1715.
25. Fréchet, J. M. J., Hawker, C. J., Gitsov, I., Leon, J. W. J. (1996). Dendrimers and hyperbranched polymers: two families of three dimensional macromolecules with similar but clearly distinct properties. *Macromol. Sci. Pure. Appl. Chem. A*, *33*, 1399–1425.
26. Emrick, T., Chang, H. T., Fréchet, J. M. J. (2000). The preparation of hyperbranched aromatic and aliphatic polyether epoxies by chloride-catalyzed proton transfer polymerization from AB(n) and A(2) + B-3 monomers. *J. Polym. Sci. Part A: Polym. Chem.*, *38*, 4850–4869.
27. Fréchet J. M. J., Tomalia D. A. *Dedrimers and Other Dendritic Polymers*. Wiley, New York, 2001.
28. Turner, S. R., Voit, B., Mourey, T. H. (1993). All aromatic hyperbranched polyesters with phenol and acetate end groups: synthesis and characterization. *Macromolecules*, *26*, 4617–4623.
29. Hölter, D., Frey, H. (1997). Degree of branching in hyperbranched polymers. 2. Enhancement of the db: scope and limitations. *Acta. Polym.*, *48*, 298–309.
30. Sunder, A., Mülhaupt, R., Haag, R., Frey, H. (2000). Hyperbranched polyether polyols: a modular approach to complex polymer architectures. *Adv. Mater.*, *12*, 235–239.
31. Gao, C., Yan, D. Y. (2001). Polyaddition of B_2 and BB'_2 type monomers to A_2 type monomer. 1. Synthesis of highly branched copoly(sulfone–amine)s. *Macromolecules*, *34*, 156–161.
32. Gao, C., Yan, D. Y., Tang, W. (2001). Hyperbranched polymers made from A_2- and BB'_2 -type monomers, 3. Polyaddition of *N*-methyl-1,3-propanediamine to divinyl sulfone. *Macromol. Chem. Phys.*, *202*, 2623–2629.
33. Gao, C., Tang, W., Yan, D. (2002). Synthesis and characterization of water-soluble hyperbranched poly(ester amine)s from diacrylates and diamines. *J. Polym. Sci. Part A: Polym. Chem.*, *40*, 2340–2349.
34. Ishida, Y., Sun, A. C. F., Jikei, M., Kakimoto, M. (2000). Synthesis of hyperbranched aromatic polyamides starting from dendrons as AB_x monomers: effect of monomer multiplicity on the degree of branching. *Macromolecules*, *33*, 2832–2838.

35. Morikawa, A., Kakimoto, M., Imai Y. (1993). Convergent synthesis of starburst poly(ether ketone) dendrons. *Macromolecules*, *26*, 6324–6329.
36. Hawker, C. J., Chu, F. (1996). Hyperbranched poly(ether ketones): manipulation of structure and physical properties. *Macromolecules*, *29*, 4370–4380.
37. Schmaljohann, D., Barratt, J. G., Komber, H., Voit, B. I. (2000). Kinetics of nonideal hyperbranched polymerizations. 1. Numeric modeling of the structural units and the diads. *Macromolecules*, *33*, 6284–6294.
38. (a) Hao, J., Jikei, M., Kakimoto, M. (2003). Synthesis and comparison of hyperbranched aromatic polyimides having the same repeating unit by AB_2 self-polymerization and $A_2 + B_3$ polymerization. *Macromolecules*, *36*, 3519–3528;
39. (a) Hawker, C. J., Fréchet, J. M. J. (1990). Preparation of polymers with controlled molecular architecture. A new convergent approach to dendritic macromolecules. *J. Am. Chem. Soc.*, *112*, 7638–7647.
40. (a) Wooley, K. L., Hawker, C. J., Fréchet, J. M. J. (1991). Hyperbranched macromolecules via a novel double-stage convergent growth approach. *J. Am. Chem. Soc.*, *113*, 4252–4261; (b) Wooley, K. L., Hawker, C. J., Fréchet, J. M. J. (1994). Branched monomer approach for rapid synthesis of dendrimers. *Angew. Chem. Int. Ed. Engl.*, *33*, 82–85; (c) Abbé, G. L., Forier, B., Dehaen, W. (1996). A fast double-stage convergent synthesis of dendritc polyethers. *Chem. Commun.*, *18*, 2143–2144.
41. (a) Percec, V., Barboiu, B., Grigoras, C., Bera, T. K. (2003). Universal iterative strategy for the divergent synthesis of dendritic macromolecules from conventional monomers by a combination of living radical polymerization and irreversible terminator multifunctional initiator (TERMINI). *J. Am. Chem. Soc.*, *125*, 6503–6516; (b) Percec, V., Grigoras, C., Kim, H. J. (2004). Toward self-assembling dendritic macromolecules from conventional monomers by a combination of living radical polymerization and irreversible terminator multifunctional initiator. *J. Polym. Sci. Part A: Polym. Chem.*, *42*, 505–513.
42. Kim, Y. M. (1992). Highly branched polymers. *Adv. Mater.*, *4*, 764–766.
43. Voit, B. I. (1995). Dendritic polymers: from aesthetic macromolecules to commercially interesting materials. *Acta Polym.*, *46*, 87–99.
44. Bosman, A. W., Janssen, H. M., Meijer, E. W. (1999). About dendrimers: structure, physical properties, and applications. *Chem. Rev.*, *99*, 1665–1688.
45. Buhleier, E. W., Wehner, W., Vögtle F. (1978). Cascade and nonskid-chain-like synthesis of molecular cavity topologies. *Synthesis*, *55*, 155–158.
46. (a) Newkome, G. R., Yao, Z. Q., Baker, G. R., Gupta, V. K. (1985). Cascade molecules: a new approach to micelles. Dendritic macromolecules: synthesis of starburst dendrimers. *J. Org. Chem.*, *50*, 2003–2004; (b) Tomalia, D. A., Barker, H., Dewald, J. R., Hall, M., Kallos, G., Martin, S., Roeck, J., Ryder, J., Smith P. (1986). *Macromolecules*, *19*, 2466–2468.
47. (a) Hawker, C. J., Fréchet, J. M. J., Grubbs, R. B., Dao, J. (1995). Preparation of hyperbranched and star polymers by a "living", self-condensing free radical polymerization. *J. Am. Chem. Soc.*, *117*, 10763–10764; (b) Jikei, M., Chon, S. H., Kakimoto, M., Kawauchi, S., Imase, T., Watanabe, J. (1999). Synthesis of hyperbranched aromatic polyamide from aromatic diamines and trimesic acid. *Macromolecules*, *32*, 2061–2064; (c) Yan, D., Gao, C. (2000). Hyperbranched polymers made from A2 and BB′2 type monomers. 1. Polyaddition of 1-(2-aminoethyl)piperazine to divinyl sulfone. *Macromolecules*, *33*, 7693–7699.

48. (a) Gitsov, I., Wooley, K. L., Hawker, C. J., Ivanova, P. T., Fréchet, J. M. J. (1993). Synthesis and properties of novel linear dendritic block-copolymers - reactivity of dendritic macromolecules toward linear-polymers. *Macromolecules*, *26*, 5621–5627; (b) Gitsov, I., Fréchet, J. M. J. (1994). Novel nanoscopic architectures. Linear-globular ABA copolymers with polyether dendrimers as A blocks and polystyrene as B block. *Macromolecules*, *27*, 7309–7315.

49. (a) Chapman, T. M., Hillyer, G. L., Mahan, E. J., Shaffer, K. A. (1994). Hydraamphiphiles: novel linear dendritic block copolymer surfactants. *J. Am. Chem. Soc.*, *116*, 11195–11196; (b) Iyer, J., Fleming, K., Hammond, P. T. (1998). Synthesis and solution properties of new linear-dendritic diblock copolymers. *Macromolecules*, *31*, 8757–8765; (c) Chang, Y., Kwon, Y. C., Lee, S. C., Kim, C. (2000). Amphiphilic linear PEO dendritic carbosilane block copolymers. *Macromolecules*, *33*, 4496–4500; (d) Lambrych, K. R., Gitsov, I. (2003). Linear-dendritic poly(ester)-block-poly(ether)-block-poly(ester) ABA copolymers constructed by a divergent growth method. *Macromolecules*, *36*, 1068–1074; (e) Namazi, H., Adeli, M. (2005). Solution proprieties of dendritic triazine/poly (ethylene glycol)/dendritic triazine block copolymers. *J. Polym. Sci. Part A: Polym. Chem.*, *43*, 28–41.

50. (a) Gitsov, I., Wu, S., Ivanova, P. T. (1997). Modular building blocks for combinatorial construction of polyether dendrimers and their hybrids. *Polym. Mater. Sci. Eng.*, *77*, 214–215; (b) Ihre, H. R., De Jesús, O. L. P., Szoka, F. C. Jr., Fréchet, J. M. J. (2002). Polyester dendritic systems for drug delivery applications: design, synthesis, and characterization. *Bioconjugate Chem.*, *13*, 443–452; (c) Newkome, G. R., Kotta, K. K., Mishra, A., Moorefield, C. N. (2004). Synthesis of water-soluble, ester-terminated dendrons and dendrimers containing internal PEG linkages. *Macromolecules*, *37*, 8262–8268.

51. (a) Roovers, J., Toporowski, P., Martin, J. (1989). Synthesis and characterization of multiarm star polybutadienes. *Macromolecules*, *22*, 1897–1903; (b) Jacob, S., Majoros, I., Kennedy, J. P. (1996). Synthesis and characterization of multiarm star polybutadienes. *Macromolecules*, *29*, 8631–8641; (c) Omura, N., Kennedy, J. P. (1997). Synthesis, characterization, and properties of stars consisting of many polyisobutylene arms radiating from a core of condensed cyclosiloxanes. *Macromolecules*, *30*, 3204–3214; (d) Inglis, A. J., Sinnwell, S., Davis, T. P., Barner-Kowollik, C., Stenzel, M. H. (2008). Reversible addition fragmentation chain transfer (RAFT) and hetero-diels–alder chemistry as a convenient conjugation tool for access to complex macromolecular designs. *Macromolecules*, *41*, 4120–4126.

52. (a) Percec, V., Ahn, C. H., Cho, W. D., Jamieson, A. M., Kim, J., Leman, T., Schmidt, M., Gerle, M., Moller, M., Prokhorova, S. A., Sheiko, S. S., Cheng, S. Z. D., Zhang, A., Ungar, G., Yeardley, D. J. P. (1998). Visualizable cylindrical macromolecules with controlled stiffness from backbones containing libraries of self-assembling dendritic side groups. *J. Am. Chem. Soc.*, *120*, 8619–8631; (b) Percec, V., Holerca, M. N., Magonov, S. N., Yeardley, D. J. P., Ungar, G., Duan, H., Hudson, S. D. (2001). Poly(oxazolines)s with tapered minidendritic side groups. The simplest cylindrical models to investigate the formation of two-dimensional and three-dimensional order by direct visualization. *Biomacromolecules*, *2*, 706–728.

53. Jin, R-H., Motokucho, S., Andou, Y., Nishikubo, T. (1998). Controlled polymerization of an AB_2 monomer using a chloromethylarene as comonomer: branched polymers from activated methylene compounds. *Macromol. Rapid. Commun.*, *19*, 41–46.

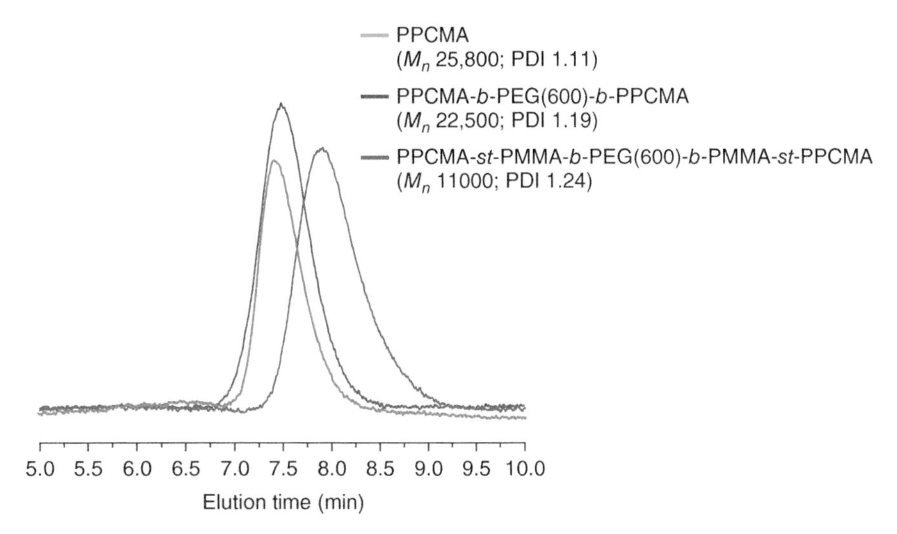

Figure 4.5 Size exclusion chromatography (SEC) of CC polymers synthesized by ATRP. (Reprinted with permission from [20]. Copyright © 2010 Wiley Periodicals Inc.)

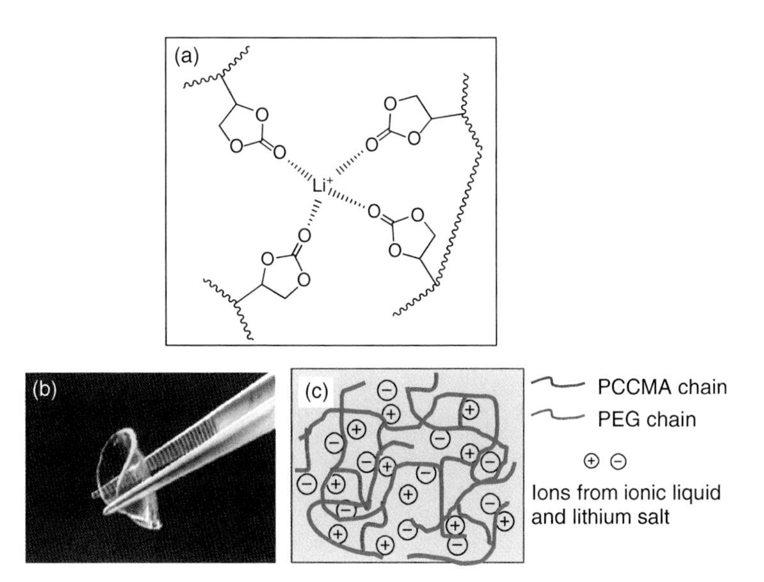

Figure 4.8 (a) Lithium-ion coordination with CC groups of polymer. (Reprinted with permission from [17d]. Copyright © 2007 American Chemical Society.) (b) Photograph and (c) schematic presentation of a transparent, flexible, and highly conductive ion gel based on pendant CC copolymer. (Reprinted with permission from [60]. Copyright © 2011 Royal Society of Chemistry.)

Synthesis and Applications of Copolymers, First Edition. Edited by Anbanandam Parthiban.

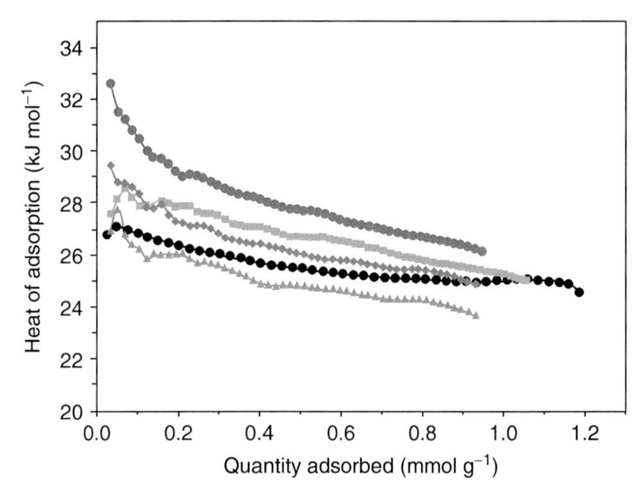

Figure 6.18 Measured isosteric heats of CO_2 for CMP networks. (See text for full caption.)

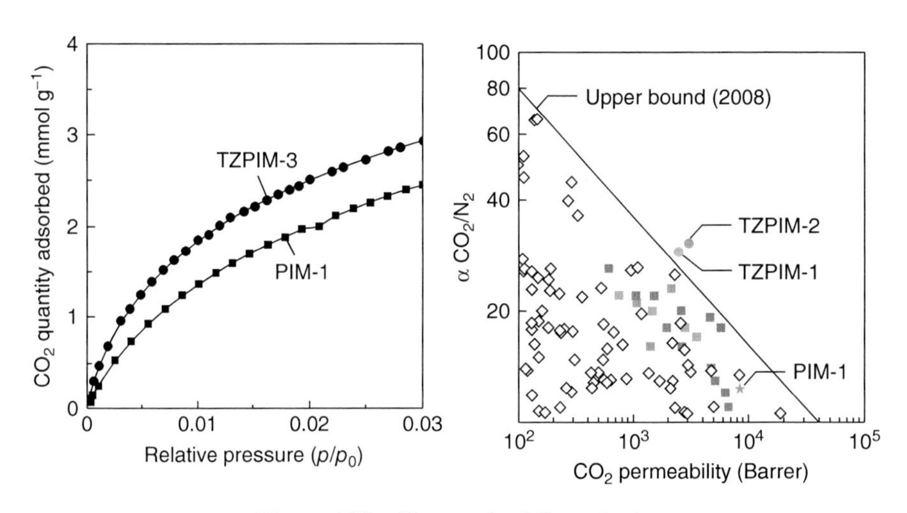

Figure 6.19 (See text for full caption.)

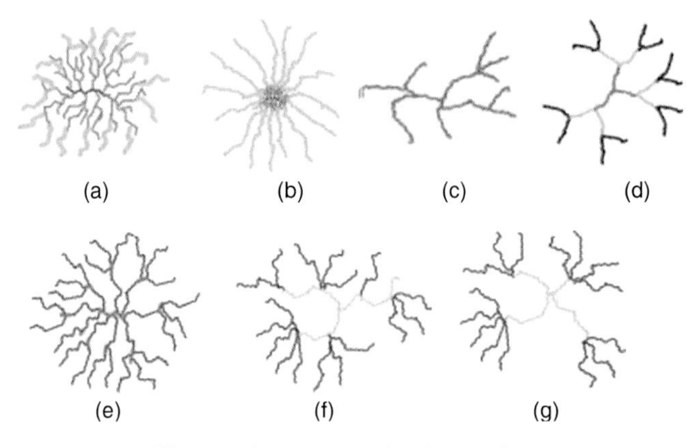

Figure 7.1 (See text for full caption.)

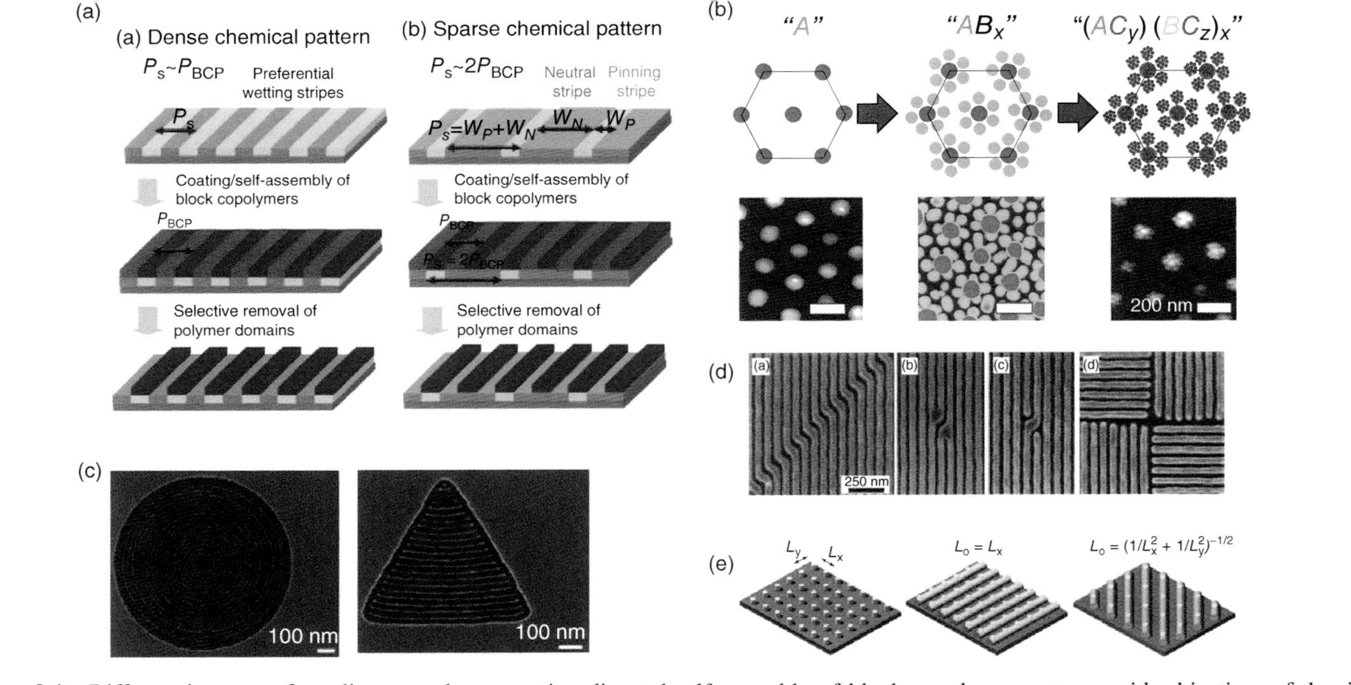

Figure 9.1 Different instances from literature demonstrating directed self-assembly of block copolymer patterns with objectives of density multiplication, hierarchical structuring, engineered pattern designs, and long-range ordering. (a) Sparse and dense chemical patterns presenting alternating preferential wetting stripes of pitch that is equal to or twice that of the block copolymer domains. (Reprinted with permission from [23]. Copyright © 2008 Wiley Periodicals Inc.) (b) Arrays of hierarchically built binary and ternary superstructures using multilevel assembly of reverse micelles. (Reprinted with permission from [31]. Copyright © 2013 American Chemical Society.) (c) Pt nanowire patterns obtained by graphoepitaxially confined polystyrene–poly(2-vinylpyridine) block copolymers within micro patterns of different shapes. (Reprinted with permission from [27]. Copyright © 2008 American Chemical Society.) (d) Directed self-assembly of PS-*b*-PMMA into patterns resembling essential elements of integrated circuit element geometries. (Reprinted with permission from [28]. Copyright © 2007 American Chemical Society.) (e) Generation of arbitrary patterns of block copolymers using electron beam lithography. (Reprinted with permission from [21]. Copyright © 2010 Nature Publishing Group.)

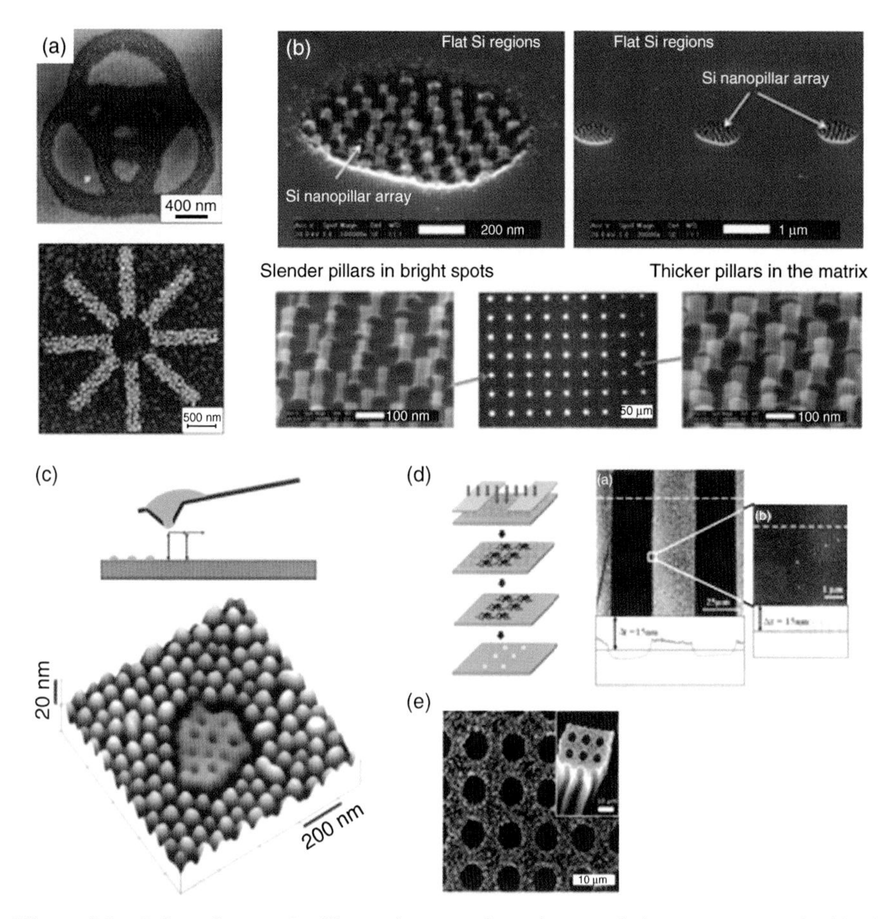

Figure 9.2 Selected examples illustrating use of top-down techniques to pattern for block copolymer nanostructures. (a) Electron beam lithography used to define selective immobilization of BCP reverse micelles (Reprinted with permission from [93,94]. Copyright © 2003, 2002 Wiley Periodicals Inc.) (b) Nanostencil lithography used to define areas for lithographic pattern-transfer of BCP templates by dry etching. (Reprinted with permission from [82]. Copyright © 2008 Wiley Periodicals Inc.) (c) Nanoscale dispensing through a hollow AFM tip (NADIS) to achieve selective morphology transformation of a reverse micelle array at pre-determined areas on surface, with resolution offered by the AFM tip. (Reprinted with permission from [84]. Copyright © 2006 Elsevier Ltd.) (Reprinted with permission from [95]. Copyright © 2004 AIP.) (d) Photolithographic patterning through selective cross-linking of an array of reverse micelles. (Reprinted with permission from [96]. Copyright © 2006 IOP Science.) (e) Microcontact printing of BCP templates used to derive patterned array of catalysts for carbon nanotube growth. (Reprinted with permission from [90]. Copyright © 2006 American Chemical Society.)

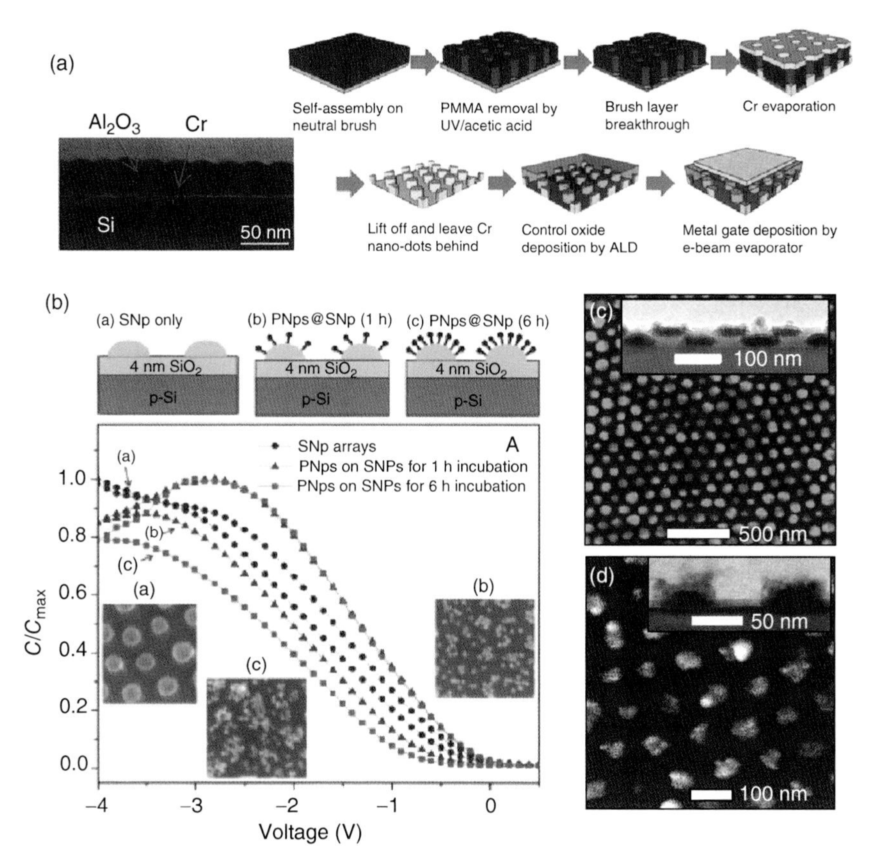

Figure 9.3 Nanopatterns for flash memory applications derived using block copolymer lithography employing (a) Cr nanocrystals. (Reprinted with permission from [109]. Copyright © 2010 American Chemical Society.) (b) Au nanoparticle clusters. (Reprinted with permission from [110]. Copyright © 2012 Royal Society of Chemistry.) (c,d) ZnO nanostructures. (Reprinted with permission from [111]. Copyright © 2012 American Chemical Society.) (Reprinted with permission from [112]. Copyright © 2012 Royal Society of Chemistry.)

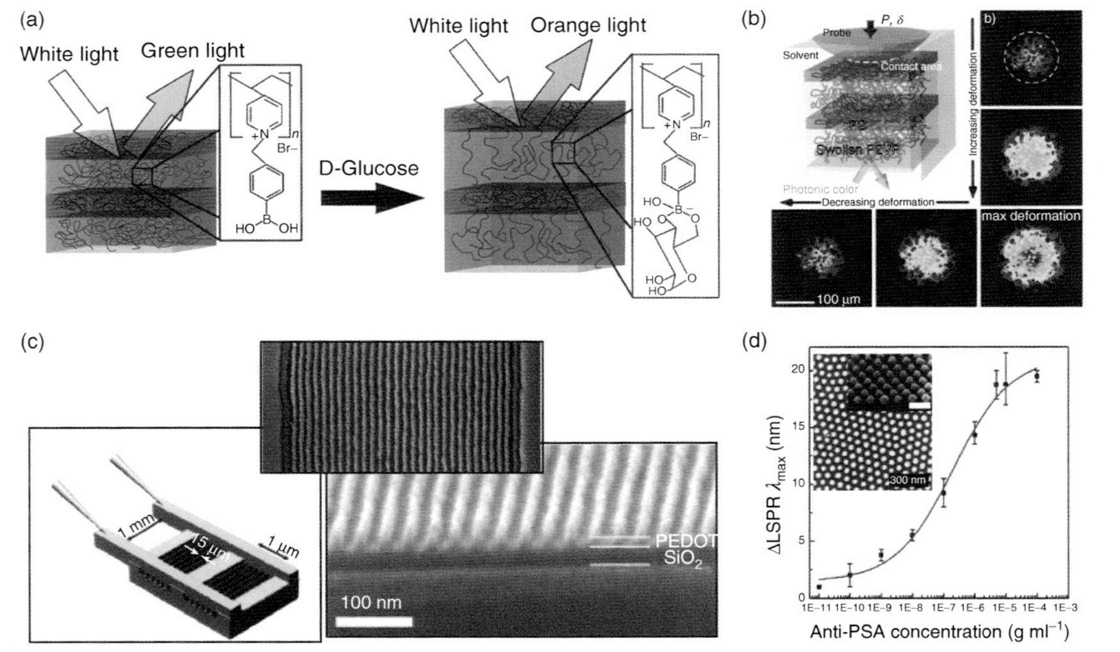

Figure 9.4 Selected instances in the use of block copolymers in sensor devices (a) Photonic gel for colorimetric sensing glucose. (Reprinted with permission from [131]. Copyright © 2011 Elsevier Ltd.) (b) Photonic sensor for transducing mechanical deformations. (Reprinted with permission from [132]. Copyright © 2011 Wiley Periodicals Inc.) (c) Gas sensing using nanowires of conjugated organic films. (Reprinted with permission from [63]. Copyright © 2008 American Chemical Society.) (d) Plasmonic sensor for quantifying Prostate Specific Antigen (PSA). (Reprinted with permission from [133]. Copyright © 2010 Royal Society of Chemistry.)

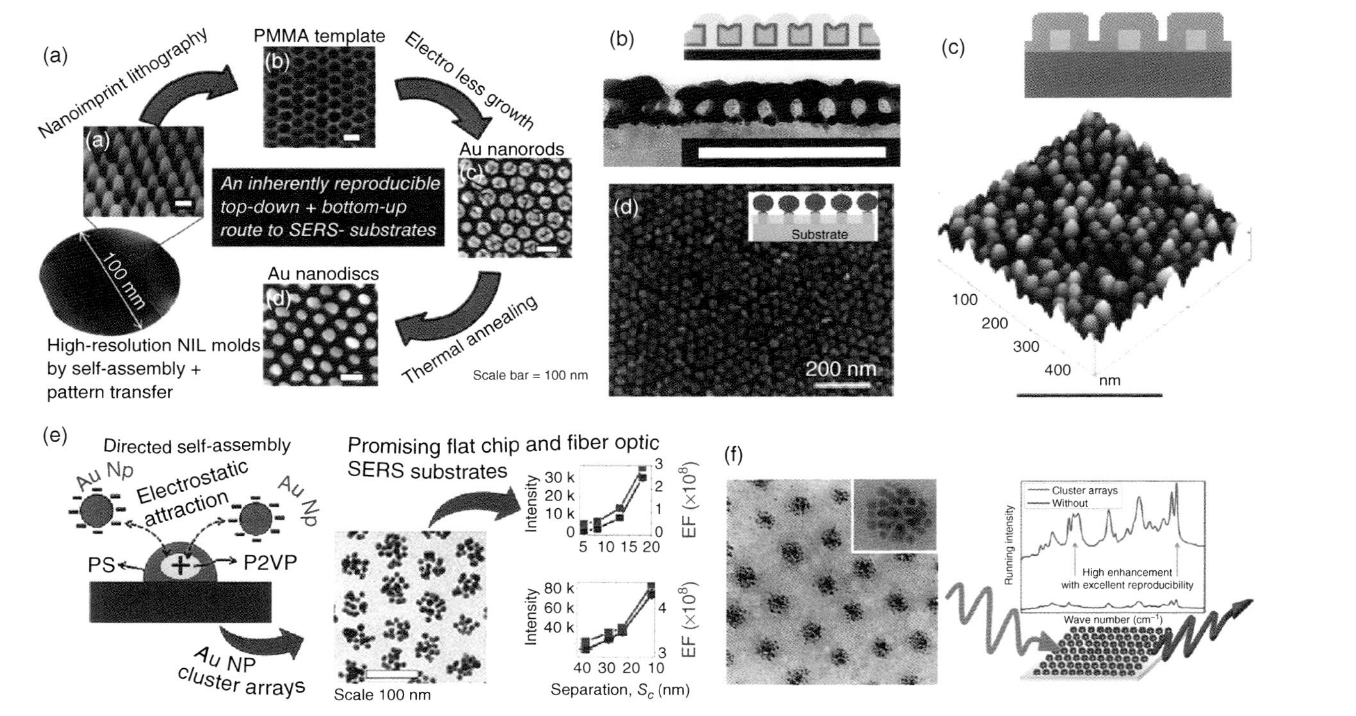

Figure 9.5 High resolution metal nanoarrays for SERS, fabricated using block copolymer thin films, using (a) BCP-NIL process to define regions for electroless Au deposition. (Reprinted with permission from [76]. Copyright © 2011 American Chemical Society.) (b) Morphology transformed PS-*b*-PVP thin films as templates for electroless Au deposition to attain mushroom-shaped structures. (Reprinted with permission from [140]. Copyright © 2009 American Chemical Society.) (c) Nanostructured silver films derived from PS-*b*-PFS thin films. (Reprinted with permission from [141]. Copyright © 2006 IOP Science.) (d) Electrostatic self-assembly of pre-formed Au nanoparticles on to BCP templates to form nanoparticle arrays with controlled separations. (Reprinted with permission from [137]. Copyright © 2011 Wiley Periodicals Inc.) (e) Electrostatic self-assembly of pre-formed Au nanoparticles on to BCP templates to form gold nanoparticle cluster arrays. (Reprinted with permission from [56]. Copyright © 2012 American Chemical Society.) (f) Silver nanocluster arrays through *in situ* synthesis within BCP reverse micelles on surface. (Reprinted with permission from [138]. Copyright © 2012 American Chemical Society.)

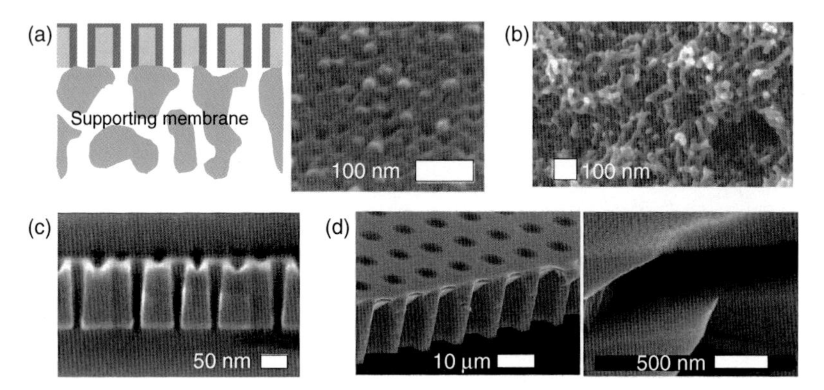

Figure 9.6 (See text for full caption.)

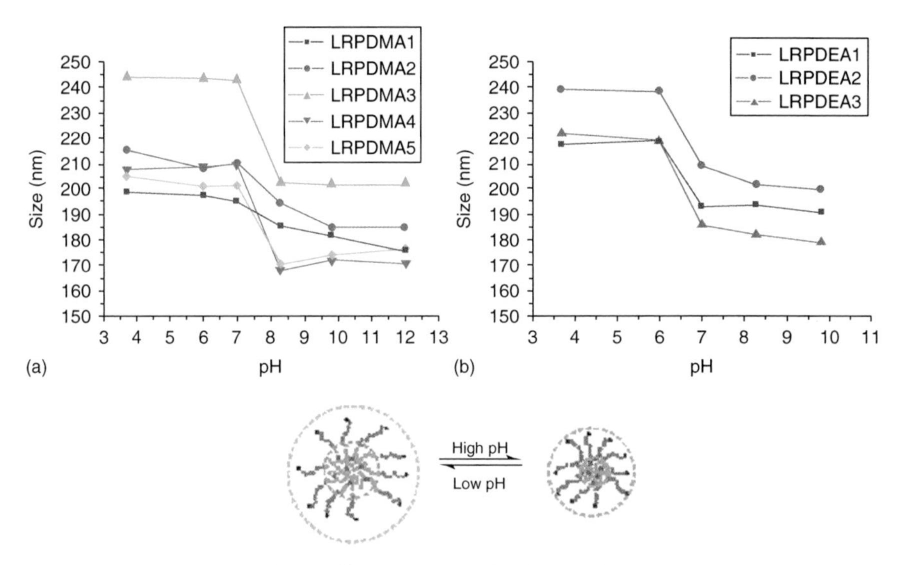

Figure 10.5 (See text for full caption.)

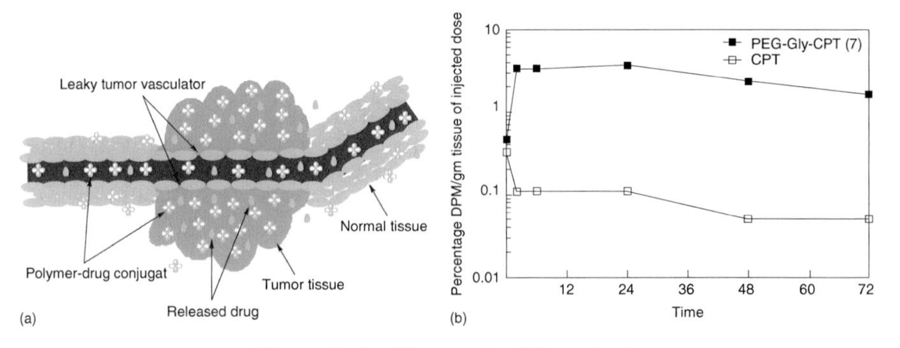

Figure 12.5 (See text for full caption.)

54. (a) Jikei, M., Fujii, K., Yang, G., Kakimoto, M. (2000). Synthesis and properties of hyperbranched aromatic polyamide copolymers from AB and AB_2 monomers by direct polycondensation. *Macromolecules*, *33*, 6228–6234; (b) Jikei, M., Fujii, K., Kakimoto, M. (2003). Synthesis and characterization of hyperbranched aromatic polyamide copolymers prepared from AB_2 and AB monomers. *Macromol. Symp.*, *199*, 223–232.

55. Sunder, A., Turk, H., Haag, R., Frey, H. (2000). Copolymers of glycidol and glycidyl ether: design of branched polyether-polyols by combination of latent AB_2 and ABR monomers. *Macromolecules*, *33*, 7682–7692.

56. Kulshrestha, A. S., Gao, W., Fu, H., Gross, R. A. (2007). Synthesis and characterization of branched polymers from lipase-catalyzed trimethylolpropane copolymerizations. *Biomacromolecules*, *8*, 1794–1801.

57. Trollsås, M., Kelly, M. A., Claesson, H., Siemens, R., Hedrick, J. L. (1999). Highly branched block copolymers: design, synthesis, and morphology. *Macromolecules*, *32*, 4917–4924.

58. (a) Gauthier, M. (2007). Arborescent polymers and other dendrigraft polymers: a journey into structural diversity. *J. Polym. Sci.: Part A: Polym. Chem.*, *45*, 3803–3810; (b) Barriau, E., Marcos, A. G., Kautz, H., Frey, H. (2005). Linear-hyperbranched amphiphilic AB diblock copolymers based on polystyrene and hyperbranched polyglycerol. *Macromol. Rapid Commun.*, *26*, 862–867.

59. (a) Sheiko, S. S., Gauthier, M., Möller, M. (1997). Film formation of arborescent graft polystyrenes. *Macromolecules*, *30*, 2343–2349; (b) Gauthier, M., Li, W., Tichagwa, L. (1997). Hard sphere behaviour of arborescent polystyrenes: viscosity and differential scanning calorimetry studies. *Polymer*, *38*, 6363–6370; (c) Gauthier, M., Chung, J., Choi, L., Nguyen, T. T. (1998). The second virial coefficient of arborescent polystyrenes and its temperature dependence. *J. Phys. Chem. B*, *102*, 3138–3142; (d) Hempenius, M. A., Zoetelief, W. F., Gauthier, M., Möller, M. (1998). Melt rheology of arborescent graft polystyrenes. *Macromolecules*, *31*, 2299–2304.

60. Gitsov, I. (2008). Hybrid linear dendritic macromolecules: from synthesis to applications. *J. Polym. Sci. Part A: Polym. Chem.*, *46*, 5295–5314.

61. Gitsov, I., Simonyan, A., Vladimirov, N. G. (2007). Synthesis of novel asymmetric dendritic-linear-dendritic block copolymers via "living" anionic polymerization of ethylene oxide initiated by dendritic macroinitiators. *J. Polym. Sci. Part A: Polym. Chem.*, *45*, 5136–5148.

62. (a) Bo, Z., Rabe, J. P., Schlüter, A. D. (1999). A poly(para-phenylene) with hydrophobic and hydrophilic dendrons: prototype of an amphiphilic cylinder with potential to segregate lengthwise. *Angew. Chem. Int. Ed.*, *38*, 2370–2372; (b) Denkewalter, R. G., Kolc, J., Lukasavage, W. J. (1981). Macromolecular highly branched homogenous compound based on lysine units. *US patent*, 4, 289, 872.

63. Hirao, A., Sugiyama, K., Tsunoda, Y., Matsuo, A., Watanabe, T. (2006). Precise synthesis of well-defined dendrimer-like star-branched polymers by iterative methodology based on living anionic polymerization. *J. Polym. Sci. Part A: Polym. Chem.*, *44*, 6659–6687.

64. Angot, S., Taton, D., Gnanou, Y. (2000). Amphiphilic stars and dendrimer-like architectures based on poly(ethylene oxide) and polystyrene. *Macromolecules*, *33*, 5418–5426.

65. Gnanou, Y., Taton, D. (2001). Stars and dendrimer-like architectures by the divergent method using controlled radical polymerization. *Macromol. Symp.*, *174*, 333–341.

66. Hou, S., Chaikof, E. L., Taton, D., Gnanou, Y. (2003). Synthesis of water-soluble star-block and dendrimer-like copolymers based on poly(ethylene oxide) and poly(acrylic acid). *Macromolecules*, *36*, 3874–3881.

67. Francis, R., Taton, D., Logan, J. L., Masse, P., Gnanou, Y., Duran, R. S. (2003). Synthesis and surface properties of amphiphilic star-shaped and dendrimer-like copolymers based on polystyrene core and poly (ethylene oxide) corona. *Macromolecules*, *36*, 8253–8259.

68. Chalari, I., Hadjichristidis, N. (2002). Synthesis of well-defined second-generation dendritic polymers of isoprene (I) and styrene (S): $(S_2I)_3$, $(SI'I)_3$, $(I''I'I)_3$, and $(I'_2I)_4$. *J. Polym. Sci. Part A: Polym. Chem.*, *40*, 1519–1526.

69. Matsuo, A., Watanabe, T., Hirao, A. (2004). Synthesis of well-defined dendrimer-like branched polymers and block copolymer by the iterative approach involving coupling reaction of living anionic polymer and functionalization. *Macromolecules*, *37*, 6283–6290.

70. (a) Hirao, A., Kato, H., Yamaguchi, K., Nakahama, S. (1986). Polymerization of monomers containing functional groups protected by trialkylsilyl groups. 5. Synthesis of poly (2-hydroxyethyl methacrylate) with a narrow molecular weight distribution by means of anionic living polymerization. *Macromolecules, 19*, 1294-1299; (b) Mori, H., Wakisaka, O., Hirao, A., Nakahama, S. (1994). Protection and polymerization of functional monomers, 23. Synthesis of well-defined poly (2-hydroxyethyl methacrylate) by means of anionic living polymerization of protected monomers. *Macromol. Chem. Phys.*, *195*, 3213–3224.

71. (a) Creutz, S., Teyssié, Ph., Jérôme, R. (1997). Anionic block copolymerization of 4-vinylpyridine and *tert*-butyl methacrylate at elevated Temperatures: influence of various additives on the molecular parameters. *Macromolecules*, *30*, 5596–5601; (b) Leemans, L., Fayt, R., Teyssié, Ph. (1990). Synthesis of new amphiphilic block copolymers. Block copolymer of sulfonated glycidyl methacrylate and alkyl methacrylate. *J. Polym. Sci. Part A: Polym. Chem.*, *28*, 1255–1262; (c) Mori, H., Hirao, A., Nakahama, S. (1994). Protection and polymerization of functional monomers. 21. Anionic living polymerization of (2, 2-dimethyl-1,3-dioxolan-4-yl)methyl methacrylate. *Macromolecules*, *27*, 35–39.

72. (a) Jung Kwon, Oh. (2008). Recent advances in controlled/living radical polymerization in emulsion and dispersion. *J. Polym. Sci. Part A: Polym. Chem.*, *46*, 6983–7001; (b) Heise, A., Nguyen, C., Malek, R., Hedrick, J. L., Frank, C. W., Miller, R. D. (2000). Starlike polymeric architectures by atom transfer radical polymerization: templates for the production of low dielectric constant thin films. *Macromolecules*, *33*, 2346–2354.

73. Zhang, K., Wang, J., Subramanian, R., Ye, Z., Lu, J., Yu, Q. (2007). Chain walking ethylene copolymerization with an ATRP inimer for one-pot synthesis of hyperbranched polyethylenes tethered with ATRP initiating sites. *Macromol. Rapid Commun.*, *28*, 2185–2191.

74. Inoue, Y., Matyjaszewski, K. (2004). Preparation of polyethylene block copolymers by a combination of postmetallocene catalysis of ethylene polymerization and atom transfer radical polymerization. *J. Polym. Sci. Part A: Polym. Chem.*, *42*, 496–504.

75. Kaneko, H., Saito, J., Kawahara, N., Matsuo, S., Matsugi, T., Kashiwa, N. (2009). Synthesis and characterization of polypropylene-based block copolymers possessing

polar segments via controlled radical polymerization. *J. Polym. Sci. Part A: Polym. Chem.*, *47*, 812–823.

76. Zhang, K., Ye, Z., Subramanian, R. (2008). Synthesis of block copolymers of ethylene with styrene and *n*-butyl acrylate via a tandem strategy combining ethylene "living" polymerization catalyzed by a functionalized Pd–diimine catalyst with atom transfer radical polymerization. *Macromolecules*, *41*, 640–649.
77. Guan, Z. B., Cotts, P. M., McCord, E. F., McLain, S. J. (1999). Chain walking: a new strategy to control polymer topology. *Science*, *283*, 2059–2062.
78. Cotts, P. M., Guan, Z. B., McCord, E., McLain, S. (2000). Novel branching topology in polyethylenes as revealed by light scattering and ^{13}C NMR. *Macromolecules*, *33*, 6945–6952.
79. Ye, Z., Zhu, S. (2003). Newtonian flow behavior of hyperbranched high-molecular-weight polyethylenes produced with a Pd–diimine catalyst and its dependence on chain topology. *Macromolecules*, *36*, 2194–2197.
80. Wang, W. J., Liu, P., Li, B. G., Zhu, S. (2010). One-step synthesis of hyperbranched polyethylene macroinitiator and its block copolymers with methyl methacrylate or styrene via ATRP. *J. Polym. Sci.: Part A: Polym. Chem.*, *48*, 3024–3032.
81. Hawker, C. J. (1994). Molecular weight control by a "living" free-radical polymerization process. *J. Am. Chem. Soc.*, *116*, 11185–11186.
82. (a) Hawker, C. J., Barclay, G. G., Orellana, A., Dao, J., Devonport, W. (1996). Initiating systems for nitroxide-mediated "living" free radical polymerizations: synthesis and evaluation. *Macromolecules*, *29*, 5245–5254; (b) Braunecker, W. A., Matyjaszewski, K. (2007). Controlled/living radical polymerization: features, developments, and perspectives. *Prog. Polym. Sci.*, *32*, 93–146.
83. Celik, C., Hizal, G., Tunca, U. (2003). Synthesis of miktoarm star and miktoarm star block copolymers via a combination of atom transfer radical polymerization and stable free-radical polymerization. *J. Polym. Sci. Part A: Polym. Chem.*, *41*, 2542–2548.
84. Tunca, U., Ozyurek, Z., Erdogan, T., Hizal G. (2004). Novel miktofunctional initiator for the preparation of an ABC-type miktoarm star polymer via a combination of controlled polymerization techniques. *J. Polym. Sci. Part A: Polym. Chem.*, *42*, 4228–4236.
85. Bussels, R., Bergman-Goettgens, C., Meuldijk, J., Koning, C. (2005). Multiblock copolymers synthesized in aqueous dispersions using multifunctional RAFT agents. *Polymer*, *46*, 8546–8554.
86. Bussels, R., Bergman-Goettgens, C., Klumperman, B., Meuldijk, J., Koning, C. (2006). Triblock copolymer synthesis via controlled radical polymerization in solution using *S-tert*-alkyl-*N,N*-alkoxycarbonylalkyldithiocarbamate RAFT agents. *J. Polym. Sci. Part A: Polym. Chem.*, *44*, 6419–6434.
87. Mori, H., Matsuyama, M., Sutoh, K., Endo, T. (2006). RAFT polymerization of acrylamide derivatives containing L-phenylalanine moiety. *Macromolecules*, *39*, 4351–4360.
88. Chong, Y. K., Krstina, J., Le, T. P. T., Moad, G., Postma, A., Rizzardo, E., Thang, S. H. (2003). Thiocarbonylthio compounds [SC(Ph)S–R] in free radical polymerization with reversible addition-fragmentation chain transfer (RAFT polymerization). Role of the free-radical leaving group (R). *Macromolecules*, *36*, 2256–2272.
89. Moad, G., Rizzardo, E., Thang, S. H. (2005). Living radical polymerization by the RAFT process. *Aust. J. Chem.*, *58*, 379–410.

90. Benaglia, M., Rizzardo, E., Alberti, A., Guerra, M. (2005). Searching for more effective agents and conditions for the RAFT polymerization of MMA: influence of dithioester substituents, solvent, and temperature. *Macromolecules*, *38*, 3129–3140.

91. Barner-Kowollik, C., Buback, M., Charleux, B., Coote, M. L., Drache, M., Fukuda, T., Goto, A., Klumperman, B., Lowe, A. B., McLeary, J. B., Moad, G., Monteiro, M. J., Sanderson, R. D., Tonge, M. P., Vana, P. (2006). Mechanism and kinetics of dithiobenzoate mediated RAFT polymerization. I. The current situation. *J. Polym. Sci. Part A: Polym. Chem.*, *44*, 5809–5831.

92. Wang, R., McCormick, C. L., Lowe, A. B. (2005). Synthesis and evaluation of new dicarboxylic acid functional trithiocarbonates: RAFT synthesis of telechelic poly(*n*-butyl acrylate)s. *Macromolecules*, *38*, 9518–9525.

93. Postma, A., Davis, T. P., Li, G., Moad, G., O'shea, M. S. (2006). RAFT polymerization with phthalimidomethyl trithiocarbonates or xanthates. On the origin of bimodal molecular weight distributions in living radical polymerization. *Macromolecules*, *39*, 5307–5318.

94. Rizzardo, E., Chen, M., Chong, B., Moad, G., Skidmore, M., Thang, S. H. (2007). RAFT polymerization: adding to the picture. *Macromol. Symp.*, *248*, 104–116.

95. Stenzel, M. H., Cummins, L., Roberts, G. E., Davis, T. P., Vana, P., Barner-Kowollik, C. (2003). Xanthate mediated living polymerization of vinyl acetate: a systematic variation in MADIX/RAFT agent structure. *Macromol. Chem. Phys.*, *204*, 1160–1168.

96. Perrier, S., Takolpuckdee, P. (2005). Macromolecular design via reversible addition–fragmentation chain transfer (RAFT)/xanthates (MADIX) polymerization. *J. Polym. Sci. Part A: Polym. Chem.*, *43*, 5347–5393.

97. Jacquin, M., Muller, P., Lizarraga, G., Bauer, C., Cottet, H., Theodoly, O. (2007). Characterization of amphiphilic diblock copolymers synthesized by MADIX polymerization process. *Macromolecules*, *40*, 2672–2682.

98. Favier, A., Charreyre, M-T. (2006). Experimental requirements for an efficient control of free-radical polymerizations via the reversible addition-fragmentation chain transfer (RAFT). *Macromol. Rapid Commun.*, *27*, 653–692.

99. Destarac, M., Brochon, C., Catala, J.-M., Wilczewska, A., Zard, S. Z. (2002). Macromolecular design via the interchange of xanthates (MADIX): polymerization of styrene with *o*-ethyl xanthates as controlling agents. *Macromol. Chem. Phys.*, *203*, 2281–2289.

100. (a) Chiefari, J., Mayadunne, R. T. A., Moad, C. L., Moad, G., Rizzardo, E., Postma, A. (2003). Thiocarbonylthio compounds (SC(Z)S–R) in free radical polymerization with reversible addition-fragmentation chain transfer (RAFT polymerization). Effect of the activating group Z. *Macromolecules*, *36*, 2273–2283; (b) Patton, D. L., Taranekar, P., Fulghum, T., Advincula, R. (2008). Electrochemically active dendritic-linear block copolymers via RAFT polymerization: synthesis, characterization, and electrodeposition properties. *Macromolecules*, *41*, 6703–6713.

101. Hong, C.-Y., You, Y.-Z., Liu, J., Pan, C.-Y. (2005). Dendrimer-star polymer and block copolymer prepared by reversible addition-fragmentation chain transfer (RAFT) polymerization with dendritic chain transfer agent. *J. Polym. Sci. Part A: Polym. Chem.*, *43*, 379–6393.

102. Jesberger, M., Barner, L., Stenzel, M. H., Eva Malmströ, M., Davis, T. P., Barner-Kowollik, C. (2003). Hyperbranched polymers as scaffolds for multifunctional

reversible addition–fragmentation chain-transfer agents: a route to polystyrene-*core*-polyesters and polystyrene-*block*-poly (butyl acrylate)-*core*-polyesters. *J. Polym. Sci. Part A: Polym. Chem.*, *41*, 3847–3861.

103. Kolb, H. C., Finn, M. G., Sharpless, K. B. (2001). Click chemistry: diverse chemical function from a few good reactions. *Angew. Chem.*, *113*, 2056–2075.
104. Binder, W. H., Kluger, C. (2006). Azide/alkyne click reactions: applications in material science and organic synthesis. *Curr. Org. Chem.*, *10*, 1791–1815.
105. Johnson, J. A., Finn, M. G., Koberstein, J. T., Turro, N. J. (2008). Construction of linear polymers, dendrimers, networks, and other polymeric architectures by copper-catalyzed azide-alkyne cycloaddition "click" chemistry. *Macromol. Rapid Commun.*, *29*, 1052–1072.
106. Lundberg, P., Hawker, C. J., Hul, A., Malkoch, M. (2008). Click assisted one-pot multi-step reactions in polymer science: accelerated synthetic protocols. *Macromol. Rapid Commun.*, *29*, 998–1015.
107. Ender, W. H., Sachsenhofer, R. (2007). 'Click' chemistry in polymer and materials science. *Macromol. Rapid Commun.*, *28*, 15–54.
108. Del Barrio, J. S., Oriol, L., Alcalá, R., Nchez, C. S. (2010). Photoresponsive poly (methyl methacrylate)-*b*-azodendron block copolymers prepared by ATRP and click chemistry. *J. Polym. Sci. Part A: Polym. Chem.*, *48*, 1538–1550.
109. Wang, G., Liu, C., Pan, M., Huang, J. (2009). Synthesis and characterization of star graft copolymers with asymmetric mixed "V-shaped" side chains via "click" chemistry on a hyperbranched polyglycerol core. *J. Polym. Sci. Part A: Polym. Chem.*, *47*, 1308–1316.
110. Durmaz, H., Dag, A., Onen, C., Gok, O., Sanyal, A., Hizal, G., Tunca, U. (2010). Multiarm star polymers with peripheral dendritic PMMA arms through Diels–Alder click reaction. *J. Polym. Sci. Part A: Polym. Chem.*, *48*, 4842– 4846.
111. Yu, D., Vladimirov, N., Fréchet, J. M. J. (1999). MALDI-TOF in the characterizations of dendritic–linear block copolymers and stars. *Macromolecules*, *32*, 5186–5192.
112. Hillenkamp, F., Karas, M., Beavis, R. C., Chait, B. T. (1991). Matrix-assisted laser desorption/ionization mass spectrometry of biopolymers. *Anal. Chem.*, *63*, 1193A–1203A.
113. Herman, L. W., Tarr, G., Kates, S. A. (1996). Optimization of the synthesis of peptide combinatorial libraries using a one-pot method. *Mol. Diversity*, *2*, 147–155.
114. (a) Barr, U., Deppe, A., Karas, M., Hillenkamp, F., Giessmann, U. (1992). Mass spectroscopy of polymers by UV-matrix assisted laser desorption/ionization. *Anal. Chem.*, *64*, 2866–2869; (b) Montaudo, G., Montaudo, M. S., Puglisi, C., Samperi, F. (1995). Characterization of polymers by matrix-assisted laser desorption ionization-time of flight mass spectrometry. End group determination and molecular weight estimates in poly(ethylene glycols). *Macromolecules*, *28*, 4562–4569; (c) Montaudo, G., Montaudo, M. S., Puglisi, C., Samperi, F. (1995). Characterization of polymers by matrix-assisted laser desorption/ionization time-of-flight mass spectrometry: molecular weight estimates in samples of varying polydispersity. *Rapid Commun. Mass Spectrom.*, *9*, 453–460; (d) Whittal, R. M., Li, L., Lee, S., Winnik, M. A. (1996). Characterization of pyrene end-labeled poly(ethylene glycol) by high resolution MALDI time-of-flight mass spectrometry. *Macromol. Rapid Commun.*, *17*, 59–64; (e) Belu, A. M., Desimone, J. M., Linton, R. W., Lange, G. W., Friedman, R. M. (1996). Evaluation of matrix-assisted laser desorption ionization mass spectrometry for polymer characterization. *J. Am. Soc. Mass Spectrom.*, *7*, 11–24.

115. (a) Leon, J. W., Kawa, M., Fréchet, J. M. J. (1996). Isophthalate ester-terminated dendrimers: versatile nanoscopic building blocks with readily modifiable surface functionalities. *J. Am. Chem. Soc.*, *118*, 8847–8859; (b) Leon, J. W., Fréchet, J. M. J. (1995). Analysis of aromatic polyether dendrimers and dendrimer-linear block copolymers by matrix-assisted laser desorption ionization mass spectrometry. *Polym. Bull.*, *35*, 449–455; (c) Kawaguchi, T., Walker, K. L., Wilkins, C. L., Moore, J. S. (1995). Double exponential dendrimer growth. *J. Am. Chem. Soc.*, *117*, 2159–2165; (d) Wu, Z. C., Biemann, K. (1997). The MALDI mass spectra of carbosilane-based dendrimers containing up to eight fixed positive or 16 negative charges. *Int. J. Mass Spectrom.*, *165*, 349–361; (e) Chessa, G., Scrivanti, A., Seraglia, R., Traldi, P. (1998). Matrix effects on matrix-assisted laser desorption/ionization mass spectrometry analysis of dendrimers with a pyridine-based skeleton. *Rapid Commun. Mass Spectrom.*, *12*, 1533–1537.

116. Holter, D., Burgath, A., Frey, H. (1997). Degree of branching in hyperbranched polymers. *Acta Polym.*, *48*, 30–35.

117. Frey, H., Holter, D. (1999). Degree of branching in hyperbranched polymers. 3 Copolymerization of AB_m-monomers with AB and AB_n-monomers. *Acta Polym.*, *50*, 67–76.

118. Kambouris, P., Hawker, C. J. (1993). A versatile new method for structure determination in hyperbranched macromolecules. *J. Chem. Soc., Perkin. Trans.*, *22*, 2717–2721.

119. Thompson, D. S., Markoski, L. J., Moore, J. S. (1999). Rapid synthesis of hyperbranched aromatic polyetherimides. *Macromolecules*, *32*, 4764–4768.

120. Jeong, M., Mackay, M. E., Vestberg, R., Hawker, C. J. (2001). Intrinsic viscosity variation in different solvents for dendrimers and their hybrid copolymers with linear polymers. *Macromolecules*, *34*, 4927–4936.

121. Bednarek, M., Jankova, K., Hvilsted, S. (2007). Novel polymers based on atom transfer radical polymerization of 2-methoxyethyl acrylate. *J. Polym. Sci. Part A: Polym. Chem.*, *45*, 333–340.

122. Poree, D. E., Giles, M. D., Lawson, L. B., He, J., Grayson, S. M. (2011). Synthesis of amphiphilic star block copolymers and their evaluation as transdermal carriers. *Biomacromolecules*, *12*, 898–906.

123. Keul, H., Möller, M. (2009). Synthesis and degradation of biomedical materials based on linear and star shaped polyglycidols. *J. Polym. Sci. Part A: Polym. Chem.*, *47*, 3209–3231.

124. Wilms, D., Stiriba, S.-E., Frey, H. (2010). Hyperbranched polyglycerols: from the controlled synthesis of biocompatible polyether polyols to multipurpose applications. *Acc. Chem. Res.*, *43*, 129–141.

125. Barrio, J. D., Oriol, L., Alcala, R., Sanchez, C. (2009). Azobenzene-containing linear–dendritic diblock copolymers by click chemistry: synthesis, characterization, morphological study, and photoinduction of optical anisotropy. *Macromolecules*, *42*, 5752–5760.

SECTION II

APPLICATIONS OF COPOLYMERS

8

A NEW CLASS OF ION-CONDUCTIVE POLYMER ELECTROLYTES: CO_2/EPOXIDE ALTERNATING COPOLYMERS WITH LITHIUM SALTS

Yoichi Tominaga

8.1 INTRODUCTION

It has been known that polyethers and salts such as poly(ethylene oxide) (PEO) and mercuric chloride [1,2] are capable of direct interactions, and structures of these complexes have been reported and widely used in organometallic chemistry. In 1973, Wright et al. [3] first reported the conducting properties of "solvent-free" PEO systems with alkali metal salts. In 1978, Armand et al. [4] highlighted the potential of these materials as a new class of solid electrolyte for energy storage applications. These new "polymer electrolytes" are recently gaining interest as solid-state alternatives to liquid electrolytes for electrochemical device applications, which range from high energy density rechargeable batteries [5,6] to solar cells [7], ion sensors, and electrochromic displays [8]. Above all, secondary lithium batteries based on polymer electrolytes have the capabilities of outstanding performance in terms of easy processibility, mechanical stability, reliability, and safety (nonflammability and leakage). There have recently been many studies on the macromolecular design of PEO-based polymers as electrolyte materials with mainly reduced degrees of crystallinity, showing good electrochemical stability and improvement in salt solubility [8–10]. However, these materials suffer from a relatively low ionic conductivity in the solid state compared with that of most

Synthesis and Applications of Copolymers, First Edition. Edited by Anbanandam Parthiban.

liquid or ceramic electrolytes. The maximum ionic conductivity of polyether-based electrolytes is lower than 10^{-4} S cm^{-1} at room temperature. Fast migration of ions in polymer can be realized by increasing the local chain mobility, since ions are transported via the segmental motion of ether chains. The localized structure that plays a crucial role for the ionic conduction is believed to involve cation–anion or cation–dipole interactions [11]. Unfortunately, the ionic interaction sometimes inhibits conduction of ions because of their strong cohesion, which increases the glass transition temperature (T_g). To overcome these problems, novel candidates for the matrix are strongly needed without the oxyethylene (OE) framework. In previous studies, some candidates without OE such as poly(acrylonitrile) [12,13], polysiloxane [14], and poly(phosphazene) [15] based polymers were evaluated as ion-conductive matrix. These polymers have low T_g because of its flexible main chains or its polar groups for salt dissociation. However, it is extremely difficult for these electrolytes to show conductivity above 10^{-4} S cm^{-1}, because both the increase in segmental motion and mobile carrier ions must be achieved at the same time.

To improve the conductivity in the solid state, processing with CO_2 under subcritical and supercritical conditions has been reported for simple polyether–salt mixtures [16–18], organically modified ceramics as polymer electrolytes [19], and polyether–clay composites [20]. Our previous reports concluded that CO_2 molecules that permeated into the treated sample can promote the dissociation of ions in the local structure and increase ionic mobility. In other words, introduction of CO_2 as a raw material into the framework of polymer may improve the conductivity of polymer electrolytes, so we accordingly focused on a CO_2/epoxide copolymer. The CO_2/epoxide alternating copolymerization was first carried out by Tsuruta et al. [21,22] in 1968, and today there are numerous reports and reviews, mainly of the development of highly active catalysts so as to yield the corresponding polymer efficiently. The copolymerization method is promising not only for the novel polymerization reaction but also in view of the potential carbon source, in environmental terms. Studies involving polycarbonates have recently been carried out for novel functional materials, including biodegradable polymers [23], nanocomposites [24], and liquid crystalline complexes [25].

Recently, novel polycarbonate-based electrolyte systems, poly(trimethylene carbonate) (p(TMC)) with lithium triflate ($LiCF_3SO_3$), perchlorate ($LiClO_4$) [26], and hexafluorophosphate ($LiPF_6$) [27] have been reported. The authors said that the combination of polar carbonate groups, linked by highly flexible methylene units, is expected to provide an attractive host matrix for ionic transport [26]. The high molecular weight p(TMC) has a T_g close to room temperature and can be an amorphous elastomer with excellent mechanical properties. Figures 8.1 and 8.2 show temperature and salt concentration dependences of ionic conductivity for the $p(TMC)_nLiCF_3SO_3$ and $LiClO_4$ (subscript n means ratio of a TMC monomer unit to the number of Li salts) system. The highest conductivity in the $LiCF_3SO_3$-based electrolytes was recorded at a composition with n close to 12. In contrast, the conductivity of $LiClO_4$-based electrolytes continued to increase with concentration up to n value of 2. The authors said that these results show the

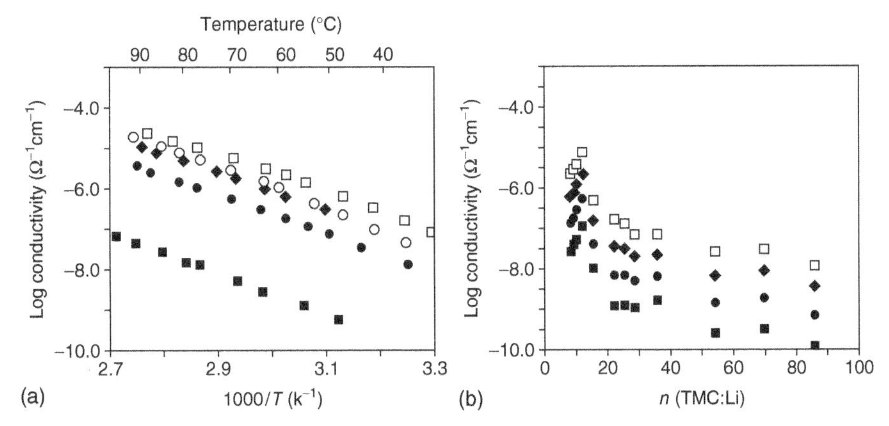

Figure 8.1 (a) Variation of conductivity with $1/T$ for electrolytes based on the $(TMC)_nLiCF_3SO_3$ system ($n = 55$ ■, 35 •, 10 ♦, 9 □, and 7 ○); (b) conductivity isotherms for $(TMC)_nLiCF_3SO_3$-based electrolytes (at 40 ■, 55 •, 70 ♦, and 85 °C □). (Reprinted with permission from [26]. Copyright © 2001 Elsevier Ltd.)

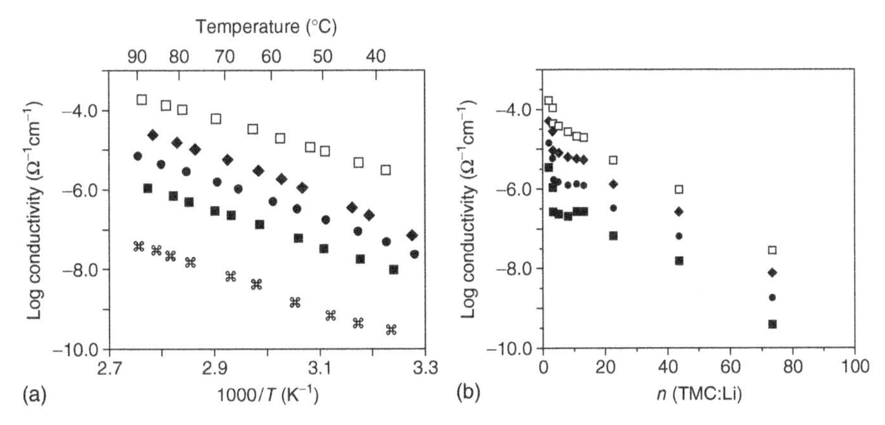

Figure 8.2 (a) Variation of conductivity with $1/T$ for electrolytes based on the $(TMC)_nLiClO_4$ system ($n = 73$ *, 43 ■, 23 •, 10 ♦, and 2 □); (b) conductivity isotherms for $(TMC)_nLiCF_3SO_3$-based electrolytes (at 40 ■, 55 •, 70 ♦, and 85 °C □). (Reprinted with permission from [26]. Copyright © 2001 Elsevier Ltd.)

substantial variation of total conductivity with the concentration of the guest salt species, greater than that found with "conventional" electrolytes based on the PEO system [26]. On the other hand, thermal behavior of both electrolytes shows a T_g at about −18 °C, which is surprisingly almost unaffected by the salt content with n between 85 and 15. As the salt content is increased for n less than 15, a minor shift of T_g to higher temperatures is perceptible. The authors said that significant interaction between the guest ionic species and the polymer host chains only may take place at these higher salt concentrations [26].

So we focused on such fascinating properties and chemical structure of the polycarbonate, which has one carbonate group (–O–(C=O)–O–) alternately in each repeating unit of the main chain. The carbonate group has large dipole moment and can dissolve many kinds of salts. Carbonate-based organic solvents such as dimethyl carbonate are used as electrolyte solution in Li-ion batteries because of their high dielectric constant. The carbonate group therefore provides a suitable structure for the polymer framework. In our recent study, we synthesized some alternating copolymers of CO_2 with glycidyl ether (GE) monomers and considered as a novel candidate for ion-conductive polymer electrolytes [28,29]. The polycarbonate is expected to weaken strong interaction with cations compared with the typical polyether electrolyte. Moreover, it has moderate polar groups on the main chain with relatively low T_g. The polycarbonate has probably no formation of complex structure, which is based on the strong coordination between cations and the polar polymer chains, as seen in the polyether-salt system. In this chapter, we review recently synthesized alternating copolymers of CO_2 with different kinds of epoxy monomers, and discuss the usefulness of these copolymers as novel candidates for the development of ion-conductive polymer electrolytes.

8.2 EXPERIMENTAL

8.2.1 Preparation of Monomers and Catalyst

Zinc oxide (ZnO, 99%, Kanto Chemical Co.), glutaric acid (GA, 98%, Kanto Chemical Co.), and CO_2 (99.99%, Toei Chemical Co.) were used as received. GE monomers possessing phenyl (side groups R=Phe, 98%), *tert*-butyl (*t*Bu, 99%), *n*-butyl (*n*Bu, 95%), ethyl (Et, 98%) and isopropyl (*i*Pr, 96%), and a non-GE-type epoxy monomer possessing ethyl (epoxy pentane, EP, 98%) were purchased from Aldrich Co. and were kept in a dry Ar gas-filled glovebox with 4 Å molecular sieves. The GE monomers possessing methoxyethyl (MeEt) group were synthesized from epichlorohydrin and 2-methoxyethanol in the presence of NaOH (Scheme 8.1). Zinc glutarate (ZnGA) as a catalyst for the polymerization was synthesized from ZnO and GA [30,31]. GA (0.99 mol) was dissolved in toluene (90 ml) in a flask equipped with a Dean–Stark trap with a reflex condenser and a drying tube. ZnO (1.00 mol) was added as a fine powder into the solution and stirred vigorously at 55 °C for 4 h, and the solution was then refluxed for 24 h. After cooling to room temperature, the mixture was filtered, washed three times with acetone, and dried under vacuum at 120 °C.

8.2.2 Copolymerization of Epoxides with CO_2

Alternating copolymerizations of CO_2 with epoxy monomers were carried out in a stainless steel reactor (Taiatsu Techno Co.). The monomer was added with ZnGA (approximately 5 mol% to the monomer) to the reactor in the glovebox. The reaction conditions were fixed at 8.2 MPa and 60 °C for 24 h. In the case of polymerization using MeEt-GE monomer, the conditions were 5.0 MPa at 60 °C for 7 days.

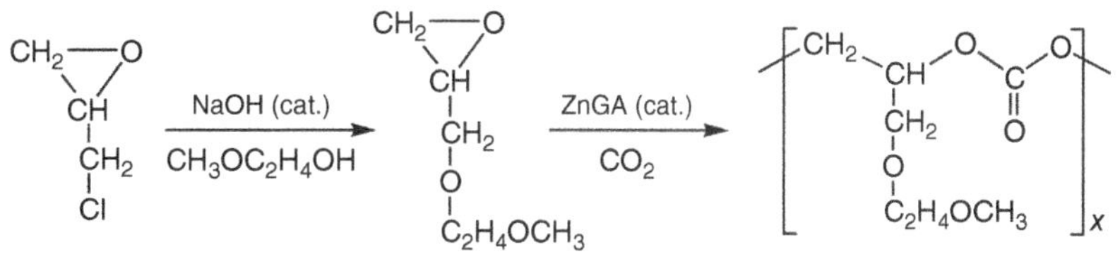

Scheme 8.1 Copolymerization of methoxyethyl glycidyl ether with CO_2.

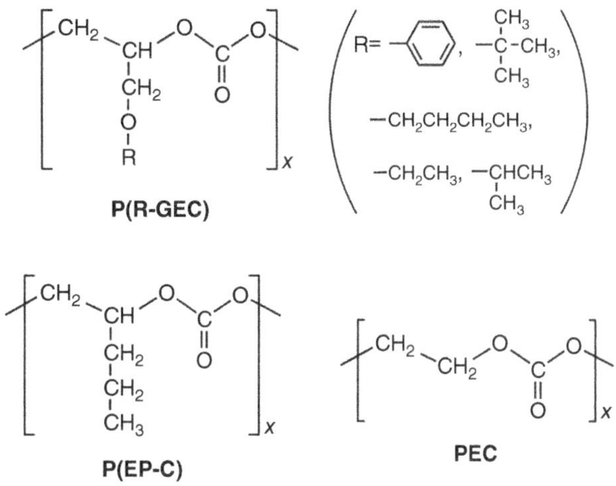

Figure 8.3 Chemical structures of P(R-GEC), P(EP-C), and PEC.

The preparation of MeEt-GE and the corresponding copolymer with CO_2 are shown in Scheme 8.1. After polymerization the reactor was cooled to room temperature, and the resulting mixture was dissolved in chloroform. The chloroform solution was filtered in order to remove ZnGA and was then concentrated to a proper volume using a rotary evaporator. The solution was dropped into excess methanol. The polymer was dissolved again in chloroform and reprecipitated at least three times. The precipitated polymer was dried under vacuum at 60 °C for 24 h. The structures of GE-type polymers, P(R-GEC), and a non-GE-type polymer, P(EP-C), are shown in Figure 8.3. Poly(ethylene carbonate), PEC, was donated by Sumitomo Seika Co. The chloroform solution of as-received PEC was added into excess methanol, and the precipitated PEC ($M_n = 3.7 \times 10^4$, $M_w/M_n = 5.9$) was dried under vacuum at 60 °C for 24 h. The ^{1}H and ^{13}C NMR spectra of all of the synthesized polycarbonates were observed using a JEOL EX-400. Molecular weights and polydispersities of all polycarbonates were estimated using a gel permeation chromatography (GPC) system (JASCO Co.), with two columns (TOSOH TSKgel GM_{HHR}-H) and chloroform (HPLC grade) as an eluent at a flow rate of 1.0 ml min^{-1} (calibrated by polystyrene standards).

8.2.3 Preparation of Electrolyte Membranes

Polycarbonate electrolytes were prepared by using the simple casting method. In case of P(R-GEC) and P(EP-C) electrolytes, the neat polycarbonate was dissolved in chloroform with lithium bis-(trifluoromethane sulfonyl) imide (LiTFSI, donated by Daiso Co.) at room temperature. LiTFSI was able to dissolve in chloroform, and the polymer/salt mixed solution was completely transparent. The LiTFSI content in the electrolyte was chosen to be 10 mol% to the repeating unit of each polycarbonate. The solution was cast onto a PTFE Petri dish, kept for several hours in a dry N_2-filled circulation chamber, and then dried under vacuum at 60 °C for 24 h. In case of the PEC system, four lithium salts, LiX (anion X= TFSI, $N(SO_2C_2F_5)_2$ (BETI), ClO_4, and CF_3SO_3), were used, and the electrolytes with the Li salt concentration from 20 to 80 wt% were prepared.

8.2.4 Measurements

The ionic conductivities of all electrolytes were measured by the complex impedance method, using an impedance/gain-phase analyzer 4194A (HP) in the frequency range 100–15 MHz or a potentiostat/galvanostat SP-150 (BioLogic) in the frequency range 100–1 MHz. The temperature was reduced from 100 to 30 °C and the cell was held constant at 10 or 20 °C intervals for at least 30 min, after which each impedance measurement was carried out in a dry Ar gas-filled glovebox. Differential scanning calorimetry (DSC) measurements of all samples were made using a DSC120 (Seiko Inst.) from −100 °C to 300 °C at a heating rate of 10 °C min^{-1} under dry N_2 gas. FT-IR spectra were recorded on a FT-IR spectrometer (FT/IR-4100, JASCO Co.) using an ATR unit (ZnSe lens) in the region from 400 to 4000 cm^{-1} with a resolution of 1 cm^{-1} under dry N_2 gas.

8.3 RESULTS AND DISCUSSION

8.3.1 NMR Characterization

The representative ^1H and ^{13}C NMR of P(MeEt-GEC) is shown in Figures 8.4 and 8.5. Figure 8.4 showed no signals in the range 3.4–3.9 ppm ($-CH_2CHO-$ main chain) corresponding to the GE homopolymer [31]. ^{13}C NMR spectrum of Figure 8.5 also revealed the presence of signal around 154 ppm because of the carbonyl group (C=O) present in the main chain. These NMR data indicate that all polycarbonates are alternating copolymers of CO_2 with epoxy monomer. In addition, no NMR signals of unreacted GE monomers were observed, except that of polycarbonate. The characterization details of all polymers are as follows:

^1H NMR ($CDCl_3$): poly(phenyl glycidyl ether carbonate) [P(Phe-GEC)], δ, ppm = 4.06 (br, H^2, CH_2O-Phe), 4.41 (m, H^2, CH_2CH), 5.16 (br, H^1, *CH*), 6.82, 6.96, 7.20 (m, H^5, Phe). ^{13}C NMR ($CDCl_3$): P(Phe-GEC), δ, ppm = 65.3 (CH_2O-Phe), 65.9 (CH_2CH), 73.8 (*C*H), 114.6, 121.3, 129.6, 157.8 (Phe), 154.2 (*C*=O).

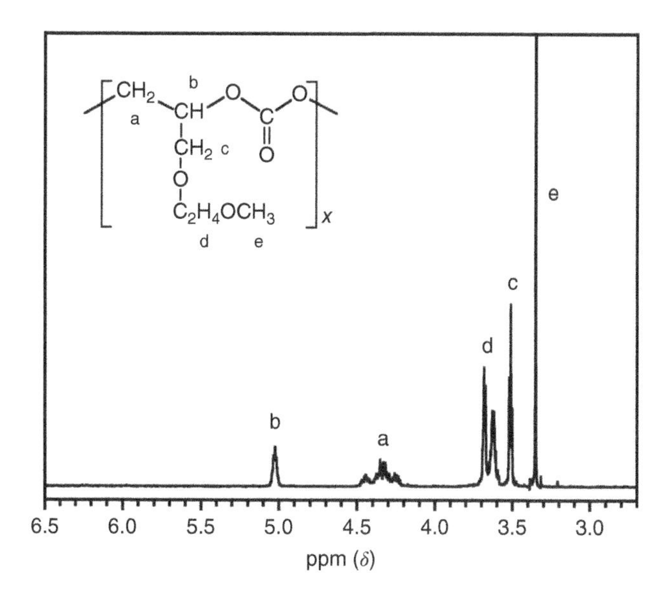

Figure 8.4 ^{1}H NMR spectrum of P(MeEt-GEC).

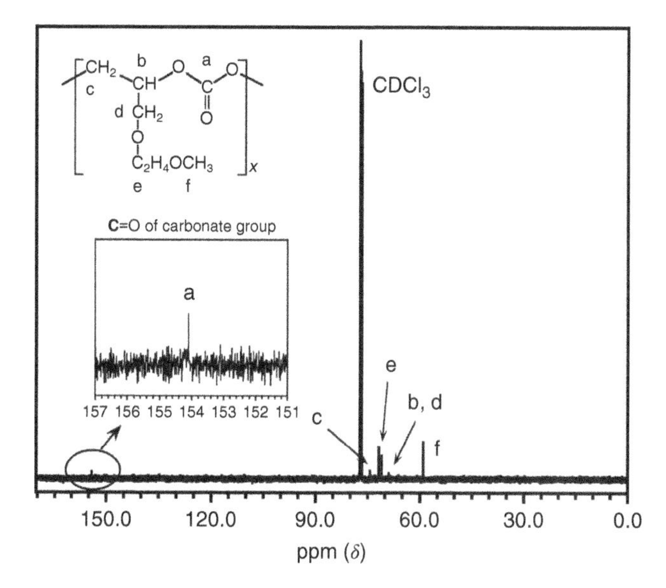

Figure 8.5 ^{13}C NMR spectrum of P(MeEt-GEC) (inset – expanded region around 154 ppm).

^{1}H NMR ($CDCl_3$): poly(*tert*-butyl glycidyl ether carbonate) [P(*t*Bu-GEC)], δ, ppm = 1.20 (s, H^9, CH_3), 3.52 (br, H^2, CH_2O-*tert*-Bu), 4.35 (br, H^2, CH_2CH), 4.95 (br, H^1, *CH*). ^{13}C NMR ($CDCl_3$): P(*t*Bu-GEC), δ, ppm = 27.4 (CH_3), 59.8 (CH_2O-*tert*-Bu), 66.4 (CH_2CH), 73.6 (*C*H), 75.1 ($C(CH_3)_3$), 154.3 (*C*=O).

^{1}H NMR ($CDCl_3$): poly(*n*-butyl glycidyl ether carbonate) [P(*n*Bu-GEC)], δ, ppm = 0.91 (br, H^3, CH_3), 1.26–1.36 (br, H^2, CH_2CH_3), 1.53–1.75 (br, H^2, $CH_2C_2H_5$), 3.39–3.46 (br, H^2, $CH_2C_3H_6$), 3.65–3.79 (m, H^2, CH_2O-*n*-Bu), 4.35 (m, H^2, CH_2CH), 5.01 (br, H^1, *CH*). ^{13}C NMR ($CDCl_3$): P(*n*Bu-GEC), δ, ppm = 13.8 (CH_3), 19.2 (CH_2CH_3), 31.5 ($CH_2C_2H_5$), 70.6–71.9 ($CH_2C_3H_6$ and CH_2O-*n*-Bu), 68.0 (CH_2CH), 74.3 (*C*H), 154.2 (*C*=O).

^{1}H NMR ($CDCl_3$): poly(ethyl glycidyl ether carbonate) [P(Et-GEC)], δ, ppm = 1.16 (m, H^3, CH_3), 3.48–3.65 (br, H^4, CH_2–O–CH_2–CH_3), 4.3–4.5 (br, H^2, CH_2CH), 5.01 (br, H^1, *CH*). ^{13}C NMR ($CDCl_3$): P(Et-GEC), δ, ppm = 15.0 (CH_3), 66.2 (*C*H), 68.2 (O–CH_2CH_3), 66.9 (CH_2CH), 74.4 (CH_2O–Et), 154.2 (*C*=O).

^{1}H NMR ($CDCl_3$): poly(*iso*-propyl glycidyl ether carbonate) [P(*i*Pr-GEC)], δ, ppm = 1.07 (br, H^6, CH_3), 3.4–3.6 (br, H^3, CH_2–O–*CH*–$(CH_3)_2$), 4.2–4.3 (br, H^2, CH_2CH), 4.9 (br, H^1, *CH*). ^{13}C NMR ($CDCl_3$): P(*i*Pr-GEC), δ, ppm = 22.0 (CH_3), 65.8 (CH_2CH), 66.0 (*C*H), 72.3 (*C*H–$(CH_3)_2$), 74.7 (CH_2O-*i*Pr), 154.0 (*C*=O).

^{1}H NMR ($CDCl_3$): poly(methoxyethyl glycidyl ether carbonate) [P(MeEt-GEC)], δ, ppm = 3.37 (s, H^3, CH_3), 3.43–3.54 (m, H^2, $CH_2OCH_2CH_2$), 3.61–3.70 (m, H^4, $CH_2CH_2OCH_3$), 4.25–4.46 (br, H^2, CH_2CH), 5.04 (br, H^1, *CH*). ^{13}C NMR ($CDCl_3$): P(MeEt-GEC), δ, ppm = 58.9 (CH_3), 66.1 ($CH_2OCH_2CH_2$), 70.9–71.8 ($CH_2CH_2OCH_3$), 69.0 (CH_2CH), 74.4 (*C*H), 154.1 (*C*=O).

^{1}H NMR ($CDCl_3$): poly(epoxy pentane carbonate) [P(EP-C)], δ, ppm = 0.87 (m, H^3, CH_3), 1.28–1.71 (br, H^4, CH_2CH_2-CH_3), 4.14 (br, H^2, CH_2CH), 4.89 (br, H^1, *CH*). ^{13}C NMR ($CDCl_3$): P(EP-C), δ, ppm = 13.6 (CH_3), 18.1 (CH_2–CH_2–CH_3), 32.2 (CH_2–CH_2–CH_3), 68.1 (CH_2CH), 75.8 (*C*H), 154.6 (*C*=O).

8.3.2 Characteristics of Polycarbonates

All polycarbonates were obtained as high molecular weight polymers (see Table 8.1), but they differed in color and stiffness. P(Phe-GEC), P(*n*Bu-GEC), and P(MeEt-GEC) were transparent polymers. P(*n*Bu-GEC), P(Et-GEC), P(*i*Pr-GEC), and P(MeEt-GEC) were jelly-like rubbers and were much softer than P(Phe-GEC) at room temperature. P(*t*Bu-GEC) was a white fibrous solid. These copolymers with 10 mol% of LiTFSI were slightly opaque and rubbery solids and were free of any precipitation of the salt. Dissociation of salt (LiTFSI) to ions (Li^+, $TFSI^-$) in polycarbonates was confirmed by the FT-IR measurement. The spectra of P(Et-GEC) and P(EP-C) samples (original copolymers and those electrolytes with LiTFSI) in the region from 550 to 650 cm^{-1} are shown in Figure 8.6. Both electrolytes showed absorption bands corresponding to the bending modes appeared at 655 cm^{-1} (ν_{SNS}), 618 cm^{-1} (ν_{aSO_2}), 604 cm^{-1} (ν_{aSO_2}), and 571 cm^{-1} (ν_{aCF_3}, $\nu_{s\ SO_2}$) [32], which are due to the existence of dissociated $TFSI^-$ in these polycarbonates. In the case of the PEC electrolytes, samples with Li salt concentrations of 20,

TABLE 8.1 Characterization of Synthesized Polycarbonates, P(R-GEC) and P(EP-C)

Polymers	Yield/%[a]	$M_n \times 10^4$	M_w/M_n	T_g (org)	T_g (Li)[b]
P(MeEt-GEC)	0.3	1.7	2.1	−45	−55
P(Et-GEC)	7	0.9	3.2	−5	−32
P(*i*Pr-GEC)	12	4.6	2.0	−9	−32
P(*n*Bu-GEC)	21	19	2.3	−24	−20
P(*t*Bu-GEC)	7	4.4	1.5	13	18
P(Phe-GEC)	9	1.5	4.5	45	41
P(EP-C)	23	2.2	4.6	−10	−4

[a]Yield (g g^{-1} of cat.) of P(R-GEC) copolymers insoluble in methanol.
[b]T_g of polycarbonate/LiTFSI (10 mol%) electrolytes.

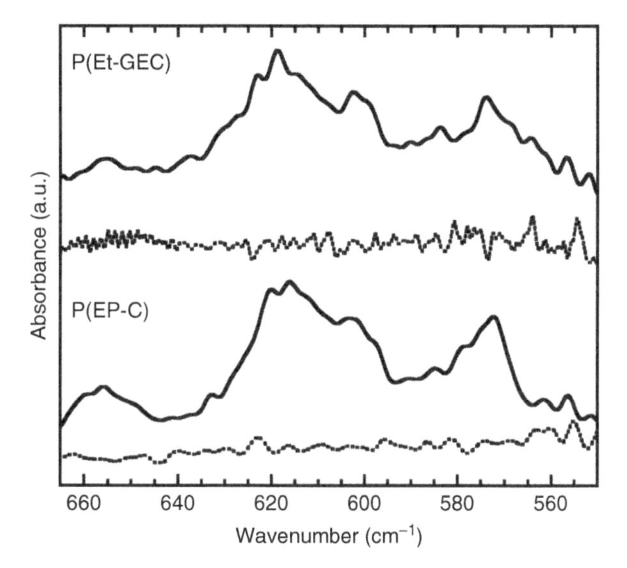

Figure 8.6 FT-IR spectra of original copolymers (dashed lines) and electrolytes with 10 mol% LiTFSI (solid lines) in the low frequency region.

40, and 60 wt% formed self-standing films, which turned to rubbery solids upon further increasing the salt concentration to 80 wt%. Copolymer electrolytes with LiTFSI, LiBETI, and $LiBF_4$ were transparent, whereas those electrolytes with $LiClO_4$ (except for the 20 and 40 wt% samples) and $LiCF_3SO_3$ formed opaque films.

8.3.3 Thermal Analysis of Polycarbonates

Figures 8.7 and 8.8 show DSC traces of the original polycarbonates and that of copolymers with LiTFSI. The glass transition temperature, T_g, of original

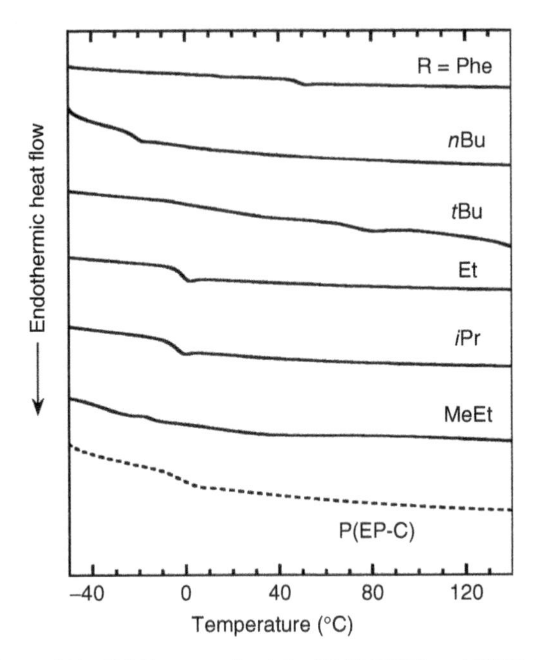

Figure 8.7 DSC traces of original P(R-GEC) and P(EP-C).

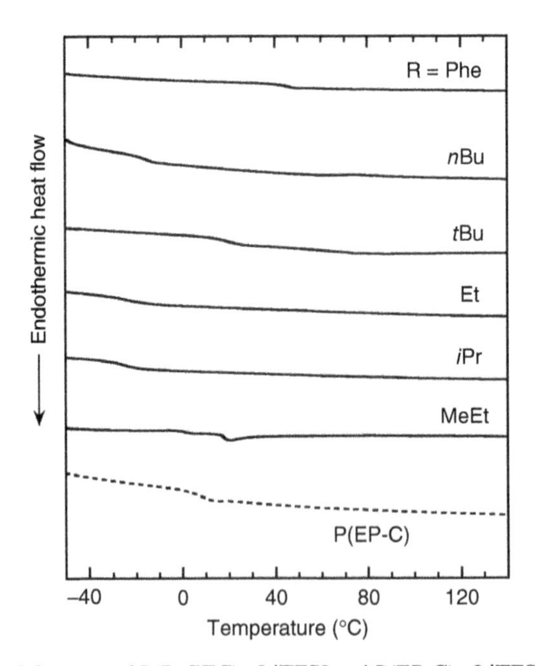

Figure 8.8 DSC traces of $P(R\text{-}GEC)_{10}LiTFSI$ and $P(EP\text{-}C)_{10}LiTFSI$ electrolytes.

copolymers and electrolytes are summarized in Table 8.1. All polycarbonates and the electrolytes were amorphous, except for the P(*t*Bu-GEC). The nature of substituents (denoted by R in Figure 8.3) determined the T_g of polycarbonate copolymers that varied from −45 to 45 °C. Polycarbonates bearing MeEt side chain formed the lowest T_g polymers and the copolymers bearing Phe group yielded highest T_g polycarbonates. For P(*n*Bu-GEC), T_g was 37 °C lower than for P(*t*Bu-GEC). This is due to the difference in mobility of the side groups, *n*- and *tert*-butyl, in the polycarbonates. Steric hindrance of the *tert*-butyl group should be very different from that of the *n*-butyl group, even though these groups have the same formula weight. For P(*t*Bu-GEC), there was a very weak glass transition because of coexistence in small amounts of crystalline domains, as indicated by the endothermic transition at 141 °C in the DSC analysis. P(Phe-GEC) is a glassy polymer because of its rigid side groups, resulting in the formation of polymers with the highest T_g of all polycarbonates studied here. P(MeEt-GEC) had the lowest T_g of all the original polymers, and the transition was also very weak (Figure 8.9). The T_g of P(Et-GEC), P(*i*Pr-GEC), and P(Phe-GEC) clearly decreased by the addition of LiTFSI, which may be caused by the plasticization effect of dissociated TFSI anions [33,34]. For P(*n*Bu-GEC) and P(*t*Bu-GEC), however, T_g clearly increased upon addition of LiTFSI. This result indicates the existence of interactions between cations or aggregated ions and main chain or pendant groups in the polycarbonate. Moreover, the small endothermic transition of P(*t*Bu-GEC) disappeared upon the addition of salts, and these electrolytes became amorphous without any thermal transitions above T_g. These results indicate that the increase in T_g is due to increased interchain interaction in amorphous regions, leading to cross-linking of polymer chains through the association of cations or aggregated ions and polar groups represented by carbonate unit of

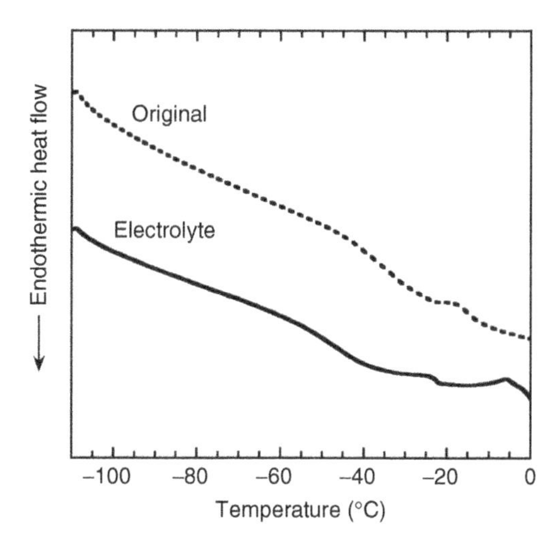

Figure 8.9 DSC traces of original P(MeEt-GEC) and electrolyte with LiTFSI on expanded scale of Figures 8.7 and 8.8 at the glass transition region.

the polycarbonate copolymer. In other words, the added salt can dissolve in the polycarbonate copolymer and dissociate into ions. The ions thus dissociated act as carrier ions in the polymer. On the basis of our study of the polar groups in polycarbonate, we believe that both ether oxygens of the side chain and the carbonate units of the main chain are involved in the interactions. The T_g values of P(Phe-GEC) and P(MeEt-GEC) changed in opposite directions upon addition of LiTFSI as compared to the corresponding original copolymers. The decrease in T_g for P(Phe-GEC) and P(MeEt-GEC) may be due to the plasticization effect of dissociated TFSI anions. In polycarbonates possessing ether side groups, the T_g tends to decrease upon the addition of salt. This indicates that the dissociated ions in polymer may disperse into both main chain and side groups, and as a result the plasticization effect on P(Et-GEC) is larger than the P(EP-C). On the other hand, the T_g of polycarbonate without ether side group, P(EP-C), increased upon the addition of salt. This probably indicates that Li ions mainly interact with polar groups of the main chain.

8.3.4 Impedance Measurement of Copolymers

The temperature dependence of the conductivity of all synthesized polycarbonate/LiTFSI electrolytes is shown in Figure 8.10. All polycarbonates gave typical Arrhenius plots similar to that of the polyether-based amorphous electrolytes, which are convex in shape throughout the temperature range measured. The conductivity of P(Phe-GEC)/LiTFSI was the lowest of all polycarbonate electrolytes, approximately 10^{-8} S cm^{-1} at 80 °C, because it was a glassy polymer with the

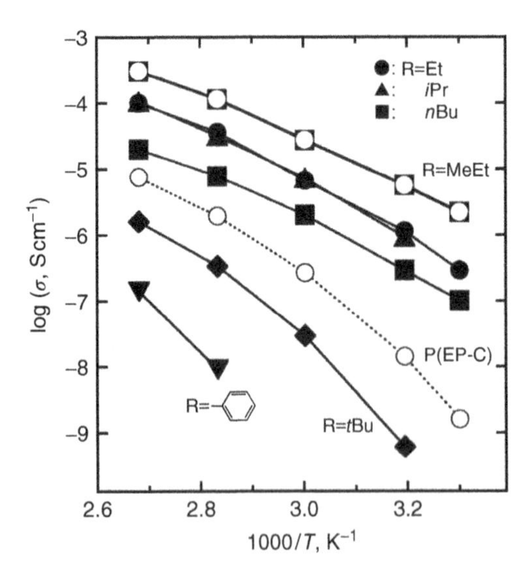

Figure 8.10 Temperature dependence of ionic conductivity for P(EP-C)$_{10}$LiTFSI and P(R-GEC)$_{10}$LiTFSI electrolytes (R=MeEt, Et, *i*Pr, *n*Bu, *t*Bu and Phe).

highest T_g among all polycarbonates studied here. The P(*t*Bu-GEC) electrolyte also had very low conductivity, of the order of 10^{-9} S cm^{-1} at 40 °C. The P(*n*Bu-GEC) electrolyte had good conductivity relative to P(Phe-GEC) and P(*t*Bu-GEC) electrolytes, because of its low T_g. This is due to the large difference in T_g because of the steric hindrance of their side groups. On the other hand, the P(Et-GEC) electrolyte had very high conductivity (2.9×10^{-7} S cm^{-1} at 30 °C), and the conductivity of a P(*i*Pr-GEC) electrolyte was almost the same as that of P(Et-GEC). As seen in Table 8.1, these two electrolytes had same T_g (−32 °C), which strongly influences the conductivity.

To clarify the influence of the structure of side groups, we compared the conductivities of P(Et-GEC) and P(EP-C) electrolytes. The conductivity of P(Et-GEC) electrolyte was approximately 100 times higher than that of the P(EP-C), and simultaneously the T_g of P(Et-GEC) electrolyte was approximately 30 °C lower than that of the P(EP-C). This means that the existence of ether side group in the polycarbonate may increase the ionic conduction, because the ether group has low rotational energy and probably shows fast relaxation. The P(MeEt-GEC) electrolyte had the highest conductivity of all polycarbonate samples (2.2×10^{-6} S cm^{-1} at 30 °C), which was 10 times lower than typical polyether-based electrolytes such as poly[oligo(oxyethylene glycol) methacrylate] (PMEO)/LiTFSI (2.4×10^{-5} S cm^{-1} at 30 °C) [35].

The Arrhenius plots of P(*n*Bu-GEC), P(*i*Pr-GEC), P(Et-GEC), and P(MeEt-GEC) electrolytes in Figure 8.10 have smaller gradient than for P(*t*Bu-GEC), P(EP-C), and P(Phe-GEC), and have similar gradient to $PMEO_{10}LiTFSI$. This suggests that there is almost no difference, in the activation energy for ionic conduction, between the polycarbonate and PMEO. The temperature dependence of amorphous polymer electrolytes is known to follow the Vogel–Tammann–Fulcher (VTF) empirical equation:

$$\sigma = \frac{A}{T^{1/2}} \cdot \exp\left[\frac{-E_a}{R\left(T-T_0\right)}\right] \tag{6.1}$$

where A (K$^{1/2}$ S cm^{-1}) is a constant that is proportional to the number of carrier ions, T_0 is an ideal T_g, R (J mol^{-1} K^{-1}) is the fundamental gas constant, and E_a (kJ mol^{-1}) is an activation energy for ionic transport via segmental motion [10]. Equation 6.1 can be rewritten by taking logarithms as

$$\ln(\sigma \cdot T^{1/2}) = -\frac{E_a}{R(T-T_0)} + \ln A \tag{6.2}$$

The parameters A and E_a can be estimated from the intercept and the gradient of each linear plot based on Equation 6.2. To make root mean square (RMS) closer to 1.0 in each plot using Equation 6.2, we took T_0 to be 50 °C lower than the value of T_g obtained from the DSC measurement. VTF plots of polycarbonate electrolytes without P(Phe-GEC) are shown in Figure 8.11. All plots appear to be lined up

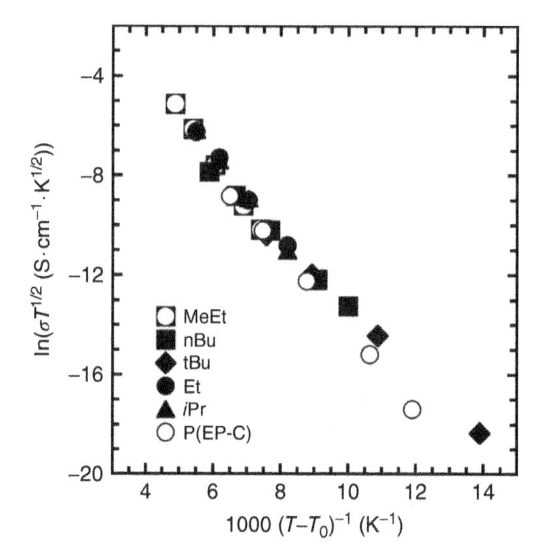

Figure 8.11 VTF plots of P(EP-C)$_{10}$LiTFSI and P(R-GEC)$_{10}$LiTFSI electrolytes (R=MeEt, Et, *i*Pr, *n*Bu and *t*Bu). (RMS of all plots >0.99).

TABLE 8.2 VTF Parameters of LiTFSI Electrolytes (10 mol%)

Polymers	T_0, K	A, K$^{1/2}$ S cm^{-1}	E_a, kJ mol^{-1}
P(MeEt-GEC)	168	109.3	16.7
P(Et-GEC)	191	27.5	14.4
P(*i*Pr-GEC)	191	36.5	14.8
P(*n*Bu-GEC)	203	1.0	11.1
P(*t*Bu-GEC)	241	0.5	10.5
P(EP-C)	219	4.7	13.1

virtually in a straight line, suggesting that ionic migration in these polycarbonates is dependent on the segmental motion of the polymer chain. The estimated VTF parameters are summarized in Table 8.2.

From our previous data [35], we estimated again parameters A and E_a of PMEO$_{10}$LiTFSI (T_g=−51 °C, T_0 = −101 °C) to be 16.1 (K$^{1/2}$ S cm^{-1}) and 11.4 (kJ mol^{-1}), respectively. The values of parameter A for P(*n*Bu-GEC), P(*t*Bu-GEC), and P(EP-C) electrolytes were quite small, and this is probably due to the small number of carrier ions. This is because the nonpolar side groups inhibit the smooth dissociation of ions. However, the values of parameter E_a for these polycarbonate electrolytes were almost the same as that for PMEO$_{10}$LiTFSI (11.4 kJ mol^{-1}). This indicates that there is no difference in the potential energy for ionic transport in polycarbonate and in PMEO, showing that carrier ions can migrate even in the polycarbonate. Shriver [36] has argued that transport in a rigid polymer electrolyte possessing a carbonate unit is decoupled from the segmental motion in the polymer. We believe that the decrease in the E_a observed in this study is a result of a similar decoupling due to the weak ion–dipole interaction. On the other hand, both A and E_a values of P(Et-GEC) and P(iPr-GEC) electrolytes were greater than those of $PMEO_{10}LiTFSI$. From the point of view of the side groups (ether or non-ether type), the A value of P(EP-C) electrolyte was clearly lower than that of the P(Et-GEC). This means that the number of carrier ions in P(EP-C) is lower than that in the P(Et-GEC) because of the difference in polarity of side groups. However, there were almost no differences in E_a between P(Et-GEC) and P(EP-C) electrolytes. This indicates that the potential energies for the ionic migration in P(Et-GEC) and in P(EP-C) are almost the same, showing that carrier ions can migrate even in such polycarbonates. The dependence for the P(MeEt-GEC) electrolyte also followed the VTF equation, but the parameters were significantly different. These unusual data are perhaps due to the decrease in T_g caused by the addition of LiTFSI. We consider that ionic migration in P(MeEt-GEC) is due to not only the segmental motion of the main chain but mainly the fast local motion of short side chains, such as γ-relaxation.

8.3.5 FT-IR Measurement

Figure 8.12 shows FT-IR spectra of original polymers [P(Et-GEC), P(EP-C)] and those electrolytes with LiTFSI in the region from 1700 to 1800 cm^{-1}. The adsorption band around 1750 cm^{-1} can be identified as stretching vibration of the carbonyl (C=O) group, $\nu_{C=O}$ of –O–(C=O)–O–, on the main chain of each polycarbonate [37]. In the case of P(Et-GEC), there was almost no significant change in the spectra between original (neat polymer) and the electrolyte, except for a minor broadness of the stretching vibration of the electrolyte, that is, the copolymer with Li salt. On the other hand, the band at this region in the P(EP-C) system was different from the original polymer, and a well-pronounced shoulder around 1720 cm^{-1} appeared upon addition of LiTFSI. It is known that the coordination of metal ions with C=O groups causes a reduction in the electron density leading to the reduced bond order of C=O, and hence a new absorption band appears at lower frequency [38]. Hence, in the P(EP-C) system, we expect that the dissociated Li ions only interact with the main chain. This is obviously related to the increase in T_g by the addition of salt, which is based on the relaxation of the main chain. Images of various modes of solvation for cations in these copolymer electrolytes are summarized in Figure 8.13. The side chain of P(EP-C) is free of any heteroatoms and consists only of hydrocarbons. Thus, it is impossible for P(EP-C) to form electrostatic interactions with Li ions through side chains (Figure 8.13c). However, in the P(Et-GEC) system, we think that the Li ions mainly interact with ether oxygens of the side groups and simultaneously through weak interactions with the C=O groups on the main chain (Figure 8.13b). Probably, the interactions with ether side chains scarcely affect the T_g, since these interactions are weak and thus cannot form the structure of strong coordinations with Li ions as seen in the PEO–salt complexes (Figure 8.13a). The existence of

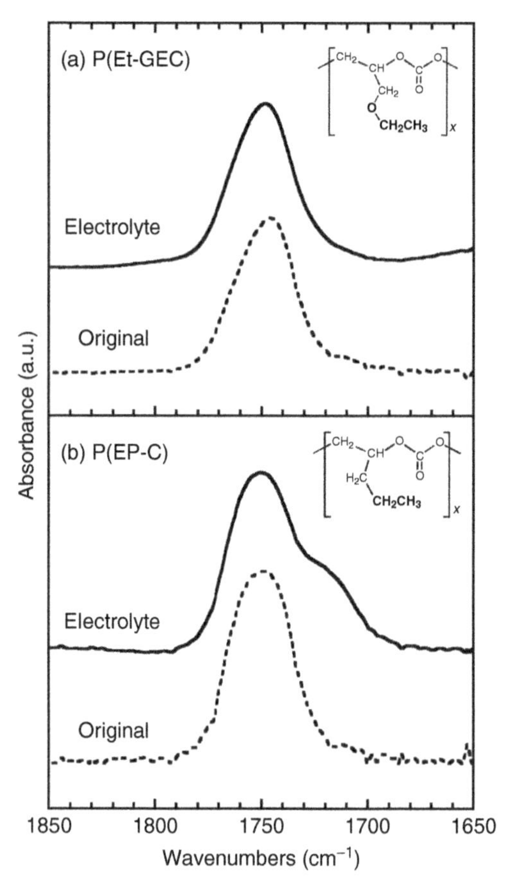

Figure 8.12 FT-IR spectra of original polycarbonates (dashed line) and electrolytes (solid line) with LiTFSI (10 mol%) in the stretching vibration region of the carbonyl group (C=O).

polar side groups such as a short OE chain in the polycarbonate may increase the flexibility of the main chain and possibly prevent large increase in T_g.

8.3.6 PEC System: Effect of Salt Concentration

Previously, GE-based polycarbonates were synthesized and its ionic conductivities were measured. However, poor yields have made it impossible to determine ionic conductivities in detail. Here, we used commercially available and thus easily accessible polycarbonate, namely, PEC, as an exemplary for a polymer matrix for studying the effect of salt concentrations on ionic conductivities. For this purpose, we prepared five kinds of Li electrolytes with salt concentrations varying from 20 to 80 wt%. The dependence of ionic conductivity on salt concentration or T_g for PEC-LiX and amorphous P(EO/EM2) (structure is inserted therein in Figure 8.14) electrolytes is shown in Figures 8.14 and 8.15. In Figure 8.14, the conductivity of

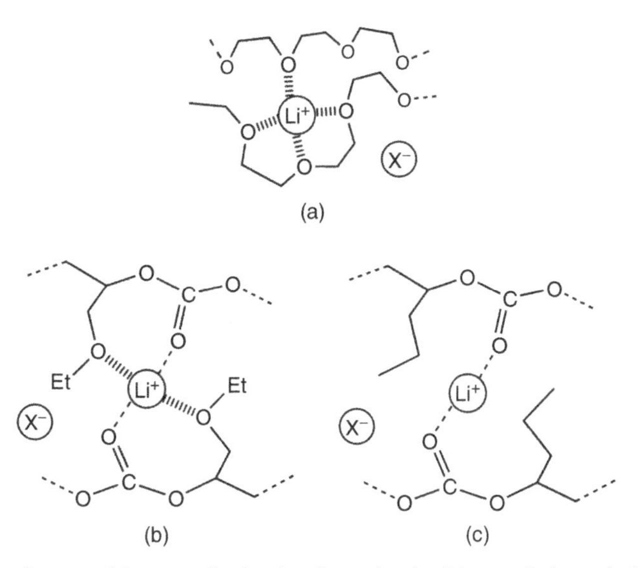

Figure 8.13 Structural images of solvation for cation by (a) oxyethylene chains, (b) ether side chains of GE type polycarbonate, and (c) carbonyl groups of P(EP-C).

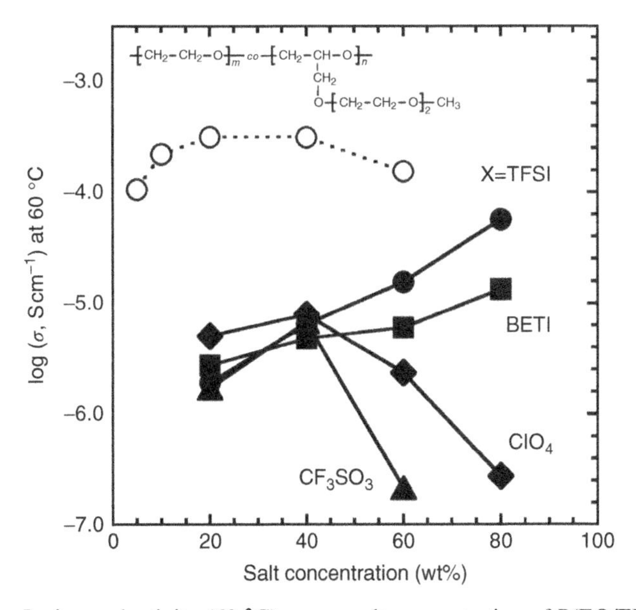

Figure 8.14 Ionic conductivity (60 °C) versus salt concentration of P(EO/EM2)-LiTFSI (○) and PEC-LiX (X=TFSI, BETI, CF_3SO_3 and ClO_4) electrolytes.

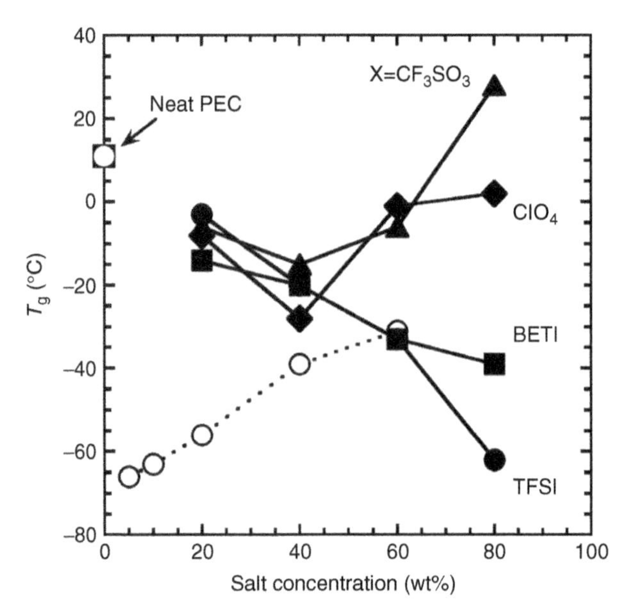

Figure 8.15 Glass transition temperature (T_g) versus salt concentration of P(EO/EM2)-LiTFSI (○) and PEC-LiX (X=TFSI, BETI, CF_3SO_3, and ClO_4) electrolytes.

polyether-based electrolyte was the highest at salt concentrations of approximately 20–30 wt% (around 5 mol% Li^+ to the OE units), and this behavior of the polyether system is well known [8–11]. The conductivity decreases slowly with increasing concentration at high salt concentrations, above 40 wt%. This is due to the increase in cross-linking structures between cations and dipoles of polyether chains. The formation of such cross-linked structures prevents segmental motion of local chains in the amorphous regions and results in large increase in T_g as can be noticed in Figure 8.15. For PEC-based electrolytes, the conduction and glass transition behavior were different. In Figure 8.14, the $LiCF_3SO_3$ and $LiClO_4$ electrolytes exhibited typical polyether-like conduction behavior, but with maximum values shifted to the higher concentration region from 20–30 to 40 wt%. Moreover, the conductivities of LiBETI and LiTFSI electrolytes increased linearly with increasing concentration, and electrolyte with 80 wt% LiTFSI had the highest conductivity of all PEC systems. The glass transition behavior of the PEC system is unique. As seen in Figure 8.15, the T_g value of PEC falls by more than 10 °C with the addition of salts of at least 20 wt%, whereas T_g for P(EO/EM2) increases with increasing salt concentration. At high salt concentrations, above 40 wt%, T_g for $LiCF_3SO_3$ and $LiClO_4$ electrolytes increased with increasing concentration. In contrast, for LiBETI and LiTFSI electrolytes T_g decreased with increasing the concentration, and the electrolyte with 80 wt% of LiTFSI resulted in the formation of polymer with the lowest T_g, a reduction of approximately 70 °C from the T_g of neat PEC. For salt concentrations higher than 40 wt%, the behavior of the PEC-based electrolytes is opposite to that of typical polyether systems. Addition of salts to polyether usually leads to an increase in T_g as a result of the strong interactions of dissociated Li^+ with ether oxygens and coordination with polymer chains [11]. For the PEC system, the dipole moment of the carbonate group in the main chain is strong enough to dissociate salts and interact with cations, but the coordinations or solvation are negligible. This behavior of salt-rich PEC electrolytes leads to the title of "polymer-in-salt" system. Angell et al. [39] have reported that polymer-in-salt electrolytes such as superionic glasses with a small amount of low M_w poly(propylene oxide) included have greater conductivities than the salt-in-polymer electrolyte, and there is a decrease in T_g. A ^7Li nuclear spin relaxation measurement was also performed on the polymer-in-salt electrolytes. The conclusion is that there are serial decoupling structures with different relaxation modes in liquids, and the conductivity of polymer-in-salt electrolytes is dominated by the migration of free Li^+ [40]. We believe that there are many "decoupled" ions (esp. Li^+) in PEC, which can migrate faster than coupled ions; the existence of these free ions may be involved in the large decrease in T_g.

8.4 CONCLUSION

To overcome the low ionic conduction of existing poly(ethylene oxide)-based polymer electrolytes, we considered polycarbonates obtained from the copolymerization of CO_2 and epoxy monomers for creation of a novel candidate for the polymer electrolytes without ethylene oxide units in the main chain. We successfully synthesized high molecular weight polycarbonates derived from CO_2 and GEs possessing phenyl, *tert*-butyl, *n*-butyl, ethyl, isopropyl, and MeEt side groups or epoxy pentane, and evaluated ionic conductivities of copolymers with Li salts like LiTFSI. Ionic conductivity of the polycarbonate electrolytes possessing MeEt side group had the highest of the order of 10^{-6} S cm^{-1} at room temperature. The activation energy (E_a) for ionic conduction in the polycarbonate electrolytes was estimated from the VTF equation, and the E_a of the electrolyte possessing *n*-butyl side groups was almost the same with the polyether-based electrolytes. The VTF parameter *A* of P(MeEt-GEC)-LiTFSI electrolyte showed the highest value of all samples, and this unusual behavior was probably due to the decrease in T_g caused by the addition of LiTFSI. The DSC measurement revealed that the series of GE-type polycarbonates almost decrease in T_g by the addition of LiTFSI, whereas the polycarbonate possessing non-ether side group P(EP-C) increase in the T_g. The FT-IR measurement indicated that a new band at 1720 cm^{-1} in the cation–dipole interaction region of carbonyl groups (C=O) on the main chain only appears in the P(EP-C) system. We expect that the polycarbonate possessing ether side group such as a short OE chain can be a good candidate for novel polymer electrolytes, because this polymer forms no strong coordination with cations as seen in the PEO-based electrolytes. Moreover, we found that the changes of conductivity and T_g versus salt concentration in the PEC-based electrolytes are opposite to that in typical polyether systems. An interesting feature of our study is that the polycarbonate is a unique candidate for ion-conductive polymers because of its flexible and hydrophobic properties.

With increasing environmental concerns about the global warming potential of CO_2 gas in the atmosphere, the results presented here can be considered as a way of mitigating CO_2 by fixing it on polymers.

ACKNOWLEDGMENT

This work was partially supported by Grant for Advanced Industrial Technology Development (Project ID: 11B01015c) from the New Energy and Industrial Technology Development Organization (NEDO), Japan.

REFERENCES

1. Iwamoto, R., Saito, Y., Ishihara, H., Tadokoro, H. (1968). Structure of poly(ethylene oxide) complexes. II. Poly(ethylene oxide)-mercuric chloride complex. *J. Polym. Sci.*, *6*, 1509–1525.
2. Yokoyama, M., Ishihara, H., Iwamoto, R., Tadokoro, H. (1969). Structure of poly(ethylene oxide) complexes. III. Poly(ethylene oxide)-mercuric chloride complex. type II. *Macromolecules*, *2*, 184–192.
3. Fenton, D. E., Parker, J. M., Wright, P. V. (1973). Complexes of alkali metal ions with poly(ethylene oxide). *Polymer*, *14*, 589.
4. Armand, M. B., Chabagno, J. M., Duclot, M. T. *Fast Ion Transport in Solids*. In: Vashishta, P., Mundy, J. N., Shennoy, G. K., editors. Proceedings of the International Conference on Fast Ion Transport in Solids, Electrodes, and Electrolytes. Elsevier, Amsterdam, 1979, p. 131.
5. Tarascon, J. M., Armand, M. (2001). Issues and challenges facing rechargeable lithium batteries. *Nature*, *414*, 359–367.
6. Scrosati, B., Garche, J. (2010). Lithium batteries: status, prospects and future. *J. Power. Sources*, *195*, 2419–2430.
7. Nogueira, A. F., Durrant, J. R., De Paoli, M. A. (2001). Dye-sensitized nanocrystalline solar cells employing a polymer electrolyte. *Adv. Mater.*, *13*, 826–830.
8. Takeoka, S., Ohno, H., Tsuchida, E. (1993). Recent advancement of ion-conductive polymers. *Polym. Adv. Technol.*, *4*, 53–73.
9. Gray, F. M. *Solid Polymer Electrolytes Fundamentals and Technological Applications*. VCH, New York, 1991.
10. Bruce, P. G. *Solid State Electrochemistry*. Cambridge University Press, Cambridge, 1995.
11. Ratner, M. A., Shriver, D. F. (1988). Ion transport in solvent-free polymers. *Chem. Rev.*, *88*, 109–124.
12. Abraham, K. M., Choe, H. S., Pasquariello, D. M. (1998). Polyacrylonitrile electrolyte-based Li ion batteries. *Electrochim. Acta*, *43*, 2399–2412.
13. Belières, J. P., Marechal, M., Saunier, J., Alloin, F., Sanchez, J. Y. (2003). Swollen polymethacrylonitrile urethane networks for lithium batteries. *J. Electrochem. Soc.*,*150*, A14–A20.

14. Zhang, Z. C., Sherlock, D., West, R., Amine, K., Lyons, L. J. (2003). Cross-linked network polymer electrolytes based on a polysiloxane backbone with oligo(oxyethylene) side chains: synthesis and conductivity. *Macromolecules, 36*, 9176–9180.
15. Allcock, H. R., Oconnor, S. J. M., Olmeijer, D. L., Napierala, M. E., Cameron, C. G. (1996). Polyphosphazenes bearing branched and linear oligoethyleneoxy side groups as solid solvents for ionic conduction. *Macromolecules, 29*, 7544–7552.
16. Tominaga, Y., Izumi, Y., Kwak, G.-H., Asai, S., Sumita, M. (2002). Improvement of the ionic conductivity for PEO-$LiCF_3SO_3$ complex by supercritical CO_2 treatment. *Mater. Lett., 57*, 777–780.
17. Tominaga, Y., Izumi, Y., Kwak, G.-H., Asai, S., Sumita, M. (2003). Effect of supercritical carbon dioxide processing on ionic association and conduction in a crystalline poly(ethylene oxide)-$LiCF_3SO_3$ complex. *Macromolecules, 36*, 8766–8772.
18. Kwak, G.-H., Tominaga, Y., Asai, S., Sumita, M. (2003). Improvement of the ionic conductivity for amorphous polyether electrolytes using supercritical CO_2 treatment technology. *Electrochim. Acta, 48*, 1991–1995.
19. Di Noto, V., Vezzu, K., Pace, G., Vittadello, M., Bertucco, A. (2005). Effect of subcritical CO_2 on the structural and electrical properties of ORMOCERS-APE systems based on Zr and Al. *Electrochim. Acta, 50*, 3904–3916.
20. Kitajima, S., Tominaga, Y. (2009). Enhanced cationic conduction in a polyether/clay composite electrolyte treated with supercritical CO_2. *Macromolecules, 42*, 5422–5424.
21. Inoue, S., Koinuma, H., Tsuruta, T. (1969). Copolymerization of carbon dioxide and epoxide. *J. Polym. Sci. Polym. Lett., 7*, 287–292.
22. Inoue, S., Koinuma, H., Tsuruta, T. (1969). Copolymerization of carbon dioxide and epoxide with organometallic compounds. *Die Makromol. Chem., 130*, 210–220.
23. Lukaszczyk, J., Jaszcz, K., Kuran, W., Listos, T. (2000). Synthesis of functional polycarbonates by copolymerization of carbon dioxide with allyl glycidyl ether. *Macromol. Rapid Commun., 21*, 754–757.
24. Tan, C. S., Juan, C. C., Kuo, T. W. (2004). Polyethercarbonate-silica nanocomposites synthesized by copolymerization of allyl glycidyl ether with CO_2 followed by sol-gel process. *Polymer, 45*, 1805–1814.
25. Yu, T., Zhou, Y., Liu, K., Chen, E., Wang, D., Wang, F. (2008). Hydrogen-bonded thermostable liquid crystalline complex formed by biodegradable polymer and amphiphilic molecules. *Macromolecules, 41*, 3175–3180.
26. Smith, M. J., Silva, M. M., Cerqueira, S., MacCallum, J. R. (2001). Preparation and characterization of a lithium ion conducting electrolyte based on poly(trimethylene carbonate). *Solid State Ion., 140*, 345–351.
27. Barbosa, P. C., Rodrigues, L. C., Silva, M. M., Smith, M. J. (2011). Characterization of $pTMC_nLiPF_6$ solid polymer electrolytes. *Solid State Ion., 193*, 39–42.
28. Tominaga, Y., Shimomura, T., Nakamura, M. (2010). Alternating copolymers of carbon dioxide with glycidyl ethers for novel ion-conductive polymer electrolytes. *Polymer, 51*, 4295–4298.
29. Nakamura, M., Tominaga, Y. (2011). Utilization of carbon dioxide for polymer electrolytes II: synthesis of alternating copolymers with glycidyl ethers as novel ion-conductive polymers. *Electrochim. Acta, 57*, 36–39.
30. Motika, S. A., Pickering, T. L., Rokicki, A., Stein, B. K. (1991). Catalyst for the copolymerization of epoxides with CO_2. *US patent*, 5026676.

31. Ree, M., Bae, J. Y., Jung, J. H., Shin, T. J. (1999). A new copolymerization process leading to poly(propylene carbonate) with a highly enhanced yield from carbon dioxide and propylene oxide. *J. Polym. Sci. Part A: Polym. Chem.*, *37*, 1863–1876.
32. Rey, I., Johansson, P., Lindgren, J., Lassègues, J. C., Grondin, J., Servant, L. (1998). Spectroscopic and theoretical study of $(CF_3SO_2)_2N^-$ (TFSI-) and $(CF_3SO_2)_2NH$ (HTFSI). *J. Physical Chem. A*, *102*, 3249–3258.
33. Armand, M., Gorecki, W., Andreani, R. (1990). Proceedings of 2nd International Symposium on Polymer Electrolytes, p. 31.
34. Sylla, S., Sanchez, J. Y., Armand, M. (1992). Eloectrochemical study of linear and cross-linked POE-based polymer electrolytes. *Electrochim. Acta*, *37*, 1699–1701.
35. Tominaga, Y., Hirahara, S., Asai, S., Sumita, M. (2005). Specific ionic conduction in poly oligo (oxyethylene glycol) methacrylate (PMEO)-Li salt complexes under high-pressure CO_2. *J. Polym. Sci. Part B: Polym. Phys.*, *43*, 3151–3158.
36. Wei, X. Y., Shriver, D. F. (1998). Highly conductive polymer electrolytes containing rigid polymers. *Chem. Mater.*, *10*, 2307–2308.
37. Chen, X. H., Shen, Z. Q., Zhang, Y. F. (1991). New catalytic-systems for the fixation of carbon-dioxide. 1. Copolymerization of CO_2 and propylene-oxide with new rare-earth catalysts-$Re(P_{204})_3$-$Al(i\text{-}Bu)_3$-$R(OH)_n$. *Macromolecules*, *24*, 5305–5308.
38. Selvasekarapandian, S., Baskaran, R., Kamishima, O., Kawamura, J., Hattori, T. (2006). Laser Raman and FTIR studies on Li^+ interaction in PVAc-$LiClO_4$ polymer electrolytes. *Spectrochim. Acta Part A - Molec. Biomolec. Spectr.*, *65*, 1234–1240.
39. Angell, C. A., Liu, C., Sanchez, E. (1993). Rubbery solid electrolytes with dominant cationic transport and high ambient conductivity. *Nature*, *362*, 137–139.
40. Fan, J., Marzke, R. F., Sanchez, E., Angell, C. A. (1994). Conductivity and nuclear-spin relaxation in superionic glasses, polymer electrolytes and the new polymer-in-salt electrolyte. *J. Non-Cryst. Solids*, *172*, 1178–1189.

9

BLOCK COPOLYMER NANOPATTERNS AS ENABLING PLATFORMS FOR DEVICE APPLICATIONS—STATUS, ISSUES, AND CHALLENGES

Sivashankar Krishnamoorthy

9.1 INTRODUCTION

Self-assembly of block copolymer (BCP) thin films on surfaces is by now a standard way of attaining nanopatterns with sub-100 nm spatial resolutions. A multitude of nanopatterns with feature and spatial resolutions typically down to sub-10 nm regime consisting of dot, line, or other complex patterns has been demonstrated. The facile realization of nanopatterns with high pattern resolutions (and therefore high feature densities) across large areas, at low cost and high throughput, makes BCP self-assembly appropriately placed in catering to different technologies. Simply by providing the right conditions for self-assembly, one could take a facile approach to create several billion features per square centimeter of surface with desired geometric characteristics. Additional benefits include manufacturing compatibility that is derived from the knowledge of processing polymers that have been mastered by semiconductor industries. Ever since the first reports on patterning using BCP thin films appeared in the 1990s, the field has progressed rapidly in many important directions. These include impressive achievements such as demonstration of long-range ordering of copolymer domains spanning complete wafers; pattern transfer down to sub-10 nm resolutions; feature density multiplication by

Synthesis and Applications of Copolymers, First Edition. Edited by Anbanandam Parthiban.

several fold; automated and scalable fabrication through nanoimprint lithography (NIL); and exploitation within nanoscale devices for data storage, sensors, microelectronics, and biology. Numerous reviews have comprehensively covered different aspects of nanopatterning using BCP thin films in great detail [1–6]. The discussion in this chapter is specifically geared toward the possibilities and challenges in relation to the exploitation of BCPs for different technologies.

9.2 BLOCK COPOLYMER TEMPLATES FOR PATTERN TRANSFER APPLICATIONS

Central to the utilization of BCP thin film-based nanopatterning are the nanoscale templates that are derived out of them. These templates are either surface-relief patterns realized by the selective removal of one of the domains from a phase-separated copolymer thin film, or the chemical/surface-energy contrast presented by a copolymer thin film even as-coated. As the template dimensions for feature width and periodicity directly determine the feature dimensions of the materials patterned, it is crucial to obtain templates with desired geometric attributes and low standard deviations. In the following discussion, the state of the art and challenges pertaining to certain key patterning specifications of interest to technologies, for example, dimensional scalability, ultrahigh densities, high resolution nanolithography, and high throughput/large-area patterning, and long-range ordering using BCP self-assembly are reviewed.

9.2.1 Dimensional Scalability and Fine-Tunability Down to Sub-10 nm Length Scales

The ultimate pattern resolution achievable using a copolymer system is determined by a high chemical incompatibility between the constituent blocks that enable phase separation even at very low molecular weights. The degree of incompatibility between the blocks is represented by the product χN, where χ signifies the Flory–Huggins interaction parameter, and N is the degree of polymerization. A sufficiently strong segregation between the blocks corresponding to $\chi N \gg 10$ is necessary for pattern formation; for $\chi N < 10$, the copolymer falls within a weak segregation limit and is disordered. Achieving high pattern resolutions would require reducing N, as the natural domain period (L_0) of the polymer scales with N as $L_0 \propto N^{\delta}$ where $0.5 < \delta < 1$ [7,8]. However, by decreasing N, the product χN also decreases and consequently a limit to pushing resolutions further is set by and can be pushed to higher values only until χN remains high enough to enable phase separation between the blocks. This places significant emphasis on the choice of copolymer systems that exhibit a high degree of incompatibility between the blocks, and thereby higher values for χ in order to scale the pattern resolutions below sub-10 nm regime. For instance, PS-*b*-PMMA, which is among the most commonly employed copolymer for BCP lithography, exhibits a χ value of $0.028 + 3.9/T$ [9]. The system is theoretically predicted to have a lower limit for $L_0 \sim 24$ nm, with experimentally observed minimum at ~27 nm [2]. The other commonly demonstrated copolymers for BCP lithography, such as PS-*b*-PI (SI) and PS-*b*-PFS (SF), share very similar values for $\chi(0.027 + 3.61/T$ for SI and $0.023 + 3.68/T$ for SF) [10]. BCPs such as PS-*b*-PEO and PS-*b*-PDMS exhibit higher χ values, allowing the realization of pattern resolutions down to sub-10 nm regime. PS-*b*-PEO patterns with feature width of 3 nm and pitch as low as 6.9 nm were demonstrated [11]. The temperature dependence of χ also poses significant impact on the ability to anneal defects [2]. The high dependence of χ values on temperature would allow annealing temperature as a useful experimental handle to tune the χ values between the strong and weak phase segregation regimes of the copolymer for a wide range of molecular weights. Furthermore, the value of χ not only determines the ability to achieve high resolutions but also the width of the interface, which eventually determines the line width roughness and the lower limit for the feature dimensions. The width of the interface shared between the copolymer domains at the limit of infinite molecular weight scales as $2\alpha/\sqrt{6\chi}$ (where "α" signifies statistical segment length of the monomer) [12], thus resulting in larger interfacial width for smaller values of χ. The above considerations point to measures that can be adopted towards achieving highest resolutions in feature width and pitch of the copolymer templates. Although the material nanopatterns produced from the copolymer templates using various pattern transfer protocols typically follow the geometric attributes of the original templates, the feature width, heights, and edge–edge separation in many cases are amenable to fine-tuning independent of the pitch through choice of processing conditions [13–18]. In general, it has been harder to achieve highest resolutions for pitch of the copolymer patterns than for feature width, aspect ratios, or edge–edge separations.

9.2.2 Directing Self-Assembly of Block Copolymers

Directed self-assembly of BCPs has been pursued aiming at varied end goals, for example, defect removal and long-range ordering, addressability of individual or group of copolymer features, density multiplication, programmed shapes, or hierarchical structuring (Figure 9.1). Long-range ordering of domains is especially sought for applications such as photonic crystals that require a periodic variation in refractive index or for high density bit-patterned magnetic media that require knowledge of the location of the individual features. For many other applications that benefit mainly from an average property of the collection of nanostructures, a polycrystalline 2D arrangement of nanostructures would suffice. Long-range ordering of copolymer domains was shown achieved through directed self-assembly using topographic or chemical guiding patterns with width (L_p) of the guiding patterns either equivalent to or several fold larger than the equilibrium periodicity (L_0) of the copolymer system. The most significant factor in determining the lateral order of the copolymer domains is the commensurability (with a typical tolerance of 10%) of the copolymer equilibrium period with respect to the width of the underlying guiding pattern. High resolution patterns made using EUV-IL and e-beam lithography

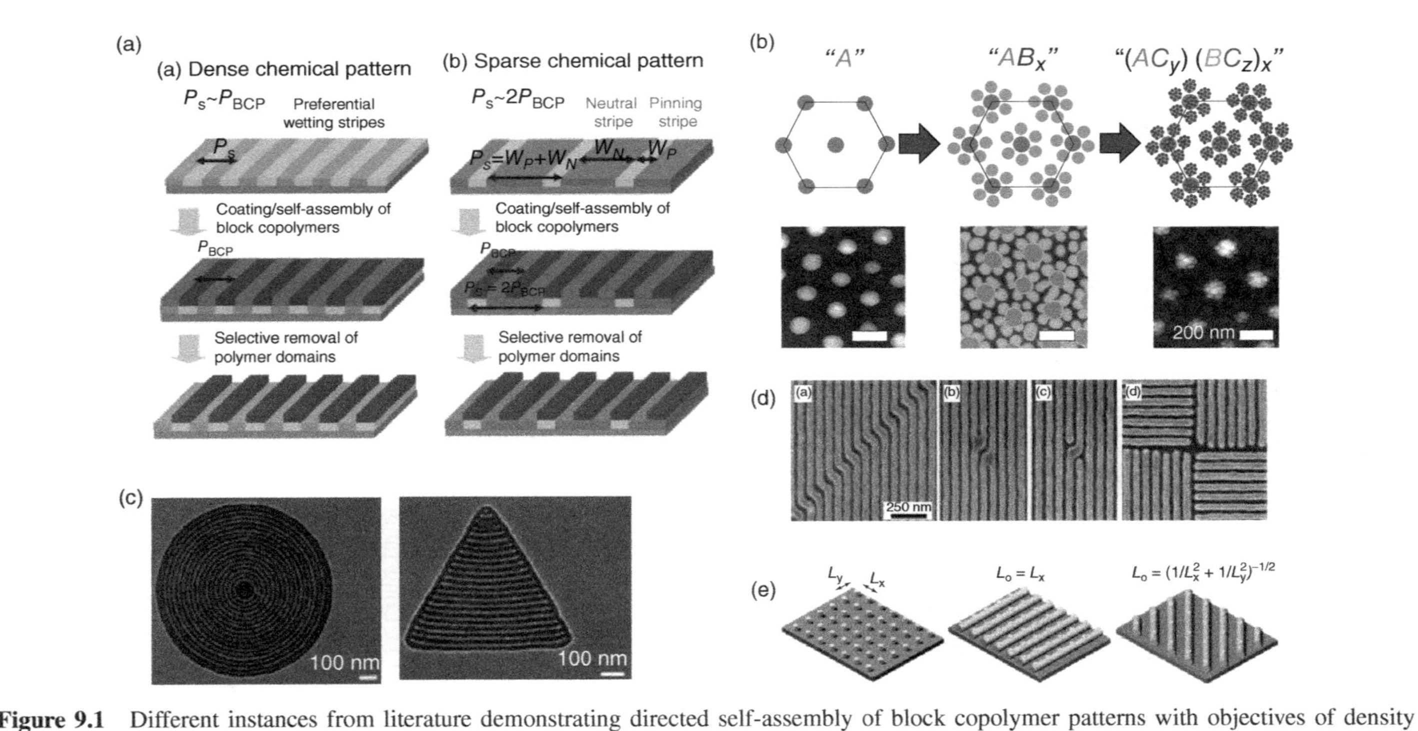

Figure 9.1 Different instances from literature demonstrating directed self-assembly of block copolymer patterns with objectives of density multiplication, hierarchical structuring, engineered pattern designs, and long-range ordering. (a) Sparse and dense chemical patterns presenting alternating preferential wetting stripes of pitch that is equal to or twice that of the block copolymer domains. (Reprinted with permission from [23]. Copyright © 2008 Wiley Periodicals Inc.) (b) Arrays of hierarchically built binary and ternary superstructures using multilevel assembly of reverse micelles. (Reprinted with permission from [31]. Copyright © 2013 American Chemical Society.) (c) Pt nanowire patterns obtained by graphoepitaxially confined polystyrene–poly(2-vinylpyridine) block copolymers within micro patterns of different shapes. (Reprinted with permission from [27]. Copyright © 2008 American Chemical Society.) (d) Directed self-assembly of PS-*b*-PMMA into patterns resembling essential elements of integrated circuit element geometries. (Reprinted with permission from [28]. Copyright © 2007 American Chemical Society.) (e) Generation of arbitrary patterns of block copolymers using electron beam lithography. (Reprinted with permission from [21]. Copyright © 2010 Nature Publishing Group.) (*See insert for color representation.*)

(EBL) [19–23] have shown to guide long-range ordering of copolymer domains using patterns with L_p that is up to fourfold greater than that of L_o (Figure 9a). In comparison, graphoepitaxy approach demonstrated long-range ordering using low resolution guiding patterns, with L_p up to 50-fold greater than L_o. In among the first examples of latter was showed by Segalman et al. [24] using a monolayer of polystyrene-*block*-poly(2-vinylpyridine) (PS-*b*-PVP) spheres spanning grooves of width $<5\,\mu m$. Long-range ordering was obtained both on mesas and within the grooves with discontinuity along the edges. Later, Cheng et al. [25,26] demonstrated systematic engineering of the number of rows of PS-*b*-PFS spheres in relation to the width of the groove. Ordering was found to be independent of the height of the grooves, as long as they were taller than the domain spacing of the copolymer system used. Chai et al. [27] employed graphoepitaxy of polystyrene-*block*-poly(2-vinylpyridine) thin films within arbitrarily shaped submicron patterns on silicon substrates, with the resulting patterns revealing significant flexibility in following the shapes of the guiding features (Figure 9c). Despite the low density multiplication factors achieved, EBL-based patterning is especially interesting to design and engineer the orientation of the BCPs [21] and to achieve complex device-oriented geometries (Figure 9d,e) [28,29]. The graphoepitaxial approaches, on the other hand, require use of less-demanding top-down approaches that can readily cater to arbitrarily large areas at low cost. In a recent work, Park et al. [11] demonstrated guidance of long-range ordering of PS-*b*-PEO copolymer domains across large areas using faceted surfaces of miscut sapphire substrates. The key advantage of the work is in being able to capitalize on high resolution guiding patterns naturally formed through surface reorganization of a miscut substrate. The lateral ordering of copolymer domains was found to be tolerant to the local defects such as dislocations in the underlying guiding patterns. In a similar approach, the guiding patterns for BCP assembly were chosen to be patterns prepared from another copolymer with larger pitch (Figure 9.1b) [30,31]. This expands the range of possibilities to obtain guiding patterns, as well as opportunities for design to realize directed self-assembly and density multiplication at scales not readily accessible by conventional lithography.

A potential alternative to lithographic guiding patterns in directing long-range ordering of copolymers is annealing through exposure to selected solvents [32–41]. The solvent annealing involves exposure of the copolymer thin films to a saturated environment of one or more solvents that impart mobility to specific blocks. Solvent exposure allows rearrangement of polymer chains between adjacent domains, and results in low standard deviation in feature width and the pitch of the patterns. One of the drawbacks of the method is the dewetting of the polymer films that accompany the process [38,42,43]. Such dewetting can often be seen when the films are inspected under an optical microscope. The localized measurements on areas spanning few microns using atomic force microscopy (AFM) do not necessarily represent the situation across entire coated area. To counter this challenge, use of lithographically patterned areas and durations of optimized solvent exposure are suggested. In spite of the economic advantages in attaining long-range ordering through solvent annealing, the reliability of the process needs significant improvement to merit technological adoption. An interesting use of the long-range ordering of solvent-annealed thin films of polystyrene-*block*-polyvinylpyridine reverse micelles was demonstrated in directing long-range ordering of electrochemically formed pores in anodized alumina (AAO) [37]. In this case, AAO technique benefits from the use of long-range ordered copolymer domains to define patterns that initiate pore formation and the simplicity and economy of the copolymers in producing such patterns.

9.2.3 Block Copolymers for Directed Nanoscale Synthesis and Self-Assembly

Numerous instances in the literature have shown the use of BCP templates to direct synthesis, attachment, or incorporation selectively within nanoscopic (<100 nm) regions [4,44–50]. The guided inorganic synthesis using BCP templates has been shown by selective incorporation of metal ions within copolymer domains, followed by their oxidation or reduction to yield the corresponding metal or metal oxide nanopatterns. The incorporation of metal ions has been carried out either in solution before films are coated on surface (pre-loading) or by exposing the copolymer-coated surface to a solution consisting of metal ions (post-loading). The latter route has the advantage of producing material arrays that closely follow the original copolymer patterns. The inclusion of ions in the solution phase, on the other hand, can result in variations in the structural attributes of the resulting nanopatterns, depending on the quantity and type of metal ion included. Templated synthesis through selective decomposition of volatile organometallic precursors within the BCP domains from vapor phase has also been shown [18,51,52]. The vapor phase methods allow for a cleaner alternative to such synthesis, and also allow for convenient handles such as pressure and duration of exposure to control the quantity of loading.

The guided attachment of pre-formed nanoparticles serves as an interesting alternative to the *in situ* synthesis option using BCP domains toward creating material patterns. Such guided attachment has also been demonstrated either by preparing solution of blends of nanoparticles with copolymers before a thin film is coated on surface [5,53,54] or by exposing copolymer thin films to a suspension of pre-formed nanoparticles [55–57]. The structural attributes of composite thin films prepared from solution blends of the nanoparticle with the copolymer is known to be influenced by the surface energy of the nanoparticle's surface in relation to the polymer domain, and the degree of loading. The post-organization of nanoparticles on a pre-existing BCP nanopattern overcomes this issue as the areas of attachment of nanoparticles are determined by the structural characteristics of the underlying BCP templates. However, the blends are much easier option if one would prefer a one-step route to nanoparticle coatings through high throughput approaches as printing or spraying, thus providing economy of time.

9.2.4 High Resolution Nanolithography

One of the significant capabilities of BCP thin film templates is their use as mask for nanolithography. Lithographic pattern transfer of copolymer templates by dry etching methods was demonstrated in the 1990s to create pillars or holes down to sub-50 nm pattern resolutions on hard substrates such as silicon nitride [58,59]. This brought in tremendous opportunities for massively parallel fabrication of several billion features spanning large areas of surface, rivaling the conventional lithography tools in resolution and throughput. However, performing nanolithography at high spatial resolutions (or very low pitch) faces serious challenges out of which the low etch contrast offered by the polymeric domains is one that is specific to the use of BCPs. Other generic challenges common to these scales exist, which include need for significant process optimization of the dry etching protocols to enable high etch specificity, anisotropy (or directionality of etch, determining how vertical are the etched features) with reasonably high etch rates and uniformity across large areas such as full wafers. Etching within confined spaces of the order of a few nanometers as encountered for sub-10 nm pore diameters or pillar separations can result in inefficient removal of trapped etchant gases resulting in undercut features and also in reduced etch rates. The low etch contrast offered by copolymer templates has been overcome by substituting the polymer templates with inorganic metal or metal oxide patterns derived out of them. This has been accomplished using different means, namely, by multi-layered pattern transfer protocols [58]; using of copolymers with organometallic blocks, for example, polystyrene-*block*-polyferrocenyldimethylsilane (PS-*b*-PFS) [60,61] or polystyrene-*block*-polydimethylsiloxane (PS-*b*-PDMS) [62,63]; selectively introducing inorganic component within one of the copolymer domains [18,64–67]; and by using template metal deposition [30,68]. Although such metal- and iron-containing particles have shown required robustness toward lithography, their use is constrained within etchers used in semiconductor processing. Silica or titania masks are better preferred in this respect.

Inclusion of titania precursors onto a pre-existing BCP pattern on surface is yet another alternative where the extent of loading of the precursors was shown recently as means of tuning the size of lithographic mask in steps of <10 nm [18]. The approach allowed BCP templates with thickness and separations below 10 nm to be successfully exploited for nanolithography resulting in Si pillars with only 8 nm separations across wafer level. In yet another recent development to high resolution nanolithography templates using pre-formed line patterns using PS-*b*-PDMS copolymer, Jung et al. [30] employed an overdeposition and etching sequence to arrive at reverse patterns of tungsten lines with 8 nm feature sizes. The latter approaches are attractive considering that the copolymer self-assembly step is decoupled from formation of the inorganic nanolithography mask, thus relieving any possible interference on the self-assembly conditions owing to such translation.

9.2.5 Nanomanufacturing Material Patterns for Applications

Use of copolymer thin films to create array of pores or posts for templated growth, deposition, or etching processes has allowed realizing nanopatterns of various metals, ceramics, and semiconducting materials. The various processes involved in such translation have been adequately covered in many earlier reviews. Here, we wish to limit the discussion to certain determining aspects of processing of templates that requires strong consideration in successful exploitation of BCP templates within applications. At the heart of such exploitation lies the ability to scale up patterns to arbitrarily large-sized surfaces that are wafer level and beyond with high throughput and capabilities to pattern 3D surfaces or flexible/plastic substrates. With such considerations in mind, the realization of template pattern alone does not guarantee one with the final pattern of chosen material on a desired substrate. This is particularly so when the resolution of the templates approaches sub-10 nm regime, and also when common pattern-transfer protocols are incompatible with the nature of substrate or templates used. In such cases, the successful realization of desired material patterns places profound emphasis on the choice of processing employed, rendering pattern transfer as significant as the template preparation itself. The pattern transfer routes should be chosen such that they do not compromise the original pattern characteristics or uniformity of the templates, and in addition preserve other advantages of the BCPs related to the processing, scalability, and manufacturing compatibility.

IBM group of workers have successfully demonstrated transfer of wafer-level patterns of PS-*b*-PMMA into silicon, silicon oxide, or metal nanopatterns, targeting microelectronic device applications [2,69–74]. The realization of templates using phase-separated thin films toward pattern transfer application typically involves a number of steps and may result in time consumption of several hours to days. Such processing may involve (i) orientation of domains by rendering the surface neutral to the different blocks of the copolymer, or by using other orienting influence such as electrical field or solvent annealing; (ii) thermal annealing at a higher temperature >100 °C to enhance lateral order; and (iii) selective removal of one of the blocks to attain a template exhibiting contrast in topography [71]. In general, the scalability of copolymer thin film-based material patterning from small chips to large areas such as full wafers demands low variation in film thickness, surface energy, and pattern transfer conditions across the entire area. Typical processes employing spin coating, wet or dry etching, and material deposition can be optimized to yield a variation lower than 5% on full wafer level. Special attention is required when processes such as solvent annealing is used, as nonuniformities can ensue because of dewetting of polymer films [38,42,43]. To have solvent-annealing conditions transferable and reproducible, it is necessary to accurately monitor and control the solvent vapor exposure parameters such as vapor pressure, composition, and exposure duration. An ability to achieve better manufacturing compatibility using significantly low thermal annealing times and the scalability of directed self-assembly on 300 mm wafers were demonstrated in separate instances by Nealey et al. [19,28,29]. Such optimization is essential in enhancing the throughput and practicality of the use of copolymers in several applications.

9.2.5.1 BCP-Derived Nanoimprint Lithography (NIL) Molds—A Step Toward Further Scalability A significant capability toward further scalability, enhanced repeatability, and automated processing of copolymer-derived nanopatterns arises in combination with NIL. NIL is a manufacturing-compatible, top-down tool that allows replication of hard molds usually formed out of one of the conventional top-down lithography. For sub-100 nm pattern resolutions, NIL faces limitation of high cost of molds or limited patterned areas. This is because high resolution NIL molds are typically fabricated by electron beam lithography which owing to the high cost and low throughput renders the molds very expensive. Pattern transfer of BCP patterns into hard substrates such as Si or SiO_2 allows facile realization of high resolution molds for NIL across large areas at significantly low costs. Park et al. [75] showed SiO_2 molds with sub-20 nm feature widths fabricated by the self-assembly of PS-*b*-PMMA cylinder forming blocks. With interesting process variations, both pillar and hole morphologies were demonstrated. Krishnamoorthy et al. [18,76] demonstrated silicon nanopillar molds with feature and spatial resolutions down to sub-50 nm length scales, using reverse micelle arrays of PS-*b*-P2VP as templates. Although these examples are applicable for chip- or wafer-level NIL molds, further scalability is feasible by step and flash imprinting and by roll-to-roll processing approaches. Conversion of the BCP pattern into a NIL mold offers additional process advantages. The height of the features in the resulting mold is limited mainly by the dry etching conditions used for pattern transfer. A well-optimized dry etching process offering high selectivity for the substrate would allow producing high aspect ratio pillars or hole features. The imprinting using these molds allows realizing templates of high aspect ratios in a single step, as compared to that of several steps of processing necessary when templates of similar geometric attributes were to be realized exclusively using copolymer self-assembly. Thus, the combination of the BCP self-assembly with NIL serves as a win-win situation for both the techniques.

9.2.5.2 Reverse Micelles Versus Phase-Separated Thin Films An effective alternative toward generating dot-array patterns of copolymer templates with a polycrystalline 2D order is the use of reverse micelles of amphiphilic copolymer coated on the surface. Reverse micelles are polymeric nanoparticles formed by aggregation of amphiphilic BCPs in selective solvents [77]. These particles can be deposited on surfaces to produce arrays of templates useful for pattern transfer applications [59]. This approach has been used to produce ordered arrays of metal or metal oxide nanoparticles and nanolithography [68,78–80]. Investigators from Centre Suisse d'Electronique et de Microtechnique (CSEM SA) had extended these capabilities to a suite of other possibilities, including lithography, responsive polymer films, surface energy patterns, and nanoporous sensors on full wafers [6,13,15,17,81–85]. The reverse micelle route to templates for pattern transfer provides noteworthy alternative to phase-separated thin films in certain respects: (i) the size of the templates can be controlled by varying the aggregation number of the micelles, which in turn can be controlled by the solvents employed; (ii) the pitch of the dot-array pattern can be controlled by varying the solution concentration and speed of evaporation (by varying spin or dip coating speeds); and (iii) the as-coated templates can act as lithographic etch masks, as they present a contrast in topography even as-coated. The approach however is best known only for dot-array patterns. Although line patterns can be realized in principle using cylindrical micelles, it is hard to obtain high density arrays with well-isolated features. Guidance of cylindrical micelles into grooves of equivalent dimensions prepared by e-beam has been shown. Currently, the approaches to extend such guided assembly of individual cylindrical micelles across large areas are still lacking.

Although the micelle arrays in their as-coated form yield only a polycrystalline order, this is not critical to most applications that can be derived out of copolymer thin films. For instance, much of the microelectronic applications shown by IBM group of workers using phase-separated thin films were demonstrated using polycrystalline 2D arrays. Typically, the lateral ordering of such polycrystalline 2D arrangement of copolymer domains spans up to a length scale of approximately 10 times the periodicity [18]. In comparison, an optimized coating of BCP micelles can already yield similar level of lateral ordering. Although the low feature heights of the micellar films pose a challenge for pattern transfer by nanolithography, this has been shown to be overcome by a first step transfer to achieve more robust inorganic masks [18,86].

9.2.5.3 High Throughput Nanometrology When patterning on large areas is concerned, one of the limitations is the measurement of quality of the resulting patterns across large areas. Typically, the nanopatterns produced by copolymers are characterized using high resolution techniques such as AFM, scanning electron microscopy (SEM), and transmission electron microscopy (TEM), which are low throughput and address very small areas. Although these techniques are indispensable for accurate measurements of geometric attributes of the patterns, it is not feasible to continuously map uniformity of the patterns across large areas, such as full wafer. Thus, it is desirable for quality control purposes, to consider alternative means of measuring uniformity of the patterns across macroscopic areas. Hexemer et al. related the characteristics of Moire patterns [87,88], namely, the periodicity and angle to the characteristics of the BCP patterns on surface, namely, orientation, grain size, and plane spacing using AFM images. The technique allows extraction of information from scan areas of AFM images that are greater by two orders of magnitude without sacrificing the measurement time [88]. An experimental demonstration of uniformity of BCP-derived Si nanopillars across full wafers was shown by Krishnamoorthy et al. [18] using optical reflectance. Variations of less than 10 nm in feature size, spacing, or height of the pillars could be measured between the spot areas spanning several square microns. The study theoretically and experimentally established the relationship between the variations in geometric variables with that of the resulting optical reflectance.

Many other techniques such as X-ray photoelectron spectroscopy (XPS), small-angle X-ray scattering (SAXS), grazing angle IR, UV–visible spectroscopy allow sampling several square microns to square millimeters of the surface at a time through averaged information derived from several million features per run. This allows them as techniques well suited to achieve high throughput measurements of variations in uniform nanoarrays spanning macroscopic areas of the surface. This is, however, possible only through adequate understanding of the correlation between the geometric variables of the nanoarrays and the spectroscopic output.

The sensitivity with which the spectroscopic technique can measure such variations would eventually determine its usefulness. Although not shown for the purpose, these techniques have been routinely used toward characterization of BCP thin films or their material nanopattern derivatives. Such high throughput nanometrology options, if developed further, can greatly save time and efforts in quality control when nanomanufacturing of BCP patterns over several batches of wafers, or sheets of substrates obtained by roll-to-roll processing, etc., is considered.

9.2.6 Top-Down Patterning of Block Copolymer Nanostructures

Top-down patterning of BCP nanostructures into features that are several microns wide to those that are only a few tens of nanometers across has been demonstrated. Such top-down patterning of BCP patterns has been pursued with varied end goals, for example, to prepare microfabricated support structures, confinement of self-assembled structures to pre-defined areas on surface, enabling long-range ordering and addressability and achieving multiple length scale surface structuring. Although techniques such as photolithography [26,89] and soft lithography [90] were used to create patterns with feature sizes of several tens of microns, nanostencils [82], NIL [91,92], E-beam [14,29,93], and AFM-based dispensing [84] have allowed patterning of BCPs down to sub-100 nm regime (Figure 9.2).

Glass et al. [93] showed arbitrary patterns of gold nanoclusters consisting of two, five, six, or eight clusters within each patterned feature, using e-beam irradiation of $HAuCl_4$-loaded BCP reverse micelles (Figure 9.2a). Using nanoscale dispensing of glycerol through a perforated AFM tip, Meister et al. demonstrated capability to selectively change morphology of a few micelles in a continuous array of PS-*b*-P2VP reverse micelles (Figure 9.2c). Glycerol is a solvent that is selective to the P2VP core, thereby causing a transformation in morphology owing to selective swelling. Such change in morphology across the entire film in response to selective solvents was shown earlier [85,97,98]. By controlling the duration of contact of the AFM tip with the surface, the authors exercise fine control over the quantity of glycerol dispensed on surface, and thereby selectively controlling the number of micelles that were eventually transformed. Krishnamoorthy et al. demonstrated capability to pattern arrays of silicon nanopillars derived out of BCP reverse micelles etch masks by performing selective protecting, removing, or etching through the pores of a nanostencil (Figure 9.2b). Nanostencil is a resistless patterning tool that is reported to perform facile patterning down to sub-100 nm regime by means of metal deposition through pores of a free standing silicon nitride membrane with a thickness of 200 nm or lesser. In addition, the authors also demonstrated patterned areas consisting of silicon nanopillars with a diameter lesser by a few nanometers in comparison to those around them. These features are optically brighter because of lower refractive index arising due to the lower volume fraction of Si nanopillars in comparison to those around the feature. Bennett et al. reported microcontact printing of $FeCl_3$-loaded PS-*b*-PAA reverse micelles using a PDMS mold to create micropatterned array of iron oxide catalysts for selective growth of carbon nanotubes (CNTs) (Figure 9.2e) [90]. The microcontact printing process

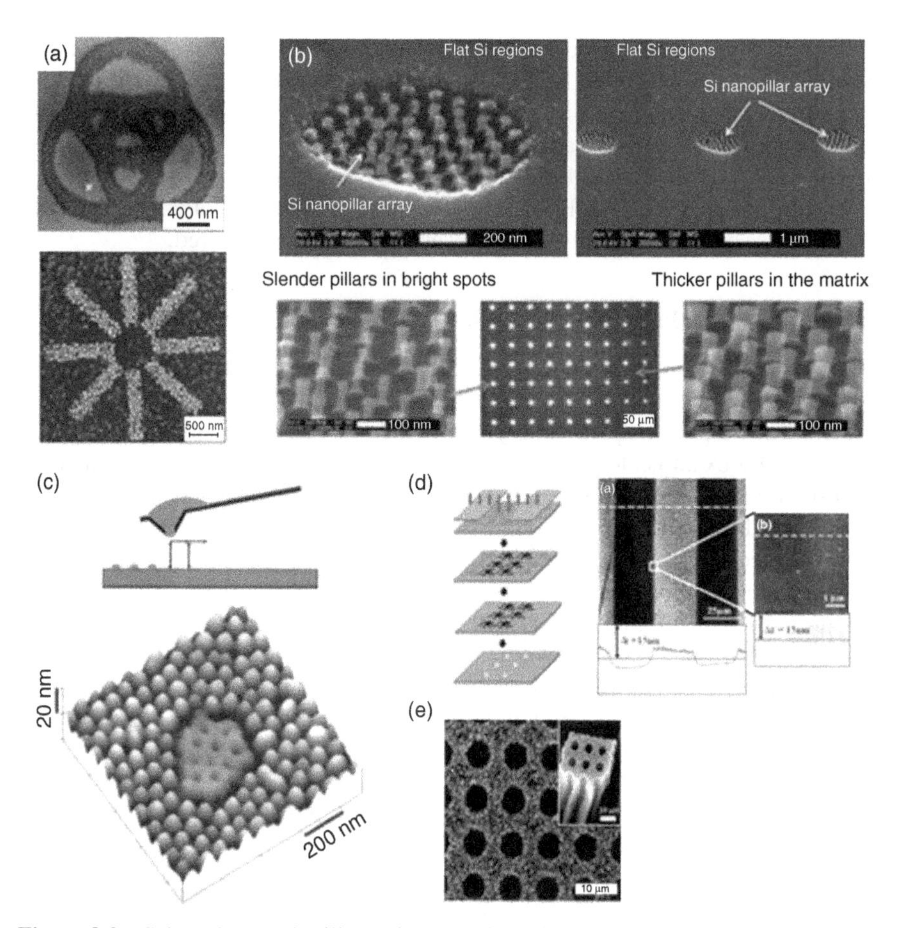

Figure 9.2 Selected examples illustrating use of top-down techniques to pattern for block copolymer nanostructures. (a) Electron beam lithography used to define selective immobilization of BCP reverse micelles (Reprinted with permission from [93,94]. Copyright © 2003, 2002 Wiley Periodicals Inc.) (b) Nanostencil lithography used to define areas for lithographic pattern-transfer of BCP templates by dry etching. (Reprinted with permission from [82]. Copyright © 2008 Wiley Periodicals Inc.) (c) Nanoscale dispensing through a hollow AFM tip (NADIS) to achieve selective morphology transformation of a reverse micelle array at pre-determined areas on surface, with resolution offered by the AFM tip. (Reprinted with permission from [84]. Copyright © 2006 Elsevier Ltd.) (Reprinted with permission from [95]. Copyright © 2004 AIP.) (d) Photolithographic patterning through selective cross-linking of an array of reverse micelles. (Reprinted with permission from [96]. Copyright © 2006 IOP Science.) (e) Microcontact printing of BCP templates used to derive patterned array of catalysts for carbon nanotube growth. (Reprinted with permission from [90]. Copyright © 2006 American Chemical Society.) (*See insert for color representation.*)

was shown to retain the micellar structure and resulting in four to five monolayers of PS-*b*-PAA in the transferred areas. In a process similar to that of the photolithography to pattern-negative resists, Gorzolnik et al. [96] reported micropatterned cross-linking of gold-salt-loaded BCP reverse micelles by selective exposure through a photo mask. The exposed regions get cross-linked thereby enabling them a high solvent resistance, and thus allowing the polymer in the unexposed regions readily removed by solvent treatment. The polymer is then removed to yield a micropatterned array of gold nanoparticles (Figure 9.2d). Cheng et al. patterned BCP thin films within shallow grooves defined by photolithography. Although process is known better for its ability to direct long-range ordering of BCP domains, it also doubles as capability to pattern BCP nanostructures or as their derivatives on addressable microarrays on surface [26]. In a similar process, Naito group of workers [91,92] have shown confinement of PS-*b*-PMMA films within circumferential pattern of grooves obtained by a NIL process. Xiang et al. [99] showed cylindrical confinement induced evolution in microdomain morphology of polystyrene-*block*-polybutadiene (PS-*b*-PBD) by patterning them within pores of AAO with pore diameters varying from 100–350 nm. The authors showed that such patterning to induce formation of concentric cylinders oriented along the nanorod axis with the number of such cylinders determined by the ratio of the nanorod diameter to the equilibrium period of the copolymer. In addition to addressability, such top-down patterning using conventional microfabrication processing has been employed to support fragile copolymer deriving ultrathin free standing nanoporous inorganic membranes [100]. This offered a patterned array of pores that permitted investigation of transport of molecules and also permitted chambers that can eventually be transformed into reservoirs for holding molecules [101,102].

9.3 SPECIFIC INSTANCES IN EXPLOITATION OF BLOCK COPOLYMERS IN DEVICE APPLICATIONS

9.3.1 Memory Devices

BCP self-assembly has immense potential toward contributing to next-generation bit-patterned media, as well as nonvolatile memory devices. The high achievable feature density greater than 1 Tb in.$^{-2}$ [11,43,103] with narrow feature size distribution has been demonstrated, amply supporting the potential of BCPs toward ultrahigh density memories. The directed self-assembly of BCPs is listed as a potential future alternative within the information technology roadmap for semiconductors (ITRS) [104]. Currently, the data storage industry carries out high volume production of magnetic disk platters with an areal density of 500 Gb in.$^{-2}$, with the next-generation hard disks aimed at 650 Gb in.$^{-2}$ (allowing 1 Tb per platter for a 3.5-in. drive). A comprehensive review highlighting advantages of patterned magnetic media to achieve high memory densities up to 1 Tbit in.$^{-2}$ and means of fabricating magnetic arrays with high pattern resolutions was published by Ross [105]. BCPs are an excellent solution for patterned media, as they allow readily accessing sub-50 nm pattern resolutions cost-effectively over large areas. Thurn-Albrecht et al. [106] showed markedly enhanced coercivities in cobalt nanowire arrays with density exceeding 1.9×10^{11} per cm^2 created by electrodeposition through BCP-derived high aspect ratio nanopores. The approach offers promising route to single-domain and ultrahigh density magnetic storage that also effectively counters superparamagnetism through the high aspect ratio of Co nanowires. Liu et al. [107] fabricated nanoporous Fe with a pore width and separation of 20 nm on continuous FeF_2 layer using PS-*b*-PMMA-derived templates. The authors show that the exchange bias can be tuned using magnetic nanostructures with sizes below typical magnetic domain sizes. Cheng et al. [108] had employed a sequential pattern transfer of PS-*b*-PFS-based copolymer template through intermediate SiO_2/W multilayer to structure underlying Co thin films. The approach is versatile and can be readily extended to a range of other thin films, especially since the pattern transfer process involves ion-beam milling, which is nonspecific to the material being patterned. Through a combination of NIL and BCL, Naito et al. [91,92] fabricated circumferential patterns of magnetic nanostructures and found an increase in perpendicular coercivity for the patterned media in comparison to their unpatterned thin film counterparts. Liu et al. [22] demonstrated directed self-assembly of copolymers on EBL-derived chevron patterns for patterned media applications.

An alternative to bit-patterned magnetic media is flash memory-based solid state disks that store information as charges (Figure 9.3) [109–112]. Flash memories offer attractive possibilities such as quick access times, better resistance to thermal shocks, low noise because of lack of mechanical moving parts, yet are more expensive compared to that of the magnetic media. Nonvolatile memories that store information even in the absence of power can be either nanocrystal flash (NCF) or charge trap flash (CTF) memories. NCF relies on storage of charges within metal or semiconductor nanocrystals, while CTF devices store charges within defect levels present in a continuous semiconductor or dielectric medium. Guarini et al. [113] reported the use of PS pore arrays derived out of PS-*b*-PMMA phase-separated thin films to produce amorphous Si nanocrystals for flash memory device applications. Si nanocrystals with a feature width of 20 nm, and center-to-center spacing of 40 nm, with a density of 6.5×10^{10} per cm^2 were achieved in this case. Excellent performance by the device, with a ΔV_{FB} >0.5 V for write voltages, $|V_w| < 4$V, high retentivity >10^6 s for a program oxide as thin as 2 nm, and an endurance >10^9 cycles were demonstrated. In an attempt to produce ultrahigh bit density metal nanocrystal-based memory devices, Hong et al. demonstrated Cr nanodot arrays using the similar approach as used by Guarini et al (Figure 9.3a). The resulting device exhibited an ultrawide memory window of 15 V at ± 18 V sweep, with high promise for multilevel cell operations [109]. Leong et al. [114,115] used arrays of Au nanoparticles synthesized within PS-*b*-P4VP reverse micelles as charge storage components within organic thin film transistor (OTFT)-based memory based on metal–pentacene–insulator–silicon (MPIS) configuration. These arrays consisted of nanoparticles with an estimated density of $2.6 - 6.7 \times 10^{10}$ cm^2. The group reported a memory window of 2.1 V, with retention of 92% over 60,000 s.

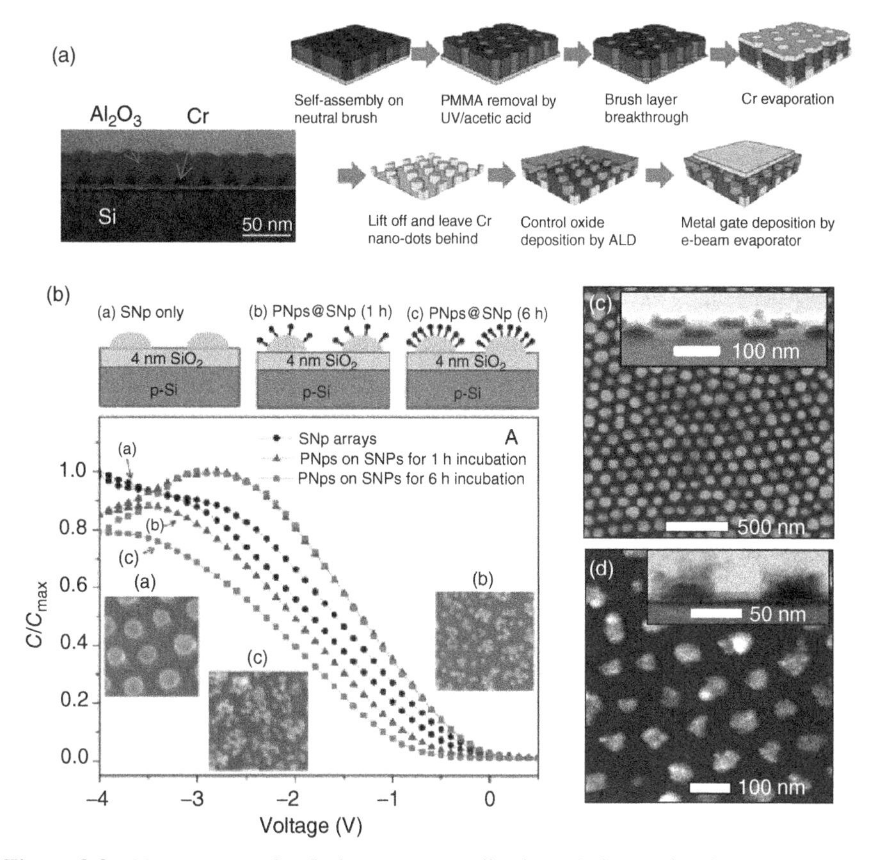

Figure 9.3 Nanopatterns for flash memory applications derived using block copolymer lithography employing (a) Cr nanocrystals. (Reprinted with permission from [109]. Copyright © 2010 American Chemical Society.) (b) Au nanoparticle clusters. (Reprinted with permission from [110]. Copyright © 2012 Royal Society of Chemistry.) (c,d) ZnO nanostructures. (Reprinted with permission from [111]. Copyright © 2012 American Chemical Society.) (Reprinted with permission from [112]. Copyright © 2012 Royal Society of Chemistry.) (*See insert for color representation.*)

Gupta et al. demonstrated hierarchical organization of citrate-stabilized gold nanoparticles around primary gold nanoparticle array templates produced using BCP reverse micelle route (Figure 9.3b). The electrical characteristics of the MIS device showed a hysteresis of 0.99 V, at a bias of ± 4 V and exhibited high retention up to 90% for 20,000 s [110]. Although the spatial isolation of charge storage centers in nanocrystal memories permits better retention because of isolation of charges and reduced stress-induced leakage currents, the CTF memories benefit from a high density of discrete trap levels that can yield high charge storage capacity. The state-of-the-art CTF memories make use of discrete nitride levels in polysilicon–oxide–nitride–oxide–silicon (SONOS) stack to store charges. The low dielectric constant of silicon nitride (7.5) coupled with small conduction band offset (1.05 eV) with SiO_2 justifies exploring alternative materials to attain better capacity and retention. Suresh et al. recently proposed nanoarrays of ZnO prepared by different BCP lithography routes as means of benefitting from both discrete charge storage centers as in NCF memories and the rich density of discrete trap levels as in CTF memories (Figure 9.3d,e). The charge storage characteristics of the resulting ZnO arrays showed memory capacities similar to that of the SONOS memories along with excellent retention [111,112].

9.3.2 Integrated Circuit Elements

BCP lithography has been found promising for exploitation within high capacity decoupling capacitors, nanowire FETs, and essential geometries for integrated circuit components. Black et al. [73] had demonstrated silicon capacitors with enhanced charge storage capacity by increasing the surface area through an array of pores created using BCP lithography. The enhanced capacitance at smaller device footprint enables shrinking of device dimensions of on-chip capacitors, which is sought after in dynamic random access memory (DRAM) storage nodes, on-chip decoupling capacitors, and ferroelectric memories. The use of BCL allows for a systematic and programmed means of achieving high capacity, as the increase in surface area is proportional to the feature width and pitch (determined by the copolymer system used) and the depth of the pores (determined by the etch duration). The work reported a 30% increase in capacitance of the metal–oxide–silicon (MOS) capacitors incorporating nanostructured electrodes in comparison to their planar counterparts. These authors later showed further improvement of the capacitance of MOS decoupling capacitors by 400% in comparison to the planar counterparts through increase in aspect ratio of the pores [116]. Later, nanolithography using directed self-assembly of cylinder forming BCPs was used to demonstrate Si nanowire FETs. By systematically varying the width of the lithographically defined guiding pattern in relation to the L_0 of the polymer domains, the work demonstrated 6, 8, 10, and 16 wires. The nanowire *n*-channel devices were found to exhibit current drives of ~five microamperes per wire and an on/off ratio of $\sim 10^5$. Stoykovich et al. [28] later demonstrated a fruitful combination of directed self-assembly of BCPs on chemically patterned surfaces to produce the entire series of essential geometries for integrated circuit layouts. The authors demonstrated nested arrays of jogs, isolated jogs, and T-array junctions using directed self-assembly of lamellar forming PS-*b*-PMMA system. Chang et al. [117] later demonstrated the use of BCP lithography to define contact holes of MOSFETs (metal oxide-semiconductor field effect transistors). Another aspect of high importance to microelectronics is the ability to attain low line-edge roughness (LER). LER signifies the quality of the lithographic pattern, with an expected specification being $3\sigma < 1.4$ nm [104]. The use of chemical patterning to direct the self-assembly of copolymers has shown to reduce LER [118]. The LER of features produced by BCL is limited by the intrinsic roughness of its domains, which in turn is related to the finite interfacial width. Increasing the Flory–Huggins parameter allows decreasing the interfacial width thereby contributing to lower LER. Thus, some of the specific capabilities of interest to microelectronic devices are (i) ability to create high density features, (ii) ability to direct self-assembly to achieve long-range ordering and registration with respect to a lithographically defined overlay, (iii) to have a desired level of density multiplication through the width of the lithographic pattern in relation to the BCP pitch, and (iv) to reduce LER of the lithographic patterns.

9.3.3 Photovoltaic and Optoelectronics Applications

BCPs consisting of semiconducting blocks or material nanopatterns derived out of BCP templates on surfaces have immense potential toward catering to optical devices through control over photonic or photophysical properties. Photovoltaic (PV) devices such as solar cells or photo detectors, on-chip wave guides, photonic crystals, and LEDs are some of the devices that can benefit from this capability. There are a large number of studies focusing on interpenetrating morphologies offered by conducting BCPs for exploitation within bulk heterojunction (BHJ) organic PV devices [119]. Such BHJ PV devices with interpenetrating architectures of semiconducting donor and acceptor features with width of the order of exciton diffusion lengths are highly attractive given their ability to split excitons. Although much of the work on BHJ devices has involved disordered architectures, there are several incentives to fabricate ordered BHJ devices [120]. An ordered BHJ allows uniformity in feature size and separation and thus puts all areas within the film to be within the exciton diffusion length of the interface between the constituent semiconductors. Furthermore, the ordered structures are readily amenable to modeling and simulation to engineer the geometric attributes to achieve high device efficiency. BCP self-assembly allows several opportunities to arrive at ordered semiconducting architectures. Few instances have however been shown using thin films of BCP or BCP-derived nanostructures as part of hybrid solar cells. Infiltration of semiconducting regioregular poly(3-hexylthiophene-2,5-diyl) (P3HT) into mesoporous titania prepared using pluronic polymers as structure-directing agents in a sol–gel process was shown [121,122]. The combination of the BCP self-assembly with sol–gel process allows facile means of realizing inorganic semiconducting nanoporous structures with uniform pore dimensions, high pore densities, along with desired values for film thickness. The study however showed that the confinement of the P3HT within the pores with a diameter of $\sim$10 nm was found to reduce the mobility of charge carriers. The fall in mobility values was attributed to the unfavorable chain conformations adopted by the P3HT chains in response to the geometric constraints imposed by the nanopores, which disrupts its π-stacking efficiency. It is however envisaged that the lower mobilities can be countered through variation in infiltration of the polymer, as well as by surface modification of the inorganic semiconductor. Many other studies showing BHJ architectures using other techniques can be strengthened using BCP self-assembly approaches. For instance, the use of high aspect ratio silicon or ZnO nanopillar arrays have been shown for solar cells [123–125]. Such nanopillars or nanowire arrays with ultrahigh densities and uniform diameters can be realized using vapor–liquid–solid (VLS) growth on catalyst particle arrays positioned on substrates using the BCP approach. Many accounts of structuring silicon using reactive ion etching processes have also been shown, with profiles that can be controlled by varying the dry etching parameters [15,18,82,85,126]. Nanopillar arrays with conical, vertical, or under-cut profiles can be realized. The conical shapes of nanopillars allow for a graded variation in refractive index and result in moth-eye like antireflection phenomenon [18]. One of the challenges in such pattern transfer is to achieve high transfer depths of the order of 200–300 nm to enable a good optical absorption.

Apart from PV devices, BCP-derived patterns are sought after as means of enhancing the luminosity or brightness of LEDs. Although LEDs with high internal quantum efficiency have been realized, a large proportion of the light produced at the donor–acceptor interface of the light-emitting diode (LED) is trapped within the device by total internal reflection caused by the large refractive index contrast between the LED material and air. This can be countered by creating surface nanostructures that contribute to enhanced light extraction efficiency through a combination of antireflection and diffraction effects. These effects can be tailored to achieve a high efficiency of light extraction through a controlled variation in the geometric attributes, namely, feature size, pitch, and height of the features by either varying copolymer or the dry etching conditions. The refractive index can be "fine-tuned" entirely through such control over geometric attributes of BCP lithography-derived semiconductor pillar or pore arrays to desired values between the index of the semiconductor material and air [127,128]. Alternatively, the BCP templates by themselves can be used for the purpose, with a limited range of tunability in index [129,130]. Furthermore, the ordered structures obtained using BCPs present facile opportunities for modeling to determine the attributes before fabrication is performed.

9.3.4 Sensors

The applications of BCP patterns for sensing applications have been gaining significance recently. Novel transduction schemes and fabrication approaches have been reported in the pursuit of realizing sensors with high sensitivity and selectivity (Figure 9.4) [131–133]. Jung et al. showed that chemiresistor devices incorporating conducting polymer nanowires (15 nm wide, 35 nm period) made of poly(3,4-ethylenedioxythiophene) : poly(styrenesulfonate) (PEDOT : PSS) exhibited higher sensitivity toward ethanol vapor as compared to that of unpatterned counterparts (Figure 9.4c). The chemiresistor device recorded conductivity of the nanowires as it reversibly swells upon exposure to the organic vapors. The nanowires were fabricated by BCP lithography using PS-*b*-PDMS-derived etch masks [63]. The authors argue that the higher sensitivity results from the combined influence of kinetic factors owing to lower characteristic diffusion lengths of the organic vapors within the nanowires as well as higher equilibrium solubility of the vapors at the surface of the nanowires, which by the virtue of their geometry exhibit a high surface-to-volume ratio. Ayyub et al. [131] demonstrated a colorimetric sensor for simple sugars by exploiting the swelling-induced variation in the interlamellar distance of boronic

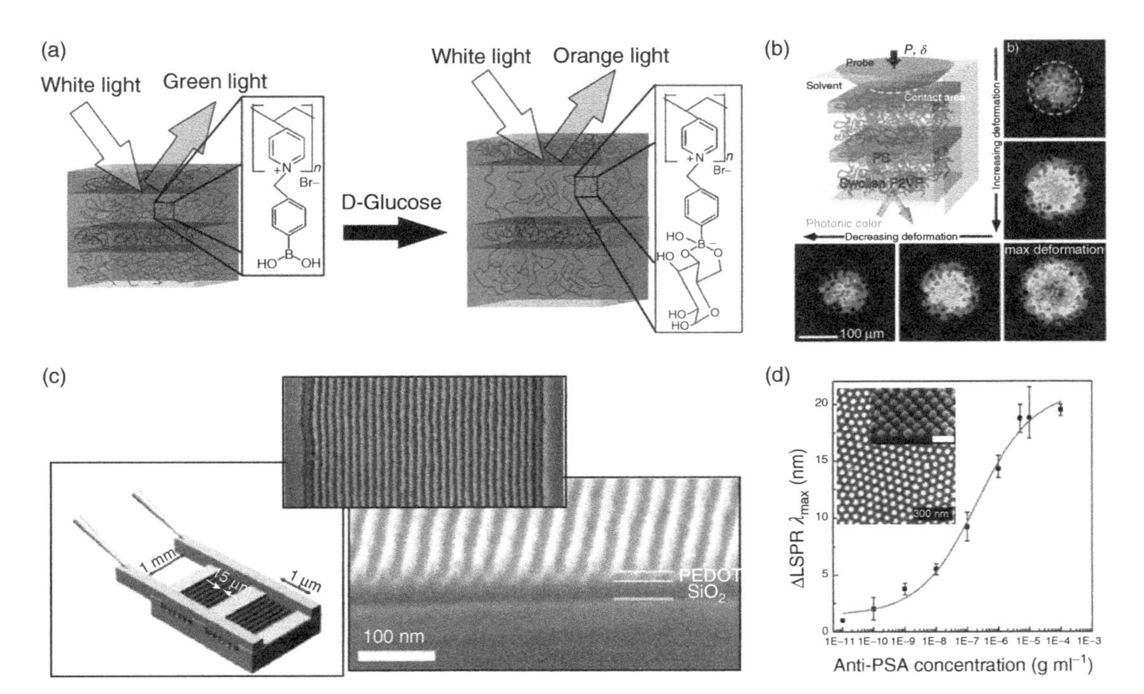

Figure 9.4 Selected instances in the use of block copolymers in sensor devices (a) Photonic gel for colorimetric sensing glucose. (Reprinted with permission from [131]. Copyright © 2011 Elsevier Ltd.) (b) Photonic sensor for transducing mechanical deformations. (Reprinted with permission from [132}. Copyright © 2011 Wiley Periodicals Inc.) (c) Gas sensing using nanowires of conjugated organic films. (Reprinted with permission from [63]. Copyright © 2008 American Chemical Society.) (d) Plasmonic sensor for quantifying Prostate Specific Antigen (PSA). (Reprinted with permission from [133]. Copyright © 2010 Royal Society of Chemistry.) (*See insert for color representation.*)

acid containing lamellar diblock copolymer system (Figure 9.4a). The selectivity of the sensor to sugars arises from the affinity of sugars to the boronic acid moiety. The difference in individual affinities allowed the sensor to exhibit different wavelength response toward glucose, fructose, and galactose. The approach is generic and can be readily extended to other chemical targets by imparting one of the copolymer domains with selectivity or specificity to the targeted analyte. Almost parallel, a similar concept on the same copolymer system was reported by Chan et al. [132]. The latter group of workers showed photonic gel for colorimetric transduction of mechanical stress (Figure 9.4b). Nanopatterning of proteins and other biomolecules through spatially selective binding to copolymer domains and studying their activity is interesting for biological sensors. Kumar et al. [134,135] investigated the immobilization of different proteins on phase-separated PS-*b*-PMMA thin films and found them to preferentially attach to the PS domains and that they retained their activity. A similar patterning of histidine-tagged proteins to norbornene-containing copolymer domains in the presence of Ni^{2+} ions was reported by Cresce et al. [136].

Recently, there has been significant interest in the use of BCP lithography to fabricate high-resolution metal nanopatterns for surface enhanced Raman spectroscopy (SERS) (Figure 9.5). SERS is a plasmonic phenomenon wherein the surface plasmon excitation of metal nanostructured surface contributes to greater than a million-fold enhancement in the Raman signals of analytes that are either in physical contact or in close vicinity to the metal surface. SERS significantly benefits from a nonlinear electromagnetic field enhancement that occurs at sharp corners or junctions of two or more metal nanostructures. BCP lithography provides excellent capabilities to cater to SERS substrates owing to their ability to deliver high density of nanostructures with feature-separations or curvatures down to sub-10 nm regime and exhibiting very low variability in nanostructure density and geometric attributes. A wide array of tools at the disposal of BCP self-assembly technique allows realizing a wide range of metal nanostructures with fine-tunable optical properties. Furthermore, the excellent definition (or narrow distribution in geometric attributes) of the metal nanostructures allows the fabrication to be rationally designed through simulations. The tunability in geometric variables of the metal arrays allows engineering their optical response towards maximizing SERS performance. A variety of metal structures have been realized, for example, metal thin films of controlled roughness, nanorods, nanodisks, and arrays of individual nanoparticles or particle aggregates with ultralow feature separations (Figure 9.5). Lu et al. employed nanostructured silver substrates for SERS by evaporating silver metal on top of iron-containing silicon oxide nanoparticles derived out of polystyrene-*b*-polyferrocenylsilane diblock copolymer (Figure 9.5c) [141]. The authors showed an SERS enhancement factor of $\sim 10^6$ at 514 nm for benzene thiol as a probe molecule. In their study, silver thickness of 15 nm on the top of the oxide nanoparticle arrays of 45 nm pitch and 30 nm diameter was found optimal for high SERS enhancements. Lee et al. showed single Au nanoparticles per domain of quarternized P4VP-containing copolymers through electrostatic attraction suitable for SERS. By varying the size of the nanoparticles, the separation between

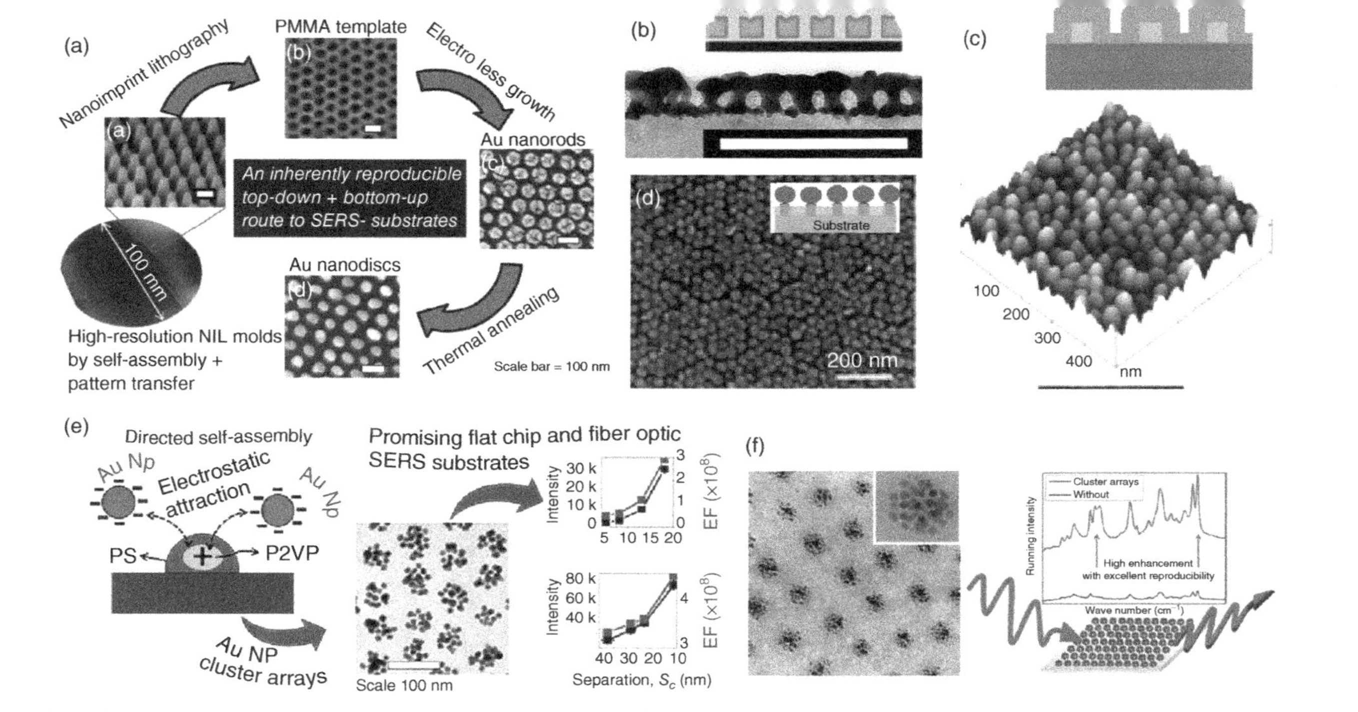

Figure 9.5 High resolution metal nanoarrays for SERS, fabricated using block copolymer thin films, using (a) BCP-NIL process to define regions for electroless Au deposition. (Reprinted with permission from [76]. Copyright © 2011 American Chemical Society.) (b) Morphology transformed PS-*b*-PVP thin films as templates for electroless Au deposition to attain mushroom-shaped structures. (Reprinted with permission from [140]. Copyright © 2009 American Chemical Society.) (c) Nanostructured silver films derived from PS-*b*-PFS thin films. (Reprinted with permission from [141]. Copyright © 2006 IOP Science.) (d) Electrostatic self-assembly of pre-formed Au nanoparticles on to BCP templates to form nanoparticle arrays with controlled separations. (Reprinted with permission from [137]. Copyright © 2011 Wiley Periodicals Inc.) (e) Electrostatic self-assembly of pre-formed Au nanoparticles on to BCP templates to form gold nanoparticle cluster arrays. (Reprinted with permission from [56]. Copyright © 2012 American Chemical Society.) (f) Silver nanocluster arrays through *in situ* synthesis within BCP reverse micelles on surface. (Reprinted with permission from [138]. Copyright © 2012 American Chemical Society.) (*See insert for color representation.*)

the particles could be controlled down to 2 nm, and plasmon resonance varied between 520 and 1000 nm (Figure 9.5d) [137]. Such electrostatic immobilization to achieve an array of close-packed nanoparticle aggregates (or clusters) with systematically tunable cluster size (defined by number of nanoparticles per cluster) and cluster separations was recently shown by Krishnamoorthy et al. (Figure 9.5e) [56]. The cluster formation occurred through electrostatic attraction experienced by Au nanoparticles toward positively charged copolymer reverse micelle features on substrate. The interparticle separations within the clusters were shown to be down to sub-5 nm regime. By tuning the self-assembly conditions, intercluster separations were shown down to sub-10 nm regime. The cluster arrays exhibited high SERS performance owing to high particle densities of 7×10^{11}per cm^2 and strong plasmonic coupling within and between the clusters. The approach was further transferable to the tip of an optical fiber, allowing high SERS performance also in remote sensing configurations. The SERS enhancements $>10^8$ on flat chips and $>10^7$ on optical fibers were observed using crystal violet (CV) as the probe molecule, at 633 nm wavelength. Almost parallel, nanoparticle clusters of silver were demonstrated by Cho et al. [138] using *in situ* synthesis within BCP templates (Figure 9.5f). These authors also showed an SERS enhancement factor as high as 10^8, using CV as a probe, at 514.5 nm. In a similar instance of *in situ* growth to create SERS substrates, Sanchez-Iglesias et al. [139] had shown the use of Au nanoparticle seeds obtained by reverse micelle route to seed growth of silver for producing SERS substrates. A novel use of BCP reverse micelle-enabled high resolution NIL was reported to create nanoporous templates for guided *in situ* electroless growth of gold nanorods [76]. These nanorods could be converted to nanodisk arrays with an ellipsoidal morphology by thermal annealing. The disk arrays proved effective for SERS, and their performance was confirmed using three different probe molecules that exhibit different optical characteristics and interaction modes on the surface. The authors found the nanorod arrays to yield a high SERS signal, yet, were found to disintegrate due to capillary forces experienced during incubation in aqueous solutions. This was discovered only by a post-investigation of the substrates after the SERS experiment, thus marking the need for such investigations to be routinely performed in order to ensure the quality of SERS substrates. The approach of combining NIL with BCP self-assembly to create SERS substrates provides a unique means of ensuring reproducibility, as identical patterns are delivered each time (Figure 9.5a). However, adequate care needs to be exercised that the electroless deposition and other pattern transfer steps involved in generating metal arrays do not introduce variability. Although there have been impressive first achievements in realizing SERS substrates using BCPs by means of novel pattern transfer routes, significant efforts are necessary before reliable SERS sensors suitable for real-life deployment become possible. In contrast to much of the plasmonic sensors using BCP have been reported for SERS applications, Shin et al. [133] demonstrated a localized surface plasmon resonance (LSPR)-based refractive index sensor for prostate cancer detection (Figure 9.4d). A detection limit in the range 0.1 – 1 ng ml^{-1} was achieved by these authors, with an observed sensitivity of 40 nm/RIU (or 40 nm shift in resonance wavelength for unit change in refractive index of the medium). Despite the excellent template uniformity readily attained through the BCP approach, the pattern transfer routes often act as sources of geometric variability. Thus, accurate batch-to-batch reproducibility of the processing conditions is critical in translating the <10% standard deviations achievable with BCP templates to that in sensing signal variability. Further investigations toward use of these sensors for detection of analytes in complex matrices such as food or biofluids (e.g., urine, saliva, and blood) are required for successful deployment in real-life scenario. This would necessitate utilization of methods beyond nanofabrication, such as data processing algorithms, microfluidic functions, and surface functionalization with designer molecules.

9.3.5 Nanoporous Membranes for Size-Exclusive Filtration or Sensing

One of the interesting means for exploitation of BCPs lies in creating nanoporous membranes to exercise size-selective control over transport of molecules across an interface for either filtration or detection purposes (Figure 9.6). The use of BCPs is promising as it allows realizing nanopores with ultrasmall feature dimensions tunable in the sub-100 nm regime, and with high feature densities (and consequently high porosities) in a low cost manner. Membranes consisting of 2D array of pores, as well as 3D interconnected porous networks have been shown in the literature. An extensive review of the nanoporous materials from BCP precursors is covered by Hillmyer [142]. BCP-derived ultrathin nanoporous membranes consisting of 2D arrays of nanopores often require a mechanical support for investigation of the

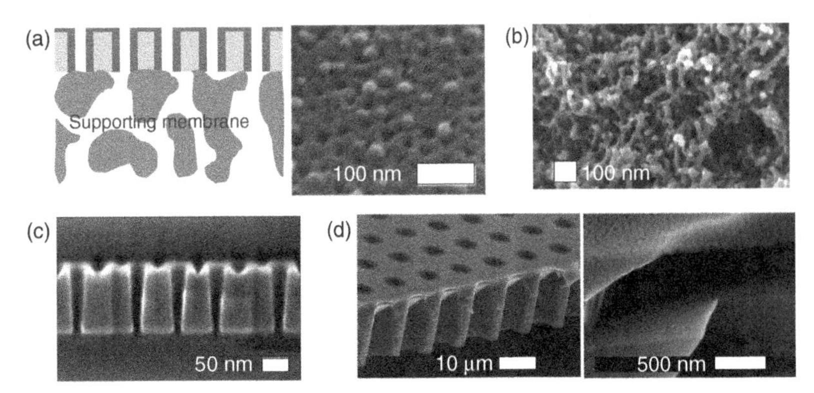

Figure 9.6 Nanoporous membranes for size-exclusive filtration or sensing obtained using block copolymers demonstrating (a) viral separation on asymmetric pores. (Reprinted with permission from [144]. Copyright © 2008 Wiley Periodicals Inc.) (b) Percolating pore structure and high mechanical stability. (Reprinted with permission from [145]. Copyright © 2007 American Chemical Society). (c) Suspended nanoporous inorganic membranes of silicon. (Reprinted with permission from [69]. Copyright © 2006 AVS Publications). (d) Suspended nanoporous inorganic membranes reinforced with silicon nitride polysilicon sandwich. (Reprinted with permission from [83]. Copyright © 2007 IEEE.) (*See insert for color representation.*)

pore characteristics, as well as their final utilization in filtration or sensing devices. Yang et al. [143] showed a double-layered nanoporous membrane suitable for filtration of human rhinovirus type 14 (HRV14). The virus has a diameter of 30 nm and is a major pathogen of common cold in humans. The membranes with a pore diameter of 15 nm diameter and 150 nm thickness were transferred and supported on an MF polysulfone (PSU) macroporous membrane with thickness of 150 μm. The study expressed promise toward concentration of viruses by filtration to enable their detection in patient samples with low viral concentrations. The flux of DI water and PBS containing 2.5×10^5 plaque forming units (PFU ml^{-1}) through an 80 nm thick BCP-derived nanoporous membrane was found to be over 10 times higher than that of a track-etched PC membrane (6.5 μm thick). The higher flux was accounted for by the ultralow thickness as well as high porosity (>20%) of the BCP films in comparison to the PC membrane (2%). Although the investigation has shown impressive capability for selective filtration through nanoporous membranes, the mechanical stability of the films is reduced and the membranes could withstand a pressure of only 0.1 bar. The authors later countered this limitation by employing a nanoporous membrane with a thickness of 160 nm, [144] consisting of a mix of parallel and perpendicular orientations of pores (Figure 9.6a). In addition, the membrane material, which was PS, was cross-linked by UV irradiation. The authors found that the mixed orientation resulted in significantly enhanced mechanical stability, with the membranes sustaining a pressure up to 2 bar, with a maximum realizable flux of 380 $l\,m^{-2}h^{-1}$. This flux is over 10 times what was achieved in their earlier work. The study further showed that the cross-linking of the PS matrix did not contribute to an increase in mechanical stability (as they made them more brittle), but resulted in excellent thermal and chemical stabilities toward various organic solvents at their boiling point for 6 h and toward strong acids and bases. Nanoporous membranes with 17 nm mean diameter and ~10% standard deviations, exhibiting a percolating pore structure, were realized by reacting a thermosetting monomer, dicyclo pentadiene with a doubly reactive BCP poly(lactide)-*b*-poly(*p*-norbornenylethylstyrene-*s*-styrene) [145] (Figure 9.6b). The membranes were very thick and exhibited a tensile strength of 32 MPa, while the stability of membranes at lower thicknesses is not known. The membranes were shown to have high porosities of ~41%, and thus exhibiting promise for high throughput separation processes. In an effort toward stable inorganic membranes, Black et al. demonstrated nanoporous Si membranes with tunable pore diameters of 20 nm (Figure 9.6c) [69]. The pore sizes could be reduced independent of pore density through stress-limited thermal oxidation of Si to 12 nm. CSEM group of workers later showed transfer of reverse micelle-based templates to create pores in ultrathin silicon nitride membranes using MEMS (microelectromechanical systems) compatible processes over complete wafers (Figure 9.6d) [146]. Porosity of 12% was demonstrated, with a promise of increasing the value by ~25% using reverse micelle patterns with shorter pitch. The same group also reported an improved processing to produce reinforced nanoporous membranes consisting of silicon nitride–polysilicon sandwich structure [83]. The special sandwich design not only allowed for pore size reduction by thermal oxidation of poly-Si layer, but also reduced the associated overall stress and contributed to better stability. Such pore size reduction can also be achieved overcoatings [69], while the roughness of such coatings has a contribution to the mechanical stability of the membrane [147]. Work reported by Tong et al. provides insights into the design and optimization of the geometry of the silicon nitride membrane, namely, pore diameter, pitch, membrane thickness, width of the silicon frame used as support, and thickness of the support so as to achieve a high mechanical stability [147]. With an appropriate design, the use of BCPs to realize silicon-based nanoporous membranes offers significant promise in enabling high throughput (or transport flux), selectivity, and robustness for filtration applications. Besides examples described above, BCP approaches can efficiently cater to other nanopore based analytical platforms reported in literature [148,149]. For the latter, it is useful if the membranes are much thinner in order to be more sensitive. For this purpose, the silicon nitride membranes are more promising in comparison to the polymeric membranes, given that an equivalent mechanical stability as polymer membranes is achieved with much thinner silicon nitride membranes.

9.4 CONCLUSIONS

The application of block copolymer (BCP) derived nanopatterns toward real-life technologies has significantly grown in the past decade. The capabilities constituting the block copolymer toolbox have evolved over years, delivering several impressive achievements recently. This includes capability toward patterning at feature and spatial resolutions down to sub-10 nm regime, across arbitrarily large areas, with freedom over choice of materials and substrates. This chapter reviewed some of the key considerations in exploitation of block copolymer patterns for technologies, discussed means of ensuring scalability, throughput and quality control in fabrication, citing concrete examples of applications in electronic, optical, sensor and filtration devices. Challenges continue to exist in ensuring integrity of the pattern-transfer processing at spatial resolutions approaching length scales of the order of a few nanometers, placing significant emphasis on process development at these scales.

REFERENCES

1. Lodge T. P. (2003). Block copolymers: past successes and future challenges. *Macromol. Chem. Phys.*, *204*, 265–273.
2. Black, C. T., Ruiz, R., Breyta, G., Cheng, J. Y., Colburn, M. E., Guarini, K. W., Kim, H. C., Zhang, Y. (2007). Polymer self assembly in semiconductor microelectronics. *IBM J. Res. Dev.*, *51*, 605–633.
3. Hamley, I. W. (2003). Nanostructure fabrication using block copolymers. *Nanotechnology*, *14*, 39–54.
4. Darling, S. B. (2007). Directing the self-assembly of block copolymers. *Prog. Polym. Sci.*, *32*, 1152–1204.

5. Forster, S. (2003). Amphiphilic block copolymers for templating applications. *Coll. Chem. 1 Top. Curr. Chem.*, *226*, 1–28.
6. Krishnamoorthy, S., Hinderling, C., Heinzelmann, H. (2006). Nanoscale patterning with block copolymers. *Materials Today*, *9*, 40–47.
7. Hamley, I. W. *The Physics of Block Copolymers*. Nineteenth ed. Oxford University Press, New York, 1998.
8. Helfand, E., Tagami, Y. (1972). Theory of the interface between immiscible polymers. *J. Chem. Phys.* *57*, 1812–1813.
9. Russell, T. P., Hjelm Jr, R. P., Seeger, P. A. (1990). Temperature dependence of the interaction parameter of polystyrene and poly (methyl methacrylate). *Macromolecules*, *23*, 890–893.
10. Eitouni, H. B., Balsara, N. P., Hahn, H., Pople, J. A., Hempenius, M. A. (2002). Thermodynamic Interactions in organometallic block copolymers: poly (styrene-block-ferrocenyldimethylsilane). *Macromolecules*, *35*, 7765–7772.
11. Park, S., Dong, H. L., Xu, J., Kim, B., Sung, W. H., Jeong, U., Xu, T., Russell, T. P. (2009). Macroscopic 10-terabit-per-square-inch arrays from block copolymers with lateral order. *Science*, *323*, 1030–1033.
12. Helfand, E., Tagami, Y. (1971). Theory of the interface between immiscible polymers. *J. Polym. Sci. B Polym. Lett.*, *9*, 741–746.
13. Krishnamoorthy, S., Pugin, R., Hinderling, C., Brugger, J., Heinzelmann, H. (2008). The systematic tunability of nanoparticle dimensions through the controlled loading of surface-deposited diblock copolymer micelles. *Nanotechnology*, *19*, 75301–75304.
14. Kastle, G., Boyen, H. G., Weigl, F., Lengl, G., Herzog, T., Ziemann, P., Riethmuller, S., Mayer, O., Hartmann, C., Spatz, J. P., Moller, M., Ozawa, M., Banhart, F., Garnier, M. G., Oelhafen, P. (2003). Micellar nanoreactors – preparation and characterization of hexagonally ordered arrays of metallic nanodots. *Adv. Funct. Mater.*, *13*, 853–861.
15. Krishnamoorthy, S., Gerbig, Y., Hibert, C., Pugin, R., Hinderling, C., Brugger, J., Heinzelmann, H. (2008). Tunable, high aspect ratio pillars on diverse substrates using copolymer micelle lithography: an interesting platform for applications. *Nanotechnology*, *19*, 85301–85307.
16. Peng, J., Gao, X., Wei, Y. H., Wang, H. F., Li, B. Y., Han, Y. C. (2005). Controlling the size of nanostructures in thin films via blending of block copolymers and homopolymers. *J. Chem. Phys.*, *122*, 114706–114712.
17. Krishnamoorthy, S., Pugin, R., Brugger, J., Heinzelmann, H., Hinderling, C. (2006). Tuning the dimensions and periodicities of nanostructures starting from the same polystyrene-block-poly(2-vinylpyridine) diblock copolymer. *Adv. Funct. Mater.*, *16*, 1469–1475.
18. Krishnamoorthy, S., Manipaddy, K. K., Yap, F. L. (2011). Wafer-level self-organized copolymer templates for nanolithography with sub-50 nm feature and spatial resolutions. *Adv. Funct. Mater.*, *21*, 1102–1112.
19. Ruiz, R., Kang, H., Detcheverry, F. A., Dobisz, E., Kercher, D. S., Albrecht, T. R., de Pablo, J. J., Nealey, P. F. (2008). Density multiplication and improved lithography by directed block copolymer assembly. *Science*, *321*, 936–939.
20. Bita, I., Yang, J. K. W., Jung, Y. S., Ross, C. A., Thomas, E. L., Berggren, K. K. (2008). Graphoepitaxy of self-assembled block copolymers on two-dimensional periodic patterned templates. *Science*, *321*, 939–943.

21. Yang, J. K. W., Jung, Y. S., Chang, J. B., Mickiewicz, R. A., Exander-Katz, A., Ross, C. A., Berggren, K. K. (2010). Complex self-assembled patterns using sparse commensurate templates with locally varying motifs. *Nat. Nanotechnol.*, *5*, 256–260.
22. Liu, G., Nealey, P. F., Ruiz, R., Dobisz, E., Patel, K. C., Albrecht, T. R. (2011). Fabrication of chevron patterns for patterned media with block copolymer directed assembly. *J. Vac. Sci. Technol. B*, *29*, 204–210.
23. Cheng, J. Y., Rettner, C. T., Sanders, D. P., Kim, H. C., Hinsberg, W. D. (2008). Dense self-assembly on sparse chemical patterns: rectifying and multiplying lithographic patterns using block copolymers. *Adv. Mater.*, *20*, 3155–3158.
24. Segalman R. A., Yokoyama, H., Kramer, E. J. (2001). Graphoepitaxy of spherical domain block copolymer films. *Adv. Mater.*, *13*, 1152–1155.
25. Cheng, J. Y., Ross, C. A., Thomas, E. L., Smith, H. I., Vancso, G. J. (2003). Templated self-assembly of block copolymers: effect of substrate topography. *Adv. Mater.*, *15*, 1599–1602.
26. Cheng, J. Y., Ross, C. A., Thomas, E. L., Smith, H. I., Vancso, G. J. (2002). Fabrication of nanostructures with long-range order using block copolymer lithography. *Appl. Phys. Lett.*, *81*, 3657–3659.
27. Chai, J., Buriak, J. M. (2008). Using cylindrical domains of block copolymers to self-assemble and align metallic nanowires. *ACS Nano*, *2*, 489–501.
28. Stoykovich, M. P., Kang, H., Daoulas, K. C., Liu, G., Liu, C. C., de Pablo, J. J., Mueller, M., Nealey, P. F. (2007). Directed self-assembly of block copolymers for nanolithography: fabrication of isolated features and essential integrated circuit geometries. *ACS Nano*, *1*, 168–175.
29. Edwards, E. W., Mueller, M., Stoykovich, M. P., Solak, H. H., de Pablo, J. J., Nealey, P. F. (2007). Dimensions and shapes of block copolymer domains assembled on lithographically defined chemically patterned substrates. *Macromolecules*, *40*, 90–96.
30. Jung, Y. S., Chang, J. B., Verploegen, E., Berggren, K. K., Ross, C. A. (2010). A path to ultranarrow patterns using self-assembled lithography. *Nano Lett.*, *10*, 1000–1005.
31. Suresh, V., Madapusi, S., Krishnamoorthy, S. (2013). Hierarchically built hetero superstructure arrays with structurally controlled material compositions. *ACS Nano*, **DOI**: 10.1021/nn400963a.
32. Kim, S. H., Misner, M. J., Xu, T., Kimura, M., Russell, T. P. (2004). Highly oriented and ordered arrays from block copolymers via solvent evaporation. *Adv. Mater.*, *16*, 226–231.
33. Freer, E. M., Krupp, L. E., Hinsberg, W. D., Rice, P. M., Hedrick, J. L., Cha, J. N., Miller, R. D., Kim, H. C. (2005). Oriented mesoporous organosilicate thin films. *Nano Lett.*, *5*, 2014–2018.
34. Albert, J. N. L., Epps III, T. H. (2010). Self-assembly of block copolymer thin films. *Mater. Today*, *13*, 24–33.
35. Bang, J., Kim, S. H., Drockenmuller, E., Misner, M. J., Russell, T. P., Hawker, C. J. (2006). Defect-free nanoporous thin films from ABC triblock copolymers. *J. Am. Chem. Soc.*, *128*, 7622–7629.
36. Chen Y., Huang, H., Hu, Z., He, T. (2004). Lateral nanopatterns in thin diblock copolymer films induced by selective solvents. *Langmuir*, *20*, 3805–3808.
37. Kim, B., Park, S., McCarthy, T. J., Russell, T. P. (2007). Fabrication of ordered anodic aluminum oxide using a solvent-induced array of block-copolymer micelles. *Small*, *3*, 1869–1872.

38. Kim, T. H., Hwang, J., Hwang, W. S., Huh, J., Kim, H. C., Kim, S. H., Hong, J. M., Thomas, E. L., Park, C. (2008). Hierarchical ordering of block copolymer nanostructures by solvent annealing combined with controlled dewetting. *Adv. Mater., 20*, 522–527.
39. Park, S., Kim, B., Yavuzcetin, O., Tuominen, M. T., Russell, T. P. (2008). Ordering of PS-b-P4VP on patterned silicon surfaces. *ACS Nano, 2*, 1363–1370.
40. Yoo, S. I., Yun, S. H., Kim, H. K., Sohn, B. H. (2010). Highly ordered hexagonal arrays of hybridized micelles from bimodal self-assemblies of diblock copolymer micelles. *Macromol. Rapid Commun., 31*, 645–650.
41. Yun, S. H., Yoo, S. I., Jung, J. C., Zin, W. C., Sohn, B. H. (2006). Highly ordered arrays of nanoparticles in large areas from diblock copolymer micelles in hexagonal self-assembly. *Chem. Mater., 18*, 5646–5648.
42. Kim, T. H., Huh, J., Hwang, J., Kim, H. C., Kim, S. H., Sohn, B. H., Park, C. (2009). Ordered arrays of PS-b-P4VP micelles by fusion and fission process upon solvent annealing. *Macromolecules, 42*, 6688–6697.
43. Xiao, S., Yang, X. M., Park, S., Weller, D., Russell, T. P. (2009). A novel approach to addressable 4 teradot/in. 2 patterned media. *Adv. Mater., 21*, 2516–2519.
44. Binder, W. H., Kluger, C., Straif, C. J., Friedbacher, G. (2005). Directed nanoparticle binding onto microphase-separated block copolymer thin films. *Macromolecules, 38*, 9405–9410.
45. Haryono, A., Binder, W. H. (2006). Controlled arrangement of nanoparticle arrays in block-copolymer domains. *Small, 2*, 600–611.
46. Tseng, Y. C., Darling, S. B. (2010). Block copolymer nanostructures for technology. *Polymers, 2*, 470–489.
47. Cohen, R. E. (1999). Block copolymers as templates for functional materials. *Curr. Opinion Solid State and Mater. Sci., 4*, 587–590.
48. Bennett, R. D., Miller, A. C., Kohen, N. T., Hammond, P. T., Irvine, D. J., Cohen, R. E. (2005). Strategies for controlling the planar arrangement of block copolymer micelles and inorganic nanoclusters. *Macromolecules, 38*, 10728–10735.
49. Cummins, C. C., Beachy, M. D., Schrock, R. R., Vale, M. G., Sankaran, V., Cohen, R. E. (1991). Synthesis of norbornenes containing Tin(II), Tin(IV), Lead(II), and Zinc(II) and their polymerization to give microphase-separated block copolymers. *Chem. Mater., 3*, 1153–1163.
50. Chan, Y. N. C., Craig, G. S. W., Schrock, R. R., Cohen, R. E. (1992). Synthesis of palladium and platinum nanoclusters within microphase-separated diblock copolymers. *Chem. Mater., 4*, 885–894.
51. Wang, Y., Qin, Y., Berger, A., Yau, E., He, C., Zhang, L., Gösele, U., Knez, M., Steinhart, M. (2009). Nanoscopic morphologies in block copolymer nanorods as templates for atomic-layer deposition of semiconductors. *Adv. Mater., 21*, 2763–2766.
52. Peng, Q., Tseng, Y. C., Darling, S. B., Elam, J. W. (2010). Nanoscopic patterned materials with tunable dimensions via atomic layer deposition on block copolymers. *Adv. Mater., 22*, 5129–5133.
53. Lin, Y., Böker, A., He, J., Sill, K., Xiang, H., Abetz, C., Li, X., Wang, J., Emrick, T., Long, S., Wang, Q., Balazs, A., Russell, T. P. (2005). Self-directed self-assembly of nanoparticle/copolymer mixtures. *Nature, 434*, 55–59.

54. Yeh S. W., Wu, T. L., Wei, K. H. (2005). Spatial position control of pre-synthesized CdS nanoclusters using a self-assembled diblock copolymer template. *Nanotechnology, 16*, 683–687.
55. Minelli, C., Hinderling, C., Heinzelmann, H., Pugin, R., Liley, M. (2005). Micrometer-long gold nanowires fabricated using block copolymer templates. *Langmuir, 21*, 7080–7082.
56. Yap, F. L., Thoniyot, P., Krishnan, S., Krishnamoorthy, S. (2012). Nanoparticle cluster arrays for high-performance SERS through directed self-assembly on flat substrates and on optical fibers. *ACS Nano, 6*, 2056–2070.
57. Wang, L., Montagne, F., Hoffmann, P., Heinzelmann, H., Pugin, R. (2011). Hierarchical positioning of gold nanoparticles into periodic arrays using block copolymer nanoring templates. *J. Colloid Interface Sci., 356*, 496–504.
58. Park M., Harrison, C., Chaikin, P. M., Register, R. A., Adamson, D. H. (1997). Block copolymer lithography: periodic arrays of $\sim 10^{11}$ holes in 1 square centimeter. *Science, 276*, 1401–1404.
59. Mansky P., Chaikin, P., Thomas, E. L. (1995). Monolayer films of diblock copolymer microdomains for nanolithographic applications. *J. Mater. Sci., 30*, 1987–1992.
60. Lammertink, R. G. H., Hempenius, M. A., Van Den Enk, J. E., Chan, V. Z. H., Thomas, E. L., Vancso, G. J. (2000). Nanostructured thin films of organic-organometallic block copolymers: one-step lithography with poly(ferrocenylsilanes) by reactive ion etching. *Adv. Mater., 12*, 98–103.
61. Lammertink, R. G. H., Hempenius, M. A., Chan, V. Z. H., Thomas, E. L., Vancso, G. J. (2001). Poly(ferrocenyldimethylsilanes) for reactive ion etch barrier applications. *Chem. Mater., 13*, 429–434.
62. Jung, Y. S., Ross, C. A. (2007). Orientation-controlled self-assembled nanolithography using a polystyrene-polydimethylsiloxane block copolymer. *Nano Lett., 7*, 2046–2050.
63. Jung, Y. S., Jung, W., Tuller, H. L., Ross, C. A. (2008). Nanowire conductive polymer gas sensor patterned using self-assembled block copolymer lithography. *Nano Lett., 8*, 3777–3780.
64. Haupt, M., Miller, S., Glass, R., Arnold, M., Sauer, R., Thonke, K., Moller, M., Spatz, J. P. (2003). Nanoporous gold films created using templates formed from self-assembled structures of inorganic-block copolymer micelles. *Adv. Mater., 15*, 829–831.
65. Haupt, M., Miller, S., Ladenburger, A., Sauer, R., Thonke, K., Spatz, J. P., Riethmuller, S., Moller, M., Banhart, F. (2002). Semiconductor nanostructures defined with self-organizing polymers. *J. Appl. Phys., 91*, 6057–6059.
66. Jagannathan, H., Deal, M., Nishi, Y., Kim, H. C., Freer, E. M., Sundstrom, L., Topuria, T., Rice, P. M. (2006). Templated germanium nanowire synthesis using oriented mesoporous organosilicate thin films. *J. Vac. Sci. Technol. B, 24*, 2220–2226.
67. Sundstrom, L., Krupp, L., Delenia, E., Rettner, C., Sanchez, M., Hart, M. W., Kim, H. C., Zhang, Y. (2006). Patterning ~20 nm half-pitch lines on silicon using a self-assembled organosilicate etch mask. *Appl. Phys. Lett., 88*, 243107–243110.
68. Glass, R., Moller, M., Spatz, J. P. (2003). Block copolymer micelle nanolithography. *Nanotechnology, 14*, 1153–1160.

69. Black, C. T., Guarini, K. W., Breyta, G., Colburn, M. C., Ruiz, R.; Sandstrom, R. L., Sikorski, E. M., Zhang, Y. (2006). Highly porous silicon membrane fabrication using polymer self-assembly. *J. Vac. Sci. Technol. B*, *24*, 3188–3191.

70. Black, C. T. (2005). Self-aligned self assembly of multi-nanowire silicon field effect transistors. *Appl. Phys. Lett.*, *87*, 1–3.

71. Guarini, K. W., Black, C. T., Yeuing, S. H. I. (2002). Optimization of diblock copolymer thin film self assembly. *Adv. Mater.*, *14*, 1290–1294.

72. Guarini K. W., Black, C. T., Zhang, Y., Kim, H., Sikorski, E. M., Babich, I. V. (2002). Process integration of self-assembled polymer templates into silicon nanofabrication. *J. Vac. Sci. Technol. B*, *20*, 2788–2792.

73. Black, C. T., Guarini, K. W., Milkove, K. R., Baker, S. M., Russell, T. P., Tuominen, M. T. (2001). Integration of self-assembled diblock copolymers for semiconductor capacitor fabrication. *Appl. Phys. Lett.*, *79*, 409–411.

74. Guarini, K. W., Black, C. T., Milkove, K. R., Sandstrom, R. L. (2001). Nanoscale patterning using self-assembled polymers for semiconductor applications. *J. Vac. Sci. Technol. B*, *19*, 2784–2788.

75. Park, H. J., Kang, M. G., Guo, L. J. (2009). Large area high density sub-20 nm SiO_2 nanostructures fabricated by block copolymer template for nanoimprint lithography. *ACS Nano*, *3*, 2601–2608.

76. Krishnamoorthy, S., Krishnan, S., Thoniyot, P., Low, H. Y. (2011). Inherently reproducible fabrication of plasmonic nanoparticle arrays for SERS by combining nanoimprint and copolymer lithography. *ACS Appl. Mater. Interfaces*, *3*, 1033–1040.

77. Riess, G. (2003). Micellization of block copolymers. *Prog. Polym. Sci.*, *28*, 1107–1170.

78. Boontongkong, Y., Cohen, R. E. (2002). Cavitated block copolymer micellar thin films: lateral arrays of open nanoreactors. *Macromolecules*, *35*, 3647–3652.

79. Lohmueller, T., Bock, E., Spatz, J. P. (2008). Synthesis of quasi-hexagonal ordered arrays of metallic nanoparticles with tuneable particle size. *Adv. Mater.*, *20*, 2297–2302.

80. Lohmueller, T., Aydin, D., Schwieder, M., Morhard, C., Louban, I., Pacholski, C., Spatz, J. P. (2011). Nanopatterning by block copolymer micelle nanolithography and bioinspired applications. *Biointerphases*, *6*, 1–12.

81. Krishnamoorthy, S., Pugin, R., Brugger, J., Heinzelmann, H., Hinderling, C. (2008). Nanopatterned self-assembled monolayers by using diblock copolymer micelles as nanometer-scale adsorption and etch masks. *Adv. Mater.*, *20*, 1962–1965.

82. Krishnamoorthy, S., van den Boogaart, M. A. F., Brugger, J., Hibert, C., Pugin, R., Hinderling, C., Heinzelmann, H. (2008). Combining micelle self-assembly with nanostencil lithography to create periodic/aperiodic micro-/nanopatterns on surfaces. *Adv. Mater.*, *20*, 3533–3538.

83. Hoogerwerf, A. C., Hinderling, C., Krishnamoorthy, S., Hibert, C., Spassov, V., Overstolz, T. (2007). Fabrication of reinforced nanoporous membranes. *Transducers '07 & Eurosensors XXI, Digest of Technical Papers, 1 and 2*, U249-U250.

84. Meister, A., Krishnamoorthy, S., Hinderling, C., Pugin, R., Heinzelmann, H. (2006). Local modification of micellar layers using nanoscale dispensing. *Microelectron. Eng.*, *83*, 1509–1512.

85. Krishnamoorthy, S., Pugin, R., Brugger, J., Heinzelmann, H., Hoogerwerf, A. C., Hinderling, C. (2006). Block copolymer micelles as switchable templates for nanofabrication. *Langmuir*, *22*, 3450–3452.
86. Bang, J., Jeong, U., Ryu, D. Y., Russell, T. P., Hawker, J. (2009). Block copolymer nanolithography: translation of molecular level control to nanoscale patterns. *Adv. Mater.*, *21*, 4769–4792.
87. Pashley, D. W., Menter, J. W., Bassett, G. A. (1957). Observation of dislocations in metals by means of Moire patterns on electron micrographs. *Nature*, *179*, 752–755.
88. Hexemer, A., Stein, G. E., Kramer, E. J., Magonov, S. Block copolymer monolayer structure measured with scanning force microscopy Moire patterns. *Macromolecules*, *38*, 7083–7089.
89. Xiao, S., Yang, X., Edwards, E. W., La, Y. H., Nealey, P. F. (2005). Graphoepitaxy of cylinder-forming block copolymers for use as templates to pattern magnetic metal dot arrays. *Nanotechnology*, *16*, 324–329.
90. Bennett, R. D., Hart, A. J., Miller, A. C., Hammond, P. T., Irvine, D. J., Cohen, R. E. (2006). Creating patterned carbon nanotube catalysts through the microcontact printing of block copolymer micellar thin films. *Langmuir*, *22*, 8273–8276.
91. Naito, K., Hieda, H., Sakurai, M., Kamata, Y., Asakawa, K. (2002). 2.5-inch disk patterned media prepared by an artificially assisted self-assembling method. *IEEE Trans. Magn.*, *38*, 1949–1951.
92. Asakawa, K., Hiraoka, T., Hieda, H., Sakurai, M., Kamata, Y., Naito, K. (2002). Nanopatterning for patterned media using block-copolymer. *J. Photopolym. Sci. Technol.*, *15*, 465–470.
93. Glass, R., Arnold, M., Blummel, J., Kuller, A., Moller, M., Spatz, J. P. Micro-nanostructured interfaces fabricated by the use of inorganic block copolymer micellar monolayers as negative resist for electron-beam lithography. *Adv. Funct. Mater.*, *13*, 569–575.
94. Spatz J. P., Chan, V. Z. H., Mossmer, S., Kamm, F. M., Plettl, A., Ziemann, P., Moller, M. (2002). A combined top-down/bottom-up approach to the microscopic localization of metallic nanodots. *Adv. Mater.*, *14*, 1827–1832.
95. Meister, A., Liley, M., Brugger, J., Pugin, R., Heinzelmann, H. (2004). Nanodispenser for attoliter volume deposition using atomic force microscopy probes modified by focused-ion-beam milling. *Appl. Phys. Lett.*, *85*, 6260–6262.
96. Gorzolnik, B., Mela, P., Moeller, M. (2006). Nano-structured micropatterns by combination of block copolymer self-assembly and UV photolithography. *Nanotechnology*, *17*, 5027–5032.
97. Peng, J., Xuan, Y., Wang, H. F., Yang, Y. M., Li, B. Y., Han, Y. C. (2004). Solvent-induced microphase separation in diblock copolymer thin films with reversibly switchable morphology. *J. Chem. Phys.*, *120*, 11163–11170.
98. Xu, T., Goldbach, J. T., Misner, M. J., Kim, S., Gibaud, A., Gang, O., Ocko, B., Guarini, K. W., Black, C. T., Hawker, C. J., Russell, T. P. (2004). Scattering study on the selective solvent swelling induced surface reconstruction. *Macromolecules*, *37*, 2972–2977.
99. Xiang, H., Shin, K., Kim, T., Moon, S. I., McCarthy, T. J., Russell, T. P. (2004). Block copolymers under cylindrical confinement. *Macromolecules*, *37*, 5660–5664.

100. Brugger, J., Berenschot, J. W., Kuiper, S., Nijdam, W., Otter, B., Elwenspoek, M. (2000). Resistless patterning of sub-micron structures by evaporation through nanostencils. *Microelectron. Eng.*, *53*, 403–405.

101. Popa, A. M., Niedermann, P., Heinzelmann, H., Hubbell, J. A., Pugin, R. (2009). Fabrication of nanopore arrays and ultrathin silicon nitride membranes by block-copolymer-assisted lithography. *Nanotechnology*, *20*, 485–303.

102. Hoogerwerf, A. C., Hinderling, C., Krishnamoorthy, S., Hibert, C., Spassov, V., Overstolz, T. (2007). Fabrication of reinforced nanoporous membranes. Solid-State Sensors, Actuators and Microsystems Conference, TRANSDUCERS 2007 International. IEEE, pp. 489–492.

103. Hellwig, O., Bosworth, J. K., Dobisz, E., Kercher, D., Hauet, T., Zeltzer, G., Risner-Jamtgaard, J. D., Yaney, D., Ruiz, R. (2010). Bit patterned media based on block copolymer directed assembly with narrow magnetic switching field distribution. *Appl. Phys. Lett.*, *96*, 052511–052513.

104. International Technology Roadmap for Semiconductors, Front End Processes (FEP), 2010 tables at http://www.itrs.net/Links/2010ITRS/Home2010.htm.

105. Ross, C. A. (2001). Patterned magnetic recording media. *Ann. Rev. Mater. Sci.*, *31*, 203–235.

106. Thurn-Albrecht, T., Schotter, J., Kastle, G. A., Emley, N., Shibauchi, T., Krusin-Elbaum, L., Guarini, K., Black, C. T., Tuominen, M. T., Russell, T. P. (2000). Ultrahigh-density nanowire arrays grown in self-assembled diblock copolymer templates. *Science*, *290*, 2126–2129.

107. Liu, K., Baker, S. M., Tuominen, M., Russell, T. P., Schuller, I. K. (2001). Tailoring exchange bias with magnetic nanostructures. *Phys. Rev. B.*, *63*, 060403–060407.

108. Cheng, J. Y., Ross, C. A., Chan, V. Z. H., Thomas, E. L., Lammertink, R. G. H., Vancso, G. J. (2001). Formation of a cobalt magnetic dot array via block copolymer lithography. *Adv. Mater.*, *13*, 1174–1178.

109. Hong, A. J., Liu, C. C., Wang, Y., Kim, J., Xiu, F., Ji, S., Zou, J., Nealey, P. F., Wang, R. L. (2010). Metal nanodot memory by self-assembled block copolymer lift-off. *Nano Lett.*, *10*, 224–229.

110. Gupta, R. K., Krishnamoorthy, S., Kusuma, D. Y., Lee, P. S., Srinivasan, M. P. (2012). Enhancing charge-storage capacity of non-volatile memory devices using template-directed assembly of gold nanoparticles. *Nanoscale*, *4*, 2296–2300.

111. Suresh, V., Huang, M. S., Srinivasan, M. P., Guan, C., Fan, H. J., Krishnamoorthy, S. (2012). Robust, high-density zinc oxide nanoarrays by nanoimprint lithography-assisted area-selective atomic layer deposition. *J. Phys. Chem. C*, *116*, 23729–23734.

112. Suresh, V., Meiyu, S. H., Srinivasan, M. P., Krishnamoorthy, S. (2012). Macroscopic high density nano disc arrays of Zinc oxide fabricated by block copolymer self-assembly assisted nanoimprint lithography. *J. Mater. Chem.*, *22*, 21871–21877.

113. Guarini, K. W., Black, C. T., Zhang, Y., Babich, I. V., Sikorski, E. M., Gignac, L. M. (2003). Low voltage, scalable nanocrystal FLASH memory fabricated by templated self assembly. Electron Devices Meeting, IEDM'03 Technical Digest. IEEE International. IEEE, pp. 22.2.1–22.2.4.

114. Leong, W. L., Mathews, N., Mhaisalkar, S., Lam, Y. M., Chen, T., Lee, P. S. (2009). Micellar poly (styrene-b-4-vinylpyridine)-nanoparticle hybrid system for non-volatile organic transistor memory. *J. Mater. Chem.*, *19*, 7354–7361.

115. Leong, W. L., Lee, P. S., Lohani, A., Lam, Y. M., Chen, T., Zhang, S., Dodabalapur, A., Mhaisalkar, G. (2008). Non-volatile organic memory applications enabled by in situ synthesis of gold nanoparticles in a self-assembled block copolymer. *Adv. Mater.*, *20*, 2325–2331.

116. Black, C. T., Guarini, K. W., Zhang, Y., Kim, H., Benedict, J., Sikorski, E., Babich, I. V., Milkove, K. R. (2004). High-capacity, self-assembled metal-oxide-semiconductor decoupling capacitors. *IEEE Electron. Dev. Lett.*, *25*, 622–624.

117. Chang, L. W., Caldwell, M. A., Wong, H. S. P. Diblock copolymer directed self-assembly for CMOS device fabrication. In Proceedings of SPIE, the International Society for Optical Engineering. Society of Photo-Optical Instrumentation Engineers, 69212M-1.

118. Nealey P. F., Edwards, E. W., Stoykovich, M. P., Mueller, M., de Pablo, J. J. (2005). Precise control over the shape and dimensions of nanostructures in block copolymer films using chemically nanopatterned substrates. Abstracts of Papers, 230th ACS National Meeting, Washington, DC, United States, August 28–September 1, 2005, POLY-266.

119. Darling, S. B. (2009). Block copolymers for photovoltaics. *Energy and Environmental Science*, *2*, 1266–1273.

120. Coakley, K. M., Liu, Y., Goh, C., McGehee, M. D. (2005). Ordered organic–inorganic bulk heterojunction photovoltaic cells. *MRS Bull.*, *30*, 37–40.

121. Coakley, K. M., Liu, Y., McGehee, M. D., Frindell, K. L., Stucky, G. D. (2003). Infiltrating semiconducting polymers into self-assembled mesoporous titania films for photovoltaic applications. *Adv. Funct. Mater.*, *13*, 301–306.

122. Coakley, K. M., McGehee, M. D. (2003). Photovoltaic cells made from conjugated polymers infiltrated into mesoporous titania. *Appl. Phys. Lett.*, *83*, 3380–3382.

123. Fan, Z., Ruebusch, D. J., Rathore, A. A., Kapadia, R., Ergen, O., Leu, P. W., Javey, A. (2009). Challenges and prospects of nanopillar-based solar cells. *Nano Res.*, *2*, 829–843.

124. Huang, J. S., Hsiao, C. Y., Syu, S. J., Chao, J. J., Lin, C. F. (2009). Well-aligned single-crystalline silicon nanowire hybrid solar cells on glass. *Sol. Energ. Mater. Sol. Cells*, *93*, 621–624.

125. Ju, X., Feng, W., Varutt, K., Hori, T., Fujii, A., Ozaki, M. (2008). Fabrication of oriented ZnO nanopillar self-assemblies and their application for photovoltaic devices. *Nanotechnology*, *19*, 435701–435706.

126. Gowrishankar, V., Miller, N., McGehee, M. D., Misner, M. J., Ryu, D. Y., Russell, T. P., Drockenmuller, E., Hawker, C. J. (2006). Fabrication of densely packed, well-ordered, high-aspect-ratio silicon nanopillars over large areas using block copolymer lithography. *Thin Solid Films*, *513*, 289–294.

127. Li R. R., Dapkus, P. D., Thompson, M. E., Jeong, W. G., Harrison, C., Chaikin, P. M., Register, R. A., Adamson, D. H. (2000). Dense arrays of ordered GaAs nanostructures by selective area growth on substrates patterned by block copolymer lithography. *Appl. Phys. Lett.*, *76*, 1689–1691.

128. Kim, H. Y., Ahn, J. H., Bang, J. (2008). Enhancement of light extraction efficiency via inductively coupled plasma etching of block copolymer templates on GaN/Al2O3. *Electron. Mater. Lett.*, *4*, 185–188.

129. Cho, Y. H., Lee, K., Kim, K., Baik, K. H., Cho, J., Kim, J., Shin, K., Bang, J. (2009). Enhanced light emission of nano-patterned GaN via block copolymer thin films. *Kor. J. Chem. Eng.*, *26*, 277–280.

130. Jung, H., Hwang, D., Kim, E., Kim, B. J., Lee, W. B., Poelma, J., Kim, J., Hawker, C. J., Huh, J., Ryu, D. Y. (2011). Three-dimensional multilayered nanostructures with controlled orientation of microdomains from crosslinkable block copolymers. *ACS Nano*, *5*, 6164–6173.

131. Ayyub, O. B., Sekowski, J. W., Yang, T. I., Briber, R. M., Kofinas, P. (2011). Color changing block copolymer films for chemical sensing of simple sugars. *Biosensors and Bioelectronics*, *28*, 349–354.

132. Chan, E. P., Walish, J. J., Thomas, E. L., Stafford, C. M. (2011). Block copolymer photonic gel for mechanochromic sensing. *Adv. Mater.*, *23*, 4702–4706.

133. Shin, D. O., Jeong, J. R., Han, T. H., Koo, C. M., Park, H. J., Lim, Y. T., Kim, S. O. (2010). A plasmonic biosensor array by block copolymer lithography. *J. Mater. Chem.*, *20*, 7241–7247.

134. Kumar, N., Hahm, J. I. (2005). Nanoscale protein patterning using self-assembled diblock copolymers. *Langmuir*, *21*, 6652–6655.

135. Kumar, N., Parajuli, O., Dorfman, A., Kipp, D., Hahm, J. (2007). Activity study of self-assembled proteins on nanoscale diblock copolymer templates. *Langmuir*, *23*, 7416–7422.

136. Cresce, A. V., Silverstein, J. S., Bentley, W. E., Kofinas, P. (2006). Nanopatterning of recombinant proteins using block copolymer templates. *Macromolecules*, *39*, 5826–5829.

137. Lee, W., Lee, S. Y., Briber, R. M., Rabin, O. (2011). Self-assembled SERS substrates with tunable surface plasmon resonances. *Adv. Funct. Mater.*, *21*, 3424–3429.

138. Cho, W. J., Kim, Y., Kim, J. K. (2012). Ultrahigh density array of silver nanoclusters for SERS substrate with high sensitivity and excellent reproducibility. *ACS Nano*, *6*, 249–255.

139. Sanchez-Iglesias, A., Aldeanueva-Potel, P., Ni, W., Perez-Juste, J., Pastoriza-Santos, I., Alvarez-Puebla, R. A., Mbenkum, B. N., Liz-Marzan, L. M. (2010). Chemical seeded growth of Ag nanoparticle arrays and their application as reproducible SERS substrates. *Nano Today*, *5*, 21–27.

140. Wang, Y., Becker, M., Wang, L., Liu, J., Scholz, R., Peng, J., Gösele, U., Christiansen, S., Kim, D. H., Steinhart, M. (2009). Nanostructured gold films for SERS by block copolymer-templated galvanic displacement reactions. *Nano Lett.*, *9*, 2384–2389.

141. Lu, J., Chamberlin, D., Rider, D. A., Liu, M. Z., Manners, I., Russell, T. P. (2006). Using a ferrocenylsilane-based block copolymer as a template to produce nanotextured Ag surfaces: uniformly enhanced surface enhanced Raman scattering active substrates. *Nanotechnology*, *17*, 5792–5797.

142. Hillmyer, M. *Nanoporous Materials from Block Copolymer precursors*. Springer, 2005, pp. 137–181.

143. Yang, S. Y., Ryu, I., Kim, H. Y., Kim, J. K., Jang, S. K., Russell, T. P. (2006). Nanoporous membranes with ultrahigh selectivity and flux for the filtration of viruses. *Adv. Mater.*, *18*, 709–712.

144. Yang, S. Y., Park, J., Yoon, J., Ree, M., Jang, S. K., Kim, J. K. (2008). Virus filtration membranes prepared from nanoporous block copolymers with good dimensional

stability under high pressures and excellent solvent resistance. *Adv. Funct. Mater.*, *18*, 1371–1377.

145. Chen, L., Phillip, W. A., Cussler, E. L., Hillmyer, M. A. (2007). Robust nanoporous membranes templated by a doubly reactive block copolymer. *J. Am. Chem. Soc.*, *129*, 13786–13787.
146. Popa, A. M., Niedermann, P., Heinzelmann, H., Hubbell, J. A., Pugin, R. (2009). Fabrication of nanopore arrays and ultrathin silicon nitride membranes by block-copolymer-assisted lithography. *Nanotechnology*, *20*, 485303–485313.
147. Tong, H. D., Jansen, H. V., Gadgil, V. J., Bostan, C. G., Berenschot, E., Van Rijn, C. J. M., Elwenspoek, M. (2004). Silicon nitride nanosieve membrane. *Nano Lett.*, *4*, 283–287.
148. Rosenstein, J. K., Wanunu, M., Merchant, C. A., Drndic, M., Shepard, K. L. (2012). Integrated nanopore sensing platform with sub-microsecond temporal resolution. *Nat. Meth.*, *9*, 487–492.
149. Howorka, S., Siwy, Z. (2009). Nanopore analytics: sensing of single molecules. *Chem. Soc. Rev.*, *38*, 2360–2384.

10

STIMULI-RESPONSIVE COPOLYMERS AND THEIR APPLICATIONS

HE TAO

10.1 INTRODUCTION

Stimuli-responsive polymers (also known as *intelligent polymers*, *smart polymers*, or *environmentally sensitive polymers*) are defined as polymers that undergo relatively large and abrupt, physical or chemical changes in response to external environment. The stimuli could be mainly classified as chemical stimuli (such as pH, chemical agents, and ionic factors), physical stimuli (such as temperature, light, electric or magnetic fields, mechanical stress, and ultrasound), and biochemical stimuli (such as glucose, enzyme, antigen, and other biochemical agents). Typically, the physical stimuli will affect the level of various energy sources and alter molecular interactions as critical onset points, while the chemical/biochemical stimuli will change the interactions of inter- or intrapolymer chains at the molecular level. Stimuli-responsive copolymers contain at least one component exhibiting stimuli-responsiveness, which are normally fabricated through copolymerizations of functional monomers with other monomers. The applications of stimuli-responsive copolymers lie in various areas such as drug delivery, DNA/gene delivery, biotechnology, chromatography, cosmetics, consumer care products, and environmental protection.

Normally, stimuli-responsive copolymers could be utilized in various forms as nanoparticles/micelles, surfaces and interfaces, hydrogels (permanently cross-linked or reverse), polymer solutions, and conjugated solutions. Generally, polymer nanoparticles/micelles are formed by aggregation of amphiphilic block copolymers or terminally modified copolymers in aqueous medium. Stimuli-responsive

Synthesis and Applications of Copolymers, First Edition. Edited by Anbanandam Parthiban.

copolymer nanoparticles/micelles could be prepared through the self-assembly of macromolecular building blocks initiated by the alternation of balance between hydrophilicity and hydrophobicity that are modulated by stimuli [1–7]. The stimuli-responsive surfaces could be fabricated after functionalizing matrix surfaces (such as polymers, silica, and metal) with stimuli-responsive copolymers [8–12]. The property of the modified interface can give a dynamic on/off switch system by changing the hydrophobic/hydrophilic surface function and the pore size of porous membranes. Permanently cross-linked hydrogels are formed with a three-dimensional network of stimuli-responsive copolymer chains, which swell but do not dissolve in aqueous environment. However, reverse hydrogels could be reversibly transformed to solutions because of environmental stimuli changes, showing solution–gelation (sol–gel) transition by altering the hydrophobic interactions of cross-linked areas in an aqueous system [13–17]. Therefore, reverse stimuli-responsive hydrogels were developed for phase change rather than dimension change, which can be exemplarily used as injectable hydrogels. As the solubility of stimuli-responsive copolymers can be controlled by changing stimuli, their conjugates can be modulated to have stimuli-responsiveness.

In this chapter, we will discuss the synthesis of different stimuli-responsive copolymer materials in various forms, together with their typical applications. We will focus on following stimuli-responsive systems: temperature, pH, biologically stimuli responsive (such as glucose, enzyme, and antigen), and field stimuli responsive (such as electric field, magnetic field, ultrasound, and light). The mechanism of stimuli-responsiveness will be investigated in detail.

10.2 TEMPERATURE-RESPONSIVE COPOLYMERS AND APPLICATIONS

Temperature-responsive polymers are the most widely studied stimuli-responsive polymer system. Temperature-responsive polymers have a critical solution temperature, at which the phases of polymers and solutions are discontinuously changed according to their composition. If polymers are completely solvated in aqueous phase below a specific temperature and are phase-separated above this temperature, these polymers generally have a low critical solution temperature (LCST) [18], the lowest temperature of the phase separation. As a comparison, some polymers have a high critical solution temperature (HCST) or upper critical solution temperature (UCST), which means that the polymers can be completely dissolved above the critical temperature and phase separated on lowering the temperature below the critical temperature. For example, the well-known poly(*N*-isopropylacrylamide) (PNIPAAM) has LCST at around 32°C, at which the polymer undergoes a reversible volume phase transition caused by the coil-to-globule transition. As copolymers with HCST (or UCST) are relatively less studied that limit their applications, this chapter will focus on temperature-responsive copolymers that have LCST.

In addition to the temperature-responsiveness that is dominated by relationship between polymers and water molecules, the intermolecular interaction in aqueous medium that could result in micelle aggregation, hydrogel shrinkage, or physical cross-link build another important temperature-responsive characteristic. For example, on lowering the temperature, two or three biopolymer chains (such as gelatin) form helix conformation that generates physical junctions to make a gel network [19]. The forming mechanism is the intermolecular association of random coil-to-helix transition based on hydrogen bonding. Temperature can also control the hydrogen bonding association/dissociation between different pendant groups, which can be applied in the preparation of reversible swelling/deswelling hydrogels. Typical example is the temperature-responsive interpenetrating polymer networks (IPNs) composed of polyacids and polyacrylamides [20,21].

10.2.1 Temperature-Responsive Copolymers Based on LCST

10.2.1.1 Functional Segments Temperature-responsive copolymers normally contain at least one responsive functional segment, which is sensitive to temperature changes. Figure 10.1 illustrates several examples of temperature-responsive monomeric blocks, which upon polymerization maintain temperature-responsiveness. Poly(*N*-substituted acrylamide (AAm)) is representative of

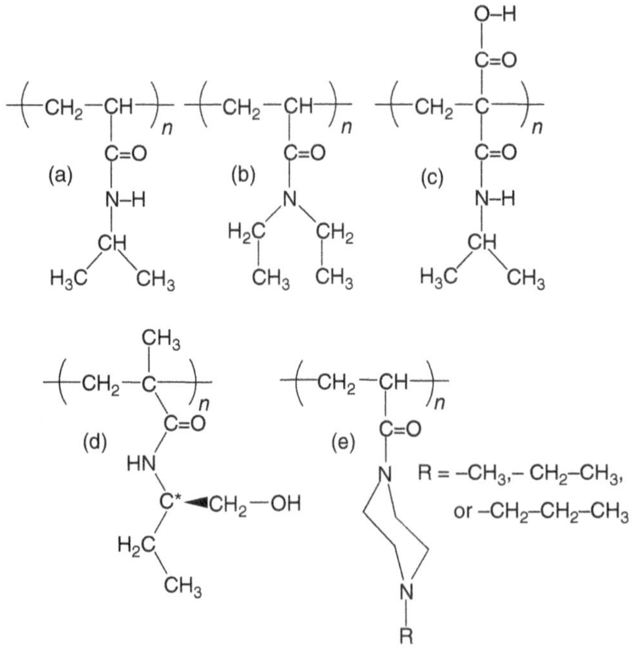

Figure 10.1 Poly(*N*-substituted acrylamide) with LCST: (a) poly(*N*-isopropylacrylamide) (PNIPAAm), (b) poly(*N*,*N*-diethylacrylamide) (PDEAAm), (c) poly(2-çarboxyisopropylacrylamide) (PCIPAAm), (d) poly(*N*-(L)-(1-hydroxymethyl) propylmethacrylamide [(L-HMPMAAm)], and (e) poly(*N*-acryloyl-*N*′-alkylpiperazine).

the group of temperature-responsive polymers that have an LCST. PNIPAAM is the most popular temperature-responsive polymer that has an LCST at around 32°C [22]. The *N*-isopropylacrylamide (NIPAAM) segment has been designed at the molecular level to control the LCST and the response kinetics. Poly(*N*-vinylcaprolactone) (PVCL), poly(*N*-(DL)-(1-hydroxymethyl)propylmethacrylamide) (P(DL)-HMPMA) [23], and poly(*N*, *N'*-diethylacrylamide) (PDEAAM) [24] have the LCST at around 32°, 37°, and 33°C, respectively. Poly(2-carboxyisopropylacrylamide) (PCIPAAM) is composed of a vinyl group, isopropylacrylamide group, and carboxyl group, which can give two benefits: the analogous temperature-responsiveness as PNIPAAM and the additional functionality in the pendant groups [25]. Poly(*N*-acryloyl-*N'*-alkylpiperazine) was reported as a temperature- and pH-responsive polymer, which show an LCST at 37°C [26].

Temperature-responsive copolymers can be synthesized from the functional segments shown above through copolymerizations, which can be fabricated into various forms such as hydrogels, nanoparticles/micelles, functional surfaces, and conjugated with biomaterials.

10.2.1.2 Temperature-Responsive Copolymers in Control of LCST and Improving Temperature Sensitivity The LCST is influenced by hydrophobic or hydrophilic moieties in the molecular chains. Copolymers of temperature-responsive monomers with other monomers can be applied in controlling the whole LCST of the copolymers. Generally, for example, copolymerizing NIPAAM monomer with hydrophilic monomers (such as AAm) would increase the final LCST, while copolymerizing with hydrophobic monomers (such as *N*-butylacrylamide) would decrease the LCST as well as improve the temperature sensitivity [27].

The influence of LCST of NIPAAM by random copolymerization with hydrophilic and hydrophobic monomers was investigated by varying the mole fractions between NIPAAM and incorporated monomers. The LCST of NIPAAM showed a discontinuous alternation or even disappeared at the pK_a of the ionizable groups when NIPAAM was copolymerized with ionizable groups such as acrylic acid (AAc) or *N,N'*-dimethylacrylamide (DMAAm) [28]. The materials with changeable LCST could be used in drug delivery. However, on being grafted with hydrophilic groups such as AAc, the LCST of NIPAAM was independent of the content of the hydrophilic groups. In other examples, the hydrolytically sensitive lactate ester groups were introduced to control the LCST of NIPAAM-based random copolymers by hydrolysis of lactate ester groups: poly(NIPAAM-*co*-HEMA-lactate) showed increased hydrophilicity after the hydrolysis of the lactate groups in which LCST was increased [5]. Terminal incorporation of hydrophilic or hydrophobic groups was reported to influence the LCST of temperature-responsive polymers such as PNIPAAM resulting in change of thermodynamics. Hydrophilic end groups such as AAc and DMAAm more pronouncedly raised the LCST of PNIPAAM and slowed down the rate of phase transition than their random copolymer analogs.

Versatile applications of temperature-responsive polymers such as biomimetic actuators require improved temperature sensitivity. The sensitivity kinetics could be controlled by molecular design. For temperature-sensitive PNIPAAM hydrogels, hydrophilic moieties could increase the deswelling rate. Hydrogels prepared from random copolymer of NIPAAM with AAc or methacrylic acid (MAAc) showed faster deswelling kinetics than NIPAAM hydrogels [29]. Another methodology to produce rapid deswelling was constructing the molecular architecture of PNIPAAM as a comb type instead of linear structure. Comb-shaped PNIPAAM hydrogel exhibited a faster acceleration of the shrinking rate than linear PNIPAAM hydrogel [30]. However, hydrophilic polyethylene oxide (PEO) grafts being introduced onto PNIPAAM backbone did not interfere with long PNIPAAM sequences. As a result, this structure could stand strong hydrophobic backbone aggregation between the long PNIPAAM segments in spite of a large amount of the hydrophilic moiety in the network. It was also found that the sensitivity of hydrogels was promoted at the synthetic steps of the hydrogel networks. When P(NIPAAM-*co*-AAc) hydrogel was synthesized at a weak alkaline condition (pH 8.8), it showed more rapid temperature sensitivity and improved oscillatory properties over small temperature cycles around the physiological temperature of 37°C [31].

10.2.1.3 Temperature-Responsive Nanoparticles/Micelles Based on LCST The LCST behavior of PNIPAAM-based copolymers has been utilized in preparation of polymer nanoparticles/micelles because of the switchable hydrophilic/hydrophobic changes of polymer chains. Normally, two types of micellar structures containing temperature-responsive segments (such as PNIPAAM) were suggested: thermosensitive outer corona and thermosensitive inner core.

Micelles will be formed below LCST, and intermolecular aggregation occurs above LCST from copolymers with temperature-responsive polymers (e.g., PNIPAAM) as corona. For example, poly(butyl methacrylate-*co*-NIPAAM) [P(BMA-*co*-NIPAAM)] and poly(styrene-*co*-NIPAAM) [P(St-NIPAAM)] would form micelles below the LCST of NIPAAM. PBMA exhibits a relatively flexible conformation compared with PNIPAAM chain and low glass transition temperature (T_g under 20°C), which results in the inner PBMA core deformation of P(BMA-NIPAAM) solution as well as the hydrophobic change of the outer PNIPAAM shell on changing the temperature during the formation of micelles [32]. Another example lies in the preparation of micelles using biodegradable hydrophobic polymer, polylactide (PLA), as the inner core with short PNIPAAM chains as the shell. PNIPAAM-*b*-PLA was synthesized through ring-opening polymerization of lactide in the presence of hydroxy-terminated PNIPAAM [33]. The micelles sizing around 40 nm formed at low temperature, and above LCST, micelles aggregated. Actually, the simplest design to prepare temperature-responsive micelles could be the utilization of alkyl-terminated PNIPAAM. Different core–shell micelles could be formed when the length of alkyl groups were varied.

Combination of hydrophilic segments with temperature-responsive polymer segments could be another method of preparing temperature-sensitive polymer micelles. These block copolymers can be completely solvated at temperature above LCST, while micelles can be formed on decreasing temperature below LCST. For example, block copolymers of PNIPAAM and PEG were synthesized by a ceric ion redox system yielding radicals at the terminal carbons of PEG, which were applied in the preparation of temperature-responsive micelles [34,35]. In another study, *N*-(2-hydroxypropyl)methacrylamide (HPMA) was selected as the hydrophilic monomer, which was copolymerized with NIPAAM to produce the block copolymer with varied molecular weights. This type of block copolymers can be used to produce temperature-sensitive polymer micelles. The size decreased with decreasing content of PHPMA in the block [36].

10.2.1.4 Temperature-Responsive Hydrogels and Surfaces Hydrogels and surfaces are two of the most common application forms of temperature-responsive copolymers. Various hydrogels and modified surfaces have been investigated.

Generally, temperature-responsive hydrogels have been prepared through two different ways: polymerization from temperature-responsive monomers in presence of cross-linkers, and cross-linking of functional groups on presynthesized polymers. In the second method, α-ray or electron beam was used to initiate the cross-linking as well. To improve the biocompatibility, swelling/deswelling rate, and mechanical properties, PMAAc component was introduced into PNIPAAM hydrogel by the copolymerization of both monomers in the presence of cross-linker. The resulting IPN structure exhibited improved properties. The coiled-coil protein, a temperature-sensitive recombinant protein, was reported to be used as cross-linking domains in synthetic polymers. This hybrid hydrogel was assembled with poly(*N*-(2-hydroxypropyl)-methacrylamide-*co*-*N*,*N*′-dicarboxymethylaminopropyl)-methacrylamide) (P(HPMAAm-*co*-DAMAAm)) and two histine-tagged coiled-coil proteins, which has transition temperature at 39°C [12]. In another report, microparticles have been fabricated from PNIPAAM copolymers of NIPAAM, *N*,*N*′-methylenebisacrylamide (BIS) and AAc or 1-vinylimidazole (VI) [37]. When the carboxyl or imidazolyl groups were ionized into the corresponding salts, the particles containing 30 mol% of AAc or VI increased tremendously from 125 to 600 nm at 25°C. However, uncharged particles varied from 125 nm at 25°C to 50 nm at 35°C (above LCST).

Recently, self-assembly between two different hydrophobically modified polymers was reported as a way of preparing temperature-responsive hydrogel particles on a nanometer scale. For example, a cholesterol-bearing pullulan (CHP) was physically mixed with a copolymer of NIPAAM and *N*-[4-(1-pyrenyl) butyl]-*N*-*n*-octadecylacrylamide] (PNIPAAm–C18Py), which made physically cross-linked networks because of the aggregation of hydrophobic moieties to nanoparticles (Figure 10.2) [38]. The radius of the nanoparticles showed a reversible change from 47 to 160 nm, where nanoparticles extended above the LCST (32°C) and recovered their original dimension upon cooling below the LCST.

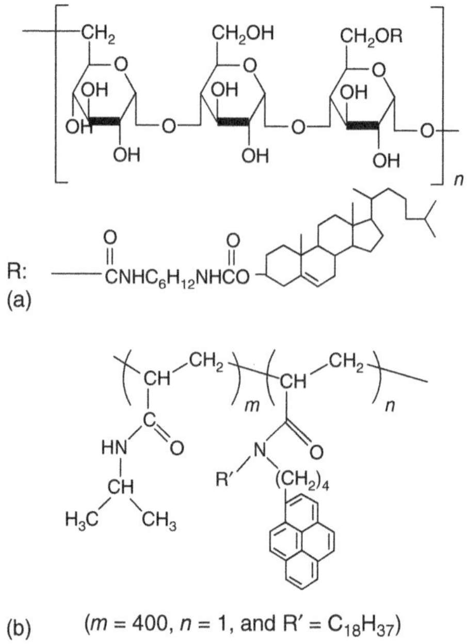

(a)

(b) ($m = 400$, $n = 1$, and $R' = C_{18}H_{37}$)

Figure 10.2 Cholesterol-bearing pullulan and *N*-[4-(1-pyrenyl)butyl]-*N*-*n*-octadecylacrylamide (PNIPAAm–C18Py).

Temperature-responsive surface or interface has potential applications in biorelated areas such as temperature-modulated membranes, chromatography, and cell culture dish. Temperature-responsive surface or interface has been designed either to modulate pores of a porous matrix as temperature-responsive gates or to control the wettability of nonporous matrix surfaces. In this chapter, we focus on the former case. For example, the modification of a membrane surface with temperature-responsive polymer chains can modulate the diffusion profiles. It is well known that PNIPAAM-grafted porous membrane shows positive control of solute diffusion, which means that the fast diffusion occurs through the opened gates at higher temperature, while a PNIPAAm-grafted nylon capsule exhibits negative control of solute diffusion by blocking the solutes from passing through the surface of the membrane above the LCST [39]. The photoreactive phenylazido group was introduced into PNIPAAM chains to fabricate a temperature-responsive porous membrane immobilized with PNIPAAM [40]. In this study, two types of azidophenyl-derivatized PNIPAAM were synthesized: (i) azophenyl group terminated PNIPAAM on one side and (ii) azidophenyl-derivatized PNIPAAM-*co*-PAAc copolymer by substituting azidophenyl group on each carboxylic group of PAAc. When a microporous membrane was filled with thick hydrogels, it showed different temperature dependences of permeation from another micoporous membrane, whose pore surfaces were covered with tethered linear chains or thin hydrogels. The latter membrane showed that the grafted chains or thin hydrogels swelled below the LCST to close the pores and deswelled above the LCST to open the pores. In contrast, the thick hydrogels filled in the former membrane swelled below the LCST to enhance the permeability and deswelled above the LCST to decrease the permeability.

Hydrophobic and ionizable hydrophilic groups were introduced in PNIPAAM chains as a random copolymer form, and this copolymer was grafted onto silica beads. In this study, poly(*N*-isopropylacrylamide-*co*-acrylicacid-*co*-*N*-*tert*-butylacrylamide) [poly(NIPAAM-*co*-AAc-*co*-*t*BAAm)] hydrogel was grafted onto silica beads and evaluated as a column matrix for cation-exchange thermoresponsive chromatography [41]. In another study, 2-carboxyisopropylacrylamide was copolymerized with PNIPAAM, and this copolymer was grafted onto a polystyrene Petri dish and exhibited an intensified cell detachment over when only PNIPAAM chains were grafted on the dish under the LCST of PNIPAAM [42].

In addition to NIPAAM derivates, temperature-responsive elastin-like polypeptides (ELPs) was also investigated as surface modifier of solid nanoparticles [12]. For example, ELP was adsorbed onto gold nanoparticles functionalized with a self-assembled monolayer (SAM) of mercaptoundecanoic acid. Reversible formation of large aggregates was observed above the LCST of ELP because of interparticle hydrophobic interaction. More recently, the ELP fusion protein (ELP-thioredoxin), which was recombinantly synthesized with a temperature-responsive polypeptide tail (ELP), was immobilized onto a patterned hydrophobic surface. The ELP moiety of this fusion protein reversibly adsorbed onto a hydrophobic SAM above its inverse phase transition temperature (e.g., LCST). In these studies, this method, where an ELP fusion protein was reversibly addressed to chemically distinct regions of a patterned surface by temperature change, was termed as *thermodynamically reversible addressing of proteins* (TRAPs).

10.2.1.5 Temperature-Responsive Copolymer Bioconjugates Various temperature-responsive bioconjugates have been synthesized to address the disadvantages of biomaterials. For example, streptavidin (SA) has been known to bear a high binding affinity for biotin, which encourages SA to be used in affinity separations, laboratory assays, and clinical diagnostics [43]. When PNIPAAM was conjugated to a specific site that is located on the loop above the biotin-binding pocket of SA, the temperature-responsiveness of PNIPAAM chains controlled the reversible binding activity with biotin and triggered the release of bound biotin as a "molecular gate" [44]. Another temperature-responsive copolymer, *N*,*N*-dimethylacrylamide (DMAAm) was also selected as site-specific SA conjugate [45]. In this study, 4-phenylazophenyl acrylate was copolymerized with DMAAm to enhance the photoresponsive control of enzyme activity of biotin. The conjugation site, molecular weight, and type of stimuli-responsive polymer should be considered to optimize the switching activity. Through the self-assembly via hybridization of two single-chain DNA sequences that were conjugated on SA at the end of a PNIPAAM chain, the location and length of the PNIPAAm chains can be controlled.

A substrate peptide of protein kinase A (PKA) was reported as a conjugate with PNIPAAM [46]. PKA forms one of the most important intracellular signals in cellular signal transduction. The conjugates containing the NIPAAM unit and the substrate peptide unit showed an increase in its LCST from 36.7°C to 40°C in response to phosphorylation by activated PKA. This conjugate was designed to form a micellar structure. Ethylene oxide blocks were used as the outer shell and a copolymer of NIPAAm with *N*-methacryloyl bearing the substrate peptide sequence was designed as inner core. The alternation of the LCST of the inner core moiety by a PKA signal suggests that encapsulation of drugs into this micelle might result in an intelligent drug delivery system that communicates with an intracellular signal.

A cell-adhesive motif was introduced into the structure of PNIPAAM-based hydrogel beads [47]. These hydrogels were fabricated by copolymerization of NIPAAM, *N*-aminoethyl methacrylate (AEMA), and cross-linkable monomer, BIS in the presence of calcium alginate that acts a temporal spherical mold. The conjugation of cell-adhesive motif, GRGDY (Gly-Arg-Gly-Asp-Tyr), was selectively incorporated into the surface region of these precollapsed hydrogels above the LCST. The temperature-responsive conjugates were proposed as cell culture substrate for chondocytes. The cell, which was attached on the bead above the LCST, was readily detached from the swollen beads below the LCST at 37°C.

10.2.1.6 Temperature-Responsive Biopolymers and Polypeptides Apart from synthesized polymers, various biopolymers such as gelatin [48], agarose [49], gellan benzyl ester [50], and modified artificial proteins [51] also exhibited temperature-responsiveness. The responsive mechanism is by forming helix conformation that resulted in physical cross-linking. For example, the helix domains on gellan and its derivatives such as gellan benzyl ester would aggregate to cross-linked structures because of the hydrophobic interaction on changing temperature, which led to gelation. Gelatin which is a thermally reversible hydrogel would form gels in aqueous solution by a cooling process as the chains transform their conformation from random coil to triple helix, during which physical junctions are promoted and gel is formed. Mixtures of pectin and chitosan could form thermoreversible gels by lowering the temperature. The gelation temperature was 30°C at 25 wt% of pectin and 48°C at 75 wt% of pectin.

Another biopolymer-recombinant artificial ELPs composed of Val-Pro-Gly-Xaa-Gly amino acid repeat units (where Xaa is a "guest residue" except proline) were also reported to undergo a thermally reversible phase transition [52,53]. A block copolypeptide bearing ELP segment and silk-like segment, where the "guest residue" Xaa was valine, was reported to undergo sol–gel transition under physiological conditions. Recently, applications of ELPs have broadened into protein purification, nanosized metal-surface modification, and temperature-responsive molecular valves in a porous membrane.

10.2.1.7 Typical Applications of Temperature-Responsive Copolymers Recently, temperature-responsive polymers have been applied for a gene carrier system. It has been recognized that introducing a stimuli-responsive property into a polymeric gene carrier can provide progressive advantages, such as nonpathogenic and nonimmunogenic problems, high transfection efficiency, and complex formation/dissociation with therapeutics controlled by stimuli. The dissociation of therapeutics from polymeric carriers within cytoplasm has been proposed by introducing a temperature-responsive mechanism produced by local hyperthermia. Two types of gene carriers have been suggested: soluble temperature-responsive polymeric carrier, which can bind anionic DNA or hydrophobic anticancer drugs at body temperature and dissociate them at hyperthermic temperatures [54,55], and the temperature-responsive micelle or nanohydrogel, which can contain anionic DNA or hydrophobic anticancer drugs in the core reservoir or the hydrogel at body temperature and release them at hyperthermic temperatures.

Typically, recently, several temperature-responsive polymeric gene carrier systems were devised to have their LCST between body temperature and 42°C, at which hyperthermic treatments of cancer patients are generally performed [55]. Targeting cancer chemotherapeutics to tumors can have benefits, such as avoiding cytotoxic chemotherapeutics, which also damage healthy tissues. Two different types of thermally responsive polymers were conjugated with rhodamine: a genetically engineered ELP and a copolymer of NIPAAM and AAm. From *in vivo* fluorescence video microscopy, it was confirmed that local hyperthermia at 42°C increased the accumulation of the conjugates in the solid tumor site. At a temperature higher than the LCST of these temperature-responsive polymers, precipitation of the polymer–drug conjugates took place. Both of the dissociation of the conjugates at intercellular cytoplasm and the accumulation of the conjugates at specific sites like tumor cells play important role in effective gene carrier system.

Temperature-responsive polymer has been extensively investigated as targeted drug delivery systems. Hydrophobic drugs, proteins, or negatively charged DNA can be loaded into the core of the micelles whose dimensions are usually less than 100 nm. Temperature-responsive micelles have an effective advantage for targeted drug delivery over general micelle structures: they have a double targeting mechanism in both active and passive manners. They also solve the problem of nonselective scavenging by the reticuloendothelial system (RES) [56,57]. Several methods have been reported to synthesize polymers that can be used in preparation of temperature-responsive micelles, such as terminal hydrophobization of NIPAAm, block or graft compolymerization of NIPAAm with hydrophobic monomers, and/or hydrophilic monomers [32,33,58].

Temperature-responsive copolymers can also be used in intelligent on/off systems for separation, permeation, actuation, and detachment control. For example, various ligands such as lectin [59], maltose [60], and antibodies [61] have been conjugated with stimuli-responsive polymers to separate targeted biomolecules. One strategy of protein purification is the precipitation of the temperature-responsive polymer after binding the targeted proteins above the LCST. By incorporating VI moiety as a metal binding site into PNIPAAM-based temperature-responsive copolymer, metal-loaded copolymers were prepared as metal affinity macroligands [62]. The temperature control of this macroligand provided the effective purification of histidine-tagged single-chain Fv-antibody fragments from the fermentation broth by metal chelate affinity precipitation. Meanwhile, the ligand-conjugated polymer chains were grafted or modified on column beads and specific proteins were separated from flowing solutions. The separated proteins, which were bound by the immobilized ligands, could be reversibly recovered from the surface of columns by adjusting temperature below the LCST. Purification of antibodies was reported by using temperature-responsive PNIPAAM and dextran-derivative conjugate as a model [63]. The main idea was the combination of the temperature sensitivity of PNIPAAM and the affinity of antibodies recognizing the polysaccharide antigen, carboxymethyl dextran benzylamide sulfonate/sulfate. This conjugate was obtained by grafting amino-terminated PNIPAAM onto this dextran derivative. The purified antibodies and the polymer conjugates could be readily separated and recycled by a thermal-dependent recovery process, which gave a rapid and sensitive procedure to separate antibodies.

Stimuli-responsive polymers could be utilized to manufacture microfluidic systems that can self-control microscale flow as well as separate, purify, analyze, and deliver biomolecules. A switchable protein trap was obtained with a surface coated with an end-tethered monolayer of PNIPAAM [64]. The monolayer temperature was controlled by a microhot plate device containing gold or platinum heater lines deposited on the thin layer. This device could adsorb and desorb the bound protein in less than 1 s because of the switchable hydrophilic/hydrophobic property of the thin layer by changing the temperature. This surprisingly rapid response time was achieved by the tenuous scale of the polymer layer, which was only 4 nm thick. An array of these devices could be utilized to purify the targeted protein on a large scale while preserving the rapid response time. Also, this device could be applied for artificial organs, for example, artificial pancreas for controlling the release of insulin. However, external power is required for operation of this device, which can restrict their use in practical systems *in vivo*.

10.3 pH-RESPONSIVE COPOLYMERS AND APPLICATIONS

pH-responsive copolymers normally contain pH-responsive segments that consist of ionizable pendants, which can donate or accept protons when the environmental pH changes. The phase transition of the copolymers happens at a specific pH value because of the alternation of the hydrodynamic volume of the polymer chain. The transition from collapsed state to expanded state can be explained by the osmotic pressure exerted by mobile counterions neutralizing the network charges. There are mainly two classes of pH-responsive segments: polyacids and polybases. Polyacids normally contain carboxylic acid group as the representative acidic pendant group, which accept protons at low pH and release protons at neutral and high pH. As a comparison, polybases such as poly(4-vinylpyridine) is protonated at high pH and positively ionized at low pH. pH-responsive copolymers can be fabricated in various forms as temperature-responsive copolymers such as nanoparticles/micelles, hydrogels, and functional surfaces, which encourage their applications in various areas especially in biorelated areas.

10.3.1 pH-Responsive Segments

As mentioned earlier, pH-responsive segments normally can be classified into polyacids and polybases. Polyacid segments undergo an ionization/deionization transition from pH 4–8. Figure 10.3 lists typical structures of polyacid segments. Poly(acrylic acid) (PAAc) [65] and poly(methacrylic acid) (PMAAc) [66] are two of the most frequently reported polyacids, which with pK_a values in the range of 5 will release protons and swell under basic pH. Compared to PAAc, PMAAc presented a more abrupt phase transition because the methyl groups in the main chain would induce stronger hydrophobic interaction as the aggregation force. Introducing more hydrophobic moieties can offer a more compact conformation in the uncharged state and more dramatically discontinuous phase. Examples lie in poly(2-ethyl acrylic acid) (PEAAc) and poly(2-propyl acrylic acid) (PPAAc), which form more compact conformational structures at low pH [67,68]. Polymer segments containing sulfonamide groups (derivatives of *p*-aminobenzene sulfonamide) was regarded as another polyacid, which showed pK_a values ranging from pH 3 to pH 11 [69,70]. The pH-responsiveness comes from different pendant substituents at the sulfonamide group acting as electron-withdrawing/electron-donating groups. This type of polyacid showed a controllable, narrow pH range and improved sensitivity compared with that of normal carboxylic-based polyacids.

Typical pH-responsive polybase segments were shown in Figure 10.4. Poly(*N,N*′-dimethyl aminoethyl methacrylate) (PDMAEMA) and poly(*N,N*′-diethyl aminoethyl methacrylate) (PDEAEMA) are two mostly reported pH-responsive polybases with pH changing point at pH 6.5 and 7.5, respectively [71]. The amino groups in the side chains would gain protons under acidic conditions and release them under basic conditions, which resulted in the phase changes. Poly(4 or 2-vinylpyridine) (PVP) is another polybase that

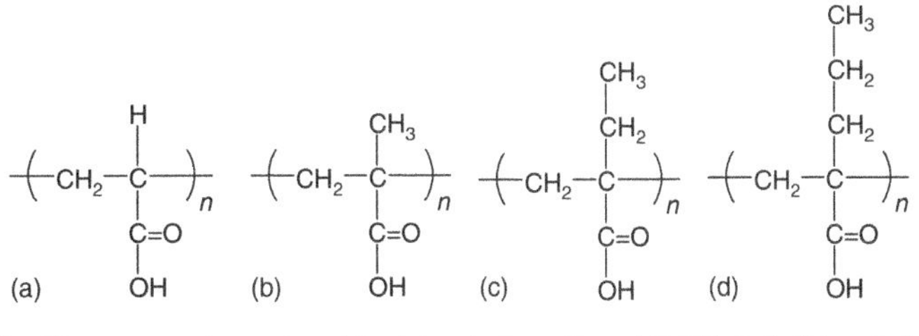

Figure 10.3 Representative polyacids: (a) poly(acrylic acid) (PAAc), (b) poly(methacrylic acid) (PMAAc), (c) poly(2-ethyl acrylic acid) (PEAAc), and (d) poly(2-propyl acrylic acid) (PPAAc).

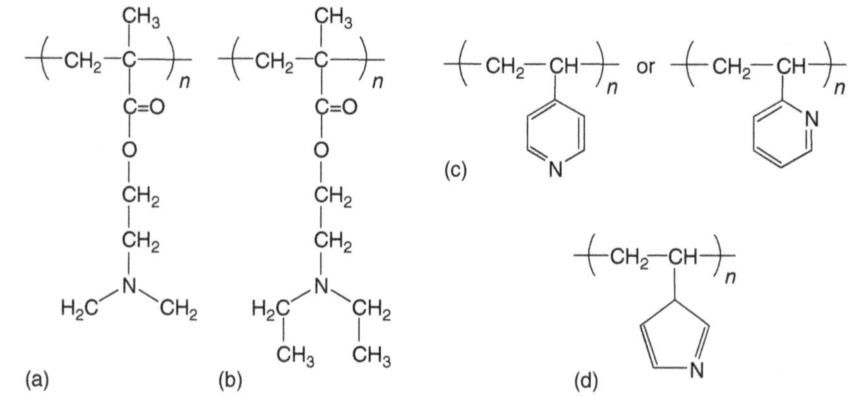

Figure 10.4 Representative pH-responsive polybases: (a) poly(*N*,*N*′-dimethyl aminoethyl methacrylate) (PDMAEMA), (b) poly(*N*,*N*-diethyl aminoethyl methacrylate) (PDEAEMA), (c) poly(4 or 2-vinylpyridine) (PVP), and (d) poly(vinyl imidazole).

undergoes phase transition at pH 5 because of the protonation of pyridine [72]. Other polybase examples include poly(vinyl imidazole) (PVI) [73] and poly(*N*-acryloyl-*N*′-alkypiperazine) [26].

In addition to polyacids and polybases, there are some other polymers/copolymers exhibiting pH responsiveness. For example, poly(ortho ester) is relatively stable at physiological pH, but degradation of this polymer happens under mildly acidic conditions [74]. This is because the ortho ester groups hydrolyze to form pentaerythritol dipropionate and diol monomer and further hydrolyze to pentaerythritol and acetic acid. This pH-responsive degradation could be used as forming hydrogel for pulsatile insulin delivery. Poly(β-amino ester) is another pH-responsive biodegradable polymer, which would be rapidly soluble at pH below 6.5. It was reported that degradation of this polymer occurred more slowly at pH 7.4 than at pH 5.1 [75].

Some biopolymers such as polysaccharides were reported to have pH-responsiveness. For example, alginate, an acidic polysaccharide bearing carboxylic acid groups would undergo phase transition around its pK_a at pH 3–4. Chitosan is the N-deacetylated derivative of chitin, which would precipitate in an aqueous solution at pH over 6–6.2 because of high hydrogen bonding strength between –OH groups and uncharged amine groups in the polymer chain [76].

Synthetic polypeptides also exhibited pH responsiveness due to the existence of amino acid groups on the polymer chain. For example, polycysteine ($pK_a = 8.4$), polyaspartic acid ($pK_a = 3.9$), poly(glutamic acid) ($pK_a = 4.1$), polyhistidine ($pK_a = 6.0$), polylysine ($pK_a = 10.5$), and polyarginine ($pK_a = 12.5$) would undergo a pH-induced phase transition around their pK_a [77]. The amino acid residues can be introduced onto synthetic polymers as side chains to produce pH-responsive copolymers.

10.3.2 Polymer Nanoparticles/Micelles Prepared from pH-Responsive Copolymers

pH-responsive copolymers have been used in the preparation of polymer nanoparticles/micelles that have hydrophobic core and hydrophilic corona. Micelles with PDEAEMA as core together with PEG, PDMAEMA, and PQDMAEMA (quaternized DMAEMA with benzyl chloride) as shell have been prepared from their corresponding diblock copolymers, respectively. The size of the micelles was controlled by the aggregation polymer number and swelling of the micelle corona.

Triblock copolymers of poly(styrene-*b*-2-vinylpyridine-ethylene glycol) [P(St-*b*-2VP-EO)] was used to prepare pH-responsive core–shell–corona (CSC) micelles [78]. The P2VP shell tuned the size of the micelles ranging from 75 nm at pH > 5 to 135 nm at pH < 5 because of the variable electrostatic repulsion forces between the charged P2VP segments. By employing different molecular weights of PEO, a series of triblock copolymers could be synthesized which could be used to fabricate various micelles. Another class of triblock copolymers of P(EO-*b*-DMAEMA-*b*-DEAEMA) was used to prepare shell cross-linked (SCL) micelles after cross-linking the shell in presence of 1,2-bis(2-iodoethoxy)-ethane (BIEE) [79]. Living radical polymerization such as atom transfer radical polymerization (ATRP) was employed to produce triblock copolymers with varied molecular weights, which in turn control the structures of the SCL micelles. Reversible swelling was observed from these SCL micelle aqueous solutions.

pH-responsive unimer micelles were also prepared from random copolymers of 50 mol% sodium 2-(acrylamido)-2-methylpropanesulfonate (NaAMPS) and a series of monomers bearing long-chain alkyl carboxyl pendants. These copolymers formed unimer micelles at pH < 5 and completely solvated at basic pH [80,81].

Well-defined pH-responsive polymer nanoparticles were prepared in water directly after a simple dialysis procedure without the process of self-assembly from branched copolymers was reported by us (Figure 10.5) [82–84]. The branched AB diblock copolymers of poly(ethylene glycol methacrylate/dimethylamino ethyl methacrylate) [P(PEGMA/DMA)] and poly(ethylene glycol methacrylate)/diethylamino ethyl methacrylate) [P(PEGMA/DEA)] have been synthesized using a one-pot ATRP approach in the presence of a bifunctional monomer, ethylene glycol dimethacrylate (EGDMA). Clear polymer nanoparticle (170–244 nm) suspensions were obtained after a dialysis process. The aqueous solution behavior of the branched and linear copolymers at different pH values (pH 3.7–12) was investigated using dynamic light scattering (DLS). pH responsiveness was observed for both branched copolymer systems. Controlled branching structure is suggested as the key point for direct synthesis of pH-responsive branched copolymer nanoparticles.

10.3.3 pH-Responsive Surfaces and Hydrogels

The simplest method to prepare pH-responsive surfaces could be immobilizing pH-responsive polymers onto a substrate surface. For example, PAAc and cationic

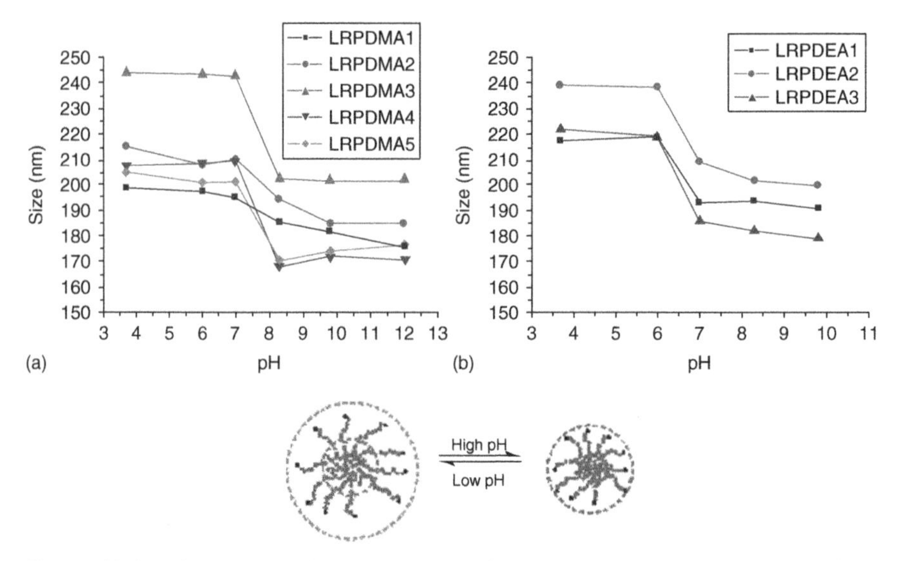

Figure 10.5 Direct synthesis of pH-responsive branched copolymer nanoparticles. (Reprinted with permission from [84]. Copyright © 2011 Royal Society of Chemistry.) (*See insert for color representation.*)

polypeptide were immobilized onto poly(vinylidene fluoride) (PVDF) and poly(tetrafluoroethylene) membrane separately [85,86]. It was reported that block copolymers of PNIPAAM-*b*-MAAc can be introduced onto porous polyethylene membrane, which could offer multi-sensitive surface properties. AAc and MAAc monomers were graft-copolymerized onto ozone-pretreated PVDF to obtain the PAAc-*g*-PVDF copolymers, which was used to construct a pH-responsive membrane. More recently, PAAc and PAAm were used to produce polyelectrolyte multilayers (PEMs) through layer-by-layer technique. At low pH, the protonated PAAs were hydrogen-bonded with PAAm, thus facilitating the fabrication of PEMs. However, the PEMs were destroyed at physiological pH because the ionized carboxyl groups could not bond hydrogen with PAAm.

pH-responsive hydrogels consist of ionic pendant groups that could ionize and develop charges on the polymeric networks, and the hydrogels would swell because of the generated electrostatic repulsion forces. For example, interpolymeric hydrogels containing *N*-vinyl pyrrolidone (NVP), polyethylene glycol diacrylate (PAC), and chitosan were prepared by free radical polymerization using azobisisobutyronitrile (AIBN) as initiator and BIS as cross-linker, followed by introducing pH-responsive chitosan [87]. The whole hydrogel exhibited strong pH sensitivity. In the other report, a semi-IPN-containing chitosan and poly(vinyl pyrrolidone) were obtained by cross-linking chitosan and poly(vinyl pyrrolidone) blend with glutaraldehyde. Air-dried and freeze-dried hydrogels were prepared and compared in this study. Porous freeze-dried hydrogels showed superior pH-dependent swelling properties over nonporous air-dried hydrogels [88].

10.3.4 Typical Applications of pH-Responsive Copolymers

pH-responsive copolymers can find their applications in various fields. Recently, applications in biorelated areas such as drug delivery and gene delivery received more and more attention. pH-responsive copolymers have been applied to drug delivery systems to protect drugs from the acidic stomach and release in neutral intestine. For example, oral delivery of polypeptide hormone salmon calcitonin was reported utilizing pH-responsive hydrogel. P(MAAc-*g*-EG) was employed to build the hydrogel film that was designed to have pH responsiveness for targeted release [66]. In another report, copolymer hydrogels from NIPAAm, butylmethacrylate (BMA), and AAc were designed to target the release of drugs into neutral intestine by pH and the LCST change of NIPAAm. The AAc groups are uncharged in the acidic environment prevailing in the stomach, leading to their hydrophobic characteristics, which not only increased the interaction between polymer chains and hydrogen bonding, but also decreased the LCST of NIPAAm below body temperature leaving the hydrogels tightly collapsed.

Intracellular delivery of genes has attracted much interest recently, as this strategy can give cells the capability of producing therapeutic proteins directly from the delivered genes. However, the delivery is hampered by two factors: effective targeting to specific sites and successful transportation of the therapeutics to cytoplasm, avoiding nonproductive intracellular trafficking. PEAAc and PPAAc were reported to enhance the disruption of endosomal membrane by undergoing pH responsive hydrophobical changes. PPAAc [68], which has the longer hydrophobic pendant, disrupted the red blood cells 15 times more efficiently than PEAAc at pH 6.1. The effect of protein binding was investigated on the pH-dependent membrane-disruptive activity of PPAAc by measuring the hemolytic activity of PPAAc–SA model complexes with different proteins. The PPAAc–SA complex showed similar hemolytic ability as that of free PPAAc. PPAAc was also incorporated into lipoplex carriers containing cationic lipid and plasmid DNA. A significant enhancement of transfections and serum stability in an *in vitro* cell culture model was observed and compared with lipoplex carriers without PPAAc. An *in vivo* murine excisional wound healing model confirmed that pH-sensitive, membrane-disruptive polymers enhanced the release of plasmid DNA for expression of therapeutics, from the acidic endosomal compartment to the cytoplasm.

Furthermore, effective DNA carriers were prepared from pseudopeptides, poly(L-lysine isophthalamide) and poly(L-lysine dodecanamide) containing carboxylic acid side groups. These polymers changed their conformations into hypercoiled structures at endosomal pH, leading to high transfection efficiency [89]. Poly(L-lysine dodecanamide) bearing a long-chain aliphatic dodecyl moiety displayed enhanced cell lysis below pH 7.0 compared with poly(L-lysine isophthalamide) bearing the aromatic isophthaloyl moiety. On the other hand, pH responsive polybase DNA carriers demonstrate a different mechanism for endosomal membrane disruption. At endosomal pH (under physiological pH), these polymers remain protonated and are positively charged, leading to increased interactions between the polymeric cations and the negatively charged membrane phospholipids. By these increased interactions, transfection efficiency of these DNA carriers through endosome lipid bilayer can be enhanced.

10.4 BIOLOGICALLY RESPONSIVE COPOLYMERS AND APPLICATIONS

Smart polymers are becoming increasingly important in the context of biomedical applications. Biologically responsive copolymers have found their applications in areas such as controlled drug delivery, biosensing/diagnostics, smart films/matrices for tissue engineering, and *in situ* construction of structural networks. This approach is a form of biomimicry, since many biomacromolecules are known to dramatically alter their conformation and degree of self-assembly in response to the presence of specific chemical species in their surroundings. Biologically responsive copolymers can be glucose-responsive, enzyme-responsive, antigen-responsive, and even redox/thiol-responsive systems. In this section, we focus on the glucose-responsive and enzyme-responsive copolymers and their applications because of their relatively wide applications compared with other biologically responsive copolymers.

10.4.1 Glucose-Responsive Copolymers and Applications

Although a variety of specific biological responses can be envisioned and have been reported in the literature, copolymers that respond to glucose have received considerable attention because of their potential applications in both glucose sensing and insulin delivery. Diabetes mellitus, commonly referred to as *diabetes*, is a chronic disease characterized by insufficient production or ineffective usage of insulin. Treatment generally involves regular monitoring of blood sugar concentrations and subcutaneously administering insulin several times per day. This need for consistent patient vigilance often leads to poor compliance with the prescribed therapy. One potential route proposed to increase patient compliance is the development and use of smart delivery systems in which insulin delivery is automatically triggered by a rise in blood glucose levels. Glucose-responsive copolymers are typically based on enzymatic oxidation of glucose by glucose oxidase (GOx), binding of glucose with concanavalin A (ConA), or reversible covalent bond formation between glucose and boronic acids.

10.4.1.1 Glucose-Responsive Copolymers Based on Glucose-GOx Glucose-responsive polymers/copolymers are mainly based on GOx-catalyzed reactions of glucose with oxygen. Actually, glucose sensitivity is not caused by direct interaction of glucose with the responsive polymer, but rather by the response of the polymer to the by-products that result from the enzymatic oxidation of glucose. The enzymatic action of GOx on glucose is highly specific and leads to by-products of gluconic acid and H_2O_2. Therefore, incorporation of a polymer that responds to either of these small molecules can indirectly lead to a glucose-responsive system. Typically, a pH-responsive polymer is loaded or conjugated with GOx, and the gluconic acid by-product that results from the reaction with glucose induces a response in the pH-responsive macromolecule. For applications specifically intended for diabetes therapy, the pH response generally causes swelling or collapse of hydrogel matrix that contains insulin.

For example, control release of insulin was archived through a GOx-conjugated PAAc-modified cellulose film [90]. At neutral and high pH levels, the carboxylate units of the PAA chains were negatively charged and extended because of electrostatic repulsion, which resulted in occlusion of the pores in the cellulose membrane. The gluconic acid that resulted from the addition of glucose led to a local pH reduction, protonation of the PAA carboxylate moieties, and concomitant collapse of the chains obscuring the membrane pores, with the latter event facilitating the release of entrapped insulin. Similar glucose-responsive gating membranes based on PAA-grafted PVDF have been reported as well.

As opposed to gluconic acid production leading to chain collapse in polyacids, the lowering of pH could also lead to chain expansion in the presence of a polybase [91–93]. For example, hydrogel containing cross-linked poly(2-hydroxyethyl methacrylate-*co*-*N*,*N*-dimethylaminoethyl methacrylate) [poly(HEMA-*co*-DMAEMA)] that contained entrapped GOx, catalase, and insulin has been prepared [94]. In this case, as the pH of the hydrogel was lowered during exposure to glucose, amines in the DMAEMA units assumed a positive charge, and electrostatic repulsion caused hydrogel swelling and insulin escape.

The need for increased biocompatibility in GOx-based responsive materials has recently led to the incorporation of poly(ethylene glycol) (PEG) grafts and other nontoxic, nonimmunogenic, biocompatible polymers, such as chitosan [95,96]. In addition, while the above-mentioned examples were largely hydrogels that rely on changes in pH resulting from production of gluconic acid during the oxidation of glucose, it is also possible to prepare other pH-responsive polymer morphologies. For example, pH-sensitive liposomes were fabricated from β-palmitoyl-γ-decyl-L-α-phosphatidylethanolamine and oleic acid [97]. The liposomes were infused with GOx and insulin, and the acidic environment that resulted from glucose oxidation destabilized the liposomes and led to insulin delivery. Moreover, it is also possible to prepare polymers that respond to the increased concentration of the H_2O_2 by-product instead of gluconic acid.

10.4.1.2 Glucose-Responsive Copolymers Based on ConA Lectins are proteins that specifically bind carbohydrates. Competitive binding of glucose with glycopolymer–lectin can be utilized in the preparation of glucose-responsive system [98]. Most lectins are multivalent, glycopolymers tend to cross-link and/or aggregate in their presence; however, this aggregation can be disrupted by introducing a competitively binding saccharide.

Monosubstituted conjugates of glucosyl-terminal PEG (G-PEG) and insulin were synthesized, which were bound to ConA that was attached pendantly along a PEG poly(vinylpyrrolidone-*co*-acrylic acid) backbone [99]. When the concentration of glucose increased, competitive binding of glucose with ConA led to displacement and release of the G-PEG–insulin conjugates. In another sample, a ConA-glycogen gel that underwent a gel–sol transition in the presence of glucose because of the preferred binding of ConA with free glucose over the glycogen-containing gel was synthesized [100]. Meanwhile, a glucose-responsive hydrogel was prepared via copolymerization of ConA vinyl macromonomers with a monomer modified with pendent glucose units [101]. The addition of free glucose caused the glycopolymer–ConA complex to dissociate and the gel to swell, with the degree of swelling depending on glucose concentration. Responsive systems in which ConA is simply entrapped in the hydrogel can lead to leakage of the protein and irreversible swelling. However, the copolymerized hydrogel described earlier prevented ConA leakage, and the hydrogels were demonstrated to be reversibly responsive. The response of these polymers was specific for glucose or mannose, while other sugars caused no response [102–104].

10.4.1.3 Glucose-Responsive Copolymers Based on Boronic Acid–Diol Complexation Normally, the glucose-responsive systems based on GOx and lectins are versatile and highly specific. However, the reliance on protein-based components may limit the applications under nonbiological conditions or over longer time spans that might promote denaturation. An alternative mechanism of glucose response relies on polymers endowed with purely synthetic components, namely boronic acid moieties. The ability of boronic acids to reversibly complex with sugars has led to their being heavily employed as glucose sensors. Boronic acids are unique because their water solubility can be tuned by changes in pH or diol concentration. In aqueous systems, boronic acids exist in equilibrium between an undissociated neutral trigonal form and a dissociated anionic tetrahedral form. In the presence of 1,2- or 1,3-diols, cyclic boronic esters between the neutral boronic acid and a diol are generally considered hydrolytically unstable. On the other hand, the anionic form is able to reversibly bind with diols to form a boronate ester, shifting the equilibria to the anionic forms. Polymers containing neutral boronic acid groups are generally hydrophobic, whereas the anionic boronate groups impart water solubility. As the concentration of glucose is increased, the ratio of the anionic forms to the neutral form increases and the hydrophilicity of the system also increases. Therefore, as a result of the interrelated equilibria discussed earlier, the solubility of boronic acid-containing polymers is dependent not only on pH but also on the concentration of compatible diols, which can be utilized in the preparation of glucose-responsive systems [105–108].

A glucose-responsive hydrogel composed of terpolymers of 3-acrylamidophenylboronic acid (APBA), (*N*,*N*-dimethylamino)propylacrylamide (DMAPA), and DMA was reported [109]. The boronic acid groups present in the terpolymer were allowed to complex with poly(vinyl alcohol) (PVA) at physiological pH. Addition of glucose led to competitive displacement of the PVA, and the resulting decrease in cross-link density led to swelling of the gel. In similar systems, responsive hydrogels were made by incorporating PVA with poly(*N*-vinyl-2-pyrrolidone-*co*-APBA) or poly(DMA-*co*-3-methacrylamidophenylboronic acid-*co*-DMAPA-*co*-butyl methacrylate). In another sample, glucose-responsive polymeric hydrogel was prepared via copolymerization of NIPAM with APBA and *N,N*-methylene-bis-acrylamide (MBA) as a cross-linker [107].

10.4.1.4 Enzyme-Responsive Copolymers and Applications Recently, there arose great demand in design of materials that undergo macroscopic property changes when triggered by the selective catalytic actions of enzymes. Sensitivity of this type is unique because enzymes are highly selective in their reactivity, are operable under mild conditions present *in vivo*, and are vital components in many biological pathways. Enzyme-responsive materials are typically composed of an enzyme-sensitive substrate and another component that directs or controls interactions that lead to macroscopic transitions. Catalytic action of the enzyme on the substrate can lead to changes in supramolecular architectures, swelling/collapse of gels, or the transformation of surface properties.

There was a report concerning the use of enzymatic dephosphorylation to induce a sol–gel transition [110,111]. The small molecule fluorenylmethyloxy-carbonyl (FMOC)-tyrosine phosphate was exposed to a phosphatase, and the resulting removal of phosphate groups led to a reduction in electrostatic repulsions, supramolecular assembly by π-stacking of the fluorenyl groups, and eventual gelation. In a similar example, the modification of a genetically engineered variant of spider dragline silk via enzymatic phosphorylation and dephosphorylation was reported.

Incorporation of functional groups that react under enzymatic conditions can also be used in fabrication of enzyme-responsive materials. Exposure of the groups to a specific enzyme can lead to the creation of new covalent linkages that cause a change in macroscopic properties. For example, self-assembly of hydrogels via reversed hydrolysis (ligation) of peptides can be initiated by proteases [112]. Trans-glutaminase, a blood clotting enzyme, has the ability to cross-link the side chains of lysine (Lys) residues with glutamine (Gln) residues within or across peptide chains. This process was exploited for the synthesis of hydrogels of cross-linked function-alized PEG and lysine-containing polypeptides. Controlled gelation kinetics was observed under rather moderate conditions, indicating this route may allow hydro-gel formation in the presence of living cells. Transglutaminase can similarly be used to cross-link naturally occurring polymers in the presence of cells. *Escherichia coli* cells entrapped in a protein–gelatin mixture continued to grow within the hydro-gel and even survived after hydrolytic gel degradation induced by exposure to a protease enzyme. Because some transglutaminase enzymes are only active in the presence of calcium ions, exposure to Ca^{2+} can also trigger enzymatic cross-linking [113,114].

10.5 FIELD-RESPONSIVE COPOLYMERS AND APPLICATIONS

Different from most of the response of traditional stimuli-responsive polymers, which is a relatively slow process, there are alternative mechanisms of stimulation overcoming this issue with applied electric and magnetic fields, or light. In addition to having the ability to be applied or removed near instantaneously, some of these stimuli have the added benefit of potentially being directional, which can give rise to anisotropic deformation. Indeed field-responsive polymers/copolymers are some of the most convenient stimuli from the point of signal control.

10.5.1 Electric-Responsive Copolymers

Electric-responsive polymers/copolymers can transform electrical energy into mechanical energy, which help to establish their promising applications in biomechanics, sensing, artificial muscle actuation, energy transduction, chemical separations, and controlled drug delivery.

Generally, electroresponsive polymers have been investigated in the form of polyelectrolyte hydrogels. Polyelectrolyte gels deform under an electric field because of anisotropic swelling or deswelling as charged ions are directed toward the anode or cathode side of the gel [115]. For instance, under an electric field, inside the hydrolyzed polyacrylamide gels, mobile H^+ ions migrate toward the cathode while the negatively charged immobile acrylate groups in the polymer networks are attracted toward the anode, creating a uniaxial stress within the gel. The region surrounding the anode undergoes the greatest stress while the area in the vicinity of the cathode exhibits the smallest stress. This stress gradient contributes to the anisotropic gel deformation under an electric field [116].

Both synthetic and natural polymers/copolymers have been applied in preparation of electric-responsive materials. For example, natural polymers such as chitosan, chondroitin sulfate, hyaluronic acid, and alginate have been employed. Meanwhile, typical synthetic polymer samples lied in polymers/copolymers synthesized from various functional monomers such as vinyl alcohol, allylamine, acrylonitrile, 2-acrylamido-2-methylpropane sulfonic acid, aniline, 2-hydroxyethyl methacrylate, MAAc, AAc, and vinyl sulfonic acids. In some cases, a combination of natural and synthetic components has been employed. For instance, a semi-IPN hydrogel of PHEMA and chitosan has been prepared as electric-responsive materials [117]. It was found that the response to an applied electric field was affected by several factors such as blending rate and angle. In another report, the electromechanical response in a direct current electric field of cross-linked strong acid hydrogels composed of sulfonated polystyrene or sulfonated poly(styrene-*b*-ethylene-*co*-butylene-*b*-styrene) in different electrolyte solutions of varying concentrations was investigated in detail [118]. Recently, an electric-responsive hydrogel based on PVA and poly(sodium maleate-*co*-sodium acrylate) was fabricated. It was observed that hydrogel deformation increased when the concentration of NaCl or the electric voltage was increased.

10.5.2 Magneto-Responsive Copolymers

Generally, magneto-responsive polymeric materials are prepared through the incorporation of inorganic magnetic particles (or nanoparticles) into a three-dimensional cross-linked polymer network, whose shape and size distortion occur reversibly and instantaneously in presence of a nonuniform magnetic field. In this case, the magnetophoretic force conferred to the polymeric material as a result of the magnetic susceptibility of the particles has led to such materials receiving significant attention for use as soft biomimetic actuators, sensors, cancer therapy agents, artificial muscles, switches, separation media, membranes, and drug delivery systems. In uniform magnetic fields, there is a lack of magnetic field–particle interactions, but particle–particle interactions arise from the creation of induced magnetic dipoles. Particle assembly within the surrounding polymer matrix can lead to dramatic transformations in material properties.

Apart from traditional methods of physically incorporating magnetic particles (e.g., Fe_3O_4) into polymer networks such as PNIPAM and PVA hydrogels, there is an increasing interest in covalent immobilization of polymer chains directly to the surface of magnetic particles. For example, nitroxide-mediated radical polymerization, one type of living radical polymerization technique, was applied to synthesize polymer surfactants that can stabilize magnetic nanoparticles [119–123]. Surface-initiated ATRP was applied in preparation of magnetic-responsive gels, during which iron nanoparticles can serve as cross-linkers, eliminating the need for a conventional cross-linking agent [124]. In another case, after ligand exchange to obtain a dispersion of polyacrylonitrile-stabilized ferromagnetic cobalt (Co) nanoparticles, a film was cast under an applied magnetic field [121]. Stabilization and pyrolysis of the resulting nanoparticle chains led to 1D carbon nanoparticle chains with cobalt inclusions.

10.5.3 Light-Responsive Copolymers

Normally, light-responsive polymers/copolymers have specific functional groups along the polymer backbone or side chains that can undergo light-induced structural transformations. Possible applications of photoresponsive polymers include reversible optical storage, polymer viscosity control, photomechanical transduction and actuation, bioactivity switching of proteins, tissue engineering, and drug delivery. An important aspect of light-sensitive polymer systems is that using irradiation as a stimulus is a relatively straightforward, noninvasive mechanism to induce responsive behavior.

Polymers/copolymers containing azobenzene groups are the mostly well-studied light-responsive polymer materials [125–127]. Azobenzene is a chromophore with an irradiation-induced *cis*-to-*trans* isomerization that is accompanied by a fast and complete change in electric structure, geometric shape, and polarity (Figure 10.6). For instance, although the more stable *trans*-azobenzene has no dipole moment, the cis form is quite polar, having a dipole moment. By incorporating azobenzene derivatives into polymer structures, materials with variable shape, polarity, and self-assembly behavior can be obtained [128]. Incorporation of azobenzene groups onto methylcellulose and methylcellulose–cyclodextrin inclusion complexes allowed phototuning of the sol–gel transition and cloud points in aqueous media. Photochromic derivatives

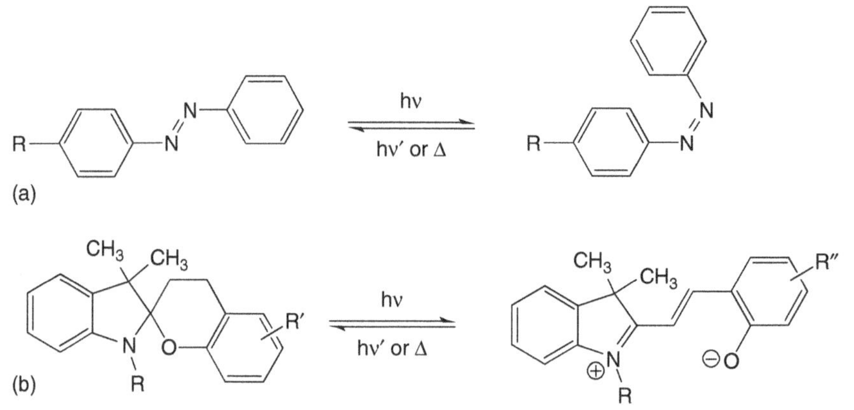

Figure 10.6 Reversible photoinduced transformations of (a) azobenzene and (b) spiropyran derivatives.

of elastin-like polypeptides [poly(VPGVG)] were prepared as well [129]. The inherent thermoresponsive nature of the polymers could be tuned by incorporating one azobenzene moiety for every 30 amino acid residues. Azobenzene-containing block copolymers were employed in preparation of photoresponsive micelles and vesicles because of the change in polarity that accompanies isomerization. Azo chromophores have been incorporated into a variety of other polymeric systems, including poly(*N*-hydroxy propyl methacrylamide) (PHPMA), PAA, PDMAEMA, and PNIPAM. Photoresponsive dendrimers based on azobenzene derivatives have also been reported.

Azobenzene groups can also be used to control the assembly of supramolecular polymers. A new photoresponsive peptide system composed of two ring-shaped cyclic peptides tethered by an azobenzene moiety was reported. When the azobenzene group was in the trans state, the cyclic peptide units demonstrated intermolecular hydrogen bonding to yield extended linear chains. UV-induced isomerization to the cis state led to intramolecular hydrogen bonding and depolymerization of the supramolecular complex. The noninvasive nature in which light can be applied to a polymer solution has led to a variety of applications of photoresponsive polymers in biological systems. For example, azobenzene-containing polymers was applied as switches to reversibly activate enzymes in response to distinct wavelengths of light and used the same concept to reversibly control biotin binding by site-specific conjugation to SA [130].

In addition to polymers/copolymers functionalized with azobenzene moieties, other chromophores have also been used to impart photoresponsive behavior. Several examples exist in which spiropyran derivatives were incorporated terminally or pendantly to establish light sensitivity [131–134]. Spiropyran groups are relatively nonpolar, but irradiation with the appropriate wavelength of light leads to the zwitterionic merocyanine isomer that has a larger dipole moment. The isomerization can be reversed by irradiating with visible light. This concept has been employed to prepare a variety of spiropyran-containing photoresponsive polymers, including PAA, PHPMA, and PNIPAM. Polymer micelles with responsive spiropyran-containing blocks were prepared. A block copolymer of PEO and a spiropyran-containing methacrylate monomer was prepared by ATRP, and the polymeric micelles formed in aqueous solution of the resulting polymer were completely disrupted by UV irradiation and regenerated by irradiation with visible light. This strategy allowed the controlled release of hydrophobic coumarin-102 from the micelle cores via UV irradiation and re-encapsulation of the dye upon irradiation with visible light.

Apart from the isomerization induced by UV irradiation, IR radiation was also employed in the preparation of light-responsive polymer materials as IR radiation can penetrate skin with less risk of damage and might be more applicable for photoactivation of drug carriers within a living system. For example, block copolymer micelles with PEG hydrophilic segments and 2-diazo-1,2-naphthoquinone hydrophobic blocks were fabricated [135]. Irradiation with IR light led to micelle dissociation as a result of the photoinduced rearrangement of the 1,2-napthoquinone units to yield anionic/hydrophilic 3-indenecarboxylate groups. Near-infrared (NIR) dissociable polymeric micelles based on diblock copolymers of PEG and poly(2-nitrobenzyl methacrylate) have also been developed where the latter undergoes photolysis and is converted to hydrophilic PMAAc.

10.6 CONCLUSION

Stimuli-responsive polymers/copolymers have been known for many years; however, their important applications especially in biorelated areas have attracted much interest only recently. The topics covered here are at various levels of maturity, but many opportunities remain to capitalize on their specificity and uniqueness. For many specific applications such as delivery and diagnostics, hybrid materials with the ability to respond to several stimuli would be extremely beneficial. Despite many advances, numerous challenges and thus opportunities remain for making an impact in the field of smart polymers. Apart from the existing stimuli-responsive polymer materials, investigation of new type of stimuli-responsive polymers/copolymers could be a more impetus task because of the new materials and properties that can be obtained.

REFERENCES

1. Jeong, B., Bae, Y. H., Lee, D. S., Kim, S. W. (1997). Biodegradable block copolymers as injectable drug-delivery systems. *Nature*, *388*, 860–862.
2. Jeong, B., Bae, Y. H., Kim, S. W. (1999). Thermoreversible gelation of PEG–PLGA–PEG triblock copolymer aqueous solutions. *Macromolecules*, *32*, 7064–7069.

3. Zhang, R., Liu, J., He, J., Han, B., Zhang, X., Liu, Z., Jiang, T., Hu, G. (2002). Compressed CO_2-assisted formation of reverse micelles of PEO–PPO–PEO copolymer. *Macromolecules*, *35*, 7869–7871.
4. Chung, J. E., Yokoyama, M., Aoyagi, T., Sakurai, Y., Okano, T. (1998). Effect of molecular architecture of hydrophobically modified poly(N-isopropylacrylamide) on the formation of thermoresponsive core–shell micellar drug carriers. *J. Control. Release*, *53*, 119–130.
5. Neradovic, D., van Nostrum, C. F., Hennink, W. E. (2001). Thermoresponsive polymeric micelles with controlled instability based on hydrolytically sensitive N-Isopropylacrylamide copolymers. *Macromolecules*, *34*, 7589–7591.
6. Kohori, F., Yokoyama, M., Sakai, K., Okano, T. (2002). Process design for efficient and controlled drug incorporation into polymeric micelle carrier systems. *J. Control. Release*, *78*, 155–163.
7. Okabe, S., Sugihara, S., Aoshima, S., Shibayama, M. (2002). Heat induced self-assembling of thermosensitive block copolymer. 1. Small-angle neutron scattering study. *Macromolecules*, *35*, 8139–8146.
8. Okubo, M., Ahmad, H., Suzuki, T. (1998). Synthesis of temperature sensitive micron-sized monodispersed composite polymer particles and its application as a carrier for biomolecules. *Colloid Polym. Sci.*, *276*, 470–475.
9. Nakayama, H., Kaetsu, I., Uchida, K., Sakata, S., Tougou, K., Hara, T., Matsubara, Y. (2002). Radiation curing of intelligent coating for controlled release and permeation. *Radiat. Phys. Chem.*, *63*, 521–523.
10. Magoshi, T., Ziani-Cherif, H., Ohya, S., Nakayama, Y., Matsuda, T. (2002). Thermoresponsive heparin coating: heparin conjugated with poly(N-isopropylacrylamide) at one terminus. *Langmuir*, *18*, 4862–4872.
11. Rama Rao, G. V., Krug, M. E., Balamurugan, S., Xu, H., Xu, Q., Lopez, G. P. (2002). Synthesis and characterization of silica-poly(N-isopropylacrylamide) hybrid membranes: switchable molecular filters. *Chem. Mater.*, *14*, 5075–80.
12. Nath, N., Chilkoti, A. (2001). Interfacial phase transition of an environmentally responsive elastin biopolymer adsorbed on functionalized gold nanoparticles studied by colloidal surface plasmon resonance. *J. Am. Chem. Soc.*, *123*, 8197–202.
13. Nordby, M. H., Kjøniksen, A., Nystrom, B., Roots, J. (2003). Thermoreversible gelation of aqueous mixtures of pectin and chitosan. Rheology. *Biomacromolecules*, *4*, 337–343.
14. Chenite, A., Chaput, C., Wang, D., Combes, C., Buschmann, M. D., Hoemann, C. D., Leroux, J. C., Atkinson, B. L., Binette, F., Selmani, A. (2000). Novel injectable neutral solutions of chitosan form biodegradable gels in situ. *Biomaterials*, *21*, 2155–61.
15. Vercruysse, K. P., Li, H., Luo, Y., Prestwich, G, D. (2002). Thermosensitive lanthanide complexes of hyaluronan. *Biomacromolecules*, *3*, 639–643.
16. Jeong, B., Kibbey, M. R., Birnbaum, J. C., Won, Y., Gutowska, A. (2000). Thermogelling biodegradable polymers with hydrophilic backbones: PEG-g-PLGA. *Macromolecules*, *33*, 8317–8322.
17. Jeong, B., Lee, K. M., Gutowska, A., An, Y. H. (2002). Thermogelling biodegradable copolymer aqueous solutions for injectable protein delivery and tissue engineering. *Biomacromolecules*, *3*, 865–868.

18. Lin, H., Cheng, Y. (2001). In-situ thermoreversible gelation of block and star copolymers of poly(ethylene glycol) and poly(N-isopropylacrylamide) of varying architectures. *Macromolecules*, *34*, 3710–3715.
19. Hatefi, A., Amsden, B. (2002). Biodegradable injectable in situ forming drug delivery systems. *J. Control. Release*, *80*, 9–28.
20. Soutar, I., Swanson, L., Thorpe, F. G., Zhu, C. (1996). Fluorescence studies of the dynamic behavior of poly(dimethylacrylamide) and its complex with poly(methacrylic acid) in dilute solution. *Macromolecules*, *29*, 918–924.
21. Garay, M. T., Llamas, M. C., Iglesias, E. (1997). Study of polymer–polymer complexes and blends of poly(N-isopropylacrylamide) with poly(carboxylic acid). 1. Poly(acrylic acid) and poly(methacrylic acid). *Polymer*, *38*, 5091–5096.
22. Schild, H. G. (1992). Poly(N-isopropylacrylamide): experiment, theory and application. *Prog. Polym. Sci.*, *17*, 163–249.
23. Aoki, T., Muramatsu, M., Torii, T., Sanui, K., Ogata, N. (2001). Thermosensitive phase transition of an optically active polymer in aqueous milieu. *Macromolecules*, *34*, 3118–3119.
24. Qiu, Y., Park, K. (2001). Environment-sensitive hydrogels for drug delivery. *Adv. Drug Deliv. Rev.*, *53*, 321–339.
25. Aoyagi, T., Ebara, M., Sakai, K., Sakurai, Y., Okano, T. (2000). Novel bifunctional polymer with reactivity and temperature sensitivity. *J. Biomater. Sci. Polym. Ed.*, *1*, 101–110.
26. Gan, L. H., Roshan D. G., Loh, X. J., Gan, Y. Y. (2001). New stimuliresponsive copolymers of N-acryloyl-N'-alkyl piperazine and methyl methacrylate and their hydrogels. *Polymer*, *42*, 65–69.
27. Chen, G., Hoffman, A. S. (1995). Graft copolymers that exhibit temperature-induced phase transition over a wide range of pH. *Nature*, *373*, 49–52.
28. Okano, T. editor. *Biorelated Polymers and Gels*. Academic Press, San Diego, CA, 1998, pp. 195–250.
29. Kaneko, Y., Nakamura, S., Sakai, K., Aoyagi, T., Kikuchi, A., Sakurai, Y., Okano, T. (1998). Rapid deswelling response of poly(N-isopropylacrylamide) hydrogels by the formation of water release channels using poly(ethylene oxide) graft chains. *Macromolecules*, *31*, 6099–6105.
30. Yoshida, R., Uchida, K., Kaneko, Y., Sakai, K., Kikuchi, A., Sakurai, Y., Okano T. (1995). Comb-type grafted hydrogels with rapid de-swelling response to temperature changes. *Nature*, *374*, 240–242.
31. Zhang, X., Yang, Y., Wang, F., Chung, T. (2002). Thermosensitive poly(N-isopropylacrylamide-co-acrylic acid) hydrogels with expanded network structures and improved oscillating swelling–deswelling properties. *Langmuir*, *18*, 2013–2018.
32. Chung, J. E., Yokoyama, M., Okano, T. (2000). Inner core segment design for drug delivery control of thermoresponsive polymeric micelles. *J. Control. Release*, *65*, 93–103.
33. Kohori, F., Sakai, K., Aoyagi, T., Yokoyama, M., Sakurai, Y., Okano, T. (1998). Preparation and characterization of thermally responsive block copolymer micelles comprising poly(N-isopropylacrylamide-b-DL-lactide). *J. Control. Release*, *55*, 87–98.
34. Topp, M. D. C., Dijkstra, P. J., Talsma, H., Feijen, J. (1997). Thermosensitive micelle-forming block copolymers of poly(ethylene glycol) and poly(N-isopropylacrylamide). *Macromolecules*, *30*, 8518–8520.

35. Topp, M. C. D., Leunen, I. H., Dijkstra, P. J., Tauer, K., Schellenberg, C., Feijen, J. (2000). Quasi-living polymerization of N-isopropylacrylamide onto poly(ethylene glycol). *Macromolecules*, *33*, 4986–4988.

36. Konak, C., Oupicky, D., Chytry, V., Ulbrich, K. (2000). Thermally controlled association in aqueous solutions of diblock copolymers of poly[N-(2-hydroxypropyl)methacrylamide] and poly(N-isopropylacrylamide). *Macromolecules*, *33*, 5318–5320.

37. Ito, S., Ogawa, K., Suzuki, H., Wang, B., Yoshida, R., Kokufuta, E. (1999). Preparation of thermosensitive submicrometer gel particles with anionic and cationic charges. *Langmuir*, *15*, 4289–4294.

38. Akiyoshi, K., Kang, E., Kurumada, S., Sunamoto, J., Principi, T., Winnik, FM. (2000). Controlled association of amphiphilic polymers in water: thermosensitive nanoparticles formed by self assembly of hydrophobically modified pullulans and poly (N-isopropylacrylamides). *Macromolecules*, *33*, 3244–3249.

39. Iwata, H., Oodate, M., Uyama, Y., Amemiya, H., Ikada, Y. (1991). Preparation of temperature sensitive membranes by graft polymerization onto a porous membrane. *J. Membr. Sci.*, *55*, 119–130.

40. Park, Y. S., Ito, Y., Imansish, Y. (1998). Permeation control through porous membranes immobilized with thermosensitive polymer. *Langmuir*, *14*, 910–914.

41. Kobayashi, J., Kikuchi, A., Sakai, K., Okano, T. (2002). Aqueous chromatography utilizing hydrophobicity-modified anionic temperature-responsive hydrogel for stationary phases. *J. Chromatogr. A*, *958*, 109–119.

42. Ebara, M., Yamato, M., Hirose, M., Aoyagi, T., Kikuchi, A., Sakai, K., Okano, T. (2003). Copolymerization of 2-carboxyisopropylacrylamide with N-isopropylacrylamide accelerates cell detachment from grafted surfaces by reducing temperature. *Biomacromolecules*, *4*, 344–349.

43. Stayton, P. S., Shimoboji, T., Long, C., Chilkoti, A., Chen, G., Harris, J. M., Hoffman, A. S. (1995). Control of protein–ligand recognition using a stimuli-responsive polymer. *Nature*, *78*, 472–474.

44. Ding, Z., Long, C. J., Hayashi, Y., Bulmus, E. V., Hoffman, A. S., Stayton, P. S. (1999). Temperature control of biotin binding and release with a streptavidin–polyNIPAAm site-specific conjugate. *Bioconjug. Chem.*, *10*, 395–400.

45. Shimoboji, T., Ding, Z., Stayton, P. S., Hoffman, A. S. (2001). Mechanistic investigation of smart polymer–protein conjugates. *Bioconjugate Chem.*, *12*, 314–319.

46. Katayama, Y., Sonoda, T., Maeda, M. (2001). A polymer micelle responding to the protein kinase A signal. *Macromolecules*, *34*, 8569–73.

47. Kim, M. R., Jeong, J. H., Park, T. G. (2002). Swelling Induced detachment of chondrocytes using RGD-modified poly(N-isopropylacrylamide) hydrogel beads. *Biotechnol. Prog.*, *18*, 495–500.

48. Kuijpers, A. J., Engbers, G. H. M., Feijen, J., De Smedt, S. C., Meyvis, T. K. L., Demeester, J., Krijgsveld, J., Zaat, S. A. J., Dankert, J. (1999). Characterization of the network structure of carbodiimide cross-linked gelatin gels. *Macromolecules*, *32*, 3325–33.

49. Ramzi, M., Rochas, C., Guenet, J. (1998). Structure–properties relation for agarose thermoreversible gels in binary solvents. *Macromolecules*, *31*, 6106–6111.

50. Dentini, M., Desideri, P., Crescenzi, V., Yuguchi, Y., Urakawa, H., Kajiwara, K. (1999). Solution and gelling properties of gellan benzyl esters. *Macromolecules*, *32*, 7109–7115.
51. Petka, W. A., Harden, J. L., McGrath, K. P., Wirtz, D., Tirrell, D. A. (1998). Reversible hydrogels from self-assembling artificial proteins. *Science*, *281*, 389–392.
52. Meyer, D. E., Chilkoti, A. (1999). Purification of recombinant proteins by fusion with thermally-responsive polypeptides. *Nat. Biotechnol.*, *17*, 1112–1115.
53. Meyer, D. E., Trabbic-Carlson, K., Chilkoti, A. (2001). Protein purification by fusion with an environmentally responsive elastin like polypeptide: effect of polypeptide length on the purification of thioredoxin. *Biotechnol. Prog.*, *17*, 720–728.
54. Kurisawa, M., Yokoyama, M., Okano, T. (2000). Gene expression control by temperature with thermo-responsive polymeric gene carriers. *J. Control. Release*, *69*, 127–137.
55. Meyer, D. E., Shin, B. C., Kong, G. A., Dewhirst, M. W., Chilkoti, A. (2001). Drug targeting using thermally responsive polymers and local hyperthermia. *J. Control. Release*, *74*, 213–224.
56. Yang, H. M., Reisfeld, R. (1998). Doxorubicin conjugated with a monoclonal antibody directed to a human melanoma-associated proteoglycan suppresses the growth of established tumor xenografts in nude mice. *Proc. Natl. Acad. Sci. U. S. A.*, *85*, 1189–1193.
57. Thedrez, P., Saccavini, J. C., Nolibe, D., Simoen, J. P., Guerreau, D., Gestin, J. F., Kremer, M., Chatal, J. F. (1989). Biodistribution of indium-111-labeled OC 125 monoclonal antibody after intraperitoneal injection in nude mice intraperitoneally grafted with ovarian carcinoma. *Cancer Res.*, *49*, 3081–3086.
58. Virtanen, J., Holappa, S., Lemmetyinen, H., Tenhu, H. (2002). Aggregation in aqueous poly(N-isopropylacrylamide)-block-poly(ethylene oxide) solutions studied by fluorescence spectroscopy and light scattering. *Macromolecules*, *35*, 4763–4769.
59. Pan, L., Chien, C. (2003). A novel application of thermo-responsive polymer to affinity precipitation of polysaccharide. *J. Biochem. Biophys. Methods*, *55*, 87–94.
60. Hoshino, K., Taniguchi, M., Kitao, T., Morohashi, S., Sasakura, T. (1998). Preparation of a new thermo-responsive adsorbent with maltose as a ligand and its application to affinity precipitation. *Biotechnol. Bioeng.*, *60*, 568–579.
61. Chen, J. P., Hoffman, A. S. (1990). Polymer–protein conjugates. II. Affinity precipitation separation of human immune gamma globulin by a poly(N-isopropylacrylamide)-protein A conjugate. *Biomaterials*, *11*, 631–634.
62. Kumar, A., Wahlund, P. O., Kepka, C., Galaev, I. Y., Mattiasson, B. (2003). Purification of histidine-tagged single-chain Fv-antibody fragments by metal chelate affinity precipitation using thermoresponsive copolymers. *Biotech. Bioeng.*, *84*, 494–503.
63. Anastase-Ravion, S., Ding, Z., Pelle, A., Hoffman, A. S., Letourneur, D. (2001). New antibody purification procedure using a thermally responsive poly(N-isopropylacrylamide)-dextran derivative conjugate. *J. Chromatogr. B*, *761*, 247–254.
64. Huber, D. L., Manginell, R. P., Samara, M. A., Kim, B., Bunker, B. C. (2003). Programmed adsorption and release of proteins in a microfluidic device. *Science*, *301*, 352–354.
65. Philippova, O. E., Hourdet, D., Audebert, R., Khokhlov, A. R. (1997). pH-responsive gels of hydrophobically modified poly(acrylic acid). *Macromolecules*, *30*, 8278–8285.

66. Torres-Lugo, M., Peppas, N. A. (1999). Molecular design and in vitro studies of novel pH-sensitive hydrogels for the oral delivery of calcitonin. *Macromolecules*, *32*, 6646–6651.
67. Tonge, S. R., Tighe, B. J. (2001). Responsive hydrophobically associating polymers: a review of structure and properties. *Adv. Drug Deliv. Rev.*, *53*, 109–122.
68. Murthy, N., Robichaud, J. R., Tirrell, D. A., Stayton, P. S., Hoffman, A. S. (1999). The design and synthesis of polymers for eukaryotic membrane disruption. *J. Control. Release*, *61*, 137–143.
69. Park, S. Y., Bae, Y. H. (1999). Novel pH-sensitive polymers containing sulfonamide groups. *Macromol. Rapid Commun.*, *20*, 269–273.
70. Kang, S. I., Bae, Y. H. (2003). A sulfonamide based glucose-responsive hydrogel with covalently immobilized glucose oxidase and catalase. *J. Control. Release*, *86*, 115–121.
71. Lee, A. S., Butun, V., Vamvakaki, M., Armes, S. P., Pople, J. A., Gast, A. P. (2002). Structure of pH-dependent block copolymer micelles: charge and ionic strength dependence. *Macromolecules*, *35*, 8540–8551.
72. Gohy, J., Lohmeijer, B. G. G., Varshney, S. K., Decamps, B., Leroy, E., Boileau, S., Schubert, U. S. (2002). Stimuli-responsive aqueous micelles from an ABC metallo-supramolecular triblock copolymer. *Macromolecules*, *35*, 9748–9755.
73. Sutton, R. C., Thai, L., Hewitt, J. M., Voycheckand, C. L., Tan, J. S. (1988). Microdomain characterization of styrene–imidazole copolymers. *Macromolecules*, *21*, 2432–2439.
74. Seymour, L. W., Duncan, R., Duffy, J., Ng, S. Y., Heller, J. (1994). Poly (ortho ester) matrices for controlled release of the antitumour agent 5-fluorouracil. *J. Control. Release*, *31*, 201–206.
75. Lynn, D. M., Langer, R. (2000). Degradable poly(b-amino esters):- synthesis, characterization, and self-assembly with plasmid DNA. *J. Am. Chem. Soc.*, *122*, 10761–10768.
76. Hudson, S., Smith, C. Chitin and chitosan: the chemistry and technology of their use as structural materials. In: Kaplan D., editor. *Biopolymers from Renewable Sources*. Springer, Heidelberg, 1998, pp. 96–118.
77. Horton, H. R., Moran, L. A., Ochs, R. S., Rawn, J. D., Scrimgeour, K. G. *Principles of Biochemistry*. Third ed. Prentice Hall, Upper Saddle River, NJ. 2002, pp. 51–80.
78. Gohy, J. F., Willet, N., Varshney, S., Zhang, J. X., Jerome, R. (2001). Core–shell-corona micelles with a responsive shell. *Angew. Chem. Int. Ed.*, *40*, 3214–3216.
79. Bütün, V., Wang, X. S., de Paz Banez, M. V., Robinson, K. L., Billingham, N. C., Armes, S. P. (2000). Synthesis of shell cross-linked micelles at high solids in aqueous media. *Macromolecules*, *33*, 1–3.
80. Yusa, S., Sakakibara, A., Yamamoto, T., Morishima, Y. (2002). Fluorescence studies of pH-responsive unimolecular micelles formed from amphiphilic polysulfonates possessing long chain alkyl carboxyl pendants. *Macromolecules*, *35*, 10182–10188.
81. Yusa, S., Sakakibara, A., Yamamoto, T., Morishima, Y. (2002). Reversible pH-induced formation and disruption of unimolecular micelles of an amphiphilic polyelectrolyte. *Macromolecules*, *35*, 5243–5249.
82. He, T., Adams, D. J., Butler, M. F., Yeoh, C. T., Cooper, A. I., Rannard, S. P. (2007). Direct synthesis of anisotropic polymer nanoparticles. *Angew. Chem. Intl. Ed.*, *46*, 9243–9247.

83. He, T., Adams, D. J., Butler, M. F., Cooper, A. I., Rannard, S. P. (2009). Polymer nanoparticles: shape-directed monomer-to-particle synthesis. *J. Am. Chem. Soc.*, *131*, 1495–1501.

84. He, T., Di Lena, F., Neo, K. C., Chai, C. L. L. (2011). Direct synthesis of pH-responsive polymer nanoparticls based on living radical polymerization and traditional radical polymerization. *Soft Matter*, *7*, 3358–3365.

85. Ying, L., Wang, P., Kang, E. T., Neoh, K. G. (2002). Synthesis and characterization of poly(acrylic acid)-graft-poly(vinylidene fluoride) copolymers and pH-sensitive membranes. *Macromolecules*, *35*, 673–679.

86. Jenkins, D. W., Hudson, S. M. (2001). Review of vinyl graft copolymerization featuring recent advances toward controlled radical based reactions and illustrated with chitin/chitosan trunk polymers. *Chem. Rev.*, *101*, 3245–3273.

87. Shantha, K. L., Harding, D. R. K. (2000). Preparation and in-vitro evaluation of poly[N-vinyl-2-pyrrolidone-polyethylene glycol diacrylate]-chitosan interpolymeric pH-responsive hydrogels for oral drug delivery. *Int. J. Pharm.*, *207*, 65–70.

88. Risbud, M. V., Hardikar, A. A., Bhat, S. V., Bhonde, R. R. (2000). pH sensitive freeze-dried chitosan–polyvinyl pyrrolidone hydrogels as controlled release system for antibiotic delivery. *J. Control. Release*, *68*, 23–30.

89. Eccleston, M. E., Kuiper, M., Gilchrist, F. M., Slater, N. K. H. (2000). pH responsive pseudo-peptides for cell membrane disruption. *J. Control. Release*, *69*, 297–307.

90. Ito, Y., Casolaro, M., Kono, K., Imanishi, Y. (1989). An insulin-releasing system that is responsive to glucose. *J. Control. Release*, *10*, 195–203.

91. Podual, K., Doyle III, F. J., Peppas, N. A. (2000). Preparation and dynamic response of cationic copolymer hydrogels containing glucose oxidase. *Polymer*, *41*, 3975–3983.

92. Podual, K., Doyle III, F. J., Peppas, N. A. (2000). Glucose-sensitivity of glucose oxidase-containing cationic copolymer hydrogels having poly(ethylene glycol) grafts. *J. Control. Release*, *67*, 9–17.

93. Podual, K., Doyle III, F. J., Peppas, N. A. (2000). Dynamic behavior of glucose oxidase-containing microparticles of poly(ethylene glycol)-grafted cationic hydrogels in an environment of changing pH. *Biomaterials*, *21*, 1439–1450.

94. Traitel, T., Cohen, Y., Kost, J. (2000). Characterization of glucose-sensitive insulin release systems in simulated in vivo conditions. *Biomaterials*, *21*, 1679–1687.

95. Ravaine, V., Ancia, C., Catargi, B. (2008). Chemically controlled closed-loop insulin delivery. *J. Control. Release*, *132*, 2–11.

96. Kashyap, N., Viswanad, B., Sharma, G., Bhardwaj, V., Ramarao, P., Kumar, M. N. V. R. (2007). Design and evaluation of biodegradable, biosensitive in situ gelling system for pulsatile delivery of insulin. *Biomaterials*, *28*, 2051–2060.

97. Kim, C. K., Im, E. B., Lim, S. J., Oh, Y. K., Han, S. K. (1994). Development of glucose triggered ph-sensitive liposomes for a potential insulin delivery. *Int. J. Pharm.*, *101*, 191–197.

98. Brownlee, M., Cerami, A. (1979). A glucose-controlled insulin-delivery system: semisynthetic insulin bound to lectin. *Science*, *206*, 1190–1191.

99. Liu, F., Song, S. C., Mix, D., Baudy, M., Kim, S. W. (1997). Glucose-induced release of glycosyl poly(ethylene glycol) insulin bound to a soluble conjugate of concanavalin a. *Bioconjugate Chem.*, *8*, 664–672.

100. Cheng, S. Y., Gross, J., Sambanis, A. (2004). Hybrid pancreatic tissue substitute consisting of recombinant insulin-secreting cells and glucose responsive material. *Biotechnol. Bioeng.*, *87*, 863–873.

101. Miyata, T., Jikihara, A., Nakamae, K., Hoffman, A. S. (2004). Preparation of reversibly glucose-responsive hydrogels by covalent immobilization of lectin in polymer networks having pendent glucose. *J. Biomater. Sci. Polym. Ed.*, *15*, 1085–1098.

102. Miyata, T., Jikihara, A., Hoffman, A. S. (1996). Preparation of poly(2-glucosyloxyethyl methacrylate)-concanavalin a complex hydrogel and its glucose-sensitivity. *Macromol. Chem. Phys.*, *197*, 1135–1146.

103. Nakamae, K., Miyata, T., Jikihara, A., Hoffman, A. S. (1994). Formation of poly(glucosyloxyethyl methacrylate)-concanavalin a complex and its glucose-sensitivity. *J. Biomater. Sci. Polym. Ed.*, *6*, 79–90.

104. Tanna, S., Taylor, M. J., Sahota, T. S., Sawicka, K. (2006). Glucose-responsive UV polymerised dextran-concanavalin a acrylic derivatised mixtures for closed-loop insulin delivery. *Biomaterials*, *27*, 586–597.

105. Lorand, J. P., Edwards, J. O. (1959). Polyol complexes and structure of the benzene boronate ion. *J. Org. Chem.*, *24*, 769–774.

106. Kataoka, K., Miyazaki, H., Okano, T., Sakurai, Y. (1994). Sensitive glucose induced change of the lower critical solution temperature of poly[n,n-(dimethylacrylamide)-co-3-(acrylamido)-phenylboronic acid] in physiological saline. *Macromolecules*, *27*, 1061–1062.

107. Kataoka, K., Miyazaki, H., Bunya, M., Okano, T., Sakurai, Y. (1998). Totally synthetic polymer gels responding to external glucose concentration: their preparation and application to on-off regulation of insulin release. *J. Am. Chem. Soc.*, *120*, 12694–12695.

108. Springsteen, G., Wang, B. (2002). A detailed examination of boronic acid–diol complexation. *Tetrahedron*, *58*, 5291–5300.

109. Hisamitsu, I., Kataoka, K., Okano, T., Sakurai, Y. (1997). Glucose-responsive gel from phenylborate polymer and poly(vinyl alcohol): prompt response at physiological pH through the interaction of borate with amino group in the gel. *Pharm. Res.*, *14*, 289–293.

110. Yang, Z., Gu, H., Fu, D., Gao, P., Lam, J. K., Xu, B. (2004). Enzymatic formation of supramolecular hydrogels. *Adv. Mater.*, *16*, 1440–1444.

111. Yang, Z., Xu, B. (2004). A simple visual assay based on small molecule hydrogels for detecting inhibitors of enzymes. *Chem. Commun.*, *2424–2425*.

112. Toledano, S., Williams, R. J., Jayawarna, V., Ulijin, R. V. (2006). Enzyme-triggered self-assembly of peptide hydrogels via reversed hydrolysis. *J. Am. Chem. Soc.*, *128*, 1070–1071.

113. Ulijin, R. V. (2006). Enzyme-responsive materials: a new class of smart biomaterials. *J. Mater. Chem.*, *16*, 2217–2225.

114. Sanborn, T. J., Messersmith, P. B., Barron, A. E. (2002). In situ crosslinking of a biomimetic peptide-peg hydrogel via thermally triggered activation of factor XIII. *Biomaterials*, *23*, 2703–2710.

115. Filipcsei, G., Feher, J., Zrinyi, M. (2000). Electric field sensitive neutral polymer gels. *J. Mol. Struct.*, *554*, 109–117.

116. Bajpai, A., Shulka, S., Bhanu, S., Kankane, S. (2008). Responsive polymers in controlled drug delivery. *Prog. Polym. Sci.*, *33*, 1088–1118.

117. Kim, S. J., Kim, H. I., Shin, S. R., Kim, S. I. (2004). Electrical behavior of chitosan and poly(hydroxyethyl methacrylate) hydrogel in the contact system. *J. Appl. Polym. Sci.*, *92*, 915–919.

118. Yao, L., Krause, S. (2003). Electromechanical responses of strong acid polymer gels in dc electric fields. *Macromolecules*, *36*, 2055–2065.

119. Benkoski, J. J., Bowles, S. E., Jones, R. L., Douglas, J. F., Pyun, J., Karim, A. (2008). Self-assembly of polymer-coated ferromagnetic nanoparticles into mesoscopic polymer chains. *J. Polym. Sci. Part B: Polym. Phys.*, *46*, 2267–2277.

120. Keng, P. Y., Shim, I., Korth, B. D., Douglas, J. F., Pyun, J. (2007). Synthesis and self-assembly of polymer-coated ferromagnetic nanoparticles. *ACS Nano*, *1*, 279–292.

121. Bowles, S. E., Wu, W., Kowalewski, T., Schalnat, M. C., Davis, R. J., Pemberton, J. E., et al. (2007). Magnetic assembly and pyrolysis of functional ferromagnetic colloids into one-dimensional carbon nanostructures. *J. Am. Chem. Soc.*, *129*, 8694–8695.

122. Benkoski, J. J., Bowles, S. E., Korth, B. D., Jones, R. L., Douglas, J. F., Karim, A, et al. (2007). Field induced formation of mesoscopic polymer chains from functional ferromagnetic colloids. *J. Am. Chem. Soc.*, *129*, 6291–6297.

123. Korth, B. D., Keng, P., Shim, I., Bowles, S. E., Tang, C., Kowalewski, T., et al. (2006). Polymer-coated ferromagnetic colloids from well-defined macromolecular surfactants and assembly into nanoparticle chains. *J. Am. Chem. Soc.*, *128*, 6562–6563.

124. Czaun, M., Hevesi, L., Takafuji, M., Ihara, H. (2008). A novel approach to magneto-responsive polymeric gels assisted by iron nanoparticles as nano cross-linkers. *Chem. Commun.*, *2124–2126*.

125. Yager, K. G., Barrett, C. J. (2006). Novel photo-switching using azobenzene functional materials. *J. Photochem. Photobiol. A*, *182*, 250–261.

126. Zhao, Y., He, J. (2009). Azobenzene-containing block copolymers: the interplay of light and morphology enables new functions. *Soft Matter*, *5*, 2686–2693.

127. Jochum, F. D., Theato, P. (2009). Temperature and light sensitive copolymers containing azobenzene moieties prepared via a polymer analogous reaction. *Polymer*, *50*, 3079–3085.

128. Viswanathan, N. K., Kim, D. Y., Bian, S., Williams, J., Liu, W., Li, L., et al. (1999). Surface relief structures on azo polymer films. *J. Mater. Chem.*, *9*, 1941–1955.

129. Strzegowski, L. A., Martinez, M. B., Gowda, D. C., Urry, D. W., Tirrell, D. A. (1994). Photomodulation of the inverse temperature transition of a modified elastin poly(pentapeptide). *J. Am. Chem. Soc.*, *116*, 813–814.

130. Shimoboji, T., Ding, Z. L., Stayton, P. S., Hoffman, A. S. (2002). Photoswitching of ligand association with a photoresponsive polymer–protein conjugate. *Bioconjugate Chem.*, *13*, 915–919.

131. Such, G. K., Evans, R. A., Davis, T. P. (2004). Control of photochromism through local environment effects using living radical polymerization (ATRP). *Macromolecules*, *37*, 9664–9666.

132. Ivanov, A. E., Eremeev, N. L., Wahlund, P. O., Galaev, I. Y., Mattiasson, B. (2002). Photosensitive copolymer of n-isopropylacrylamide and methacryloyl derivative of spirobenzopyran. *Polymer*, *43*, 3819–3823.

133. Konak, C., Rathi, R. C., Kopeckova, P., Kopecek, J. (1997). Photoregulated association of water-soluble copolymers with spirobenzopyran containing side chains. *Macromolecules*, *30*, 5553–5556.

134. Lee, H-i., Wu, W., Oh, J. K., Mueller, L., Sherwood, G., Peteanu, L., et al. (2007). Light induced reversible formation of polymeric micelles. *Angew. Chem. Int. Ed.*, *46*, 2453–2457.

135. Goodwin, A. P., Mynar, J. L., Ma, Y., Fleming, G. R., Frechet, J. M. J. (2005). Synthetic micelle sensitive to IR light via a two-photon process. *J. Am. Chem. Soc.*, *127*, 9952–9953.

11

PHARMACEUTICAL POLYMERS

Natarajan Venkatesan
Hideki Ichikawa

11.1 INTRODUCTION TO PHARMACEUTICAL POLYMERS

Among the various biomaterials studied, both natural and synthetic polymers have been studied the most for various medical and biomedical applications. For nearly two centuries, polymers have been in use even before their nature and characteristics were understood. In simple terms, polymers are repeat units of monomeric units contributing to a large molecule. Polymers have found applications in our everyday life in one form or the other starting from commuting (car tires) to healthcare. In this chapter, the readers will be introduced to the role of polymers in pharmaceutical industry and about the recent development of novel polymers for various pharmaceutical applications.

Polymers can be either made of single backbone structure referred to as *homochains* or multiple backbone structure referred to as *heterochains* (Figure 11.1). Polymers can either be made of single monomer units called as *homopolymers* or those that contain two or more monomers called as *copolymers*.

Polymers represent an important constituent of the pharmaceutical dosage forms. They have been used in conventional dosage forms such as tablets, capsules, suspensions, and microcapsules; in controlled drug delivery dosage forms such as extended-release tablets, implants, microspheres, and transdermal drug delivery. Polymers have also been studied for novel applications such as drug–polymer conjugates, thermal, and pH-sensitive drug release. Polymers have been widely used in pharmaceutical industry because of their flexibility. Researchers have the ability to use the same polymer for different applications

Synthesis and Applications of Copolymers, First Edition. Edited by Anbanandam Parthiban.

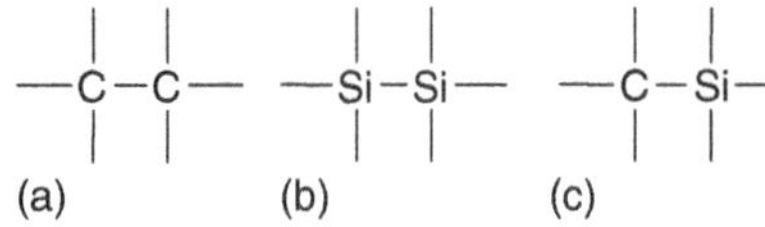

$$-\overset{|}{\underset{|}{C}}-\overset{|}{\underset{|}{C}}- \quad -\overset{|}{\underset{|}{Si}}-\overset{|}{\underset{|}{Si}}- \quad -\overset{|}{\underset{|}{C}}-\overset{|}{\underset{|}{Si}}-$$

(a) (b) (c)

Figure 11.1 (a) and (b) Single backbone structure and (c) multiple backbone structure.

based on modification of their physicochemical properties. For example, some copolymers of polymethylmethacrylates are soluble in certain pH, and by changing their copolymer composition it is possible to make it soluble in different pH values.

The use of polymers for pharmaceutical applications is determined by four major characteristics, namely, functionality, mechanical strength, biocompatibility, and toxicity. Polymers may have good functional and mechanical properties, however, to be applicable for human or veterinary use; the polymer needs to be biocompatible and nontoxic (in the prescribed dose). This important factor puts the researchers in a tough position as there are lot of synthetic polymers developed everyday keeping in mind the growing needs in the medical and biomedical fields. However, upon prolonged exposure to human or veterinary application, they turn out to be toxic that limits their use for the desired application.

The pharmaceutical application of polymers are not limited to act as a tablet binder or a viscosity building agent in a suspension, they have now been widely used to control the release of the drug and have played a crucial role in improving the quality of life. Without the use of polymers, it would have not been possible for development of dosage forms, which can release the drug at a desired rate over a day, week, month, or year. This significantly changed how the drug is being delivered. Not only did it change the way the drug delivery was carried out, but also played a significant role in reducing the dose in many cases, thereby decreasing the exposure of the patient to any untoward side-effects, thus improving the quality of life. The objective of this chapter is to introduce the role of polymers in pharmaceutical formulations.

11.2 APPLICATIONS OF PHARMACEUTICAL POLYMERS

Polymers used in pharmaceutical industry can be classified based on their chemical structure, physicochemical properties, or their application. Keeping in mind the benefit of the readers, in this chapter, we would like to classify polymers based on their application. The authors have made every attempt to provide the readers with the most commonly used polymers. However, it is beyond the scope of this chapter to list all the polymers studied for various pharmaceutical applications.

11.2.1 Polymers as Excipients

Polymers have become an integral part of pharmaceutical dosage forms. The most common use of polymer in a dosage form is in the form of an excipient.

Excipients are the substances that provide certain properties in designing a dosage form, but do not have any known pharmacological activity of their own. A variety of polymers have been used for this purpose. Polymers under this category are mostly derivatives of cellulose. The cross-linking of these cellulose derivatives provides polymeric materials with different physicochemical properties. On the basis of their properties, the applications of these polymers differ.

11.2.1.1 Carboxymethylcellulose Sodium Carboxymethylcellulose sodium (CMC sodium) is cellulose-ether-based long-chain polymer. It is an anionic polyelectrolyte, readily soluble in hot or cold water. It is widely used in oral and topical pharmaceutical formulations. The viscosity of the polymer differs based on the degree of polymerization. The higher the degree of polymerization, the higher will be the viscosity. CMC sodium has been widely used as viscosity building agent in suspensions and some topical preparations. It acts as a binder and a disintegrant in tablets and capsules [1,2]. Different grades are available depending on their viscosity for different pharmaceutical applications. Higher concentrations (3–6% w/w) of a medium viscosity grade are typically used as a base for pastes.

11.2.1.2 Ethylcellulose Ethylcellulose (EC) is an ethyl ether of cellulose. It is a long-chain polymer and has the structural formula as shown in Figure 11.2. EC is widely used in oral controlled-release formulations. The polymeric solution or dispersion can be applied onto the tablet, pellet, or other oral dosage forms as a thin film that controls the release of the drug by diffusion mechanism [3]. EC, in itself is poorly water soluble; however, the solubility can be altered by addition of hypermellose [4] or a plasticizer [5]. EC coating has also been used in taste masking as well as to improve stability of the drug. EC has also been used in modified-release (MR) dosage forms, wherein the polymer is used as a matrix-forming agent in order to modify the release of the drug [6]. The powder form of EC, Ethocel™ (Dow Chemical Company) is available in a wide range to provide different solution viscosity. Most common example for an aqueous dispersion of EC is Aquacoat® ECD (FMC Biopolymer) and Surelease® (Colorcon).

In addition to their abilities as controlled-release agents, EC is also used as binder in tablet formulations, thickening agent in creams, lotions and gels for topical applications, suspensions, vaginal preparations, etc. EC is also commonly used

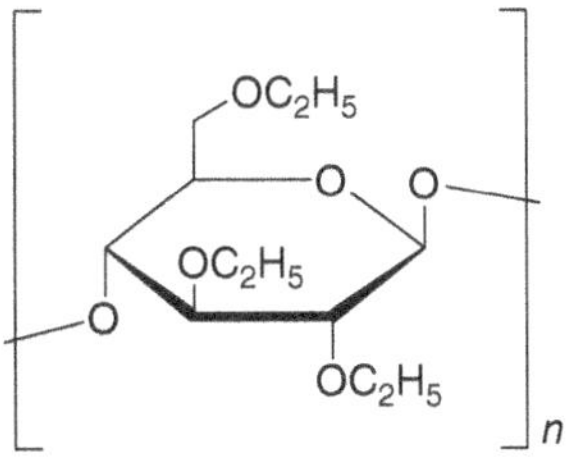

Figure 11.2 Structure of ethylcellulose.

in the food and cosmetic industry. EC has also been used as a backing membrane on mucoadhesive patches intended for buccal [7] and gastrointestinal administration [8]. Though a large number of studies have been carried out on EC as a polymer for preparation of microspheres, no commercial products have been reported.

11.2.1.3 Methylcellulose Methylcellulose (MC) is methyl ether-substituted long-chain cellulose. Depending on the degree of polymerization or molecular weight, viscosity of the polymer may change. The degree of substitution of methoxyl groups affects the physical properties of MC, such as its solubility and viscosity. MC is used a tablet binder, tablet and capsule disintegrant, coating agent, emulsifying agent, viscosity building agent, and emulsifying agent. MC is also used as a bulk laxative (Citrucel®). MC with low to medium viscosity grades have been used in tablet formulations either as binder (1–5% w/w) or disintegrant (2–10% w/w) or as sustained release agent (5–75% w/w). Low viscosity MC has been used as an emulsifying agent (1–5% w/w). MC has been used as a vehicle in ophthalmic preparations (0.5–1% w/w). MC can be dispersed slowly in cold water and they are insoluble in hot water.

11.2.1.4 Hydroxypropyl Cellulose Hydroxypropyl cellulose (HPC) is a partially substituted poly(hydroxypropyl)ether of cellulose. Figure 11.3 shows the structure of the common chain for HPC. Like other cellulose derivatives, this too imparts different physical properties depending on the degree of substitution. Similar to their predecessors, this polymer is also used as a tablet binder, thickening agent, suspending agent, coating agent, and controlled-release agent. HPC at a concentration of 15–35% w/w is used in tablets to form a matrix and thereby control the release of the drug. It is freely soluble in water below 38 °C and insoluble in hot water. Mucoadhesive HPC microspheres have been reported for inhalation applications [9].

11.2.1.5 Hydroxypropyl Methylcellulose or Hypromellose Hydroxypropyl methylcellulose (HPMC) is officially known as *hypromellose* in various pharmacopoeias. Hypromellose, as defined in the *United States Pharmacopeia-36*

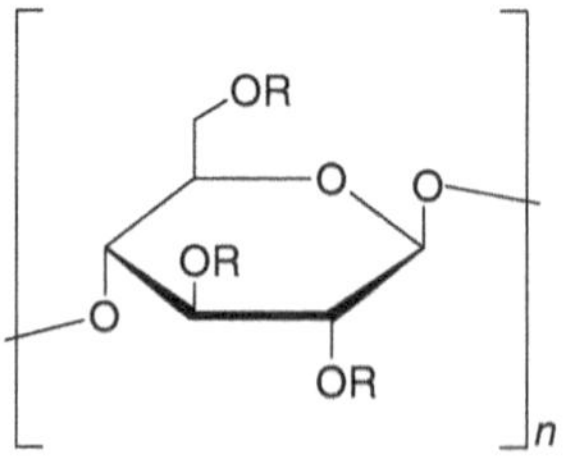

R = H or $CH_2CH(OH)CH_3$

Figure 11.3 Structure of HPC.

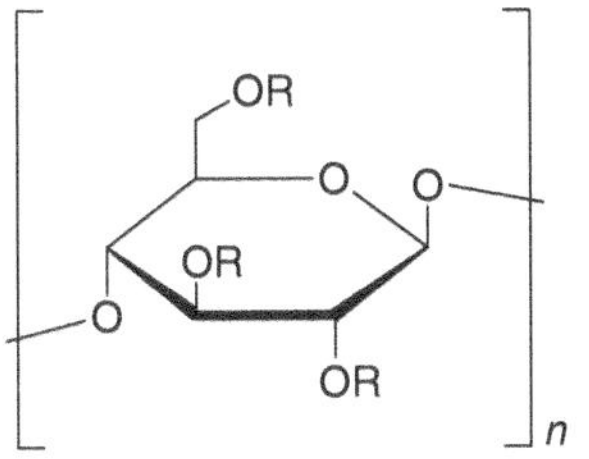

R = H, CH_3 or $CH_2CH(OH)CH_3$

Figure 11.4 Structure of HPMC.

(*USP-36*) (*Pharmacopoeia* is a book describing drugs, chemicals, and excipients used in medicinal preparations. The book serves as a standard, approved officially by recognized authority or agency) [10] is a methyl and hydroxypropyl mixed ether of cellulose. It has the structural formula as shown in Figure 11.4. Hypromellose is available in several grades that vary in viscosity. In the *USP*, the substitution type is used as a four-digit identification code for the grade. For example, in hypromellose 1828, the first two digits refer to the approximate percentage content of the methoxy group ($-OCH_3$). The last two digits refer to the approximate percentage content of the hydroxypropoxy group ($-OC_3H_6OH$) calculated on a dried basis. Hypromellose is widely used as a tablet binder [11] and in film coating [12]. It is also one of the main excipient in development of extended-release tablet formulations, wherein they form a matrix system to control the release of the drug [13]. Similar to other cellulose derivatives, HPMC is also used as a viscosity building agent in liquid formulations. Hypromellose ophthalmic solution is also official in the USP [10]. It is also used as a viscosity building agent in topical formulations. Other applications of hypromellose that have been studied include action as bioadhesive material, nasal formulations, and stabilizing agents.

11.2.1.6 Hydroxypropyl Methylcellulose Acetate Succinate or Hypromellose Acetate Succinate Hydroxypropyl methylcellulose acetate succinate (HPMC-AS) is a mixture of acetic acid and monosuccinic acid esters of hypromellose. It has a chemical structure as shown in Figure 11.5. HPMC-AS has been widely used as a coating agent, controlled-release agent, and solubility-enhancing agent. This polymer is not soluble in gastric pH and hence been used as an enteric coating agent (discussed later in this chapter under functional polymers). This polymer is available in several grades according to the pH in which the polymer dissolves. HPMC-AS has also been studied in melt extrusion technique.

11.2.1.7 Povidone Povidone is a synthetic polymer obtained by starting from butyrolactone and ammonia. The product obtained is then subjected to reaction with acetylene to get the monomer vinylpyrrolidone that is subsequently polymerized to yield povidone. Povidone is also known as *polyvinylpyrrolidone (PVP)*. Povidone is widely used in solid dosage forms. Its application as a binder in tablets is well

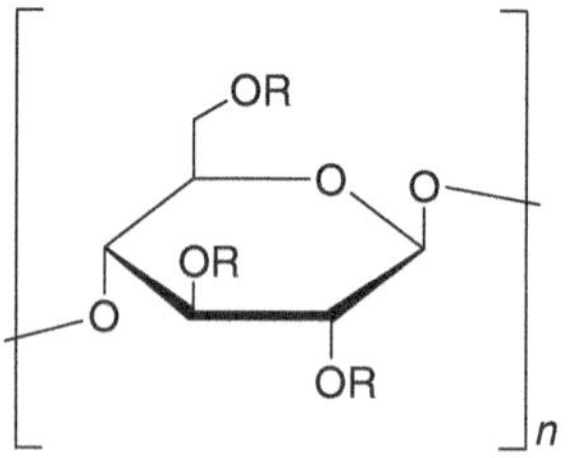

R = H, CH_3,$CH_2CH(OH)CH_3$
$COCH_3$ or $COCH_2CH_2CO_2H$

Figure 11.5 Structure of HPMC-AS.

known in the pharmaceutical industry. The polymer is either added as a dry powder or is dissolved in a solvent to be used in wet-granulation techniques. The use of povidone as a solubilizing agent is also well known in the industry. Povidone has found applications as a coating agent as well. The use of povidone in liquid dosage form has been widely attributed to its viscosity. It has been used as a suspending, viscosity building, or stabilizing agents for solutions and suspension dosage forms.

11.2.1.8 Copovidone Copovidone is prepared by free-radical copolymerization of vinylpyrrolidone and vinyl acetate in the ratio of 6 : 4. It has a structural formula as shown in Figure 11.6. Copovidone is widely used as a tablet binder and as matrix-forming agent in controlled-release dosage forms. Other common application is as a film-forming agent. Copovidone has excellent adhesion, elasticity, and hardness. One has to be mindful of the use of this agent as a binder as higher quantities may lead to a very hard tablet that may not disintegrate within the required time to release the drug for absorption.

11.2.1.9 Polyvinyl Alcohol Polyvinyl alcohol (PVA) is water-soluble synthetic polymer produced through the hydrolysis of polyvinyl acetate. PVA comes in various grades depending on their molecular weight. Viscosity is directly proportional to the molecular weight of the polymer. PVA has been used as a viscosity building agent, stabilizing agent for emulsions, ophthalmic formulations, and topical lotions.

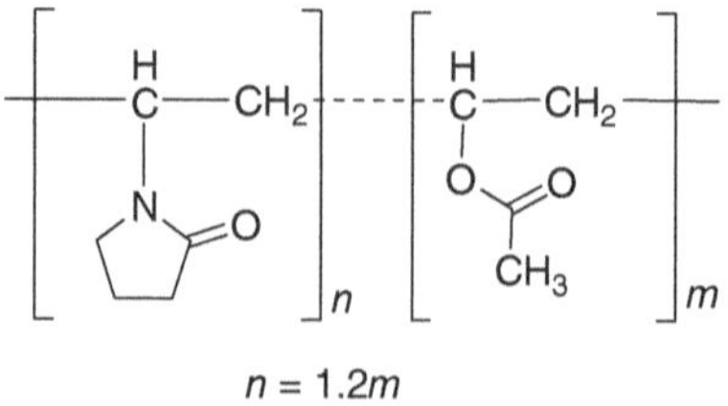

$n = 1.2m$

Figure 11.6 Structure of copovidone.

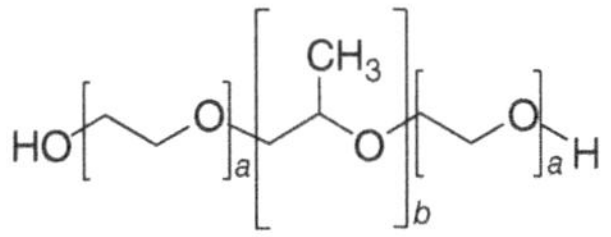

Figure 11.7 Structure of poloxamer.

This polymer is used in artificial tears and contact lens solutions for lubrication purposes. PVA has also been studied for controlled-release formulations [14].

11.2.1.10 Poloxamers Poloxamers are nonionic polyoxyethylene–polyoxypropylene copolymers. Poloxamers are prepared by a reaction between propylene oxide and propylene glycol to form polyoxypropylene glycol to which is added ethylene oxide to obtain the block copolymer. They have a structural formula as shown in Figure 11.7. They comprise a central block of relatively hydrophobic polypropylene oxide (PPO) surrounded on both sides by the blocks of relatively hydrophilic polyethylene oxide (PEO). Owing to the PEO/PPO ratio of 2:1, when these molecules are immersed into the aqueous solvents, they form micellar structures above critical micellar concentration [15]. Poloxamers find a variety of applications in pharmaceutical formulations. They are primarily used as solubilizing or emulsifying agents [16]. Poloxamers have also been used as wetting agents in ointments and gels. They have been administered orally as a stool lubricant in the treatment of constipation. Poloxamers are also commonly used in contact lens care.

Poloxamers are available in different grades and forms (liquid, gel, and solid). Poloxamer grades are designated by a three-digit number, for example, poloxamer 124. The three-digit number indicates the molecular mass and percent polyoxyethylene present in that grade. In this case, the first two digits (12) when multiplied by 100 gives the molecular mass (1200 in this case) and the last digit (4) when multiplied by 10 gives the polyoxyethylene content (40% in this case). Commercial grades are designated differently based on the form they are available and the molecular mass of the hydrophobe and polyoxyethylene contents. To compare this, poloxamer 124 is commercially available as Lutrol® L-44; "L" stands for liquid, the first digit (4) multiplied by 300 gives the hydrophobic molecular mass (1200 in this case), the second digit (4) multiplied by 10 gives the polyoxyethylene content (40% in this case). Although the number designation between nonproprietary name and commercial grade may be different, they use the same values for that particular grade. Table 11.1 lists some of the commonly used poloxamer grades and their HLB value.

11.2.1.11 Polyethylene Glycols Polyethylene glycols (PEGs) are formed by a reaction between ethylene oxide and water in the presence of a catalyst under pressure. PEGs are structurally represented as in Figure 11.8, where m represents the number of oxyethylene groups. PEGs are available in varying molecular weights and their physical form varies depending on their molecular weight as

TABLE 11.1 List of Some Commonly Used Poloxamers

Nonproprietary Name	Physical Form	Commercial Grade	HLB Value
Poloxamer 124	Liquid	L-44	12–18
Poloxamer 188	Solid flake	F-68	>24
Poloxamer 237	Solid flake	F-87	>24
Poloxamer 338	Solid flake	F-108	>24
Poloxamer 407	Solid flake	F-127	18–23

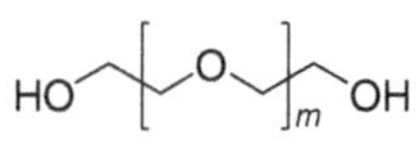

$$HO-[CH_2CH_2O]_m-CH_2CH_2-OH$$

Figure 11.8 Structure of polyethylene glycol.

well. Low molecular weight PEGs are liquids (PEG 200–600). As the molecular weight increases, they turn into solids. One good property about PEGs is that irrespective of molecular weight they are all soluble in water. PEGs also do not support microbial growth nor do they turn rancid. The numerous advantages of PEGs have shown the pharmaceutical industry a wide range of application to choose from. Low molecular weight PEGs have been used as suspending agent, solubilizing agent, and water-miscible solvent base. Higher molecular weight PEGs have been used as tablet binders, coating agents, plasticizers in coating composition, as a base for topical application, suppository base, parenteral, and ocular applications.

The use of PEGs to improve aqueous solubility of pharmaceutically active ingredient is well known. The use of PEGs in parenteral product development has been phenomenal. The hydrophilic property of PEGs has found application in the novel delivery of drug molecules.

11.2.1.12 Polyethylene Oxide PEO is a nonionic homopolymer of ethylene oxide prepared by the polymerization of ethylene oxide with a suitable catalyst. It is represented by the molecular formula of $(CH_2CH_2O)_n$, where "n" represents the average number of oxyethylene groups. They are structurally very similar to PEGs. The difference being that PEOs are very high molecular weight polymers. PEO is widely used as a tablet binder. They have excellent mucoadhesive property [17]. PEOs at low concentrations have been used as thickening agents. Though PEOs are water soluble, when incorporated in a matrix tablet, they act as controlled-release agent. PEOs at low concentrations have also been used as film formers.

11.2.1.13 Poly(Methyl Vinyl Ether/Maleic Anhydride) Poly(methyl vinyl ether/maleic anhydride) is manufactured by reacting methyl vinyl ether and maleic anhydride. A variety of copolymers are manufactured by dispersing them in different solvent or salt solutions. It has found wide application as a coating

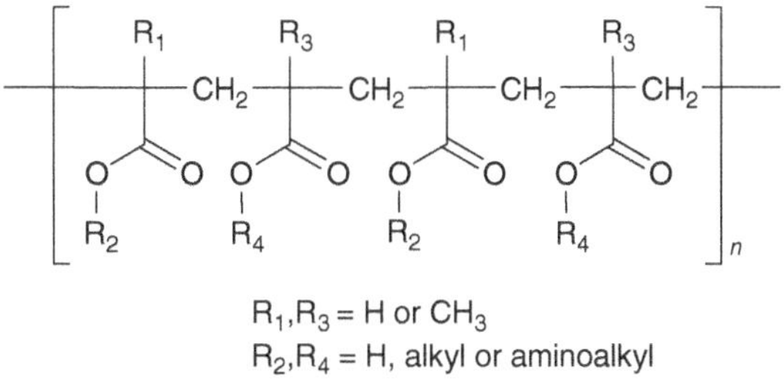

R_1,R_3 = H or CH_3
R_2,R_4 = H, alkyl or aminoalkyl

Figure 11.9 Structure of polymethacrylates.

agent in controlled release and enteric coating. This polymer has also found application in transdermal patches and as denture adhesive base. Some other applications include emulsion stabilizer, viscosity building agent, and complexing agent.

11.2.1.14 Polymethacrylates Polymethacrylates are synthetic cationic and anionic polymers of dimethylaminoethylmethacrylates, methacrylic acid, and methacrylic acid esters in varying ratios. They are prepared by the polymerization of acrylic and methacrylic acids or their esters. Their general structural formula is depicted in Figure 11.9. Polymethacrylate polymers have found various applications in the pharmaceutical industry. This polymer is most commonly referred to in the pharmaceutical industry by the name *Eudragit*® polymers (from Evonik Industries). The most common application of this polymer has been as a coating agent. It has also found various other applications such as diluent, tablet binder, and in sustained release formulations (Table 11.2).

11.2.1.15 Polyvinyl Alcohol/Acrylic Acid/Methyl Methacrylate Copolymer Copolymer of PVA/acrylic acid/methyl methacrylate (POVACOAT®) is a novel pharmaceutical polymer recently developed by Daido Chemical Corporation in Japan [18]. POVACOAT is a synthetic polymer obtained by an emulsion polymerization technique of two acrylic monomers, that is, acrylic acid and methyl methacrylate (MMA), with PVA (Figure 11.10). Currently two grades with different molecular weights are available commercially. POVACOAT® is classified into a water-soluble polymer, but one should note that this polymer is not dissolved in water completely as its liquid form usually gives a translucent appearance possibly because of the presence of hydrophobic unit (MMA) in their molecular structure. A striking feature of the physicochemical properties of this polymer is its low permeability against oxygen in its film form and also oil resistance property [18]. Because of this nature, POVACOAT has been used as a substrate for hard shell capsule and as a coating agent for solid dosage forms. Several other applications including as binder for wet granulation and as matrix agent for a solid dispersion product and/or as solubilizing agent for poorly water-soluble drugs are being investigated [18–20].

TABLE 11.2 Various Grades of Polymethacrylates and Their Application in Pharmaceutical Dosage Form Design

Chemical Name	Trade Name	Solubility/ Permeability	Application
Poly(butyl methacrylate, (2-dimethylaminoethyl) methacrylate, methyl methacrylate) 1 : 2 : 1	Eudragit E	Gastric fluid up to pH 5.0	Film coating
Poly(ethyl acrylate, methyl methacrylate) 2 : 1	Eudragit NE	Insoluble, swellable, low permeability	Sustained release
Poly(methacrylic acid, methyl methacrylate) 1 : 1	Eudragit L	Intestinal fluid pH >6.0	Enteric coating
Poly(methacrylic acid, ethyl acrylate) 1 : 1	Acryl-EZE; Eudragit L 30 D-55; Kollicoat MAE 30 DP; Eastacryl 30D	Intestinal fluid pH >5.5	Enteric coating
Poly(methacrylic acid, methyl methacrylate) 1 : 2	Eudragit S	Intestinal fluid pH >7.0	Enteric coating
Poly(methyl acrylate, methyl methacrylate, methacrylic acid) 7 : 3 : 1	Eudragit FS 30D	Intestinal fluid pH >7.0	Enteric coating
Poly(ethyl acrylate, methyl methacrylate, trimethylammonioethyl methacrylate chloride) 1 : 2 : 0.2	Eudragit RL	Insoluble, swellable, high permeability	Sustained release
Poly(ethyl acrylate, methyl methacrylate, trimethylammonioethyl methacrylate chloride) 1 : 2 : 0.1	Eudragit RS	Insoluble, swellable, low permeability	Sustained release

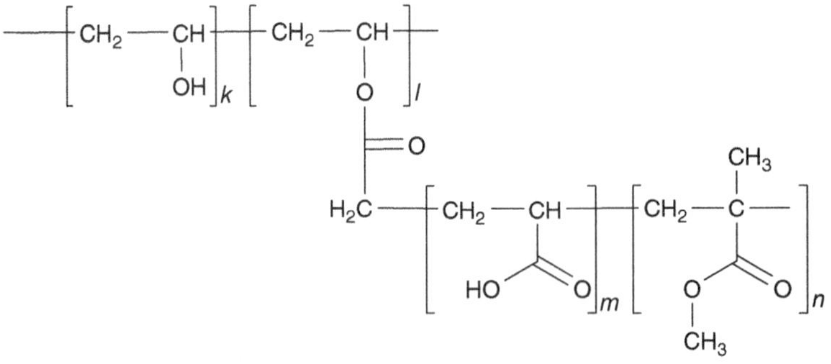

Figure 11.10 Structure of POVACOAT.

11.2.2 Functional Excipients

As we have discussed in the earlier section, polymers have been extensively used in pharmaceutical industry as an excipient. Although each excipient has a certain function in dosage form design, in this section we would like to classify polymers that have been used with a specific function. By specific function, we mean that the use of polymer is an integral part of the formulation to carry out the required function. For example, the use of polymethacrylates for enteric coating can be categorized as a functional excipient as the specific coating will protect the drug from a certain pH and will release the drug at a desired pH. This way, the polymer carries out a certain "function" in delivering the drug. In this section, we will discuss some polymers which have been used as functional excipients. Some polymers that have been discussed in Section 11.2.1 will be dealt in here with the specific function and may not be duplicated under individual titles.

11.2.2.1 Cellulose Acetate Cellulose acetate is obtained by treating cellulose with acetic anhydride under acid catalysis. Various grades are available based on the percent acetylation, chain length, and molecular weight. Cellulose acetate has been widely used in sustained release formulations and taste masking. The semipermeable nature of cellulose acetate has been used in the development of osmotic pump-type dosage forms. The semipermeable nature controls the flow of body fluid into the osmotically driven dosage form and thereby controlling the release of drug from the dosage form. Figure 11.11 schematically presents the role of a semipermeable membrane in osmotic drug delivery. The semipermeable membrane allows the biological fluid to pass through the delivery system, and once the fluid enters the delivery system, it dissolves the osmogen present in the delivery system which then dissolves and pushes the drug layer, which delivers the drug through the laser-drilled orifice.

Cellulose acetate has long been used as a taste-masking agent. Tablets containing bitter drugs have been delivered following a coating with the taste-masking agent. It also found wide application in transdermal drug delivery.

11.2.2.2 Cellulose Acetate Butyrate Cellulose acetate butyrate (CAB) is a reaction product of cellulose, acetic acid or acetic anhydride, and butyric acid or butyric

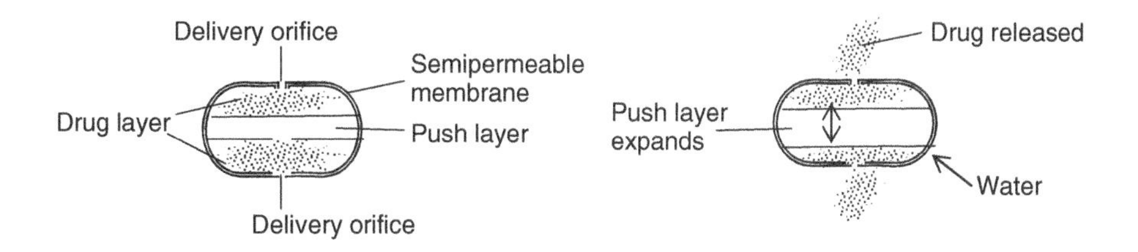

Figure 11.11 Schematic of an osmotic drug delivery system containing a semipermeable membrane.

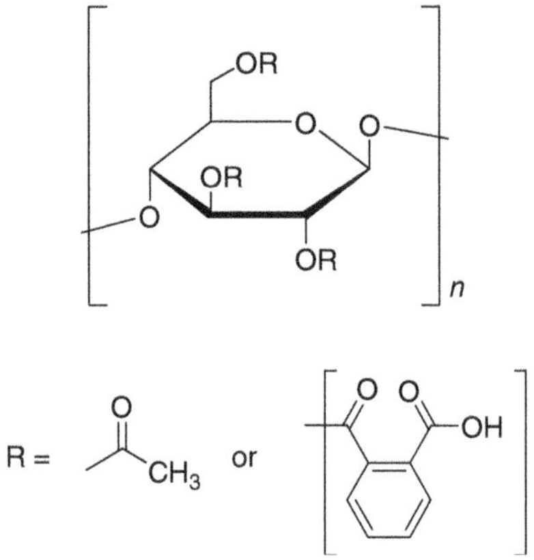

Figure 11.12 Structure of CAP.

anhydride. CAB is used as a sustained release agent incorporated either into the matrix tablet or as a coating agent. CAB is also used as a semipermeable membrane in sustained drug delivery [21,22]. This polymer is listed in the National Formulary (NF) as *Cellaburate*.

11.2.2.3 Cellulose Acetate Phthalate Cellulose acetate phthalate (CAP) (Figure 11.12) is obtained by reacting partial acetate ester of cellulose with phthalic anhydride in the presence of a strong acid or organic base. CAP has been widely used as an enteric coating material. Enteric coating is a protective coating provided over the desired solid dosage form to protect the drug from harsh gastric pH condition. Enteric coating is not only used to protect acid-labile drugs, but also used in sustained release dosage form or in targeted delivery of the drug to the small intestine or large intestine.

11.2.2.4 Hydroxypropyl Methylcellulose Phthalate or Hypromellose Phthalate Hydroxypropyl methylcellulose phthalate (HPMC-P) or hypromellose phthalate is a cellulosic polymer wherein the hydroxyl groups are replaced with methyl ethers, 2-hydroxypropyl ethers, or phthalyl esters (Figure 11.13). HPMC-P is prepared by the esterification of hypromellose with phthalic anhydride. HPMC-P is widely used as an enteric coating polymer for oral use. Being a tasteless polymer, it has found application as a taste-masking agent. HPMC-P has also been studied for sustained release dosage forms. The release of the drug from formulation depends on the pH of the environment. HPMC-P is available in different grades. The commonly used brand is the Shin-Etsu polymer, which comes in three different grades HP-50, HP-55, and HP-55S. All the three grades of polymer are used as enteric coating agents. HP-50 dissolves at a biological pH of >5.0 while HP-55 grade dissolves at a pH of >5.5. HP-55S is similar to that of HP-55 in solubility but with higher film strength. Some drugs are having a narrow absorption window and even a small difference in pH can cause big change in the absorption. This is true

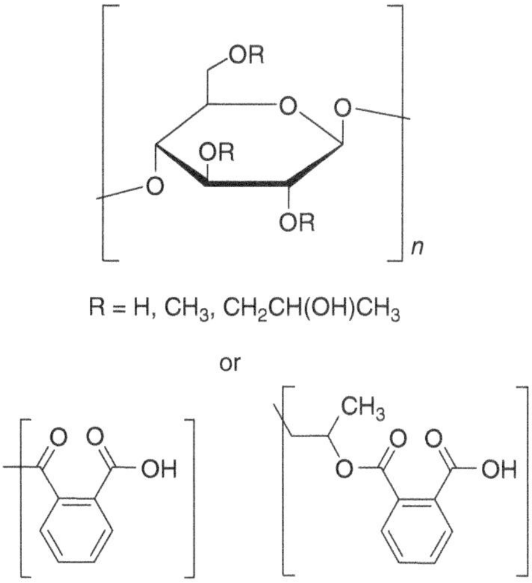
OR
O
O
OR
O
OR
n
R = H, CH3, CH2CH(OH)CH3
or
O O
OH
CH3
O O
O
OH

Figure 11.13 Structure of HPMC-P.

with drugs that are well absorbed from the duodenal region of the small intestine as compared to other parts of the small intestine. The duodenal region being a very small portion of the small intestine and if the drug needs to be protected from the harsh pH of the stomach, one will have to use an enteric polymer that will protect the drug from the acidic pH and at the same time dissolve at the site of absorption.

11.2.2.5 Polyvinyl Acetate Phthalate Polyvinyl acetate phthalate (PVA-P) is obtained by reacting partially hydrolyzed low molecular weight PVA with phthalic anhydride and sodium acetate. The general structural formula of PVA-P is depicted in Figure 11.14. PVA-P is another commonly used enteric coating polymer. This polymer acts similar to some of the polymethacrylate grades and CAP. It is also used as a seal coating agent for sugar spheres.

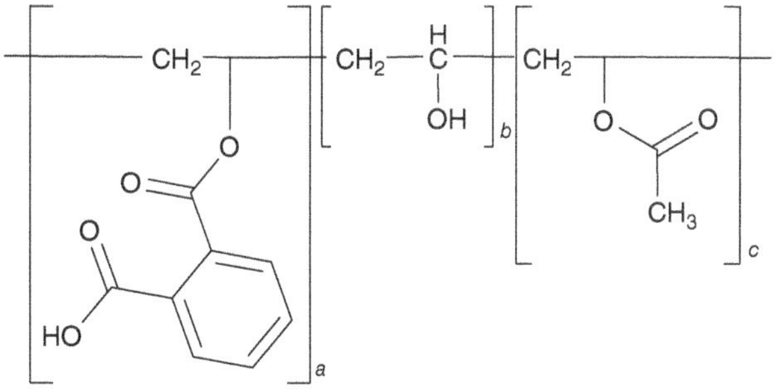
CH2
CH2
H
C
OH
b
CH2
O
O
CH3
c
O
O
O
HO
a

Figure 11.14 Structure of PVA-P.

TABLE 11.3 List of Some Commonly Used Enteric Polymers and Their Solubility in Biological Fluid

Polymer Category	Make	Trade Name	Solubility
Cellulose acetate phthalate	Eastman	C-A-P	Intestinal fluid pH >6.0
Hypromellose acetate succinate	Shin-Etsu	AQOAT-LF/LG	Intestinal fluid pH >5.5
		AQOAT-MF/MG	Intestinal fluid pH >6.0
		AQOAT-HF/HG	Intestinal fluid pH >6.8
Hypromellose phthalate	Shin-Etsu	HP-50	Intestinal fluid pH >5.0
		HP-55	Intestinal fluid pH >5.5
		HP-55S	Intestinal fluid pH >5.5
Polymethacrylates	Evonik	Eudragit L	Intestinal fluid pH >6.0
		Eudragit L 30 D-55	Intestinal fluid pH >5.5
		Eudragit S	Intestinal fluid pH >7.0
		Eudragit FS 30D	Intestinal fluid pH >7.0
Polyvinyl acetate phthalate	Colorcon	Opadry Enteric	Intestinal fluid pH >6.0

11.2.2.6 Other Enteric Coating Polymers The most commonly used enteric coating polymer is polymethacrylates. Basic characteristics of this polymer have been discussed in Section 11.2.1.14. The enteric coating property of polymethacrylates has found wide application in oral drug delivery. These polymers have enabled the delivery of acid-labile drug to small intestine; targeted delivery of drug to the small and large intestine; and helped improve the oral bioavailability of certain drugs because of their protective coating. Another hypromellose derivative used as an enteric coating polymer is the hypromellose acetate succinate (AQOAT®). Different grades of this polymer are soluble under different pH conditions. This polymer is soluble in the pH range of 5.5–6.8 depending on the grade used for enteric coating (Table 11.3).

11.2.3 Drug Delivery Agents

11.2.3.1 Aliphatic Polyesters Aliphatic polyesters are polymers that can be synthesized as homopolymers or copolymers. Poly(L-lactide) and poly(L-lactide-*co*-glycolide) are common examples of this class of polymer. They are nontoxic and can be easily fabricated into a variety of shape and size as desired. These polymers have been widely studied in drug delivery. The most important being their application in development of controlled drug delivery over an extended period of time first as implant [23], as microspheres (Leupron Depot®) [24], and as injectable liquid controlled-release systems (Eligard®) [25]. The advantage of aliphatic polyesters is that they are easily susceptible to hydrolysis in the biological system that makes them biodegradable. Poly(lactide), poly(glycolide),

TABLE 11.4 Chemical Names and Composition for Some of the Commonly Studied Aliphatic Polymers

		Composition (%)		
Generic Name	Synonym	Lactide	Glycolide	Caprolactone
Poly(L-lactide)	L-PLA	100	0	0
Poly (DL-lactide)	DL-PLA	100	0	0
Poly(DL-lactide-*co*-glycolide)	PLGA (85 : 15)	85	15	0
Poly(DL-lactide-*co*-glycolide)	PLGA (75 : 25)	75	25	0
Poly(DL-lactide-*co*-glycolide)	PLGA (65 : 35)	65	35	0
Poly(DL-lactide-*co*-glycolide)	PLGA (50 : 50)	50	50	0
Poly-ε-caprolactone	PCL	0	0	100
Poly(DL-lactide-*co*-ε-caprolactone)	DL-PLCL	85	0	15

poly(lactide-*co*-glycolide), and polycaprolactone (PCL) are used in parenteral pharmaceutical formulations. Table 11.4 shows a list of aliphatic polymers that have been used or studied for various pharmaceutical applications.

The use of biodegradable polymers such as PLA and PLGA has been well understood and their potential has been well studied. A brief about the applications of these polymers will be discussed here. Both PLA and PLGA polymers have found profound applications in controlled drug delivery. The use of these polymers is well documented in the form of the commercially successful drug delivery of leuprolide acetate as microspheres for delivery of the drug over 1-, 3-, and 6-month period, marketed in the United States under the brand name Lupron Depot®. Other drugs that have been delivered as microspheres in the United States include human growth hormone (now withdrawn), naltrexone, risperidone, and exenatide. Drug has also been incorporated into polymer and injected as a suspension. Upon injection, suspension forms an *in situ* depot and releases the drug over an extended period of time. The best example for this is the commercially available Eligard® (refer to Table 11.5). The readers should be mindful that the microsphere dosage form is also injected as a suspension upon reconstitution and before injection. However, the drug is encapsulated into the microspheres, whereas in the case of Eligard®, the drug is dispersed in the polymeric solution prior to administration. Although the type of polymer may be similar, the dosage form is different.

11.2.3.2 Carbomer Carbomers are high molecular weight polymers of acrylic acid. It is prepared by cross-linking acrylic acid with allyl sucrose or allyl pentaerythritol. Carbomers are also commonly referred to as *carbopol*. Carbomers are not soluble in water. When dispersed in water, they take up water and swell. This swelling property of carbomers has been used in controlled drug delivery [26]. Various grades of carbomers are available for pharmaceutical application. They are also used in tablets as binder, suspending agent, emulsifying agent, and stabilizing agent. Upon swelling, these polymers exhibit a bioadhesive property; hence, it has been widely studied as a bioadhesive material [27].

TABLE 11.5 List of Some Commercially Available Polymer-Based Controlled Drug Delivery System

Drug	Trade Name	Polymer	Dosage Form	Duration of Action
Leuprolide acetate	Lupron Depot®	PLGA	Microspheres	1 month
		PLA	Microspheres	3, 4, and 6 months
Naltrexone	Vivitrol®	PLGA (75 : 25)	Microspheres	1 month
Risperidone	Risperidal® Consta®	PLGA (75 : 25)	Microspheres	2 weeks
Exenatide	Bydureon®	PLGA (50 : 50)	Microspheres	1 week
Leuprolide acetate	Eligard®	PLGA (50 : 50)	Suspension	1 month
		PLGA (75 : 25)	Suspension	3 months
		PLGA (75 : 25)	Suspension	4 months
		PLGA (85 : 15)	Suspension	6 months

11.2.3.3 ***Polycarbophil*** Polycarbophils are similar to that of carbomer. They too are synthetic high molecular weight cross-linked polymers of acrylic acid. However, the difference is in the cross-linking agent, which in this case is divinyl glycol. Similar to carbomer, polycarbophils are used as a bioadhesive material, controlled-release agent in tablet, and capsule formulations. They also act as an emulsifying, suspending, and thickening agent. The gel formulation containing this polymer has been studied for local application to oropharyngeal, periodontal, and gingival areas [28–30]. Polycarbophil has also been studied as a bioadhesive agent for delivery of drugs to the nasal, rectal, vaginal, ophthalmic, and intestinal areas.

11.2.4 Solubility and Bioavailability Enhancement

In order for a drug to be effective, the drug needs to be well absorbed, which in turn depends on the solubility of the drug in biological fluid. As a general rule, poor solubility leads to poor absorption. Drugs have been classified based on their solubility and absorption behavior. This classification is known as the *Biopharmaceutical Classification System (BCS)* introduced by Amidon et al. [31]. According to this system, drugs fall under any of the four different categories from BCS-I to BCS-IV (Table 11.6).

Most of the newly developed drugs either fall in class II or class IV. This turns out to be a difficult situation where the bioavailability of the drug is seriously affected because of its poor solubility. By solubility, we hereby mean solubility in water. In order to improve the solubility and absorption of the drug, various approaches have been made. This includes prodrug formation, solubilizing the drug in oil or lipid vehicles, size reduction, nanoparticle formation, self-emulsifying drug delivery system, use of surfactants, and polymers. In this section, we will know how

TABLE 11.6 BCS Classification

Class	Solubility	Permeability
I	High	High
II	Low	High
III	High	Low
IV	Low	Low

polymers have played an important role in improving the solubility and bioavailability of the drug. Polymers have been used in improving solubility starting with simple solution formation of the drug in a specific polymer, for example, PEGs, preparation of solid dispersion/hot melt extrusion, and solid solutions.

PEG is a polymer that has been studied for dissolving poorly soluble drugs. They have also been used for dissolving complex protein molecules into solution. In the case of solid dispersion, the hydrophobic drug is mixed with a hydrophilic polymer. First, the polymer is melted and brought to liquid state, the drug is dispersed into the molten polymer and upon bringing the drug to room temperature, a solid dispersion is obtained that may have better aqueous solubility as compared to the hydrophobic drug alone. Second, the polymer is melted to a liquid and the hydrophobic drug is added to the solution where in the drug dissolves at a higher temperature; however, when brought to room temperature, some of the drug may separate because of the poor thermodynamic stability. Third, the polymer is melted into liquid and the drug is dissolved in the molten polymer, which upon cooling to room temperature still holds the drug in a thermodynamically stable state (Figure 11.15).

Preparation of solid dispersions has been reported using various techniques such as hot-melt extrusion, spray congealing, and melt granulation. Hot-melt extrusion process comprises the following steps: melting, mixing, and shaping. Hot-melt extrusion instrument is equipped with an external heat source that allows the polymer to melt; and then the drug is added to the molten polymer that may melt or get dispersed depending on its melting point; and finally the product is extruded with a desired shape and size. The whole process involves the movement of screw, which may be single, twin, or multiple screws. For a detailed review on hot-melt extrusion, the readers may refer to References 32–34.

Kaletra® is a coformulation of lopinavir and ritonavir. Lopinavir is an inhibitor of the HIV-1 protease while ritonavir inhibits the CYP3A-mediated metabolism

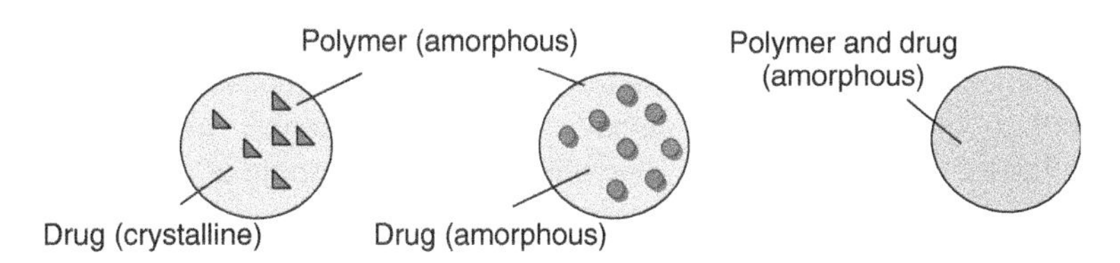

Figure 11.15 Schematic representation of solid dispersion.

of lopinavir, thereby providing increased plasma levels of lopinavir. By increasing the bioavailability, the dosing frequency of these drugs has been reduced that brings in a significant improvement in the quality of life for patients [35]. Various polymers have been studied for use in hot-melt extrusion technique. This includes Eudragit, hypromellose, HPC, copovidone, and polyvinyl caprolactam–polyvinyl acetate–PEG graft copolymer (Soluplus®).

11.2.5 Transdermal Drug Delivery

Delivery of drugs via the transdermal route has shown significant success in the last two decades. Drugs that are delivered across the intact skin are called *transdermal delivery*. This route of administration provides several advantages over the oral and other routes of administration. First, delivery by transdermal route avoids first-pass metabolism of the drug. Second, continuous and controlled delivery of the drug can be achieved via this route. Transdermal drug delivery systems have been designed in several different ways. The basis of all design is to have a drug reservoir, a backing membrane to provide the support for the reservoir, and an adhesive layer to stay in contact with the intact skin. Most of the components in the design of a transdermal drug delivery system involve the use of polymeric film. Various polymers that have found application for use in transdermal drug delivery include polyester, polyurethane, polyolefin, polyethylene as backing membrane, fluoropolymer-coated polyesters as liners, ethylene vinyl acetate as rate-controlling membrane, polyethylene foam, and polyurethane foam as tapes. Table 11.7 provides a list of transdermal patches that was approved for use by the United States Food and Drug Administration (US FDA).

11.2.6 Novel Polymeric Hydrogels for Drug Delivery Applications

Research in the field of polymer has led to the advancement of drug delivery. This includes delivery of drug in a safe and efficient way, controlled drug delivery, and targeted drug delivery. In some cases, the drug has been conjugated to one or more polymer to achieve the desired action. In this section, we will be discussing about novel polymers that have been reported for drug delivery applications. Importance of polymers in the field of drug delivery has grown with the innovation of biodegradable polymers. Biodegradable polymers are those polymers that can be degraded *in vivo* by biological enzymes. These polymers are also having the capability of delivering the drug to the desired site (targeting) with suitable modification and to deliver the drug in a controlled fashion.

Novel polymers have been engineered with an objective to make them respond according to the environmental stimulus. This stimuli-responsive smart polymer responds to the environmental signals such as temperature, pH, light, sound, magnetic/electric field, enzyme, and other chemical responses. In this section, we will be discussing about the use of novel polymers for microencapsulation, functional coating, and hydrogel applications. Hydrogels are results of the swelling of polymeric network in the presence of water. Hydrogels maintain

TABLE 11.7 List of Approved Transdermal Patches

Drug	Disease	Brand Name
Scopolamine	Motion sickness	Transderm-Scop®
Nitroglycerin	Angina pectoris	Transderm-Nitro®
Clonidine	Hypertension	Catapres-TTS
Estradiol	Menopausal symptoms	Estraderm®
Fentanyl	Chronic pain	Duragesic®
Nicotine	Smoking cessation	Nicoderm® CQ®
Testosterone	Testosterone deficiency	Androderm®
Estradiol/norethidrone	Menopausal symptoms	Combipatch®
Lidocaine	Postherpetic neuralgia pain	Lidoderm®
Norelgestromin/ethinyl estradiol	Contraception	Ortho Evra®
Estradiol/levonorgestrel	Menopausal symptoms	Climara®, Climara Pro®
Estradiol	Postmenopausal osteoporosis	Menostar®
Oxybutynin	Overactive bladder	Oxytrol®
Methylphenidate	Attention deficit hyperactivity disorder	Daytrana®
Selegiline	Major depressive disorder	Emsam®
Rotigotine	Parkinson's disease	Neupro®
Rivastigmine	Dementia	Exelon®
Sumatriptan	Migraine	Zecuity®

their three-dimensional integrity [36]. Fukumori et al. [37,38] reported the synthesis of an acrylic copolymer latex of copoly[ethylacrylate (EA)-MMA-2-hydroxy-ethylmethacrylate (HEMA)] as a coating material that can be used for microencapsulation purpose using the Wurster coating technique [39–41]. In a subsequent study, Ichikawa et al. [42] reported the development of a thermosensitive microcapsule for drug delivery application. In this microcapsule, water-soluble core particles (crystalline lactose) were coated with the composite acrylic latex of copoly(EA-MMA-HEMA) containing the shell of the thermosensitive poly(*N*-isopropylacrylamide (NIPAAm)). The swelling behavior of poly(NIPAAm) at temperature below its lower critical solution temperature (LCST) or lower gel collapse temperature (LGCT), which is 32 °C, has been explored for various applications including controlled release of drugs and enzymes, and extraction (also discussed in Chapter 10). By utilizing appropriate polymer combination and particle structure design of microcapsules, another class of thermosensitive drug release microcapsules with different thermosensitive release modes can be fabricated [43,44]. Figure 11.16 presents drug-layered calcium carbonate ($CaCO_3$) core particle being coated with a mixture of EC pseudolatex and the core-shell-like composite latex of copoly(EA-MMA) with poly(NIPAAm). By fabricating this type of microcapsule membrane structure,

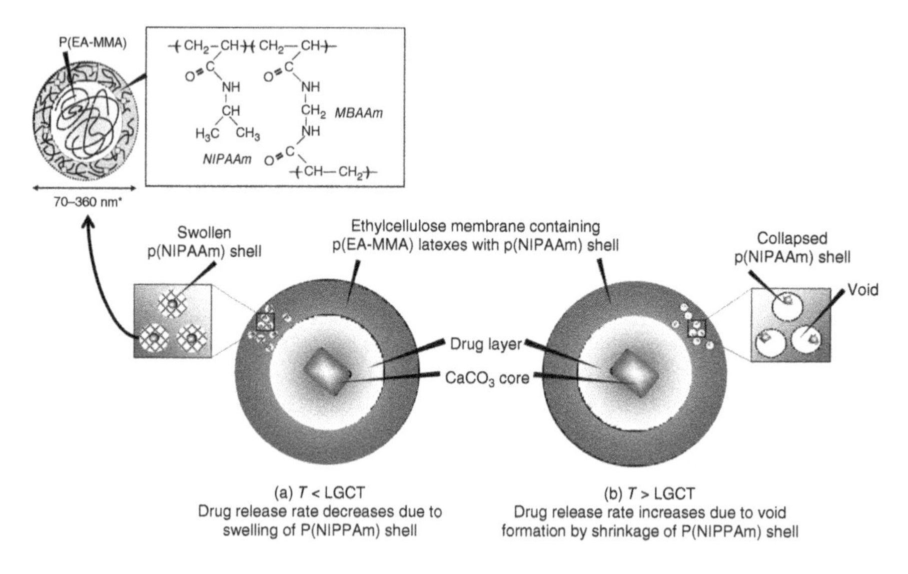

EA, ethyl acrylate (EA); MBAAm, methylene bisacrylamide; MMA, methyl methacrylate; NIPAAm, N-isopropylacrylamide.
*Indicates typical particle size changes in response to environmental temperature.

Figure 11.16 Schematic representation of a calcium carbonate core particle microencapsulated with blend of ethylcellulose pseudolatex and composite latex particles with a thermosensitive hydrogel shell. (Reprinted with permission from Reference 43. Copyright ©2000 Elsevier Ltd.)

one can obtain a thermosensitive drug release with pulsatile "on-off" switching response [43] (Figure 11.17). As poly(NIPAAm) should shrink at temperature above LGCT, this shrinkage creates many void in the microcapsule membrane and hence aqueous permeability of the membrane increases, leading to faster drug release at the higher temperature. This may be suitable for a particular application. In drug delivery, deviation of body temperature from 37 °C, because of the presence of pathogens or pyrogens, may serve as useful stimulus to induce the release of therapeutic agents from the thermosensitive microcapsular device. Physically controlled temperature by a heat source such as microwave from outside the body may also be useful for temperature-activated oscillatory release of endogeneous hormones such as insulin. For these applications, however, thermosensitivity of poly(NIPAAm) must be tuned very precisely within the narrow range of temperature deviation, and this is a critical issue remaining unsolved.

Novel polymers have been studied to encapsulate pharmaceuticals into them at lower temperature and pH conditions. One such approach is the development of novel composite nanoparticles consisting of a poly(NIPAAm) core having a layer of poly(methacrylic acid) with grafted PEG (p(MAA-*g*-EG)) (Figure 11.18) [45,46]. As discussed earlier, poly(NIPAAm) swells at temperature below 32 °C and shrinks at temperature above 32 °C. Poly(MAA-*g*-PEG) is a pH-sensitive polymer that collapses in acidic media and swells in neutral media [45–48]. In addition

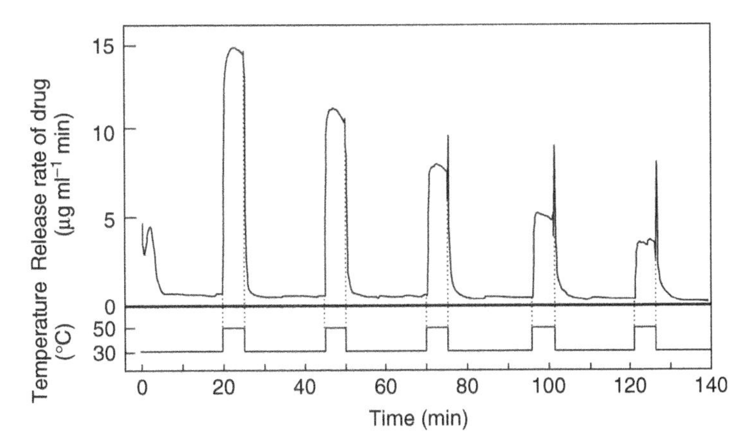

Figure 11.17 Drug release from temperature-sensitive microcapsules. (Reprinted with permission from [43]. Copyright © 2000 Elsevier Ltd.)

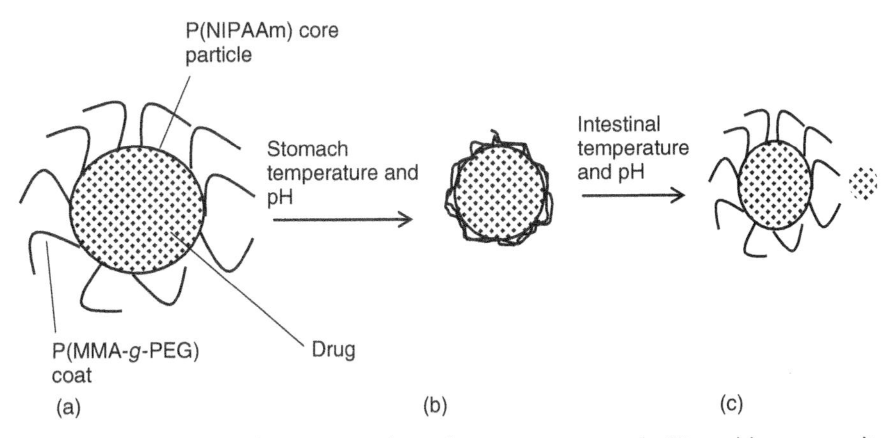

Figure 11.18 Schematic representation of a temperature and pH-sensitive composite nanoparticles: (a) swelling condition where the drug is loaded; (b) shrinking condition where the drug is kept safely inside as the pH-sensitive polymer protects; and (c) drug is released at intestinal pH as the polymer layer loosens up.

to pH sensitivity, this polymer also exhibits some mucoadhesive property [48,49]. One can use a combination of these polymers to load thermosensitive protein and peptide drugs [45,46].

In the past decade, hydrogels have gained wide attention as a drug delivery system [49]. More attention has been paid to *in situ* forming hydrogels. These are polymeric systems that are in liquid (sol) state below body temperature. The minute they are injected into the body, they turn into a gel state because of the change (increase) in temperature. This type of thermosensitive sol–gel systems have been widely studied for controlled delivery of drugs, biologicals, cells, and tissue engineering [50,51].

Poly(*N*-isoproylacrylamide)-based hydrogels have created interest in development of other PNIPAAm copolymers. One such polymer that has been developed and studied for drug delivery application is the poly(*N*-isopropylacrylamide-*co*-acrylic acid) (PNIPAAm-AA). This polymer showed similar LCST as PNIPAAm; however, the advantage of this polymer is that the aqueous solution of this polymer does not precipitate at LCST such as PNIPAAm in water [52]. Similarly, thermoreversible polymers with terminal PEG chains have been reported. The use of PEG-PNIPAAm-PEG and PEG-PNIPAAm as thermoreversible micelles has also been reported [53,54]. Another modified copolymer of PNIPAAm that has been reported for ocular delivery is the PEG diacrylate (PEG-DA) cross-linked PNIPAAm hydrogel [55]. The application of thermoreversible hydrogels has been targeted toward immobilization of cell. Modified PNIPAAm copolymer was synthesized with petaerythritol diacrylate monostearate (PEDAS), acrylamide (AAm), and hydroxyethyl acrylate (HEA) and studied for their theromresponsive behavior [56,57]. They found that by varying the concentration of comonomers, they could bring in thermally induced physical gelation around body temperature (36.2–40.5 °C). They also found that by adding calcium ions to the solution, the sol–gel transition temperature could be decreased to a significant level (27.6–34.4 °C) [58].

Among the various polymers used for drug delivery, polylactic acid and polylactide-*co*-glycolide have found special place owing to their application in controlled and prolonged delivery of drugs. Development of a copolymer using PEG-PLLA-PEG was reported by Jeong and coworkers [50]. The triblock copolymer synthesized was able to form a hydrogel. However, the formation of hydrogel with this polymer was gel–sol. That is, the polymer formed an initial gel state, which at higher temperature turned into a solution, although this was different from other hydrogel-forming polymers, which were typically turning sol–gel state. They did explore some applications for this type of polymers. To overcome the drawback of the thermoreversibile polymer and to take advantage of the poly(lactide) polymer, they further investigated and synthesized a PEG-PLGA-PEG triblock copolymer [59]. This triblock copolymer was the solution at ambient temperature and turned into a gel at higher temperature. This can perfectly play as an *in vivo* controlled-release agent [60]. Subsequently, PLGA-PEG-PLGA triblock copolymer was developed as a thermosensitive polymer for drug delivery applications [61]. This polymer is commercially available under the name ReGel®. ReGel has been investigated for solubilization of hydrophobic drugs, delivery of interleukin-2 for cancer treatment [62], paclitaxel-loaded delivery system named OncoGel™ [63], and various other applications. Another biodegradable polymer picked for development of a thermosensitive-block copolymer was the PCL. PEG-PCL and PEG-PCL-PEG polymers were reported with thermosensitive applications [64,65]. The application of all these thermosensitive polymers has been to deliver drug at a controlled rate or encapsulate cells or enzymes for drug delivery.

11.3 SUMMARY

In pharmaceutical field, polymers have played an important role for the design and manufacturing of pharmaceutical dosage forms for a long period of time. Nowadays, a wide variety of polymers including both natural and synthetic polymers are available as FDA-approved excipients. Traditionally, polymers have been widely used as excipients such as fillers, binders, thickening agents, dispersing agents, disintegrating agents, and coating agents for solid, semisolid, and liquid dosage forms. Recent progress in controlled drug delivery requires more sophisticated functional polymeric systems so as to rationalize highly functional controlled-release systems as well as novel cutting-edge drug carriers. This chapter has described fundamental aspects of representative pharmaceutical polymers and also surveyed their applications including not only typical uses as excipients/vehicles but also novel roles in drug delivery.

ACKNOWLEDGMENT

A part of this work was supported by MEXT-supported Program for the Strategic Research Foundation at Private Universities, 2012–2016 (S1201010).

REFERENCES

1. Khan, K. A., Rhodes, C. T. (1975). Evaluation of different viscosity grades of sodium carboxymethylcellulose as tablet disintegrants. *Pharm. Acta Helv.*, *50*, 99–102.
2. Singh, J. (1992). Effect of sodium carboxymethylcelluloses on the disintegration, dissolution and bioavailability of lorazepam from tablets. *Drug Dev. Ind. Pharm.*, *18*, 375–383.
3. Porter, S. C. (1989). Controlled-release film coatings based on ethylcellulose. *Drug Dev. Ind. Pharm.*, *15*, 1495–1521.
4. Rowe, R. C. (1986). The prediction of compatibility/incompatibility in blends of ethycellulose with hydroxypropyl methylcellulose or hyroxypropyl cellulose using 2-dimensional solubility parameter maps. *J. Pharm. Pharmacol.*, *38*, 214–215.
5. Saettone, M. F., Perini, G., Rijli, P., Rodriguez, L., Cini, M. (1995). Effect of different polymer-plasticizer combinations on '*in vitro*' release of theophylline from coated pellets. *Int. J. Pharm.*, *126*, 83–88.
6. Katikaneni, P., Upadrashta, S. M., Neau, S. H., Mitra, A. K. (1995). Ethylcellulose matrix controlled-release tablets of a water-soluble drug. *Int. J. Pharm.*, *123*, 119–125.
7. Sharma, P., Hamsa, V. (2001). Formulation and evaluation of buccal mucoadhesive patches of terbutalinesulphate. *STP Pharm. Sci.*, *11*, 275–281.
8. Eaimtrakarn, S., Itoh, Y., Kishimoto, J., Yoshikawa, Y., Shibata, N., Murakami, M., Takada, K. (2001). Gastrointestinal mucoadhesive patch system (GI-MAPS) for oral administration of G-CSF, a model protein. *Biomaterials*, *23*, 145–152.

9. Sakagami, M., Sakon, K., Kinoshita, W., Makino, Y. (2001). Enhanced pulmonary absorption following aerosol administration of mucoadhesive powder microspheres. *J. Control. Release*, *77*, 117–129.
10. United States Pharmacopoeia 36, National Formulary 31, 2013.
11. Chowhan, Z. T. (1980). Role of binders in moisture-induced hardness increase in compressed tablets and its effect on in vitro disintegration and dissolution. *J. Pharm. Sci.*, *69*, 1–4.
12. Rowe, R. C. (1980). The molecular weight and molecular weight distribution of hydroxypropyl methylcellulose used in the film coating of tablets. *J. Pharm. Pharmacol.*, *32*, 116–119.
13. Hogan, J. E. (1989). Hydroxypropylmethylcellulose sustained release technology. *Drug Dev. Ind. Pharm.*, *15*, 975–999.
14. Thanoo, B. C., Sunny, M., Jayakrishnan, A. (1993). Controlled release of oral drugs from cross-linked polyvinyl alcohol microspheres. *J. Pharm. Pharmacol.* *45*, 16–20.
15. Alexandridis, P., Hatton, T. A. (1995). Poly(ethylene oxide)-poly(propylene oxide)-poly(ethylene oxide) block copolymer in aqueous solutions and at interfaces: thermodynamics, structure, dynamics, and modeling. *Colloid Surf. A.*, *96*, 1–46.
16. Mata, J. P., Majhi, P. R., Guo, C., Liu, H. Z., Bahadur, P.(2005). Concentration, temperature and salt induced micellization of a triblock copolymer Pluronic L64 in aqueous media. *J. Colloid Interface Sci.*, *292*, 548–556.
17. Bottenberg, P., Cleymaet R., de Muynck, C., Remon, J. P., Coomans, D., Michotte, Y., Slop, D. (1991). Development and testing of bioadhesive, fluoride containing slow-release tablets for oral use. *J. Pharm. Pharmacol.*, *43*, 457–464.
18. Daido Chemical Corporation, Technical Information of POVACOAT®, http://www.daido-chem.co.jp/english/techinfo/pdf/pova-e.pdf. Accessed on 07/17/2013.
19. Uramatsu, S., Shinike, H., Kida, A., Uemura, T., Ichikawa, H., Fukumori, Y. (2007). Investigation of POVACOAT® as a novel carrier for solid dispersions. *J. Pharm., Sci., Tech., Japan*, *67*, 266.
20. Xu, M., Zhang, C., Luo, Y., Xu, L., Tao, X., Wang, Y., He, H., Tang, X. (2014). Application and functional characterization of POVACOAT®, a hydrophilic co-polymer poly(vinyl alcohol/acrylic acid/methyl methacrylate) as a hot-melt extrusion carrier. *Drug Dev. Ind. Pharm.*, *40*, 126–135. **DOI**: 10.3109/03639045.2012.752497.
21. Yuan, J., Wu, S. H. (2000). Sustained release tablets via direct compression: a feasibility study using cellulose acetate and cellulose acetate butyrate. *Pharm. Tech.*, *24*, 92–106.
22. Yuan, J., Shang, P. P., Wu, S. H. (2001). Effects of polyethylene glycol on morphology, thermomechanical properties and water vapor permeability of cellulose acetate free films. *Pharm. Tech.*, *25*, 62–74.
23. Shim, I. K., Yook, Y. J., Lee, S. Y., Lee, S. H., Park, K. D., Lee, M. C., Lee, S. J. (2008). Healing of articular cartilage defects treated with a novel drug-releasing rod-type implant after microfracture surgery. *J. Control. Release*, *129*, 187–191.
24. Okada, H., Heya, T., Ogawa, Y., Shimamoto, T. (1988). One-month release injectable microcapsules of a luteinizing hormone-releasing hormone agonist (leuprolide acetate) for treating experimental endometriosis in rats. *J. Pharmacol. Exp. Ther.*, *244*, 744–750.
25. Ravivarapu, H. B., Moyer, K. L., Dunn, R. L. (2000). Sustained suppression of pituitary-gonadal axis with an injectable, *in situ* forming implant of leuprolide acetate. *J. Pharm. Sci.*, *89*, 732–740.

26. Huang, L. L., Schwartz, J. B. (1995). Studies on drug release from a carbomer tablet matrix. *Drug Dev. Ind. Pharm.*, *21*, 1487–1501.
27. Singh, B., Ahuja, N. (2002). Development of controlled-release buccoadhesive hydrophilic matrices of diltiazem hydrochloride: optimization of bioadhesion, dissolution and diffusion parameters. *Drug Dev. Ind. Pharm.*, *28*, 431–442.
28. Jones, D. S. et al. (2000). Design, characterisation and preliminary clinical evaluation of a novel mucoadhesive topical formulation containing tetracycline for the treatment of periodontal disease. *J. Control. Release*, *67*, 357–368.
29. Jones, D. S. et al. (1998). Viscoelastic properties of bioadhesive, chlorhexidine containing semi-solids for topical application to the oropharynx. *Pharm. Res.*, *15*, 1131–1136.
30. Jones, D. S. Physicochemical characterization and preliminary *in vivo* efficacy of bioadhesive, semisolid formulations containing flurbiprofen for the treatment of gingivitis. *J. Pharm. Sci.*, *88*, 592–598.
31. Amidon, G. L., Lennernas, H., Shah, V. P., Crison, J. R. (1995). A theoretical basis for biopharmaceutic drug classification: The correlation of in vitro drug product dissolution and in vivo bioavailability. *Pharm. Res.*, *12*, 413–420.
32. Crowly, M. M., Zhang, F., Repka, M. A., Thumma, S., Upadhye, S. B., Battu, S. K., McGinity, J. W., Martin, C. (2007). Pharmaceutical applications of hot-melt extrusion: part I. *Drug Dev. Ind. Pharm.* *33*, 909–926.
33. Repka, M. A., Battu, S. K., Upadhye, S. B., Thumma, S., Crowley, M. M., Zhang, F., Martin, C., McGinity, J. W. (2007). Pharmaceutical applications of hot-melt extrusion: part I. *Drug Dev. Ind. Pharm.* *33*, 1043–1057.
34. Wilson, M., Williams, M. A., Jones, D. S., Andrews, G. P. (2012). Hot-melt extrusion technology and pharmaceutical application. *Ther. Deliv.*, *3*, 787–797.
35. Kelin, C. E., Chiu, Y. L., Awni, W., Zhu, T., Heuser, R. S., Doan, T., Breitenbach, J., Morris, J. B., Brun, S. C., Hanna, G. J.(2007). The tablet formulation of lopinavir/ritonavir provides similar bioavailability to the soft gelatin capsule formulation with less pharmacokinetics variability and diminished food effect. *J. Acq. Imm. Def. Synd.*, *44*, 401–410.
36. Hoffman, A. (2002). Hydrogels for biomedical applications. *Adv. Drug Deliv. Rev.*, *54*, 3–12.
37. Fukumori, Y., Yamaoka, Y., Ichikawa, H., Takeuchi, Y., Fukuda, T., Osako, Y. (1988). Coating of pharmaceutical powders by fluidized bed process. III. Aqueous coating with ethyl acrylate-methyl methacrylate-2-hydroxyethyl methacrylate copolymer and the dissolution properties of the products. *Chem. Pharm. Bull.* *36*, 3070–3078.
38. Fukumori, Y., Yamaoka, Y., Ichikawa, H., Takeuchi, Y., Fukuda, T., Osako, Y. (1988). Coating of pharmaceutical powders by fluidized bed process. IV. Softening temperature of acrylic copolymers and its relation to film-formation in aqueous coating. *Chem. Pharm. Bull.* *36*, 4927–4932.
39. Fukumori, Y., Ichikawa, H., Yamaoka, Y., Akaho, E., Takeuchi, Y., Fukuda, T., Kanamori, R., Osako, Y. (1991). Microgranulation and encapsulation of pulverized pharmaceutical powders with ethylcellulose by the Wurster process. *Chem. Pharm. Bull.* *39*, 1806–1812.
40. Ichikawa, H., Jono, K., Tokumitsu, H., Fukuda, T., Fukumori, Y. (1993). Coating of pharmaceutical powders by fluidized bed process-V. Agglomeration and efficiency in coating with aqueous latices of copoly (ethyl acrylate-methyl methacrylate-2-hydroxyethyl methacrylate). *Chem. Pharm. Bull.*, *41*, 1132–1136.

41. Ichikawa, H., Tokumitsu, H., Jono, K., Fukuda, T., Osako, Y., Fukumori, Y. (1993). Coating of pharmaceutical powders by fluidized bed process-VI. Microencapsulation using blend and composite latices of copoly (ethyl acrylate-methyl methacrylate-2-hydroxyethyl methacrylate). *Chem. Pharm. Bull.*, *42*, 1308–1314.
42. Ichikawa, H., Kaneko, S., Fukumori, Y. (1996). Coating performance of aqueous composite latices with N-isopropylacrylamide shell and thermosensitive permeation properties of their microcapsule membrane. *Chem. Pharm. Bull.*, *44*, 383–391.
43. Ichikawa, H., Fukumori, Y. (2000). A novel positively thermosensitive controlled release microcapsule with membrane of nano-sized poly(*N*-isopropylacrylamide) gel dispersed in ethylcellulose matrix. *J. Control. Release*, *63*, 107–119.
44. Ichikawa, H. and Fukumori, Y. (1999). Negatively thermosensitive release of drug from microcapsules with hydroxypropyl cellulose membrane prepared by the Wurster process. *Chem. Pharm. Bull.*, *47*, 1102–1107.
45. Lowman, A. M., Peppas, N. A. (1997). Analysis of the complexation/decomplexation phenomena in graft copolymer networks. *Macromolecules*, *30*, 4959–4965.
46. Lowman, A. M., Morishita, M., Kajita, M., Nagai, T., Peppas, N. A. (1999). Oral delivery of insulin using pH-responsive complexation gels. *J. Pharm. Sci.*, *88*, 933–937.
47. Ichikawa, H., Yamasaki, Y., Fukumori, Y. *New Trends in Polymers for Oral and Parenteral Administration from Design to Receptors*. Barratt, G., Duchene, D., Fattal, F., Legendre, J. Y., editors. Editions DeSante, Paris, 2001, pp. 257–260.
48. Iijima, M., Yoshimura, M., Tsuchiya, T., Tsukada, M., Ichikawa, H., Fukumori, Y., Kamiya, H. (2008). Direct measurement of interactions between stimulation-responsive drug delivery vehicles and artificial mucin layers by colloid probe atomic force microscopy. *Langmuir*, *24*, 3987–3992.
49. Peppas, N. A., Bures, P., Leobundung, W., Ichikawa, H. (2000). Hydrogels in pharmaceutical formulations. *Eur. J. Pharm. Biopharm.*, *50*, 27–46.
50. Jeong, B., Bae, Y. H., Lee, D. S., Kim, S. W. (1997). Biodegradable block copolymers as injectable drug-delivery systems. *Nature*, *388*, 860–862.
51. He, C., Kim, S., Lee, D. (2008). In situ gelling stimuli-sensitive block copolymer hydrogels for drug delivery. *J. Control. Release*, *127*, 189–207.
52. Han, C., Bae, Y. (1998). Inverse thermally-reversible gelation of aqueous N-isopropylacrylamide copolymer solutions. *Polymer*, *39*, 2809–2814.
53. Topp, M., Dijkstra, P., Talsma, H., Feijen, J. (1997). Thermosensitive micelle-forming block copolymers of poly(ethylene glycol) and poly(N-isoproylacrylamide). *Macromolecules*, *30*, 8518–8520.
54. Zhang, W., Shi, L., Wu, K., An, Y. (2005). Thermoresponsive micellization of poly(ethylene glycol)-b-poly(N-isopropylacrylamide) in water. *Macromolecules*, *38*, 5743–5747.
55. Turturro, S. B., Guthrie, M. J., Appel, A. A., Drapala, P. W., Brey, E. M., Perez-Luna, V. H., Mieler, W. F., Kang-Mieler, J. J. (2011). The effects of cross-linked thermo-responsive PNIPAAm-based hydrogel injection on retinal function. *Biomaterials*, *32*, 3620–3626.
56. Hacker, M. C., Klouda, L., Ma, B. B., Kretlow, J. D., Mikos, A. G. (2008). Synthesis and characterization of injectable, thermally and chemically gelable amphiphilic poly (N-isopropylacrylamide) based macromers. *Biomarcomolecules*, *9*, 1558–1570.

57. Klouda, L., Hacker, M. C., Kretlow, J. D., Mikos, A. G. (2009). Cytocompatibility evaluation of amphiphilic, thermally responsive and chemically crosslinkable macromers for in situ forming hydrogels. *Biomaterials*, *30*, 4558–4566.

58. Kretlow, J. D., Hacker, M. C., Klouda, L., Ma, B. B., Mikos, A. G. (2010). Synthesis and characterization of dual stimuli responsive macromers based on poly(N-iospropylacrylamide) and poly(vinylphosphonic acid). *Biomacromolecules*, *11*, 797–805.

59. Jeong, B., Bae, Y. H., Kim, S. W. (1999). Thermoreversible gelation of PEG-PLGA-PEG triblock copolymer aqueous solutions. *Macromolecules*, *32*, 7064–7069.

60. Jeong, B., Chooi, Y. K., Bae, Y. H., Zentner, G., Kim, S. W. (1999). New biodegradable polymers for injectable drug delivery systems. *J. Control. Release*, *62*, 109–114.

61. Shim, M., Lee, H., Shim, W., Park, I., Lee, H., Chang, T., Kim, S., Lee, D. (2002). Poly(D,L-lactic acid-co-glycolic acid)-b-poly(ethylene glycol)-b-poly(D,L-lactic acid-co-glycolic acid) triblock copolymer and thermoreversible phase transition in water. *J. Biomed. Mater. Res. A*, *61*, 188–196.

62. Samlowski, W. E., McGregor, J. R., Jurek, M., Baudys, M., Zentner, G. M., Fowers, K. D. (2006). ReGel® polymer based delivery of interleukin-2 as a cancer treatment. *J. Immunother.*, *29*, 524–535.

63. Elstad, N., Fowers, K. (2009). OncoGel (ReGel/paclitaxel) clinical applications for a novel paclitaxel delivery system. *Adv. Drug Deliv. Rev.* *61*, 785–794.

64. Kim, M., Seo, K., Khang, G., Cho, S., Lee, H. (2004). Preparation of poly(ethylene glycol)-block-poly(caprolactone) copolymers and their applications as thermo-sensitive materials. *J. Biomed. Mater. Res. A*, *70*, 154–158.

65. Gong, C. Y., Qian, Z. Y., Liu, C. B., Huang, M. J., Gu, Y. C., Wen, Y. J., Kan, B., Wang, K., Dai, M., Li, X. Y., Gou, M. L., Tu, M. J., Wei, Y. Q. (2007). A thermosensitive hydrogel based on biodegradable amphiphilic poly(ethylene glycol)-polycaprolactone-poly(ethylene glycol) block copolymers. *Smart Mater. Struct.*, *16*, 927–933.

12

POLYMER CONJUGATES OF PROTEINS AND DRUGS TO IMPROVE THERAPEUTICS

PARIJAT KANAUJIA

AJAZUDDIN

12.1 INTRODUCTION

The advent of new techniques in biotechnology has led to a number of proteins and peptides being approved as therapeutic agents. These proteins and peptides are highly specific therapeutic agents for the target but suffer from high immunogenicity owing to their microbial origin, degradation by proteases, short shelf-life, and short plasma circulation half-life. The small organic active pharmaceutical ingredients (APIs) suffer from low aqueous solubility, toxic side effects, enzymatic degradation, and rapid clearance from the circulation [1]. Various drug delivery technologies have been explored to overcome these deficiencies and improve their therapeutic performance. The polymer conjugation of drugs and proteins is one of the ways to improve the solubility, stability, and permeation; and reduce immunogenicity, toxicity, and side effects. Polymer–drug conjugates, also referred as *Polymer Therapeutics*, include chemical conjugation of hydrophilic polymer with small organic API, proteins and peptides, and polymeric micelles [2]. Polymers are widely used in the pharmaceutical formulation to improve the therapeutic performance of products and acceptability by the patients. The polymers used in the preparation of conjugate are generally water soluble, and thus are useful in drug administration.

Synthesis and Applications of Copolymers, First Edition. Edited by Anbanandam Parthiban.

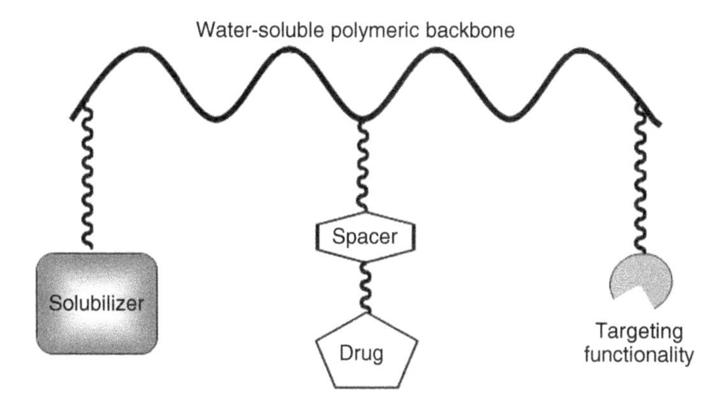

Figure 12.1 The Ringsdorf rational model for drug–polymer conjugate.

The synthesis of polyvinylpyrrolidone conjugate of glycyl-L-leucyl-mescaline as a depot formulation by Jatzkewits [3] was considered as one of the first reports in the field. The major landmark in the area of drug polymer conjugation came in 1975 when Ringsdorf [4,5] proposed rational model for pharmacologically active polymers. This model (Figure 12.1) essentially consists of a biocompatible polymer chemically coupled with a solubilizer, a drug, and a targeting functionality.

In 1977, Davis and colleagues [6] published the effect of poly(ethylene glycol) (PEG) conjugation on the immunogenicity and circulation half-life of bovine liver catalase. This formed the basis for PEGylation of therapeutically important proteins and peptides. In 1981, anticancer enzyme L-asparaginase was conjugated with several units of PEG 5000 to reduce its immunogenicity and increase blood circulation half-life [7]. This conjugate was the first to receive market approval from US Food and Drug Administration (FDA) in 1994 for use in combination chemotherapy for the treatment of patients with acute lymphoblastic leukemia (ALL). Poly(styrene-co-maleic acid) derivatives of the antitumor protein neocarzinostatin were synthesized by Maeda and coworkers [8] and reported significant improvements in pharmacological properties. This conjugate was approved for human use in 1990 in Japan for the treatment of hepatocellular carcinoma. Doxorubicin conjugated to *N*-(2-hydroxypropyl)methacrylamide (HPMA) was the first drug–polymer conjugate tested clinically in 1994 [9–11]. Some major milestones in the development of drug–polymer conjugates are represented in Figure 12.2. Currently, there are several products available in the market based on this technology and many are under development.

12.2 POLYMERS FOR THERAPEUTIC CONJUGATION

Majority of therapeutic polymer conjugates have been synthesized using water-soluble linear polymers. PEG and HPMA copolymers are the most widely

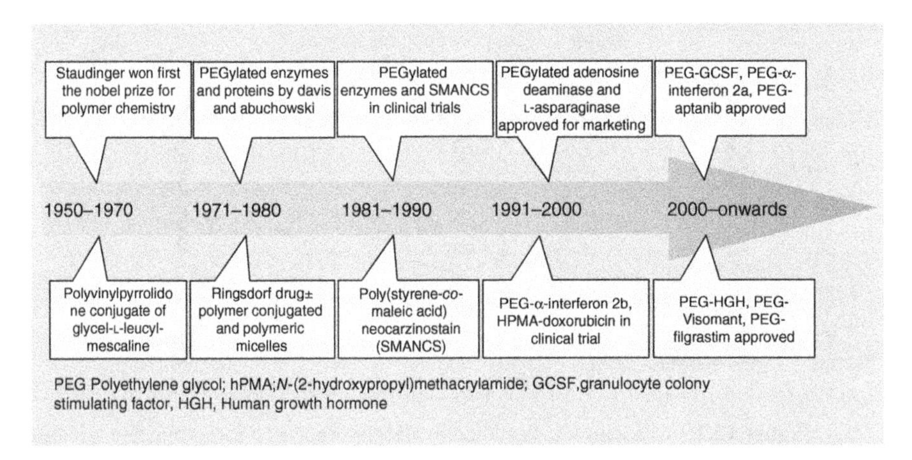

Figure 12.2 Historical development of drug–polymer conjugates. (Modified with permission from [12]. Copyright © 2006 Nature Publishing Group.)

investigated polymers for therapeutic conjugation with protein, peptides, and drugs. Poly(L-glutamic acid) and polysialic acid (PSA) are also investigated by conjugation with proteins and drugs. Highly branched star-shaped dendrimers are catching attention of researchers owing to the large density of functional group present for conjugation.

12.2.1 Poly(ethylene Glycol) Protein Conjugate

PEG is one of the most prevalent polymers exploited in recent years for the delivery of proteins, peptides, and low molecular weight bioactives. The modification of protein(s), drugs(s), and drug-carrier system by chemical conjugation with PEG is known as *PEGylation*. PEG-conjugated proteins and bioactives are more effective than their corresponding unmodified parent compounds.

PEGs are amphiphilic polymers comprising of repeated ethylene oxide subunits, whose number is represented by whole integer n. PEG contains two terminal hydroxyl groups that can be chemically activated [13].

$$HO{-}(CH_2CH_2O)_n{-}H$$

PEG is synthesized by anionic ring-opening polymerization of ethylene oxide initiated by nucleophilic attack of a hydroxide ion such as methanol or derivatives such as methoxyethoxy ethanol resulting in monoalkyl-capped PEG such as methoxy poly(ethylene) glycol (mPEG).

$$CH_3O{-}(CH_2CH_2O)_n{-}H$$

12.2.2 Significance of PEG

The PEG possesses an array of properties ideally required for any polymeric carrier to be suitable for therapeutic use.

- *Solubility and hydration*: PEG is soluble in both aqueous and organic solvents, hence making it suitable for conjugation with biological molecules under mild physiological conditions. As an amphiphilic polymer, it can solubilize poorly water-soluble compounds [14]. In aqueous solution, it typically binds with two to three water molecules per ethylene oxide unit, which gives rise to large exclusion volume and it acts as if it is 5–10 times as large as a soluble protein of comparable molecular weight.
- *Toxicity*: PEGs are nontoxic to humans even at higher doses. The chronic administration of PEG is also well tolerated in humans, making it suitable for the long-term use. It is a nonbiodegradable polymer but cleared rapidly from the biological system without any structural changes. It is eliminated from the body by a combination of renal and hepatic routes depending on molecular weight of PEG [15]. US FDA has approved PEG for human intravenous, intramuscular, oral, and topical administration.
- *Polymer chemistry*: The polymerization reaction of PEG can be manipulated and controlled to form a variety of molecular weights from 1000–50,000 Da with very low polydispersity ($M_w/M_n \leq 1.1$). The use of mPEG for conjugation to biologically relevant materials minimizes the possibilities of cross-linking and improves homogeneity of the conjugates [13,16].
- *Pharmacokinetics and biodistribution*: PEG exhibits excellent pharmacokinetic required for a long circulatory carrier. The distribution and tissue uptake of I^{125}-labeled PEGs with different molecular weights were studied after intravenous administration to mice and results showed that high molecular weight PEGs have longer half-life than low molecular weight PEGs. The terminal half-life of PEG in the circulation extends from 18 min to 24 h as PEG molecular weight increases from 6 to 90 kDa. The PEGs of low molecular weight freely translocate from circulation to extra vascular tissues, whereas PEGs of high molecular weight translocate more slowly to extra vascular tissues [17]. The biofate of different water-soluble polymers administered by intraperitoneal (ip), subcutaneous (sc), and intramuscular (im) routes to mice revealed that the elimination rate is influenced not only by molecular weight but also by route of injection. PEG of molecular weight 6000 disappeared from injection site rapidly, whereas PEG of molecular weight 50,000 was absorbed slowly. The elimination rate increases in order of ip>sc>im injections irrespective of molecular weight of PEG [18].
- *Immunogenicity*: PEGs are very weak antigens and exhibit negligible immunogenic response even at higher molecular weights. Conjugation of mPEG to proteins prevents the recognition by the body's immune system, making them nonimmunogenic. PEG conjugation also increases apparent

molecular size of the protein, reducing its renal clearance and increasing its circulation half-life [19].

- *Ease of sterilization*: Aqueous solutions can be sterilized by filtration and autoclaving and are needed to be stored in airtight containers.

12.2.3 Chemistry of Protein–PEG Conjugation

To couple a PEG moiety to a protein, peptides, and small organic compounds, it is necessary to activate the hydroxyl group(s) of PEG by converting to some functional groups capable of reacting with the functional groups found on the compound of interest to be coupled with PEG. Lysine and N-terminal amino group conjugation is the most common route of protein conjugation with PEG. Lysine is one of the most abundant amino acids present in protein sequence and constitutes 10% of the total amino acid sequence. Broadly, PEGylation chemistry can be divided into two major categories based on the molecular weight and type of activated PEG used.

12.2.3.1 Chemistry of First Generation of PEG The first-generation PEG derivates were mostly mPEG-based electrophiles with PEG molecular weight between 2000 and 5000 Da. These activated mPEGs are mainly arylating or acylating agents containing a reactive aryl chloride or acyl chloride residue, which is displaced by a nucleophilic alpha or epsilon amino group upon reaction with peptides, proteins, and organic molecules. Various first-generation PEG reagents used for PEGylation of proteins and peptides include (i) mPEG dichlorotriazine, (ii) $mPEG_2$ chlorotriazine (iii) mPEG succinimidyl carbonate, (iv) mPEG benzotriazole carbonate, (v) mPEG carbonylimidazole, (vi) mPEG *p*-nitrophenyl carbonate, (vii) mPEG trichlorophenyl carbonate, (viii) PEG tresylate, and (ix) PEG succinimidyl succinate.

The first synthesis of PEG–protein conjugate was reported by Davis and coworkers using cyanuric chloride to activate mPEG to form mPEG dichlorotriazine (**1**) (Scheme 12.1) for attaching to primary amines present on the bovine serum albumin (BSA). The conjugation of PEG chains reduced the antigenicity and immunogenicity of BSA significantly [20]. The mPEG dichlorotriazine was lacking selectivity and also able to react with multiple nucleophilic functional groups such as hydroxyl and sulfhydral groups present in the amino acid sequence and resulted in the loss of protein activity of phenylalanine-ammonia lyase [21]. To overcome the lack of selectivity of mPEG dichlorotriazine, Inada and coworkers synthesized 2,4-bis(methoxypolyethylene glycol)-6-chloro-*s*-triazine (**2**) (Scheme 12.1) by replacing two of the most reactive chlorides with mPEG. The low reactivity of the remaining chloride prevents side reaction and cross-linking [22].

Various acylating agents have been developed for mPEG conjugation with proteins via amide bonds in the early 1990s. These acylating mPEG derivatives displayed good selectivity for amino groups and retained biological activity of the proteins. Succinimidyl ester of carboxymethylated PEG (SC-mPEG, **3**) [23] and

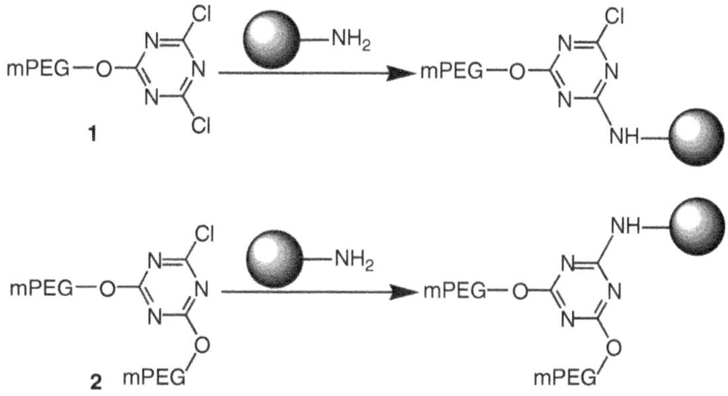

Scheme 12.1 PEGylation by chlorotriazine.

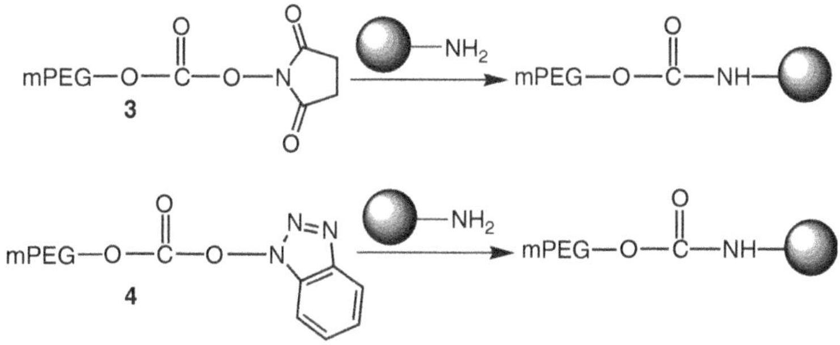

Scheme 12.2 PEGylation by acylating agents.

benzotriazole ester of carboxymethylated PEG (BTC-mPEG, **4**) (Scheme 12.2) reacted selectively with lysine residues to form amides [24].

The SC-mPEG and BTC-mPEG are able to produce protein conjugates under mild physiological conditions in a short period of time. These acylating agents are prone to hydrolysis and have a half-life of 15–20 min at pH 8.0.

The use of urethane linkage to form mPEG-conjugated proteins was achieved by using immidazolylcarbonyloxy derivative of mPEG **5** [25], *p*-nitrophenyl carbonate derivative of mPEG **6**, and trichlorophenyl carbonate derivative of mPEG **7** [26]. These reagents have much lower reactivity than SC-mPEG or BTC-mPEG (Scheme 12.3). These reagents are very easy to prepare and form chemically stable bonds with amines present on the protein.

Nilsson and Mosbach synthesized tresylate of mPEG **8** (Scheme 12.4) for protein conjugation. The conjugation reaction can be accomplished by the alkylation of the amino and thiol groups under mild physiological conditions at pH 8–9 [27].

Succinimidyl succinate ester of mPEG (SS-mPEG) **9** (Scheme 12.5) is prepared by reacting succinic anhydride with mPEG and activating the carboxylic acid to the

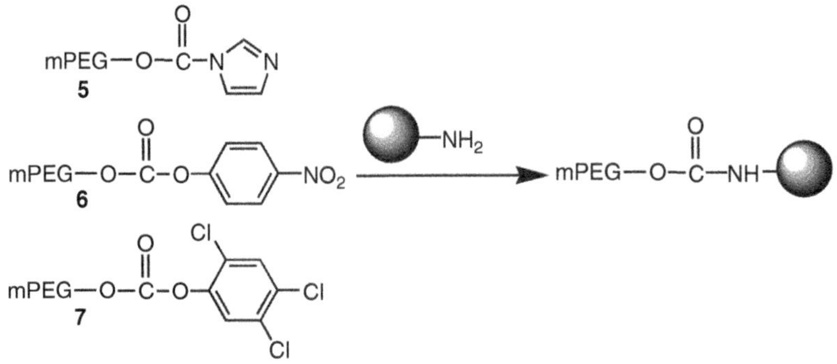

Scheme 12.3 PEGylation by urethane linkers.

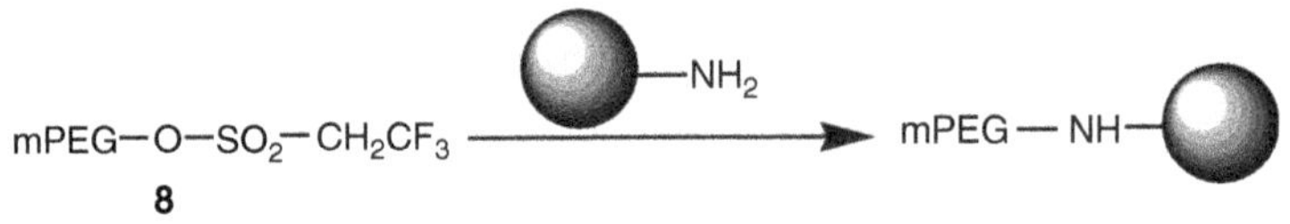

Scheme 12.4 PEGylation by alkylating agent.

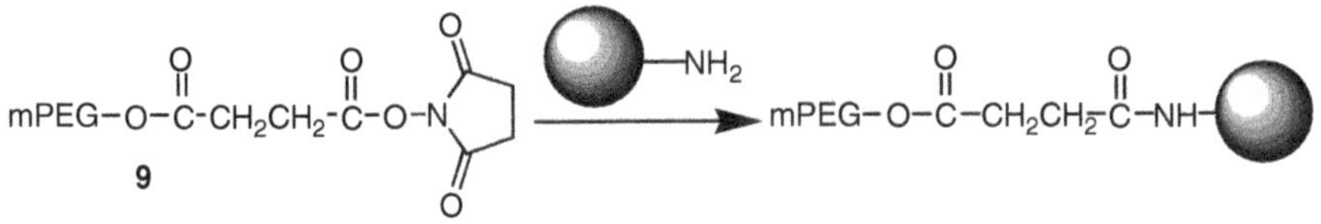

Scheme 12.5 PEGylation by succinimidyl succinate.

succinimidyl ester. Owing to the presence of an ester linkage, the protein–mPEG conjugate is highly prone to hydrolysis leaving a succinate tag on the protein which leads to an increase in immunogenicity [28].

The first-generation PEG reagents are generally simple and prepared by reacting the free hydroxyl group of mPEG with anhydrides, carbonates, chlorides, and chloroformates. The derivatives of low molecular weight mPEG are highly unselective because the relatively small PEG chains can penetrate into, otherwise poorly, accessible regions in the protein structure. The PEGs used in the first-generation PEGylation are of generally low molecular weight because high molecular weight mPEG contains high amount of diol ($>15\%$) resulting from the presence of water during polymer synthesis. In addition to various associated limitations such as low selectivity, side reactions, cross-linking, and aggregation from bifunctional contaminant, hydrolysis of PEG conjugate makes first-generation PEGylation chemistry inefficient for protein conjugation.

12.2.3.2 Chemistry of Second-Generation PEG The second-generation PEG reagents were designed to prevent the problem faced in the first-generation reagents such as high diol contamination, lack of selectivity toward functional groups, unstable linkages, and side reactions. These reagents have PEG chain with molecular weights between 5 and 20 kDa.

Activated mPEG for Amine Conjugation Aldehydes are one of the most preferred groups for modification of amino groups of proteins because of its selectivity toward primary amino groups of proteins and formation of stable conjugates with polymer chains without any change in the net charge of proteins [29]. The first activated PEG reagent of the second-generation chemistry is mPEG propionaldehyde **10** [30]. Under the acidic conditions ($pH < 5$), aldehydes are highly selective for the N-terminal α-amino group because of low pK_a of the terminal amino group [31]. The coupling reaction gives a permanent linkage after Schiff base formation followed by cyanoborohydride reduction, but this reaction is very slow and sometimes a day is needed for completion of reaction (Scheme 12.6).

Another approach using mPEG aldehyde is the derivatization of mPEG aldehyde to form acetal derivative [32]. The *in situ* acid hydrolysis generates aldehyde hydrate of the acetal derivative, and the reaction proceeds by the Schiff base formation and reduction to form protein conjugate. The acetal derivatives are having longer storage stability and high purity compared to those of their free aldehyde counterparts.

The activated esters of mPEG carboxylic acids are the most popular second-generation derivatives for amine PEGylation of proteins. Reaction between lysine and terminal amines and the active esters of mPEG produces a stable amide linkage under mild physiological conditions [33]. Succinimidyl succinate ester of mPEG (mPEG-SS) is one of the oldest and the most widely used PEG derivatives, but it possesses an ester linkage in its backbone and thus has the property of undergoing hydrolysis *in vivo*. Harris and coworkers have prepared mPEG-succinimidyl propionic acid **11** (mPEG-SPA) and mPEG-succinimidyl butanoatic acid **12** (mPEG-SBA) (Scheme 12.7), which do not have an ester

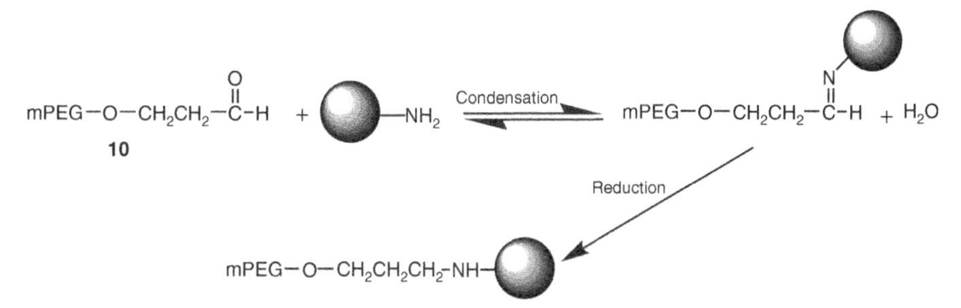

Scheme 12.6 PEGylation by reductive amination using mPEG propionaldehyde. (Reprinted with permission from [31]. Copyright © 2002 Elsevier Ltd.)

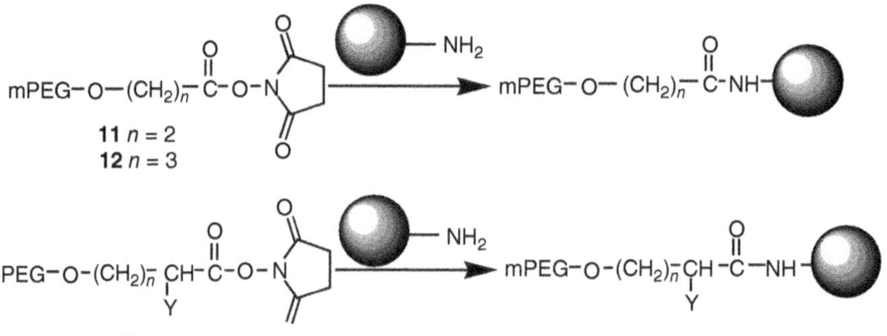

Scheme 12.7 PEGylation by NH ester of mPEG succinimidyl acid and α-branched NHS ester of mPEG succinimidyl acid. (Reprinted with permission from [31]. Copyright © 2002 Elsevier Ltd.)

linkage in the backbone. They generate stable linkages and have nearly ideal reactivity for protein modification. The hydrolytic half-life of mPEG-SPA is 17 min at pH 8, whereas the hydrolytic half-life of mPEG-SBA increased by 23 min because of the presence of two extra methylene units [34]. To reduce activity further, mPEG was attached to an α-branched acid. Activation with *N*-hydroxysuccinimide gave the active ester **13** having a hydrolytic half-life of 33 min at pH 8.

A branched PEG having two polymer chains has been synthesized by coupling mPEG to the primary amines of lysine, whereas the carboxylic acid is activated by NHS ester to bind the protein [35]. The mPEG esters are purely monofunctional because the intermediate acid is purified by ion exchange chromatography. The $mPEG_2$-NHS **14** (40,000 kDa) reagent is being used for attachment to several proteins, such as Hoffmann–La Roche's PEG-ASYS (PEGylated-interferon α-2-b), which is approved for human use by US FDA (Scheme 12.8).

12.2.3.3 Site-Specific PEGylation The ε-amino conjugated proteins are heterogeneous mixtures containing proteins having different number of PEG chains attached to different sites. First few approved products such as PEG asparaginase and PEG adenosine deaminase were produced by nonspecific PEGylation. Unlike

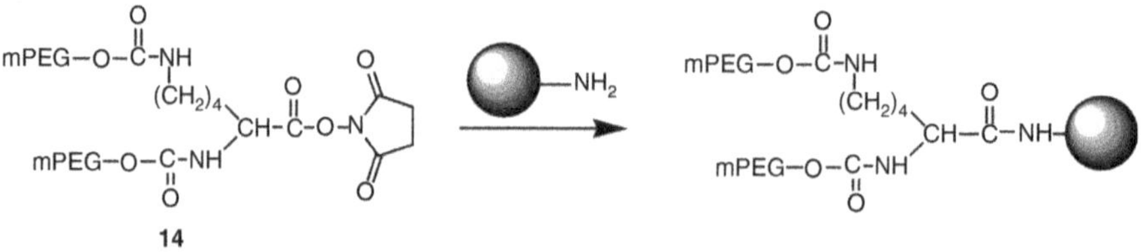

Scheme 12.8 PEGylation by branched PEG_2–NHS ester.

that of the random PEGylation of proteins in site-specific PEGylation, suitable functionalized mPEG chains are attached to a specific group or site to produce polymer conjugate without any steric hindrance into the binding site of proteins. This is achieved by either attaching polymer chains to free thiol group present on cysteine residues or attaching PEG chain to N-terminal amino group.

The regulatory requirements for biologicals are very stringent and require thorough characterization to meet the specifications. Site-specific PEGylated products are reproducible and characterization is straightforward, therefore they are preferred by the industry.

Cysteine Modification Site-specific protein PEGylation is achieved by attaching PEG chain to free thiol groups present on the cysteine residues, but free thiol is rarely available because it is involved in disulfide linkage with other thiol groups. The cysteine-specific-activated mPEG reagents such as mPEG maleimide **15**, mPEG vinylsulfone **16**, mPEG iodoacetamide **17**, and mPEG orthopyridyl disulfide **18** have been developed for thiol-specific PEGylation. The main characteristics of these reagents are given in Table 12.1.

TABLE 12.1 Thiol-Specific Activated PEG Reagents

No.	Name	Structure	Reaction Conditions	Ref.
15	mPEG maleimide	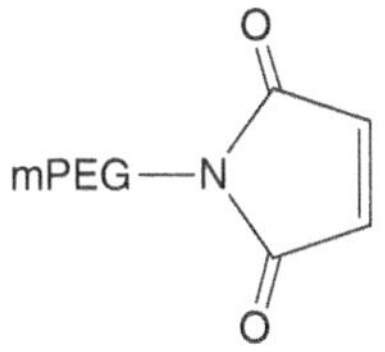 mPEG—N (maleimide ring, two C=O)	Forms stable bond under acidic pH; more reactive to thiol but unstable in water	[36,37]
16	mPEG vinylsulfone	$mPEG{-}S(=O)_2{-}CH{=}CH_2$	Forms stable thioether bond at pH 7–8. At higher pH it reacts with amines	[38]
17	mPEG iodoacetamide	$mPEG{-}NH{-}C(=O){-}CH_2{-}I$	Forms stable thioether linkage. Reaction is slow and should be performed in dark	[39]
18	mPEG orthopyridyl disulfide	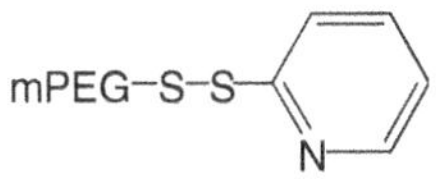 mPEG–S–S–(2-pyridyl)	Reacts specifically to thiol groups under wide pH range (3–10) to form stable disulfide bond but can be cleaved by reducing agents	[40]

The PEGylation of thiol group of cysteine is usually performed in physiological buffers in order to prevent the denaturation of protein. Veronese and coworkers conjugated granulocyte colony stimulating factor (G-CSF) containing an unpaired cysteine buried in a hydrophobic cleft. Under physiological conditions, negligible PEG chain could attach with thiol groups. When urea and guanidine were added to the reaction buffer, partial denaturation and a significant degree of PEGylation were achieved because buried cysteine residues were exposed. The native structure and protein activity of the conjugate was restored after the removal of urea and guanidine [41].

Bridging PEGylation The disulfide bonds of any protein are considered unsuitable for polymer conjugation because disulfide bridges are usually important for tertiary structure and biological activity. The disulfide linkages are usually present in small numbers in therapeutic proteins yielding homogenous polymer protein conjugate after PEGylation reaction. PEGylation at disulfide linkage was first proposed by Brocchini's research group by using a thiol-specific, cross-functionalized PEG monosulfone [42].

This PEGylation reaction was completed in two steps: First, the disulfide linkage was reduced to generate free thiol groups and then reforming disulfide between the two sulfur atoms with three-carbon bridge containing PEG. This process denatures protein reversibly and protein activity is recovered (same as marketed PEGylated interferon α-2-b) after the PEG conjugation as shown by PEGylation of interferon α-2-b (Figure 12.3) [43].

PEGylation of Glycoproteins For modification of glycoproteins, reactivity of oligosaccharide residues is used to attach PEG chains as shown in Scheme 12.9. Glycoproteins such as ovalbumin, glucose oxidase, horseradish peroxidase, and immunoglobulins are oxidized with periodate to generate aldehyde groups on the carbohydrate residues that can be reacted with either mPEG hydrazide to form a permanent hydrazone linkage or with mPEG amine to produce reversible Schiff base. If necessary, this hydrazone linkage can be reduced with $NaCNBH_3$ to a more stable alkyl hydrazide [16,44].

PEGylation of N-Terminal α-Amino Group The reductive alkylation of mPEG amine in the presence of protein causes cross-linking because the amino groups of a protein have similar reactivity toward aldehydes as mPEG amine [45]. The pK_a of ε-amino residues of lysine is about 9.3–10, whereas the pK_a of α-amino group present at the terminal is about 7.6–8. Under mild acidic conditions (pH 5.5–6), all the ε-amino residues of lysine will be protonated and will not compete for hydrazone formation with mPEG aldehyde. Interleukin-8 was PEGylated at N-terminal by oxidizing N-terminal serine to form glyoxyl derivative, which was conjugated to mPEG amine or hydrazide derivative [46]. G-CSF was PEGylated at N-terminal methionine residue using aldehyde-activated PEG under reductive condition to yield almost 90% mono-PEGylated protein (Scheme 12.10). The dosing frequency of G-CSF has reduced from daily sc injection to once-a-week

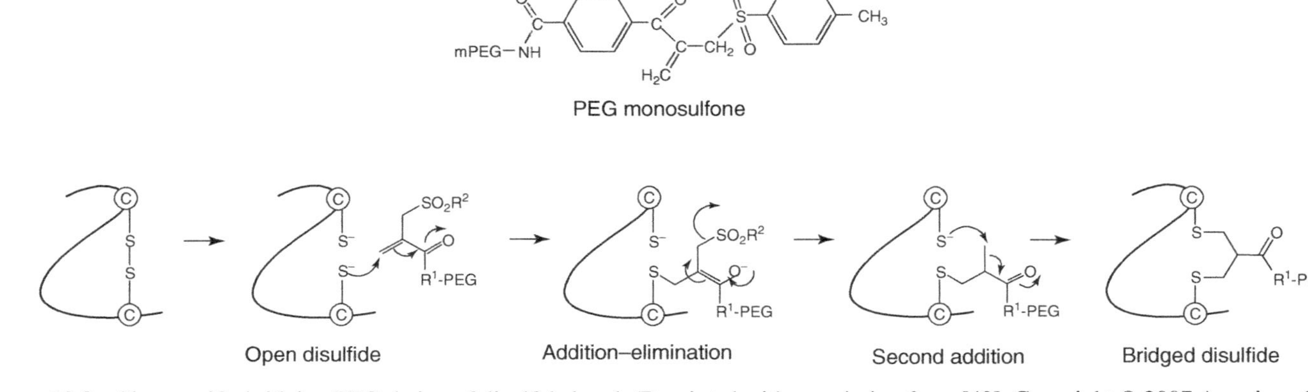

Figure 12.3 Site-specific bridging PEGylation of disulfide bond. (Reprinted with permission from [43]. Copyright © 2007 American Chemical Society.)

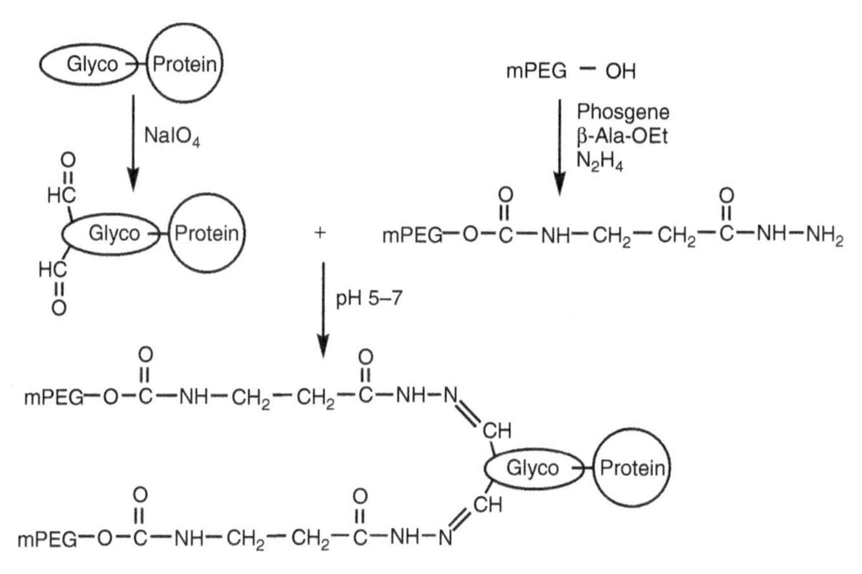

Scheme 12.9 PEGylation of glycoproteins by oxidation of carbohydrate. (Reprinted with permission from [16]. Copyright © 1995 Elsevier Ltd.)

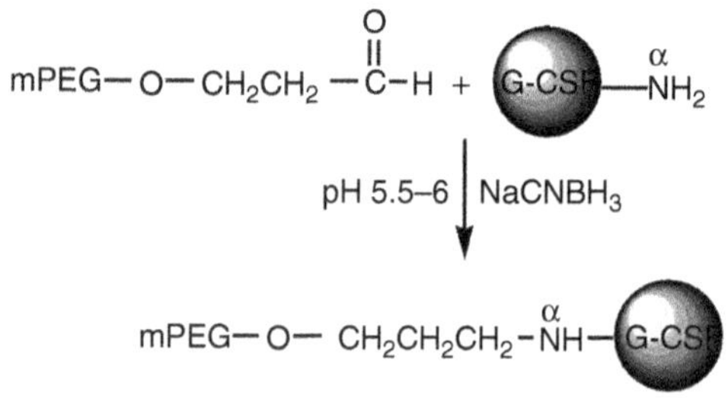

$$mPEG-O-CH_2CH_2-C(=O)-H + \text{G-CSF}-\alpha NH_2 \xrightarrow[\text{pH 5.5–6}]{NaCNBH_3} mPEG-O-CH_2CH_2CH_2-\overset{\alpha}{N}H-\text{G-CSF}$$

Scheme 12.10 N-terminal PEGylation of G-CSF using aldehyde-activated PEG.

injection because of reduced renal clearance, long-circulation half-life, and increased stability against enzymatic degradation [47].

12.2.3.4 Releasable PEGylation The reduced biological activity of the protein after polymer conjugation is often observed. This is caused by the interference of the PEG chain at the protein/receptor binding site or enzyme/substrate recognition process. Site-specific PEGylation can prevent this when the polymer conjugation site is not embedded or close to the active site of the protein.

Various approaches to restore the lost protein activity owing to PEGylation have been designed. These approaches are based on the use of PEG linkers that can release native protein over time by enzymatic or hydrolytic degradation of the protein polymer conjugate.

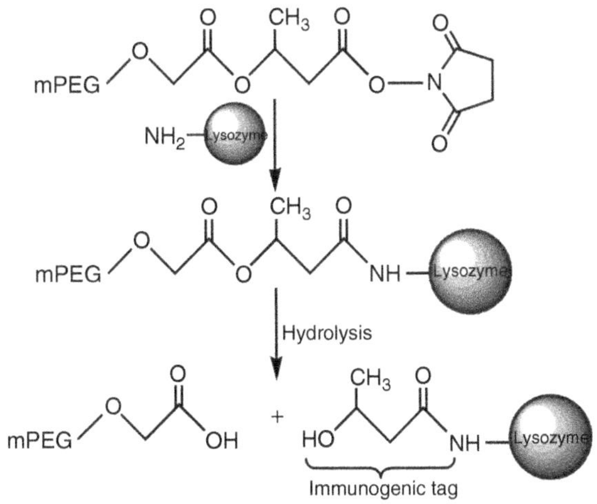

Scheme 12.11 PEGylation of lysozyme using double-ester-activated PEG.

Double Ester Linkers Succinimidyl succinate ester of mPEG **9** was the first releasable PEG linker as described in Section 12.2.3.1. Other reagents containing double ester such as hydroxyl acids attached to carboxylic acids of PEG (carboxymethyl, propionic, or butanoic) have been used to attach PEG chains to proteins and could regenerate approximately 60% of the native activity of lysozyme *in vitro* [1]. This double ester PEG reagent leaves a tag on the protein that could increase the immunogenicity (Scheme 12.11).

Maleic Anhydride Linker In order to release the protein without any "tag" and loss of activity, Garman et al. [48] used mPEG maleic anhydride to PEGylate urokinase and tissue plasminogen activator (Scheme 12.12) and showed the regeneration of native protein under physiological conditions with slow clearance from the body.

1,6-Benzyl Elimination and Trimethyl Lock Lactonization Enzon Pharm, Inc., developed releasable PEG linkers on a double prodrug system that generated native protein upon hydrolysis. The ester bonds between polymer and linker and PEG are hydrolyzed to release protein coupled with linker, which upon rearrangement of the linker generates the native protein as shown in Scheme 12.13 [49,50].

Bicin Linkers The bicin series of releasable linkers are developed by linking PEG chain through an ester bond with one or both hydroxyl groups of bicin. The protein is conjugated to carboxyl group of bicin through an amide bond. The native protein is released by the rapid hydrolysis of acetal group, triggering the elimination reaction to release the PEG. The release of native protein occurs via intramolecular cyclization (Scheme 12.14) [51].

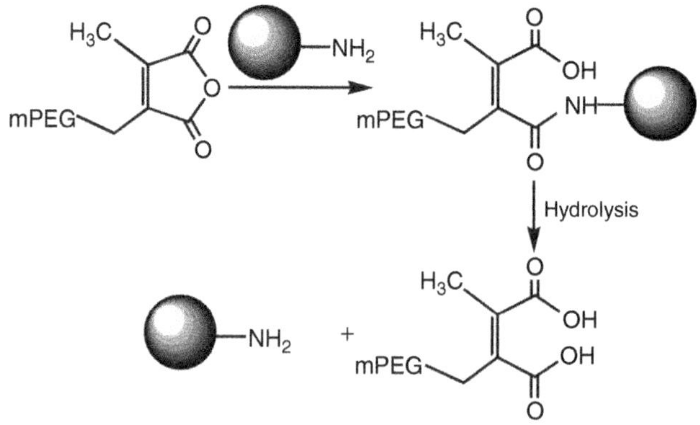

Scheme 12.12 Releasable PEGylation using mPEG maleic anhydride. (Reprinted with permission from [31]. Copyright © 2002 Elsevier Ltd.)

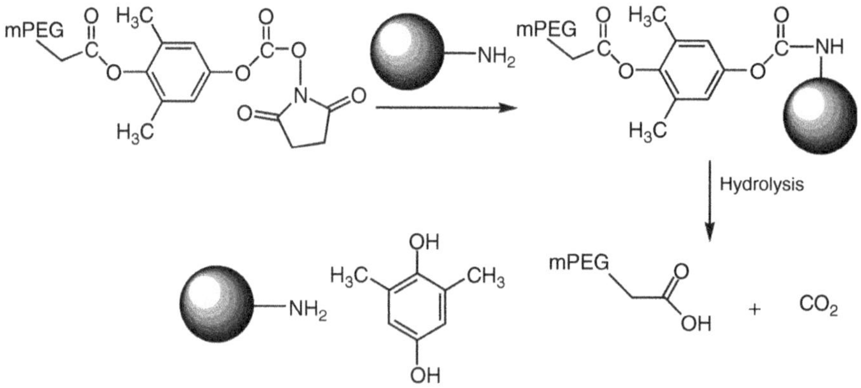

Scheme 12.13 Releasable PEGylation using mPEG phenyl carbonate. (Reprinted with permission from [31]. Copyright © 2002 Elsevier Ltd.)

Histidine PEGylation Enzon developed PEGylated interferon α-2-b (Pegintron®) by using a hydrolyzable link between the PEG and $N^{\delta 1}$ position of the imadazole ring in histidine to form a carbamate link (Scheme 12.15). The interferon α-2-b is reacted with SC-mPEG (12 kDa) at pH 5.0 to get a mixture of positional mono-PEGylated isomers with His34 accounting for 48%. The His34-PEGylated isomer is most active because the histidine residue is located away from receptor recognition site [52].

12.2.4 Biofate of PEGylated Proteins

12.2.4.1 Increased Circulation Half-Life The PEGylation significantly improved the pharmacokinetic profile of proteins and peptides when injected to

mPEG O N H O O O N N O O N O O O O NH$_2$

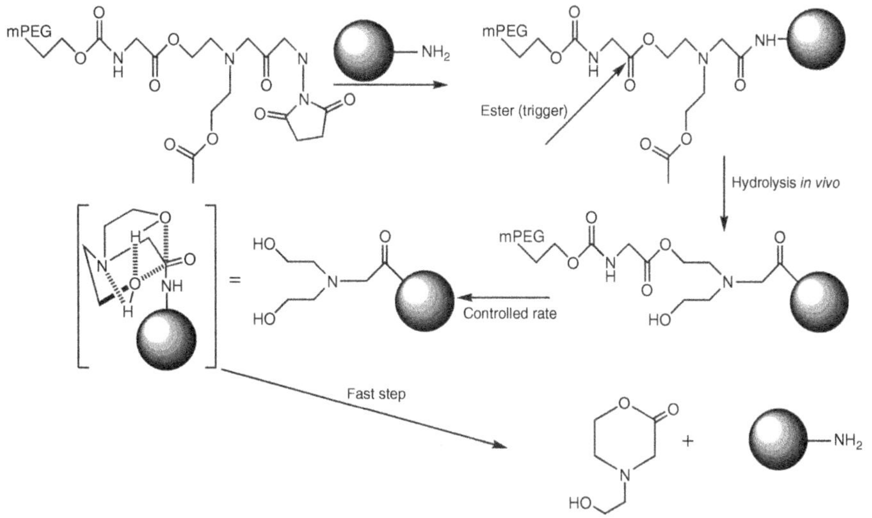

mPEG O N H O O O N NH O O O O

Ester (trigger)

Hydrolysis *in vivo*

mPEG O N H O O O N O HO

HO N O HO

Controlled rate

N H O H O NH O =

Fast step

O O N HO + NH$_2$

Scheme 12.14 Releasable PEGylation using bicin linker. (Reprinted with permission from [51]. Copyright © 2006, American Chemical Society.)

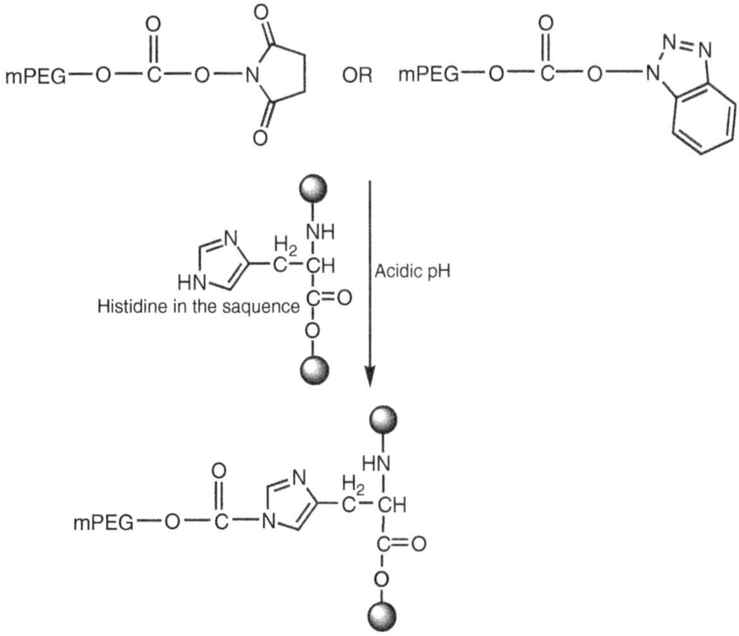

mPEG—O—C(=O)—O—N (succinimide) OR mPEG—O—C(=O)—O—N (benzotriazole, N=N)

Acidic pH

HN N H_2 C–CH NH C=O O

Histidine in the saquence

mPEG—O—C(=O)—N N H_2 C–CH HN C=O O

Scheme 12.15 Histidine PEGylation using benzotriazole or succinimidyl carbonate-activated PEG. (Reprinted with permission from [31]. Copyright © 2002 Elsevier Ltd.)

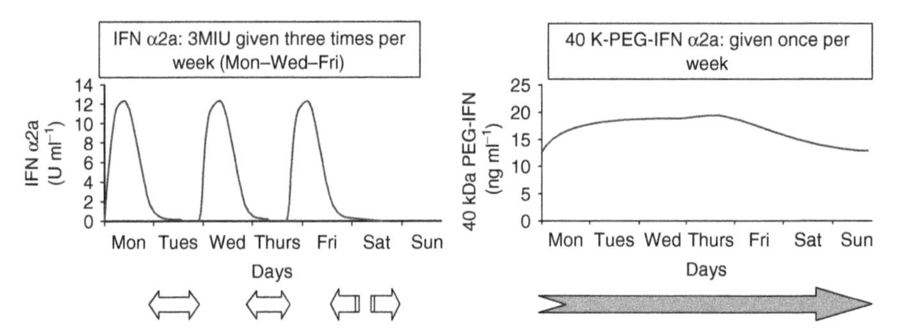

Figure 12.4 Pharmacokinetic profile of interferon α-2a and PEGylated (40 kDa) interferon α-2a in humans after subcutaneous injection. (Reprinted with permission from [55]. Copyright © 2008 Wiley-Liss, Inc.)

body. The hydrodynamic volume of PEG is 5- to 10-fold higher than the protein of equivalent molecular weight because of its strong affinity to water. This causes a dramatic increase in the effective molecular weight of the PEG–protein conjugate [53]. PEG and PEGylated proteins are mainly excreted by renal clearance, and rate of clearance is significantly low for conjugate with molecular weights above 40 kDa [17]. The PEGylated interferon α-2-a with branched 40 kDa PEG has increased the plasma circulation half-life from 3.8 to 65 h because of the slowed clearance by more than 100-fold (Figure 12.4) [54,55].

PEGylation protects therapeutic proteins from rapid proteolytic degradation caused by protease and peptidases by steric hindrance. The continuous mobility of the PEG chains provides sufficient flexibility to enable high-affinity protein–receptor interaction, thus exhibiting a biological effect. The steric hindrance caused by the large PEG chains often result in the reduced biological activity as compared to that of the native protein because of the reduced accessibility of binding site caused by PEGylation. But the prolonged circulation half-life together with reduced proteolytic degradation compensates for the reduced biological activity and creates an overall improved pharmacological profile with improved efficacy and reduced dosing frequency [55].

12.2.4.2 Low Immunogenicity The highly flexible and hydrated PEG chains of PEGylated protein that prevents enzymatic degradation by steric hindrance are also responsible for the reduced immunogenicity of the conjugate. The PEG chain impairs the recognition by the antigenic determinants, thereby reducing or preventing the generation of antibodies against the injected protein.

L-Asparaginase isolated from *Escherichia coli* with a molecular weight of 138–141 kDa or isolated from *Erwinia chrysanthemi* with a molecular weight of 138 kDa is an important chemotherapeutic agent in the treatment of ALL and other lymphoid malignancies. These two forms of L-asparaginase are although similar but do not present antigenic cross-reactivity. But both forms suffer from limitations such as clinical hypersensitivity in 3–78% patients, acute allergic reaction, and rapid clearance from the body [56]. The PEGylated L-asparaginase developed by Enzon (Oncaspar®) was granted FDA approval in 1994 for human treatment of ALL. The polymer coupling was a random PEGylation, where several PEG chains of 5000 Da molecular weight were coupled to the protein surface, thus leading to a mixture of multi-PEG–asparaginase conjugates with different degrees of PEG modification. It has very low hypersensitivity reaction and can be given to patients who are hypersensitive to native protein. The circulation half-life is increased from 1 day to 15 days, thereby reducing the dosing frequency and increasing patient compliance [57].

12.3 PEGYLATED PROTEINS IN CLINICAL PRACTICE

Ever since Davis and coworkers have reported the improved immunogenic and pharmacokinetic properties of PEGylated liver catalase [6], significant research work has been done to demonstrate the improved therapeutic performance of the PEGylated proteins as compared to that of the native protein. The new recombinant biotechnological techniques also helped to produce many natural biomolecules such as cytokines, hormones, and antibodies at a large scale. PEGylated form of these proteins has been approved by regulatory bodies for human use to treat diseases that are summarized in Table 12.2.

12.3.1 PEG Conjugate with Low Molecular Weight Drugs

After the success of the PEG–protein conjugate, PEGylation technologies are investigated to solve the problem encountered with small molecule drugs. The large number of these drugs suffers from (i) poor aqueous solubility, (ii) short blood circulation half-life, (iii) low accumulation at the site of action, and (iv) severe side effect. PEGylation of drugs is one of the promising approaches to overcome some of these problems. Generally, the properties of PEG are conveyed to the conjugated drugs and their body fate reflects that of the polymer. The PEG conjugation of low molecular weight drugs is easy and straightforward as compared to that of the biological macromolecules because of the reduced number of functional group for PEG conjugation, absence of conformational constraints, independence of choosing reaction medium and solvents, easier purification, and characterization. The coupling methods originally developed for the proteins and peptide PEGylation found a wide application in the synthesis of PEG drugs [58].

12.3.2 PEG Structures for Small-Molecule PEGylation

The PEGylation of drug has been extensively researched for the last two to three decades resulting in hundreds of interesting publications and patents. However, it has not so far yielded commercially available PEG small-molecule products. The main limitation of PEGylation of drugs is the low drug loading as the PEG possesses only one or two hydroxyl groups available for conjugation with drugs.

TABLE 12.2 US-FDA-Approved PEG–Protein Conjugate with Their Pharmacokinetic Parameters

Protein–PEG Conjugate	PEG Linker	Route of Administration	Pharmaco-kinetics	*In Vitro* Activity Retained (%)	Indication	Company	Year of Approval
Adagen® mPEG adenosine deaminase	mPEG-SS 5 kDa Random	IM	30 min to 3–6 days		Severe combined immunodeficiency disease	Enzon Inc.	1990
Oncaspar® m-PEG-L-asparaginase	mPEG-SS 5 kDa Random	IM/IV	24–360 h		Acute lymphoblastic leukemia (ALL)	Enzon Inc. (United States)/ RPR (Europe)	1994
Pegintron® mPEG Interferon α-2-b	mPEG-SC 12 kDa RandomHistidine (47%)	SC	3.8 h to 27–37 h	28%	Chronic hepatitis C	Schering-Plough Corp.	2000
Pegasys® mPEG interferon-α-2-a	$mPEG_2$-NHS ester, 20 kDa, lysine	SC	3.8–65 h	7%	Chronic hepatitis C	Hoffmann-La Roche	2002
Neulasta® mPEG granulocyte colony stimulating factor	mPEG aldehyde, 20 kDa, N-terminal methionine	SC	3.5–42 h	41%	Febrile neutropenia	Amgen	2002

Somavert® mPEG human growth hormone antagonist	mPEG-SS 5 kDa random	SC	0.25–100 h	24%	Acromegaly	Pfizer	2002
Macugen® mPEG antivascular endothelial growth factor-RNA aptamer	$mPEG_2$-NHS ester, 20 kDa, selective at cytosine	IV	12–142 h	25%	Neovascular (wet) age-related macular degeneration	Pfizer/OSI Pharma	2004
Mircera® mPEG epoetin β	mPEG-SBA, 30 kDa, random	IV/SC	12–134 h		Anemia associated with chronic kidney disease	Hoffmann-La Roche	2007
Cimzia® mPEG antitumor necrosis factor Fab'	$mPEG_2$ maleimide, 20 kDa, thiol specific	SC	1–14 days	100%	Rheumatoid arthritis and Crohn's disease	UCB Pharma	2008
Krystexxa® mPEG uricase	mPEG nitrophenyl carbonate, 10 kDa, random	IV	6.5–56 h	90%	Refractory chronic gout	Savient Pharmaceuticals	2010

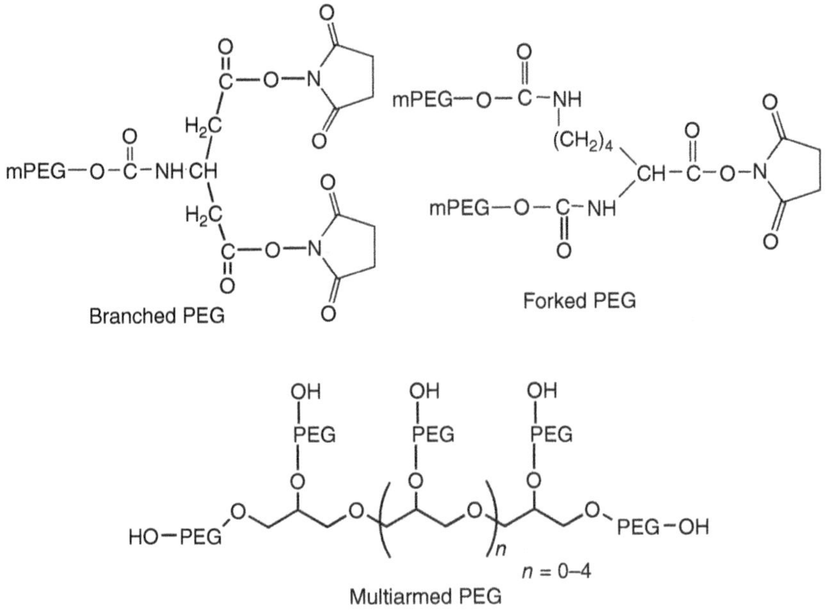

Branched PEG

Forked PEG

Multiarmed PEG

$n = 0–4$

Scheme 12.16 PEG structures used in PEGylation of small organic molecules.

This problem is evident in the need to administer high amounts of polymer to achieve the therapeutic level of drug [59]. New types of PEG structures in different shape having a number of functional groups for coupling have been synthesized to overcome these problems (Scheme 12.16).

12.3.2.1 Branched PEG A branched PEG having two PEG chains attached to lysine core was synthesized by PEGylating both amino groups of lysine [35,60]. This branched PEG allows conjugation of larger molecular weight (60 kDa) and can be obtained in highly pure form with a single reactive end group. This reagent is preferred in protein and peptide PEGylation and PEGYSY is commercially produced using this reagent.

12.3.2.2 Forked PEG Forked PEG has two functional groups either at one end of a single PEG chain (linear forked) or at branched PEG (branched forked) as shown in Scheme 12.16. It is synthesized by attaching the terminal hydroxyl group of mPEG to a trifunctional linker like serinol or β-glutamic acid [61]. This structure increases the drug loading because of two coupling points.

12.3.2.3 Multiarmed PEG Multiarmed PEG is a dendron-like structure having many PEG chains protruding from a common core having a free hydroxyl group with increased active sites for conjugation. These are prepared by the ethoxylation of various polyols (derived from glycerol condensation) having 4–8 arms

(Scheme 12.16). This is widely used in conjugation of various small-molecule drugs, and conjugates of multiarmed PEG with irinotecan, docetaxel, and SN-38 [62] are in different stages of clinical development.

12.3.3 Advantages of PEGylated Drugs

12.3.3.1 Improved Solubility Characteristics The most commonly observed effect of PEGylation on poorly soluble drugs is a significant increase in solubility because of the coupling of bulky hydrophilic PEG chains. The SN-38 (active metabolite of camptothecin-11) is 100–1000 times more potent than camptothecin-11. However, clinical utility of SN-38 has been hampered by its poor aqueous solubility. The PEG conjugate of SN-38 containing four-armed PEG (10 kDa each) increases the solubility of SN-38 about 1000-fold. It also stabilizes the E-ring in closed and active forms. PEG–SN-38 (ENZ 2208) has shown remarkable *in vivo* efficacy in preclinical models of solid tumors, hematological cancers, and even in camptothecin-11 refractory models (Scheme 12.17) [62].

The aqueous solubility of amphotericin B (an antifungal antibiotic) has been increased to 250-fold, when it is conjugated with PEG having a molecular weight 40 kDa using releasable benzyl elimination approach in order to maintain antifungal activity [63].

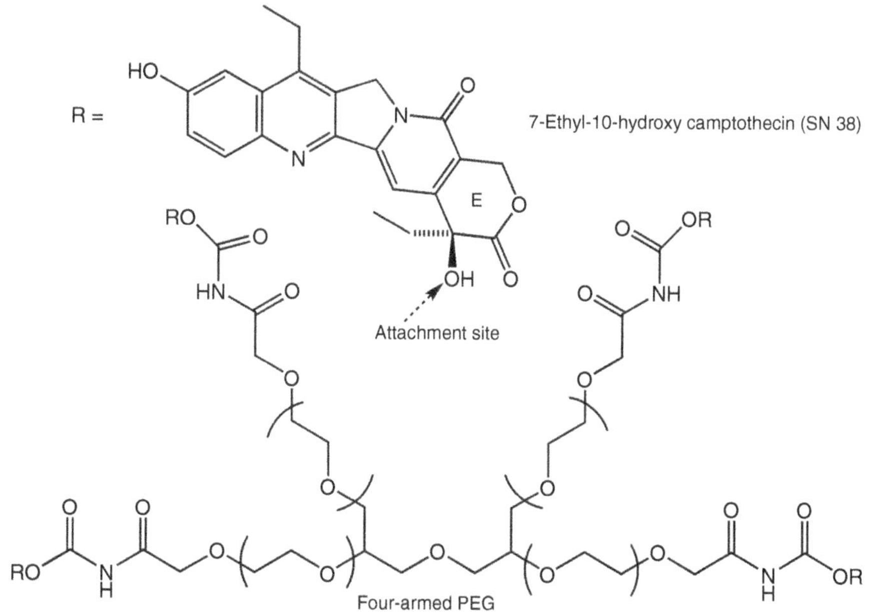

R = 7-Ethyl-10-hydroxy camptothecin (SN 38)

Attachment site

Four-armed PEG

Scheme 12.17 Four-armed PEG conjugate of SN-38 (ENZ 2208).

12.3.3.2 Long-Circulation Half-Life and Enhanced Permeation and Retention (EPR) Effect Majority of the injected dose of low molecular weight drugs generally cleared from the circulation within minutes, requiring frequent administration of the drug in order to maintain the minimum effective concentration of the drug in the blood. The polymer conjugates of small-molecule drugs circulate for several hours because of a reduced renal clearance caused by an increase in overall molecular weight. The hepatic clearance of PEG conjugate by the mononuclear phagocytic systems also impaired due to the presence of PEG chains which makes conjugate more hydrophilic. This effect is more pronounced in the case of PEGylated liposomes and nanoparticles [64]. Irinotecan is a semisynthetic derivative of camptothecin having a short-circulation half-life of 1–2 days, and it is not detectable in the plasma after 5–6 days of injection. After PEGylation of irinotecan (NKTR 102) using four-armed PEG, the half-life of conjugate was increased to 15 days and it was detected in plasma even after 50 days of administration [65].

The drug–polymer conjugate can be passively targeted to the tumor site, thereby reducing the exposure of other organs to the anticancer drug and its related toxicity. This is also known as *enhanced permeation and retention* (EPR) effect, first described by Maeda and coworkers [66]. The blood vessels of tumor have higher permeability with minimum lymphatic drainage than that of the normal vessels (angiogenic effect) of the body. The long circulatory drug–polymer conjugate permeates from tumor blood vessels and accumulates in the tumor as shown in Figure 12.5a. This accumulation of drug–polymer conjugate in the tumor increases the drug concentration very high and drug enters inside the cells because of the concentration gradient [67]. The PEG–camptothecin conjugate with a glycine linker showed 30-fold greater tumor accumulation of drug in mice bearing sc tumors compared to that of the unmodified CPT as shown in Figure 12.5b [68].

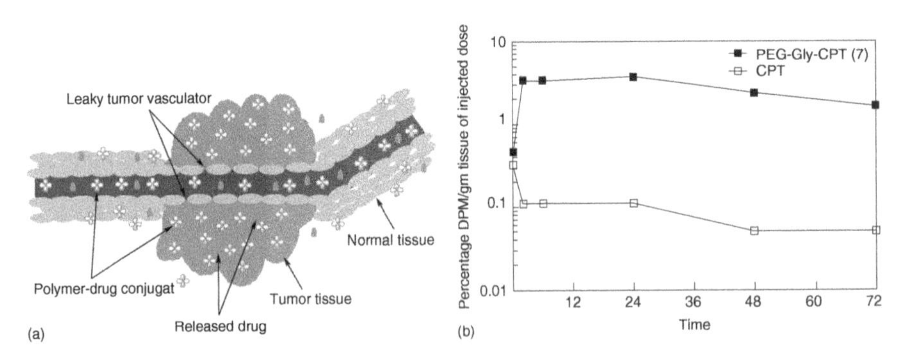

Figure 12.5 (a) Passive targeting of PEGylation drugs by EPR effect. (*See insert for color representation*). (b) Accumulation of CPT in HT-29 tumor of nude mice after single IV injection of PEGylated CPT and CPT solution. (Reprinted with permission from [59]. Copyright © 2003 Elsevier Ltd.)

12.3.3.3 Reduction in Toxic Side Effects The polymer–drug conjugates have been proved to reduce the toxic side effects of drugs when administered to the body. This is especially seen in the case of polymer conjugation with anticancer agents. When injected to the body as normal solution or taken as a tablet, the anticancer agents are nonselectively toxic to both cancer cells and normal cells of the body. These drugs have a very low therapeutic index, and since they are generally used at the maximum tolerated dose, the appearance of systemic toxicities is a common occurrence.

Bioconjugation to polymers significantly decreased the uptake of camptothecin by the immune system organs. Following IV injection in tumor-bearing animals, camptothecin mainly accumulated in the liver. Polymer conjugation of camptothecin with high molecular weight (40 kDa) PEG has reduced the accumulation of drug in the liver and spleen significantly, so also the related toxicity [68].

Amphotericin B remains the gold standard for the treatment of life-threatening fungal infections despite its severe side effects, hemolytic, and renal toxicity. The PEG conjugation was employed to increase its aqueous solubility by reduced aggregation. As the primary amine present in mycosamine sugar is essential for its antifungal activity, 1,6-benzyl elimination linker was used to PEGylate amphotericin B (Scheme 12.18), which will release the parent drug upon hydrolysis. The PEGylated drug was well tolerated with 100% survival rate in *Candida albicans* infection animal model. The PEG amphotericin B conjugate showed 3% hemolysis of rabbit's red blood cells up to a concentration of 1 mg ml^{-1}, whereas free drug in the form of deoxycholate micelles caused 70% hemolysis at concentrations greater than 10 μg ml^{-1} in 3 h [63].

NKTR-118 (PEG-naloxol) (Scheme 12.19) is a new oral peripheral opioid antagonist (POA) under clinical development for the treatment of opioid-induced constipation (OIC) and other manifestations of opioid bowel dysfunction (OBD). Naloxol

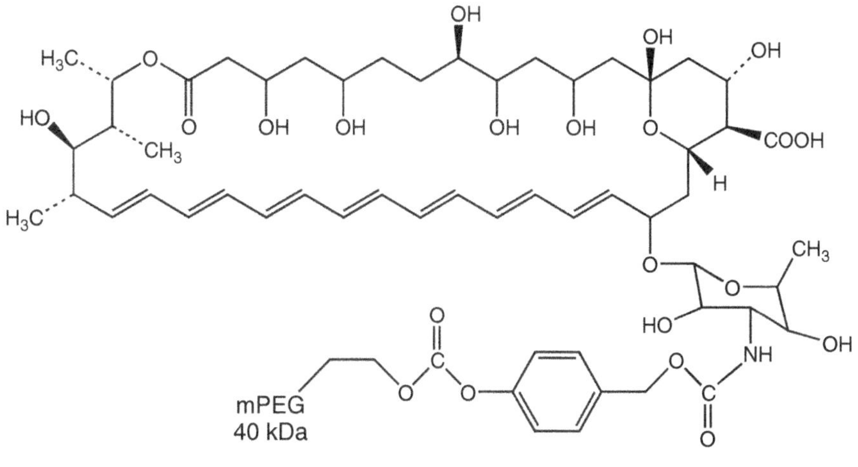

Scheme 12.18 PEG prodrug of amphotericin B conjugated using 1,6 benzyl linker.

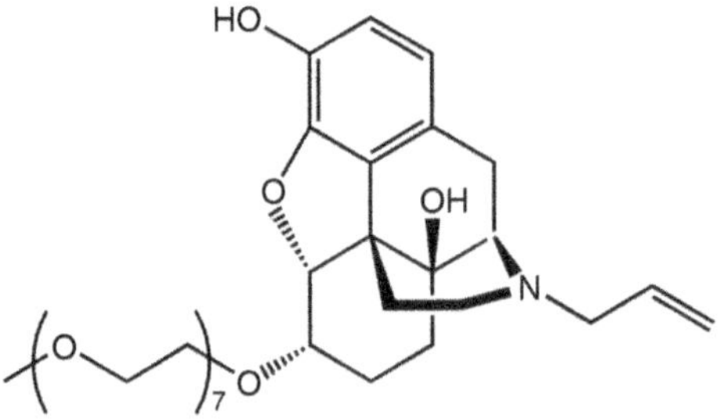

Scheme 12.19 Structure of PEGylated naloxol (NKTR-118).

conjugate with PEG reduces the ability of NKTR-118 to enter the central nervous system (CNS) by 15-fold, so that the central analgesic effect of opioid therapy is maintained [69].

Currently, no approved product based on PEGylation of small organic molecule has reached the market but four products are undergoing clinical trials. Table 12.3 gives a list of different small organic molecules, which are under clinical trial and preclinical investigation.

12.4 *N*-(2-HYDROXYPROPYL) METHACRYLAMIDE (HPMA) COPOLYMER CONJUGATE

HPMA is a hydrophilic linear homopolymer designed to act as a plasma expander in the early 1070s [85]. After PEG, HPMA is the most investigated and advanced polymer used in polymer conjugation because of its versatility as a vehicle, water solubility, nontoxicity, and biocompatibility. The general structure of the HPMA copolymer is shown in Scheme 12.20. This polymer can conjugate high drug payload unlike that of PEG where generally only two terminal groups are available for

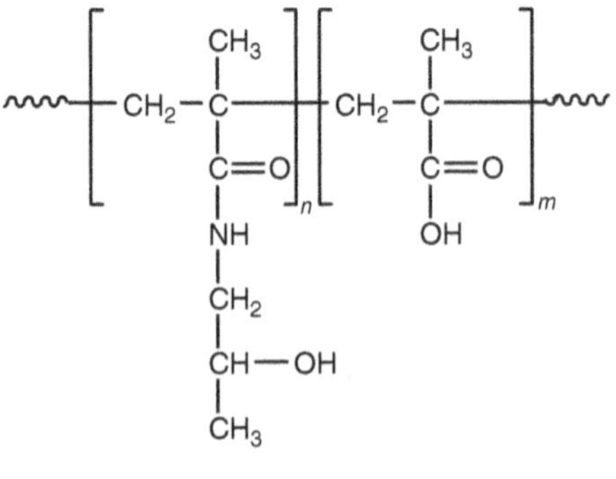

Scheme 12.20 General structure of the HPMA copolymers.

TABLE 12.3 Small Organic Molecule PEG Conjugates

PEG Conjugate	PEG Size	Indication	Clinical Stage	Ref
Under Clinical Trial				
PEG–SN-38 (EZN 2208 Enzon Inc.)	Four-armed PEG, 40 kDa	Solid tumor	Phase II	[62]
PEG irinotecan (NKTR 102 Nektar therapeutics)	Four-armed PEG	Solid tumor, colorectal cancer, and metastatic breast cancer	Phase III/Phase II	[65]
PEG naloxol (NKTR 118 Nektar therapeutics)	Linear PEG	Opoid-induced constipation	Phase III	[69]
PEG docetaxel (NKTR 105 Nektar therapeutics)	Four-armed PEG	Solid tumor	Phase I (suspended)	[70]
PEG paclitaxel (Enzon Inc.)	30 kDa PEG	Solid tumor	Phase II (suspended)	[71]
PEG camptothecin (Prothecan Enzon Inc.)	Diol, 40 kDa	Gastric and esophageal adenocarcinoma	Phase II (suspended)	[68]
Under Clinical Trial				
PEG amphotericin B	Linear, 40 kDa, 1,6-benzyl elimination Linear PEG with four–six ethylene oxide attached to COOH	Antifungal antibiotic	Preclinical	[63] [72]
PEG cisplatin	Linear Folate PEG 3000	Anticancer for testicular cancer	Preclinical	[73]

(continued)

TABLE 12.3 ***(Continued)***

PEG Conjugate	PEG Size	Indication	Clinical Stage	Ref
PEG doxorubicin	Linear PEG	Anticancer drug	Preclinical	[74]
PEG gemcitabine	Folate PEG with two drug molecule	Small-cell lung cancer	Preclinical	[75]
PEG wortmannin	Linear PEG	Anticancer drug	Preclinical	[76]
PEG methotrexate	Linear PEG 40 kDa	Leukemia	Preclinical	[77]
PEG melphalan	Linear PEG, 5 kDa	Anticancer drug	Preclinical	[78]
PEG daunorubicin	Linear PEG 1, 6-benzyl elimination	Anticancer drug	Preclinical	[59]
PEG podophyllotoxin	Linear PEG	Lymphocytic leukemia	Preclinical	[79]
PEG acyclovir	Linear PEG	Anti viral agent	Preclinical	[80]
PEG saquinavir	Linear PEG with leucine spacer	Anti HIV agent	Preclinical	[81]
PEG tacrolimus	Linear PEG	Immunosuppressive agent	Preclinical	[82]
PEG sirolimus	Linear and branched	Immunosuppressive agent	Preclinical	[83]
PEG gentamicin	Linear PEG-SH with labile linker	Anti microbial agent	Preclinical	[84]

conjugation. The initial studies evaluated the effect of molecular weight on endocytosis and biodistribution, and an optimal HPMA copolymer M_w of ~30 kDa was chosen to ensure slow clearance from the body [86]. The polymer–drug loading of 10% w/w was chosen to be optimal. HPMA drug conjugate can be synthesized with targeting ligands (sugars, antibodies, peptides) to realize fully the Rigsdorf's model for drug–polymer conjugate. HPMA drug conjugate without a targeting functionality show EPR-mediated passive accumulation at the tumor site.

The first drug–HPMA conjugate that reached clinical stage was prepared with an anticancer agent doxorubicin (PK1) (Scheme 12.21) using a lysosomally degradable peptidyl linker (Gly-Phe-Leu-Gly), which is stable in the circulation but is cleaved by lysosomal proteolytic enzyme like cathepsin B to release the drug [87]. The drug pay load of the polymer–doxorubicin conjugate was 8.5% w/w, and it was found to be less toxic than the free drug and can accumulate inside solid tumor models [88]. This was the first synthetic polymer–anticancer drug conjugate that entered clinical trials in 1994. In phase II clinical study, PK1 showed

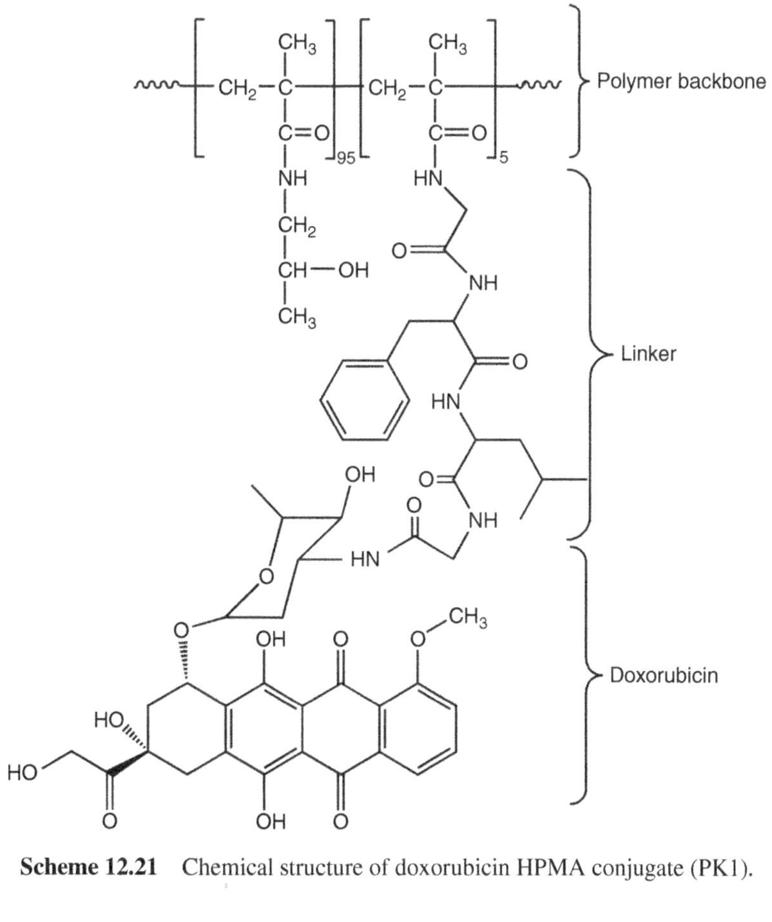

Scheme 12.21 Chemical structure of doxorubicin HPMA conjugate (PK1).

partial responses in patients suffering with from breast, nonsmall-cell lung cancer (NSCLC) but showed no activity in patients with colon cancer [89]. This study confirmed that HPMA is a safe polymeric carrier and can be administered up to a dose of 20 g m^{-2} without immunogenicity and polymer toxicity.

Another HPMA–doxorubicin conjugate (PK2) was synthesized containing 27 kDa polymer backbone, 6.5% mol/wt drug, 2% w/w free doxorubicin, and 2% mol/wt galactose, an efficient targeting moiety for asialoglycoprotein receptors expressed only on hepatocytes and hepatomas for treatment of liver cancers [90]. In clinical studies, the maximum tolerated dose of PK2 was 160 mg m^{-2} doxorubicin equivalent with neutropenia being the dose-limiting toxicity. The galactosamine containing conjugate was found to be more toxic than that of without galactosamine. The normal liver accumulation of the doxorubicin was 15–20% of the administered dose and hepatoma accumulation was 3.2% of the administered dose of PK2, which was 15- to 20-fold higher than free doxorubicin [91]. The phase I clinical trials of HPMA-conjugated camptothecin (MAG-CPT) [92] and paclitaxel (PNU166945) [67] failed to show any response. The HPMA–carboplatinum analog conjugates have shown better results in clinical trials. The conjugate containing a malonate ligand (AP5280) [93] and 1,2-diaminocyclohexane (DACH) palatinate (AP5346) [94] showed reduced platinum toxicity and greater anticancer efficacy. The pharmacokinetics of AP5346 indicated a prolonged half-life (mean terminal half-life) of 72 h and evidence of antitumor activity.

12.5 POLY(L-GLUTAMIC ACID) CONJUGATES

Poly(L-glutamic acid) (PGA) is a linear polymer of the naturally occurring amino acid, that is, L-glutamic acid, linked together through amide bonds. The free γ-carboxylic acid in each monomer is negatively charged at neutral pH, rendering polymer highly water soluble and also providing functionality for the drug conjugation. The PGA has a peptidyl structure with C–N bond that is easily breakable by the enzymes, hence biodegradable. The PGA has been conjugated to many anticancer drugs of different classes, which are reviewed by Li [95,96].

The PGA–paclitaxel conjugate (CT-2103, XYOTAX™, Cell Therapeutics Inc) reached to phase III clinical trials. The paclitaxel is conjugated to PGA with 17 kDa molecular weight using an ester linkage with γ-carboxylic acid side chain. The conjugation efficiency is approximately 10%, meaning only one out of 10 glutamic acid monomer unit is coupled with paclitaxel molecule. The final conjugate has a molecular weight of 49 kDa with 37% w/w paclitaxel loading (Scheme 12.22). The paclitaxel is released slowly by the hydrolysis of the ester bond (up to 14% in 24 h), but after the endocytic uptake, the release was increased because of the lysosomal degradation of the polymeric backbone [97–99]. In phase I/II studies, CT-2103 demonstrated a significant number of partial responses or stable disease in patients with mesothelioma, renal cell carcinoma, and NSCLC, and in paclitaxel-resistant ovarian cancer [100,101]. This conjugate has completed phase III trials but still not

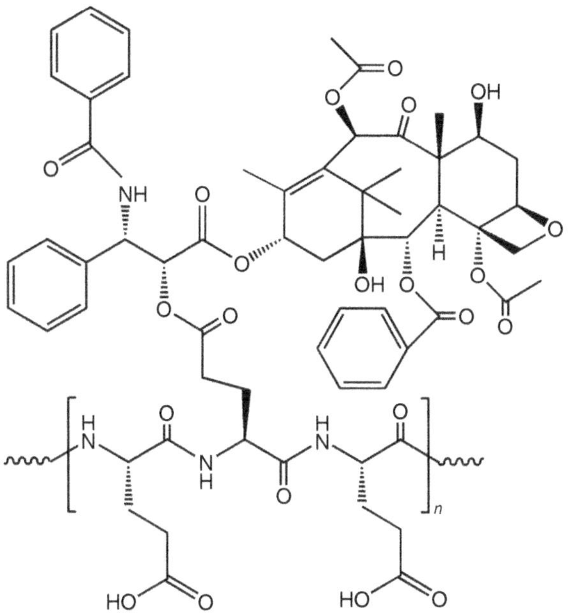

Scheme 12.22 Structure of paclitaxel conjugated with PGA conjugate.

approved by US FDA for human use. A PGA conjugate of camptothecin (CT2106) with 50 kDa polymer having a glycine linker is also under clinical investigation [102].

12.6 POLYSIALIC ACID (PSA) CONJUGATES

PSA is a linear polymer of repeating units of *N*-acetyl neuraminic acid (NeuSAc; sialic acid), a sugar present on the cell surface. PSA is also a normal constituent of the body, facilitating neural tissue development by modulating cell-to-cell contact inhibition or helping cancer cells to metastasize, by making them less adherent and more prone to migration. The α(2→8)-linked B capsular polysaccharide (also known as *colominic acid*) obtained from *E. coli* and its shorter derivative are most appropriate for the protein conjugation because of their structural similarity with PSA, nonimmunogenicity, and biodegradability in the cells but stable in blood [103].

The activation of the polymer is done by periodate oxidation of the nonreducing sugar to generate aldehyde followed by reaction with ε-amino groups or the N-terminal of the protein and reductive amination to form polymer conjugate as shown in Scheme 12.23 [104].

Gregoriadis and coworkers conjugated several proteins based on this technology and evaluated for long-circulation half-life and reduced immunogenicity.

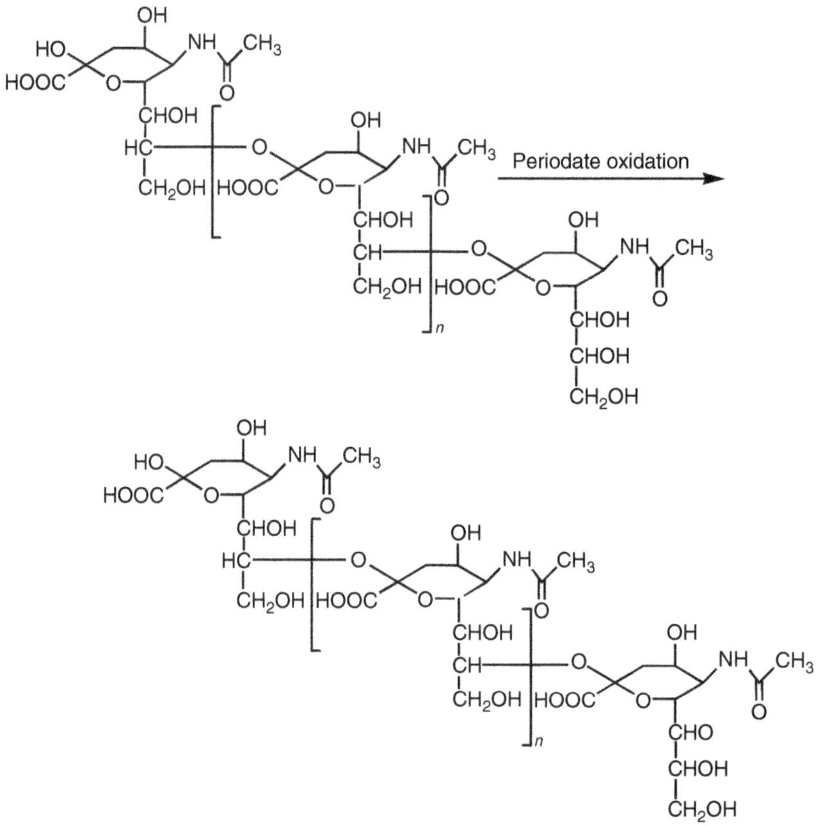

Scheme 12.23 Structure of polysialic acid (colominic acid) and its activated form.

The polysialated L-asparaginase showed 100% retention of activity after polymer conjugation and terminal half-life have been prolonged from 15 to 37 h [105]. Similar results were reported with other therapeutic proteins such as insulin [106] and antibody Fab fragment [107]. Currently, PSA-erythropoietin (EREPOXen®, Xenetic Biosciences) and PSA-insulin (SuliXen®, Xenetic Biosciences) are under different stages of clinical development. Xenetic Biosciences is also developing the PSA conjugate of recombinant clotting factor VIII (rFVIII) for hemophilia and DNase I for cystic fibrosis.

12.7 CONCLUSION

In the last two decades, the field of polymer conjugation to proteins and drugs has grown enormously with an increasing number of approved products and candidates entering into clinical trials especially proteins. Development of new polymers with unique architecture such as multiarmed PEG and dendrimer, innovative-coupling chemistry, and increasing number of approved antibodies and proteins for therapeutic use will drive the field of polymer therapeutics.

REFERENCES

1. Roberts, M. J., Harris, J. M. (1998). Attachment of degradable poly(ethylene glycol) to proteins has the potential to increase therapeutic efficacy. *J. Pharm. Sci.*, *87*, 1440–1445.
2. Duncan, R., Spreafico, F. (1994). Polymer conjugates. Pharmacokinetic considerations for design and development. *Clin. Pharmacokinet.*, *27*, 290–306.
3. Jatzkewits, H. (1955). Peptamin (glycel-L-leucyl-mescaline) bound to blood plasma expander (polyvinylpyrrolidone) as a new depot form of a biologically active primary amine (mescaline). *Z. Naturforsch.*, *10*, 27–31.
4. Ringsdorf, H. (1975). Structure and properties of pharmacologically active polymers. *J. Polym. Sci.: Polym. Symp.*, *51*, 135–153.
5. Gros, L., Ringsdorf, H., Schupp, H. (1981). Polymeric antitumour agents on a molecular and cellular level. *Angew. Chem. Int. Ed.*, *20*, 301–323.
6. Abuchowski, A., McCoy, J. R., Palczuk, N. C., van Es T., Davis, F. F. (1977). Effect of covalent attachment of polyethylene glycol on immunogenicity and circulating life of bovine liver catalase. *J. Biol. Chem.*, *252*, 3582–3586.
7. Park, Y. K., Abuchowski, A., Davis, S., Davis, F. (1981). Pharmacology of Escherichia coli-L-asparaginase polyethylene glycol adduct. *Anticancer Res.*, *1*, 373–376.
8. Maeda, H., Ueda, M., Morinaga, T., Matsumoto, T. (1985). Conjugation of poly(styrene-co-maleic acid) derivatives to the antitumor protein neocarzinostatin: pronounced improvements in pharmacological properties. *J. Med. Chem.*, *28*, 455–461.
9. Yeung, T. K., Hopewell, J. W., Simmonds, R. H., Seymour, L. W., Duncan, R., Bellini, O., Grandi, M., Spreafico, F., Strohalm, J., Ulbrich, K. (1991). Reduced cardiotoxicity of doxorubicin given in the form of N-(2-hydroxypropyl)methacrylamide conjugates: and experimental study in the rat. *Cancer Chemother. Pharmacol.*, *29*, 105–111.
10. Duncan, R. (1992). Drug-polymer conjugates: potential for improved chemotherapy. *Anticancer Drugs*, *3*, 175–210.
11. Duncan, R., Seymour, L. W., O'Hare, K. B., Flanagan, P. A., Wedge, S., Hume, I. C., Ulbrich, K., Strohalm, J., Subr, V., Spreafico, F., Grandi, M., Ripamonti, M., Farao, M., Suarato, A. (1992). Preclinical evaluation of polymer-bound doxorubicin. *J. Control. Release*, *19*, 331–346.
12. Duncan, R. (2006). Polymer conjugates as anticancer nanomedicines. *Nat. Rev. Cancer*, *6*, 688–701.
13. Zalipsky, S., Gilon, C., Zilkha, A. (1983). Attachment of drugs to polyethylene glycols. *Eur. Polym. J.*, *19*, 1177–1183.
14. Powell, G. M. Polyethylene glycol. In: Davidson RL, editor. *Handbook of Water Soluble Gums and Resins*. McGraw-Hills, New York, 1980. pp. 1–31.
15. Friman, S., Leandersson, P., Tagesson, C., Svanvik, J. (1990). Biliary excretion of different sized polyethylene glycols in the cat. *J. Hepatol.*, *11*, 215–220.

16. Zalipsky, S. (1995). Chemistry of polyethylene glycol conjugates with biologically active molecules. *Adv. Drug Deliv. Rev.*, *16*, 157–182.
17. Yamaoka, T., Tabata, Y., Ikada, Y. (1994). Distribution and tissue uptake of poly(ethylene glycol) with different molecular weights after intravenous administration to mice. *J. Pharm. Sci.*, *83*, 601–606.
18. Yamaoka, T., Tabata, Y., Ikada, Y. (1995). Comparison of body distribution of poly(vinyl alcohol) with other water-soluble polymers after intravenous administration. *J. Pharm. Pharmacol.*, *47*, 479–486.
19. Delgado, C., Francis, G. E., Fisher, D. (1992). The uses and properties of PEG-linked proteins. *Crit. Rev. Ther. Drug Carrier Syst.*, *9*, 249–304.
20. Abuchowski, A., van Es, T., Palczuk, N. C., Davis, F. F. (1977). Alteration of immunological properties of bovine serum albumin by covalent attachment of polyethylene glycol. *J. Biol. Chem.*, *252*, 3578–3581.
21. Wieder, K. J., Palczuk, N. C., van Es, T., Davis, F. F. (1979). Some properties of polyethylene glycol:phenylalanine ammonia-lyase adducts. *J. Biol. Chem.*, *254*, 12579–12587.
22. Matsushima, A., Nishimura, H., Ashihara, Y., Yakata, Y., Inada,Y. (1980). Modification of E. coli asparaginase with 2,4-bis(o-methoxypolyethylene glycol)-6-chloro-s-triazine (activatedPEG2); disappearance of binding ability towards anti-serum and retention of enzymatic activity. *Chem. Lett.*, *9*, 773–776.
23. Zalipsky, S., Seltzer, R., Menon-Rudolph, S. (1992). Evaluation of a new reagent for covalent attachment of polyethylene glycol to proteins. *Biotechnol. Appl. Biochem.*, *15*, 100–114.
24. Dolence, E. K., Hu, C., Tsang, R., Sanders, C. G., Osaki, S. (1997). Electrophilic polyethylene oxides for the modification of polysaccharides, polypeptides (proteins) and surfaces. *US patent*, 5,650,234.
25. Beauchamp, C. O., Gonias, S. L., Menapace, D. P., Pizzo, S. V. (1983). A new procedure for the synthesis of polyethylene glycol-protein adducts; effects on function, receptor recognition, and clearance of superoxide dismutase, lactoferrin, and alpha 2-macroglobulin. *Anal. Biochem.*, *131*, 25–33.
26. Veronese, F. M., Largajolli, R., Boccu, E., Benassi, C. A., Schiavon, O. (1985). Surface modification of proteins. Activation of monomethoxy-polyethylene glycols by phenylchloroformates and modification of ribonuclease and superoxide dismutase. *Appl. Biochem. Biotechnol.*, *11*, 141–152.
27. Nilsson, K., Mosbach, K. (1984). Immobilization of ligands with organic sulfonyl chlorides. *Methods Enzymol.*, *104*, 56–69.
28. Abuchowski, A., Kazo, G. M., Verhoest, C. R., Jr.,, Van Es, T., Kafkewitz, D., Nucci, M. L., Viau, A. T., Davis, F. F. (1984). Cancer therapy with chemically modified enzymes. I. Antitumor properties of polyethylene glycol-asparaginase conjugates. *Cancer Biochem. Biophys.*, *7*, 175–186.
29. Chamow, S. M., Kogan, T. P., Venuti, M., Gadek, T., Harris, R. J., Peers, D. H., Mordenti, J., Shak, S., Ashkenazi, A. (1994). Modification of CD4 immunoadhesin with monomethoxypoly(ethylene glycol) aldehyde via reductive alkylation. *Bioconjug. Chem.*, *5*, 133–140.
30. Harris, J. M., Herati, R. M. (1993). Preparation and use of polyethylene glycol propionaldehyde. *US patent*, 5,252,714.

31. Roberts, M. J., Bentley, M. D., Harris, J. M. (2002). Chemistry for peptide and protein PEGylation. *Adv. Drug Deliv. Rev.*, *54*, 459–76.

32. Bentley, M. D., Harris, J. M. (1999). Poly(ethylene glycol) aldehyde hydrates and related polymers and applications in modifying. *US patent*, 5,990,237.

33. Zalipsky, S., Barany, G. (1986). Preparation of polyethylene glycol derivatives with two different functional groups at the termini. *Polym. Prep.*, *27*, 1–2.

34. Harris, J. M., Kozlowski, A. (1997). Polyethylene glycol and related polymers monosubstituted with propionic or butanoic acids and functional derivatives thereof for biotechnical applications. *US patent* 5,672,662.

35. Monfardini, C., Schiavon, O., Caliceti, P., Morpurgo, M., Harris, J. M., Veronese, F. M. (1995). A branched monomethoxypoly(ethylene glycol) for protein modification. *Bioconjug. Chem.*, *6*, 62–69.

36. Goodson, R. J., Katre, N. V. (1990). Site-directed pegylation of recombinant interleukin-2 at its glycosylation site. *Biotechnol. (N Y)*, *8*, 343–346.

37. Kogan, T. P. (1992). The synthesis of substituted methoxy-poly(ethylene glycol) derivatives suitable for selective protein modification. *Synth. Commun.*, *22*, 2417–2424.

38. Morpurgo, M., Veronese, F. M., Kachensky, D., Harris, J. M. (1996). Preparation and characterization of poly(ethylene glycol) vinyl sulfone. *Bioconjug. Chem.*, *7*, 363–368.

39. Gard, F. R. N. (1972). Carboxymethylation. *Methods Enzymol.*, *B25*, 424–449.

40. Woghiren, C., Sharma, B., Stein, S. (1993). Protected thiol-polyethylene glycol: a new activated polymer for reversible protein modification. *Bioconjug. Chem.*, *4*, 314–318.

41. Veronese, F. M., Mero, A., Caboi, F., Sergi, M., Marongiu, C., Pasut, G. (2007). Site-specific pegylation of G-CSF by reversible denaturation. *Bioconjug. Chem.*, *18*, 1824–1830.

42. Shaunak, S., Godwin, A., Choi, J. W., Balan, S., Pedone, E., Vijayarangam, D., Heidelberger, S., Teo, I., Zloh, M., Brocchini, S. (2006). Site-specific PEGylation of native disulfide bonds in therapeutic proteins. *Nat. Chem. Biol.*, *2*, 312–313.

43. Balan, S., Choi, J. W., Godwin, A., Teo, I., Laborde, C. M., Heidelberger, S., Zloh, M., Shaunak, S., Brocchini, S. (2007). Site-specific PEGylation of protein disulfide bonds using a three-carbon bridge. *Bioconjug. Chem.*, *18*, 61–76.

44. Urrutigoity, M., Souppe J. (1989). Biocatalysis in organic solvents with a polymer-bound horseradish peroxidase. *Biocatalysis*, *2*, 145–149.

45. Kinstler, O. B., Brems, D. N., Lauren, S. L., Paige, A. G., Hamburger, J. B., Treuheit, M. J. (1996). Characterization and stability of N-terminally PEGylated rhG-CSF. *Pharm. Res.*, *13*, 996–1002.

46. Gaertner, H. F., Offord, R. E. (1996). Site-specific attachment of functionalized poly(ethylene glycol) to the amino terminus of proteins. *Bioconjug. Chem.*, *7*, 38–44.

47. Kinstler, O., Molineux, G., Treuheit, M., Ladd, D., Gegg, C. (2002). Mono-N-terminal poly(ethylene glycol)-protein conjugates. *Adv. Drug Deliv. Rev.*, *54*, 477–485.

48. Garman, A. J., Kalindjian, S. B. (1987). The preparation and properties of novel reversible polymer-protein conjugates. 2-omega-Methoxypolyethylene (5000) glycoxymethylene-3-methylmaleyl conjugates of plasminogen activators. *FEBS Lett.*, *223*, 361–365.

49. Greenwald, R. B., Yang, K., Zhao, H., Conover, C. D., Lee, S., Filpula, D. (2003). Controlled release of proteins from their poly(ethylene glycol) conjugates: drug delivery systems employing 1,6-elimination. *Bioconjug. Chem.*, *14*, 395–403.
50. Greenwald, R. B., Choe, Y. H., Conover, C. D., Shum, K., Wu, D., Royzen, M. (2000). Drug delivery systems based on trimethyl lock lactonization: poly(ethylene glycol) prodrugs of amino-containing compounds. *J. Med. Chem.*, *43*, 475–487.
51. Zhao, H., Yang, K., Martinez, A., Basu, A., Chintala, R., Liu, H. C., Janjua, A., Wang, M., Filpula, D. (2006). Linear and branched bicin linkers for releasable PEGylation of macromolecules: controlled release in vivo and in vitro from mono- and multi-PEGylated proteins. *Bioconjug. Chem.*, *17*, 341–351.
52. Wylie, D. C., Voloch, M., Lee, S., Liu, Y. H., Cannon-Carlson, S., Cutler, C., Pramanik, B. (2001). Carboxyalkylated histidine is a pH-dependent product of pegylation with SC-PEG. *Pharm. Res.*, *18*, 1354–1360.
53. Harris, J. M., Chess, R. B. (2003). Effect of pegylation on pharmaceuticals. *Nat. Rev. Drug Discov.*, *2*, 214–221.
54. Zeuzem, S., Welsch, C., Herrmann, E. (2003). Pharmacokinetics of peginterferons. *Semin. Liver Dis.*, *23*, 23–28.
55. Fishburn, C. S. (2008). The pharmacology of PEGylation: balancing PD with PK to generate novel therapeutics. *J. Pharm. Sci.*, *97*, 4167–4183.
56. Whelan, H. A., Wriston Jr.,, J. C. (1969). Purification and properties of asparaginase from Escherichia coli B. *Biochemistry*, *8*, 2386–2393.
57. Graham, M. L. (2003). Pegaspargase: a review of clinical studies. *Adv. Drug Deliv. Rev.*, *55*, 1293–1302.
58. Zalipsky, S., Gilon, C., Zilkha, A. (1983). Attachment of drugs to polyethylene glycol. *Eur. Poly. J.*, *19*, 1177–1183.
59. Greenwald, R. B., Choe, Y. H., McGuire, J., Conover, C. D. (2003). Effective drug delivery by PEGylated drug conjugates. *Adv. Drug Deliv. Rev.*, *55*, 217–50.
60. Yamasaki, N., Matsuo, A., Isobe, H. (1988). Novel polyethylene glycol derivatives for modification of proteins. *Agric. Biol. Chem.*, *52*, 2125–2127.
61. Harris, J. M., Kozlowski, A. (2002). Poly(ethylene glycol) derivatives with proximal reactive groups. *US patent*, 6362254.
62. Zhao, H., Rubio, B., Sapra, P., Wu, D., Reddy, P., Sai, P., Martinez, A., Gao, Y., Lozanguiez, Y., Longley, C., Greenberger, L. M., Horak, I. D. (2008). Novel prodrugs of SN38 using multiarm poly(ethylene glycol) linkers. *Bioconjug. Chem.*, *19*, 849–859.
63. Conover, C. D., Zhao, H., Longley, C. B., Shum, K. L., Greenwald, R. B. (2003). Utility of poly(ethylene glycol) conjugation to create prodrugs of amphotericin B. *Bioconjug. Chem.*, *14*, 661–666.
64. Senior, J., Delgado, C., Fisher, D., Tilcock, C., Gregoriadis, G. (1991). Influence of surface hydrophilicity of liposomes on their interaction with plasma protein and clearance from the circulation: studies with poly(ethylene glycol)-coated vesicles. *Biochim. Biophys. Acta*, *1062*, 77–82.
65. Awada, A., Chan, S., Jerusalem, G. H. M., Coleman, R. E., Huizing, M. T., Mehdi, A., O'Reilly, S. M., Hamm, J. T., Garcia, P. A., Perez, E. A. (2012). Significant antitumor activity in a randomized phase 2 study comparing two schedules of etirinotecan pegol (NKTR-102). IMPAKT Breast Cancer Conference, May 3–5, 2012.

66. Maeda, H. (2012). Macromolecular therapeutics in cancer treatment: the EPR effect and beyond. *J. Control. Release*, *164*, 138–144.

67. Meerum Terwogt, J. M., ten Bokkel Huinink, W. W., Schellens, J. H., Schot, M., Mandjes, I. A., Zurlo, M. G., Rocchetti, M., Rosing, H., Koopman, F. J., Beijnen, J. H. (2001). Phase I clinical and pharmacokinetic study of PNU166945, a novel water-soluble polymer-conjugated prodrug of paclitaxel. *Anticancer Drugs*, *12*, 315–323.

68. Conover, C. D., Greenwald, R. B., Pendri, A., Gilbert, C. W., Shum, K. L. (1998). Camptothecin delivery systems: enhanced efficacy and tumor accumulation of camptothecin following its conjugation to polyethylene glycol via a glycine linker. *Cancer Chemother. Pharmacol.*, *42*, 407–414.

69. Eldon, M. A., Song, D., Neumann, T. A., Wolff, R., Cheng, L., Viegas, T. X., Bentley, M. D., Fishburn, C. S., Kugler, A. R. (2007). NKTR-118 (oral PEG-naloxol), a PEGylated derivative of naloxone: demonstration of selective peripheral opioid antagonism after oral administration in preclinical models. American Academy of Pain Management, 18th Annual Clinical Meeting, 27–30 September, Las Vegas, NV.

70. Harada, M., Saito, H., Kato, Y. (2011). Polymer derivative of docetaxel, method of preparing the same and uses thereof. *US patent appl. pub.*, 2011/0136990 A1.

71. Greenwald, R. B., Gilbert, C. W., Pendri, A., Conover, C. D., Xia, J., Martinez, A. (1996). Drug delivery systems: water soluble taxol 2′-poly(ethylene glycol) ester prodrugs-design and in vivo effectiveness. *J. Med. Chem.*, *39*, 424–431.

72. Yamashita, K., Janout, V., Bernard, E. M., Armstrong, D., Regen, S. L. (1995). Micelle/monomer control over the membrane-disrupting properties of an amphiphilic antibiotic. *J. Am. Chem. Soc.*, *117*, 6249–6253.

73. Aronov, O., Horowitz, A. T., Gabizon, A., Gibson, D. (2003). Folate-targeted PEG as a potential carrier for carboplatin analogs. Synthesis and in vitro studies. *Bioconjug. Chem.*, *14*, 563–574.

74. Veronese, F. M., Schiavon, O., Pasut, G., Mendichi, R., Andersson, L., Tsirk, A., Ford, J., Wu, G., Kneller, S., Davies, J., Duncan, R. (2005). PEG-doxorubicin conjugates: influence of polymer structure on drug release, in vitro cytotoxicity, biodistribution, and antitumor activity. *Bioconjug. Chem.*, *16*, 775–784.

75. Pasut, G., Canal, F., Dalla Via, L., Arpicco, S., Veronese, F. M., Schiavon, O. (2008). Antitumoral activity of PEG-gemcitabine prodrugs targeted by folic acid. *J. Control. Release*, *127*, 239–248.

76. Zhu, T., Gu, J., Yu, K., Lucas, J., Cai, P., Tsao, R., Gong, Y., Li, F., Chaudhary, I., Desai, P., Ruppen, M., Fawzi, M., Gibbons, J., Ayral-Kaloustian, S., Skotnicki, J., Mansour, T., Zask, A. (2006). Pegylated wortmannin and 17-hydroxywortmannin conjugates as phosphoinositide 3-kinase inhibitors active in human tumor xenograft models. *J. Med. Chem.*, *49*, 1373–1378.

77. Riebeseel, K., Biedermann, E., Loser, R., Breiter, N., Hanselmann, R., Mulhaupt, R., Unger, C., Kratz, F. (2002). Polyethylene glycol conjugates of methotrexate varying in their molecular weight from MW 750 to MW 40000: synthesis, characterization, and structure-activity relationships in vitro and in vivo. *Bioconjug. Chem.*, *13*, 773–785.

78. Ajazuddin, Alexander, A., Amarji, B., Kanaujia, P. (2013). Synthesis, characterization and in vitro studies of pegylated melphalan conjugates. *Drug Dev. Ind. Pharm.*, *39*, 1053–1062.

79. Greenwald, R. B., Conover, C. D., Pendri, A., Choe, Y. H., Martinez, A., Wu, D., Guan, S., Yao, Z., Shum, K. L. (1999). Drug delivery of anticancer agents: water soluble 4-poly (ethylene glycol) derivatives of the lignan, podophyllotoxin. *J. Control. Release*, *61*, 281–294.

80. Zacchigna, M., Di Luca, G., Maurich, V., Boccu, E. (2002). Syntheses, chemical and enzymatic stability of new poly(ethylene glycol)-acyclovir prodrugs. *Farmaco*, *57*, 207–214.

81. Gunaseelan, S., Debrah, O., Wan, L., Leibowitz, M. J., Rabson, A. B., Stein, S., Sinko, P. J. (2004). Synthesis of poly(ethylene glycol)-based saquinavir prodrug conjugates and assessment of release and anti-HIV-1 bioactivity using a novel protease inhibition assay. *Bioconjug. Chem.*, *15*, 1322–1333.

82. Chung, Y., Cho, H. (2004). Preparation of highly water soluble tacrolimus derivatives: poly(ethylene glycol) esters as potential prodrugs. *Arch. Pharm. Res.*, *27*, 878–883.

83. Fawzi, M., Gu, J., Ruppen, M., Zhu, T. (2009). Processes for preparing water-soluble polyethylene glycol conjugates of macrolide immunosuppressants. *US patent*, 7605257.

84. Marcus, Y., Sasson, K., Fridkin, M., Shechter, Y. (2008). Turning low-molecular-weight drugs into prolonged acting prodrugs by reversible pegylation: a study with gentamicin. *J. Med. Chem.*, *51*, 4300–4305.

85. Kopecek, J., Bazilova, H. (1973). Poly(N-(hydroxypropyl)-methacrylamide) 1. Radical polymerization and copolymerization. *Eur. Polym. J.*, *9*, 7–14.

86. Seymour, L. W., Miyamoto, Y., Maeda, H., Brereton, M., Strohalm, J., Ulbrich, K., Duncan, R. (1995). Influence of molecular weight on passive tumour accumulation of a soluble macromolecular drug carrier. *Eur. J. Cancer*, *31A*, 766–770.

87. Duncan, R., Cable, H. C., Lloyd, J. B., Rejmanova, P., Kopecek, J. (1982). Degradation of side-chains of N-(2-hydroxypropyl)methacrylamide copolymers by lysosomal thiol-proteinases. *Biosci. Rep.*, *2*, 1041–1046.

88. Seymour, L. W., Ulbrich, K., Steyger, P. S., Brereton, M., Subr, V., Strohalm, J., Duncan, R. (1994). Tumour tropism and anti-cancer efficacy of polymer-based doxorubicin prodrugs in the treatment of subcutaneous murine B16F10 melanoma. *Br. J. Cancer*, *70*, 636–641.

89. Bilim, V. (2003). Technology evaluation: PK1, Pfizer/Cancer Research UK. *Curr. Opin. Mol. Ther.*, *5*, 326–330.

90. Seymour, L. W., Ferry, D. R., Anderson, D., Hesslewood, S., Julyan, P. J., Poyner, R., Doran, J., Young, A. M., Burtles, S., Kerr, D. J. (2002). Hepatic drug targeting: phase I evaluation of polymer-bound doxorubicin. *J. Clin. Oncol.*, *20*, 1668–1676.

91. Julyan, P. J., Seymour, L. W., Ferry, D. R., Daryani, S., Boivin, C. M., Doran, J., David, M., Anderson, D., Christodoulou, C., Young, A. M., Hesslewood, S., Kerr, D. J. (1999). Preliminary clinical study of the distribution of HPMA copolymers bearing doxorubicin and galactosamine. *J. Control. Release*, *57*, 281–290.

92. Schoemaker, N. E., van Kesteren, C., Rosing, H., Jansen, S., Swart, M., Lieverst, J., Fraier, D., Breda, M., Pellizzoni, C., Spinelli, R., Grazia Porro, M., Beijnen, J. H., Schellens, J. H., ten Bokkel Huinink, W. W. (2002). A phase I and pharmacokinetic study of MAG-CPT, a water-soluble polymer conjugate of camptothecin. *Br. J. Cancer*, *87*, 608–614.

93. Gianasi, E., Buckley, R. G., Latigo, J., Wasil, M., Duncan, R. (2002). HPMA copolymers platinates containing dicarboxylato ligands. Preparation, characterisation and in vitro and in vivo evaluation. *J. Drug Target*, *10*, 549–556.
94. Campone, M., Rademaker-Lakhai, J. M., Bennouna, J., Howell, S. B., Nowotnik, D. P., Beijnen, J. H., Schellens, J. H. (2007). Phase I and pharmacokinetic trial of AP5346, a DACH-platinum-polymer conjugate, administered weekly for three out of every 4 weeks to advanced solid tumor patients. *Cancer Chemother. Pharmacol.*, *60*, 523–533.
95. Li, C. (2002). Poly(L-glutamic acid)--anticancer drug conjugates. *Adv. Drug Deliv. Rev.*, *54*, 695–713.
96. Melancon, M. P., Li, C. (2011). Multifunctional synthetic poly(L-glutamic acid)-based cancer therapeutic and imaging agents. *Mol. Imaging*, *10*, 28–42.
97. Li, C., Yu, D. F., Newman, R. A., Cabral, F., Stephens, L. C., Hunter, N., Milas, L., Wallace, S. (1998). Complete regression of well-established tumors using a novel water-soluble poly(L-glutamic acid)-paclitaxel conjugate. *Cancer Res.*, *58*, 2404–2409.
98. Li, C., Price, J. E., Milas, L., Hunter, N. R., Ke, S., Yu, D. F., Charnsangavej, C., Wallace, S. (1999). Antitumor activity of poly(L-glutamic acid)-paclitaxel on syngeneic and xenografted tumors. *Clin. Cancer Res.*, *5*, 891–897.
99. Singer, J. W., Baker, B., De Vries, P., Kumar, A., Shaffer, S., Vawter, E., Bolton, M., Garzone, P. (2003). Poly-(L)-glutamic acid-paclitaxel (CT-2103) [XYOTAX], a biodegradable polymeric drug conjugate: characterization, preclinical pharmacology, and preliminary clinical data. *Adv. Exp. Med. Biol.*, *519*, 81–99.
100. Singer, J. W., Shaffer, S., Baker, B., Bernareggi, A., Stromatt, S., Nienstedt, D., Besman, M. (2005). Paclitaxel poliglumex (XYOTAX; CT-2103): an intracellularly targeted taxane. *Anticancer Drugs*, *16*, 243–254.
101. Singer, J. W. (2005). Paclitaxel poliglumex (XYOTAX, CT-2103): a macromolecular taxane. *J. Control. Release*, *109*, 120–126.
102. Bhatt, R., de Vries, P., Tulinsky, J., Bellamy, G., Baker, B., Singer, J. W., Klein, P. (2003). Synthesis and in vivo antitumor activity of poly(l-glutamic acid) conjugates of 20S-camptothecin. *J. Med. Chem.*, *46*, 190–193.
103. Muhlenhoff, M., Eckhardt, M., Gerardy-Schahn, R. (1998). Polysialic acid: three-dimensional structure, biosynthesis and function. *Curr. Opin. Struct. Biol.*, *8*, 558–564.
104. Fernandes, A. I., Gregoriadis, G. (1996). Synthesis, characterization and properties of sialylated catalase. *Biochim. Biophys. Acta*, *1293*, 90–96.
105. Fernandes, A. I., Gregoriadis, G. (1997). Polysialylated asparaginase: preparation, activity and pharmacokinetics. *Biochim. Biophys. Acta*, *1341*, 26–34.
106. Jain, S., Hreczuk-Hirst, D. H., McCormack, B., Mital, M., Epenetos, A., Laing, P., Gregoriadis, G. (2003). Polysialylated insulin: synthesis, characterization and biological activity in vivo. *Biochim. Biophys. Acta*, *1622*, 42–49.
107. Epenetos, A. A., Hreczuk-Hirst, D. H., McCormack, B., Gregoriadis, G. (2002). Polysialylated proteins: a potential role in cancer therapy. *Clin. Pharm.*, *21*, 2186.

INDEX

Synthesis and Applications of Copolymers, First Edition. Edited by Anbanandam Parthiban.